IB	IIB	IIIA	IVA	VA	VIA	VIIA	VIIIA
							2 4.00260 -272 **He** $1s^2$
		5 10.81 2300 **B** (He) $2s^22p$	6 12.011 -3500 **C** (He) $2s^22p^2$	7 14.0067 -209.9 **N** (He) $2s^22p^3$	8 15.9994 -218.4 **O** (He) $2s^22p^4$	9 18.998403 -219.6 **F** (He) $2s^22p^5$	10 20.179 -248.7 **Ne** (He) $2s^22p^6$
		13 26.98154 660 **Al** (Ne) $3s^23p$	14 28.0855 1415 **Si** (Ne) $3s^23p^2$	15 30.97376 44.1 **P** (Ne) $3s^23p^3$	16 32.06 112.8 **S** (Ne) $3s^23p^4$	17 35.453 -100.9 **Cl** (Ne) $3s^23p^5$	18 39.948 -189.2 **Ar** (Ne) $3s^23p^6$
29 63.546 1083 **Cu** (Ar) $3d^{10}4s$	30 65.38 419.6 **Zn** (Ar) $3d^{10}4s^2$	31 69.72 29.8 **Ga** (Ar) $3d^{10}4s^24p$	32 72.59 937 **Ge** (Ar) $3d^{10}4s^24p^2$	33 74.926 **As** (Ar) $3d^{10}4s^24p^3$	34 78.96 217.4 **Se** (Ar) $3d^{10}4s^24p^4$	35 79.904 -7.2 **Br** (Ar) $3d^{10}4s^24p^5$	36 83.80 -156.6 **Kr** (Ar) $3d^{10}4s^24p^6$
47 107.868 962 **Ag** (Kr) $4d^{10}5s$	48 112.41 321 **Cd** (Kr) $4d^{10}5s^2$	49 114.82 156.6 **In** (Kr) $4d^{10}5s^25p$	50 118.69 232 **Sn** (Kr) $4d^{10}5s^25p^2$	51 121.75 630 **Sb** (Kr) $4d^{10}5s^25p^3$	52 127.60 449 **Te** (Kr) $4d^{10}5s^25p^4$	53 126.9045 113.5 **I** (Kr) $4d^{10}5s^25p^6$	54 131.29 -111.9 **Xe** (Kr) $4d^{10}5s^25p^7$
79 196.965 1064 **Au** (Xe) $4f^{14}5d^{10}6s$	80 200.59 -36.9 **Hg** (Xe) $4f^{14}5d^{10}6s^2$	81 204.37 308.5 **Tl** (Xe) $4f^{14}5d^{10}6s^26p$	82 207.2 327.5 **Pb** (Xe) $4f^{14}5d^{10}6s^26p^2$	83 208.9804 271 **Bi** (Xe) $4f^{14}5d^{10}6s^26p^3$	84 (209) 254 **Po** (Xe) $4f^{14}5d^{10}6s^26p^4$	85 (210) **At** (Xe) $4f^{14}5d^{10}6s^26p^5$	86 (222) -71 **Rn** (Xe) $4f^{14}5d^{10}6s^26p^6$

HW4
pag 342

65 158.9254 1356 **Tb** (Xe) $4f^95d6s^2$	66 162.50 1412 **Dy** (Xe) $4f^{10}5d6s^2$	67 164.9304 1474 **Ho** (Xe) $4f^{11}5d6s^2$	68 167.26 1529 **Er** (Xe) $4f^{12}5d6s^2$	69 168.9342 1545 **Tm** (Xe) $4f^{12}5d6s^2$	70 173.04 819 **Yb** (Xe) $4f^{14}5d6s^2$	71 174.967 1663 **Lu** (Xe) $4f^{14}5d6s^2$
97 (247) **Bk** (Rn) $5f^86d7s^2$	98 (251) **Cf** (Rn) $5f^86d7s^2$	99 (252) **Es**	100 (257) **Fm**	101 (258) **Md**	102 (259) **No**	103 (260) **Lr**

Atomic Number — 26
Atomic Weight — 55.847
Element — **Fe**
Melting point (°C) — 1535
Electron Configuration — (Ar)$3d^64s^2$

Principles of Growth and Processing of Semiconductors

S. Mahajan
Arizona State University

K. S. Sree Harsha
San Jose State University

Boston Burr Ridge, IL Dubuque, IA Madison, WI New York San Francisco St. Louis
Bangkok Bogotá Caracas Lisbon London Madrid Mexico City Milan
New Delhi Seoul Singapore Sydney Taipei Toronto

WCB/McGraw-Hill

A Division of The ***McGraw-Hill*** *Companies*

PRINCIPLES OF GROWTH AND PROCESSING OF SEMICONDUCTORS

This book is printed on acid-free paper.

1 2 3 4 5 6 7 8 9 0 DOC/DOC 9 0 9 8 7

ISBN 0-07-039605-1

Vice president and editorial director: *Kevin T. Kane*
Publisher: *Tom Casson*
Senior administrative assistant: *Jean Turrise*
Marketing manager: *John T. Wannemacher*
Senior project manager: *Denise Santor-Mitzit*
Production supervisor: *Michael R. McCormick*
Designer: *Jennifer Hollingsworth*
Compositor: *Interactive Composition Corporation*
Typeface: 10/12 *Linotype/Adobe Times Roman*
Printer: *R. R. Donnelley & Sons Company*

Library of Congress Cataloging-in-Publication Data

Mahajan, Subhash.
Principles of growth and processing of semiconductors / S.
Mahajan, K.S. Sree Harsha.
p. cm.
Includes index.
ISBN 0-07-039605-1
1. Semiconductors—Design and construction. I. SreeHarsha, K. S.
II. Title.
TK7871.85.M295 1998
621.3815'2—dc21 98-3385
CIP

http://www.mhhe.com

To our families and parents

PREFACE

Fabricating state-of-the-art integrated circuits is a joint endeavor between chemists, chemical engineers, electrical engineers, materials scientists, and physicists. Each group's expertise is not broad enough to appreciate all the interdisciplinary issues in the microelectronics industry. For example, electrical engineers and physicists, though familiar with semiconducting materials and devices, are not exposed to defects in solids or crystal growth. This book builds bridges across various disciplines. We begin with the physics of semiconductors and devices, and follow with the growth and processing of semiconductors. We emphasize how defects arise during growth and processing and what effects these defects have on the device behavior. This approach will help prepare students for the eclectic microelectronics industry.

In planning this book, we asked ourselves the following question: What is a suitable background for a student intending to join the microelectronics industry? It became apparent that chemists, chemical engineers, and materials scientists need exposure to the physics of semiconductors and principles of semiconducting devices, so we decided to discuss these topics in Chapters 1 and 2. Even though materials scientists are familiar with defects in solids, they do not learn specific characteristics of defects in semiconductors. We cover these characteristics in Chapter 3, which also includes the necessary background on various types of defects for students from other disciplines. Since real materials contain defects and since semiconducting behavior is affected by impurities, evaluation of semiconductors is essential. Therefore, we cover structural, chemical, and electrical evaluations of semiconductors in Chapter 4. To illustrate the salient features of each technique and its limitations, we have included one or two examples in each case. Furthermore, semiconducting devices require doped single crystals because grain boundaries act as carrier recombination centers. We cover crystal growth in Chapter 5. In particular, we emphasize the reduction of dislocation densities in as-grown crystals, the precipitation of oxygen in Czochralski silicon, and the formation of impurity striations. For efficient operation, most devices require epitaxial growth, a topic we cover in Chapter 6, along with the introduction of defects during epitaxy and heteroepitaxy.

To convert the crystal into a device requires several fabrication steps. These steps include oxidation, diffusion, ion implantation, metallization, lithography, and etching. The growth of a thermal oxide on silicon forms the backbone of ULSI technology. We cover oxide growth kinetics, thermodynamics, structure and oxidation-induced stacking faults in Chapter 7. The fabrication of several devices requires a local change in carrier concentration and conductivity type using diffusion and ion implantation. We cover these processes in Chapters 8 and 9. External and inter-device communication requires metal contacts and interconnects. We consider the techniques available for depositing contacts and interconnects and other relevant issues in Chapter 10. Circuit fabrication requires transferring a circuit pattern on a wafer using lithography and etching. We discuss the principles of these two technologies and their limitations in Chapter 11. In Chapter 12, we cover some of the future challenges in growth and processing of semiconductors.

To flesh out the concepts being developed, we have provided problems and their solutions within each chapter, as well as problem material at the end of each chapter. Furthermore, the approach underlying this book has been tested at Carnegie Mellon University and San Jose University, and the student response has been very encouraging.

The authors are grateful to Professor D. W. Greve, Professor M. E. McHenry, and Professor H. Temkin for their feedback on some of the chapters. To assess the student's reaction, some of the chapters were critiqued by Sanjoy, Sunit, and Ashish Mahajan, and their contribution is much appreciated. The authors are also very much obliged to Mrs. Valerie Thompson for her impressive word processing effort and to Mr. Kelly Young for his meticulous illustrations. Finally, they are very grateful to their families for their support and patience through this arduous endeavor.

S. Mahajan
K. S. Sree Harsha

ABOUT THE AUTHORS

Subhash Mahajan received a B.Sc. degree in physical sciences from Panjab University, Chandigarh, a B.E. in metallurgy from the Indian Institute of Science, Bangalore, and a Ph.D. from the University of California, Berkeley, in materials science and engineering. He has worked at the University of Denver, the Atomic Energy Research Establishment, Harwell, England; AT&T Bell Laboratories, Murray Hill; and Carnegie Mellon University, Pittsburgh. He is currently a Professor of Electronic Materials in the Department of Chemical, Bio and Materials Engineering, Arizona State University, Tempe, Arizona.

Dr. Mahajan has lectured and published extensively on origins of defects in semiconductors and their influence on device behavior and deformation behavior of solids. He has received several honors, such as the John Bardeen Award of the Minerals, Metals and Materials Society (TMS) (and the Albert Sauveur Achievement Award of the American Society at Metals (ASM) International. In addition, Dr. Mahajan consults nationally and internationally and is a member of several leading materials societies.

K. S. Sree Harsha is a Professor of Materials Engineering in the Department of Materials Engineering of San Jose State University at San Jose, California. He received a B.Sc. degree from the University of Mysore (and also received undergraduate education in metallurgy from the Indian Institute of Science), an M.S. degree from the University of Notre Dame, and a Ph.D. from Pennsylvania State University. Dr. Sree Harsha has been teaching at San Jose State University since 1968, and has served as the chair person of the Department of Materials Engineering for twenty-one years. He has taught extensively in the areas of materials analysis, thermodynamics, semiconductor processing, and fracture mechanics. His research work includes publications in all these areas of his interest. He is a member of the Metals, Minerals Society (TMS), the Materials Research Society, and the American Society for Materials (ASM Int), and has served on several committees. He has spent several summers in industrial research, and two one-year sabbaticals at Bell Laboratories.

CONTENTS

CHAPTER 1

Semiconductors: An Introduction

This chapter develops an introductory framework for understanding the behavior of semiconductors. It introduces the concepts of band gaps and charge carriers in semiconductors, that is, electrons and holes, discusses the changes in carrier concentration due to the addition of dopants, and correlates the conductivity of a semiconductor with the mobilities of the carriers. These concepts underlie the operation of semiconducting devices covered in Chapter 2.

1.1 BEHAVIOR OF FREE ELECTRONS

We show later in this chapter that the conduction in semiconductors occurs by the migration of two types of charge carriers, one of them being electrons. The presence of two types of carriers produces interesting effects in semiconductors. Therefore, we first discuss the properties of free electrons, that is, electrons that exist outside a solid—and progressively add more realism to this model so that it represents a semiconductor.

1.1.1 Particle-Wave Duality

Free electrons exhibit particle-wave duality. Figure 1.1 shows a setup to demonstrate the particle-like behavior. The electrons from a hot cathode overcome the surface potential barrier when a suitable potential is applied between the cathode and an anode. The anode has a pinhole that collimates the free-electron beam emitted from the cathode. When this beam hits a target metal, it ejects core-shell electrons from the atoms. An outer-shell electron rapidly fills the resulting vacancy in the core shell. The difference in the energies between the two electronic levels is given off as an X-ray photon. This behavior is consistent with the particle-like nature of electrons.

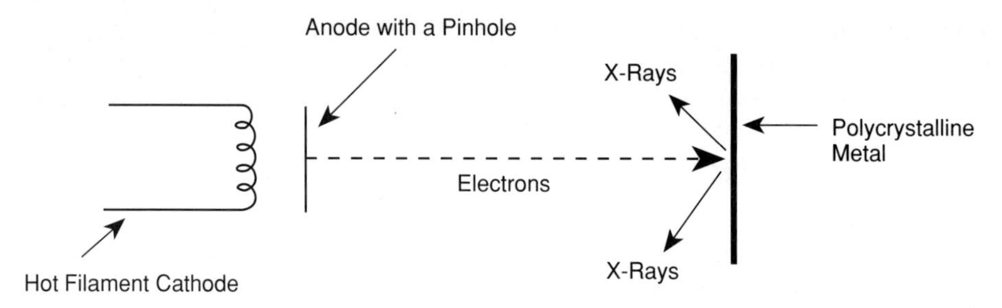

FIGURE 1.1
Schematic of a setup that can demonstrate the particle-like behavior of free electrons.

Now consider the setup shown in Figure 1.2. The electron beam defined by magnetic lenses traverses a very thin, single crystal sample and is Bragg diffracted. The diffraction produces a large number of diffraction spots on a fluorescent screen as shown in Figure 1.2. The pattern is best understood if the electrons behave as waves.

The equivalence between the particle- and wavelike behaviors of free electrons is provided by the Planck and de Broglie relations. According to Planck, electron energy E is related to its frequency ν by

$$\nu = \frac{E}{h}, \tag{1.1}$$

where h is Planck's constant. This relation is applicable to all types of electromagnetic radiation. On the other hand, de Broglie hypothesized that the wavelength λ of a wave associated with an electron is related to its momentum p by

$$\lambda = \frac{h}{p}. \tag{1.2}$$

Every particle can exhibit the particle- and wavelike behaviors. Whether the wavelike nature of a particle is experimentally discernible depends on the wave-

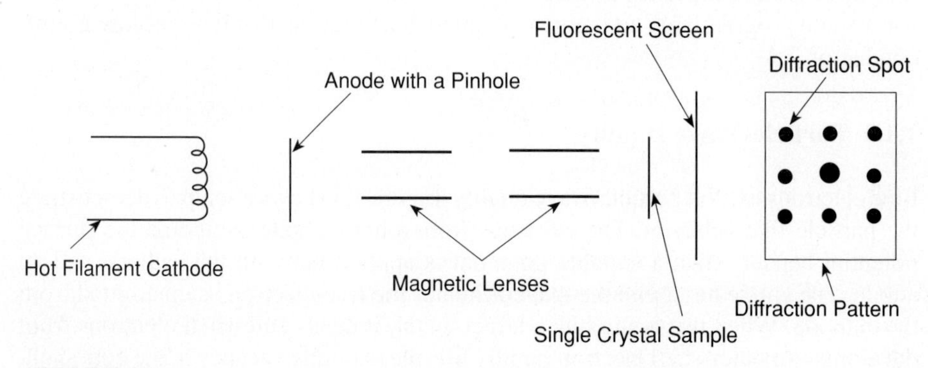

FIGURE 1.2
Schematic of a setup that can demonstrate the wave-like behavior of free electrons.

length of the wave associated with the particle relative to the dimension of the experiment. For electrons the mass is extremely small (9.11×10^{-31} kg). Therefore, for reasonable values of velocities, as shown in Example 1.1, the values of momentum are fairly small, resulting in associated wavelengths of waves that can be discerned by their diffraction from a grating consisting of lattice planes within a crystal.

EXAMPLE 1.1. Electrons are excited from a hot wire cathode at a potential of 100 kV. Calculate the wavelength of the electrons.

Solution

$$\begin{aligned} \mathrm{E} = \text{the kinetic energy of the accelerated electrons} &= \frac{1}{2}\mathrm{mv}^2 \\ &= 100\ \mathrm{keV} \\ &= 1.6 \times 10^{-14}\ \mathrm{J}. \end{aligned}$$

$\mathrm{E} = \frac{1}{2}\mathrm{mv}^2$ is reasonable because E is much less than the rest energy mc^2 (500 keV), where c is the velocity of light. For higher energies you need to use the relativistic formulas.

p = momentum of the electrons $= \sqrt{2\mathrm{mE}}$, where m is the mass of the electron (9.11×10^{-31} kg).

Thus
$$\begin{aligned} \mathrm{p} &= \sqrt{2 \times 9.11 \times 10^{-31} \times 1.6 \times 10^{-14}}\ \text{kg-m/sec} \\ &= 1.7 \times 10^{-22}\ \text{kg-m/sec}. \end{aligned}$$

According to de Broglie

$$\lambda = \frac{\mathrm{h}}{\mathrm{p}}.$$

Substituting for p and $\mathrm{h} = 6.6 \times 10^{-34}$ J sec, we obtain

$$\lambda = \frac{6.6 \times 10^{-34}}{1.7 \times 10^{-22}} = 0.004\ \mathrm{nm}.$$

Waves having the preceding wavelength can be diffracted from various lattice planes within silicon and gallium arsenide whose lattice parameters are 0.543 and 0.565 nm. Diffraction occurs because the electron wavelength is considerably smaller than the separations between different gratings formed by different crystal planes.

1.1.2 Uncertainty Principle

We cannot describe the events involving atomic particles with absolute precision. Instead, we must think of the average values of position, momentum, and energy of a particle such as an electron. According to Heisenberg, the uncertainties in the measurements of position Δx and momentum Δp are related by the uncertainty relation

$$(\Delta \mathrm{x})(\Delta \mathrm{p}) \geq \hbar, \tag{1.3}$$

where $\hbar = h/2\pi$. Similarly, the uncertainty in an energy measurement ΔE is related to the uncertainty in the time Δt at which the measurement is made by the following expression:

$$(\Delta E)(\Delta t) \geq \hbar. \tag{1.4}$$

The preceding limitations, implicit in the uncertainty principle, manifest experimentally with atomic particles such as electrons.

EXAMPLE 1.2. A mass of 1 μg has a velocity of 1000 cm/sec. If its velocity is uncertain by 1 percent, what is the order of magnitude of the minimum uncertainty in its position?

Solution

$$\begin{aligned}
m &= \text{Mass of the particle} = 1 \times 10^{-6} \times 10^{-3} = 1 \times 10^{-9}\ \text{kg.}\\
v &= \text{Velocity of the particle} = 10^{3} \times 10^{-2} = 10\ \text{m/sec.}\\
\Delta v &= \text{Uncertainty in the velocity} = 10^{-2} \times 10 = 10^{-1}\ \text{m/sec.}\\
\Delta p &= \text{Uncertainty in the momentum} = 1 \times 10^{-9} \times 10^{-1}\\
&= 1 \times 10^{-10}\ \text{kg-m/sec.}\\
\Delta x &= \text{Uncertainty in the position can be determined by Equation (1.3)}\\
&\geq \frac{\hbar}{\Delta p} = \frac{6.63 \times 10^{-34}}{2\pi \times 1 \times 10^{-10}}\\
&\approx 1.00 \times 10^{-24}\ \text{m.}
\end{aligned}$$

Δx is extremely small in this case because the mass is fairly large.

1.1.3 Quantum Mechanical Treatment

Classical mechanics dealt successfully with large objects whose de Broglie wavelengths are very small in comparison to their sizes. At atomic and subatomic scales, the wavelengths become comparable to the size of the system being studied. Then classical mechanics no longer applies. To address this problem, Heisenberg developed matrix mechanics in 1925, which was followed by Schrödinger's wave mechanics approach in 1926. The two approaches are equivalent because they make the same predictions, but Schrödinger's formulation, emphasizing the wave nature of matter and based on the familiar classical wave equation, is easier to conceptualize. We therefore approach quantum mechanics from Schrödinger's viewpoint.

1.1.3.1 Energy continuum of free electrons

Free electrons have applications in electron microscopes, televisions, X-ray generators, and electron-beam evaporators. Furthermore, understanding the behavior of free electrons is important because many problems can be treated as small perturbations to a free-electron model. For example, simple theories of conductivity that work surprisingly well are based on a nearly free-electron model.

The behavior of standing waves associated with electrons can be described by the Schrödinger equation:

$$\nabla^2 \psi + \frac{2m}{\hbar^2}(E - V)\,\psi = 0, \tag{1.5}$$

where $\nabla^2\psi = \partial^2\psi/\partial x^2 + \partial^2\psi/\partial y^2 + \partial^2\psi/\partial z^2$, ψ is the wave function associated with an electron, m is the mass of the electron, E is the total energy (kinetic + potential), and V is the potential energy. In Equation (1.5) ψ represents the amount of matter at a particular location. Its exact meaning remained unclear until Born postulated that $|\psi|^2 = \psi\psi^*$ is the probability density, where ψ^* is the complex conjugate of ψ. In one dimension Equation (1.5) reduces to

$$\frac{d^2\psi}{dx^2} + \frac{2m}{\hbar^2}(E - V)\psi = 0. \tag{1.6}$$

Now consider electrons that propagate freely, that is, in a potential free space in the positive x direction. The potential energy V is zero, and Equation (1.6) becomes

$$\frac{d^2\psi}{dx^2} + \frac{2m}{\hbar^2}E\psi = 0. \tag{1.7}$$

The solution of Equation (1.7) is

$$\psi(x) = Ae^{ikx} \tag{1.8}$$

with

$$k = \sqrt{\frac{2mE}{\hbar^2}}. \tag{1.9}$$

Rewriting Equation (1.9), we find

$$E = \frac{\hbar^2k^2}{2m}. \tag{1.10}$$

Since E in this case is also kinetic energy (V = 0), it can be written as

$$E = \frac{p^2}{2m}. \tag{1.11}$$

Substituting for E from Equation (1.11) into Equation (1.9), we have

$$k = \sqrt{\frac{2mE}{\hbar^2}} = \frac{p}{\hbar}. \tag{1.12}$$

We know from Equation (1.2) that $p/\hbar$ is equal to $2\pi/\lambda$. Therefore, by making these substitutions, we have

$$k = \frac{2\pi}{\lambda}. \tag{1.13}$$

From Equation (1.12) we know that k is proportional to the momentum p and is also proportional to the velocity of electrons because $\vec{p} = m\vec{v}$. Since both momentum and velocity are vector quantities, it follows that k is also a vector. Therefore, we should write k as

$$|\vec{k}| = \frac{2\pi}{\lambda}. \tag{1.14}$$

Since $\vec{k}$ is inversely proportional to λ, $\vec{k}$ is usually called the wave vector. The wave vector points in the direction of propagation of the wave, and its magnitude is the rate of change of phase of the wave with the distance of propagation. Furthermore, as Equation (1.10) is valid for all values of k, it implies that E can vary continuously; that is, free electrons do not have discrete energy levels.

The probability density associated with the wave function in Equation (1.8) is A^2; that is, it has a constant magnitude everywhere. Therefore, a free electron is equally likely to be found anywhere in space. This situation is not surprising, since we have a solution with definite momentum, $\vec{p} = \hbar\vec{k}$, so that $\Delta\vec{p} = 0$. By the uncertainty principle $\Delta x = \infty$, the particle is spread over all space.

1.2 ELECTRONS IN POTENTIAL WELLS

1.2.1 Infinite Potential Well

Now let us consider the case of an electron confined between two infinitely high-potential barriers schematically shown in Figure 1.3. These barriers prevent the electron from escaping from the well. We need to understand this situation before addressing the issue of a finite barrier that represents practical cases such as quantum-well structures or electrons confined in a solid. In this case $\psi = 0$ for $x \leq 0$ and $x \geq a$. Furthermore, $V = 0$ inside the well. Since the electron is reflected from the walls, it travels in both directions. As before, the Schrödinger equation describing this situation can be written as

$$\frac{d^2\psi}{dx^2} + \frac{2m}{\hbar^2}E\psi = 0. \tag{1.15}$$

Equations (1.7) and (1.15) are the same, but the boundary conditions are different.

The general solution of Equation (1.15) is $\psi = Ae^{ikx} + Be^{-ikx}$. As the electron can propagate in either direction, the equivalent solution is

$$\psi(x) = A\sin(kx) + B\cos(kx), \tag{1.16}$$

where

$$k = \sqrt{\frac{2mE}{\hbar^2}}. \tag{1.17}$$

Applying the boundary conditions that $\psi = 0$ for $x \leq 0$ and $x \geq a$, we obtain

$$\psi(0) = B = 0 \tag{1.18a}$$

and

$$\psi(a) = 0 = A\sin(ka). \tag{1.18b}$$

Except for the trivial result $\psi = 0$ obtained by setting $A = 0$, Equation (1.18*b*) is satisfied only when (ka) is an integer multiple of π. Therefore, k is restricted to the values

$$k = \pm\frac{n\pi}{a}, n = 1, 2, 3, \ldots \tag{1.19}$$

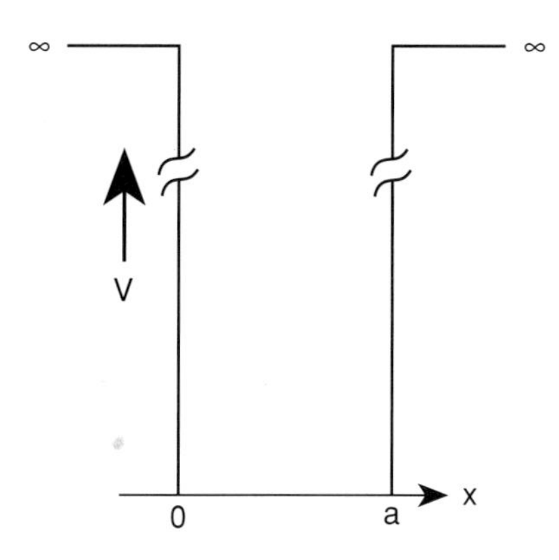

FIGURE 1.3
One-dimensional potential well; the walls provide infinitely high-potential barriers.

with the wave function corresponding to a given k (or n) being

$$\psi_n(x) = A_n \sin\left(\frac{n\pi x}{a}\right) \text{ for } 0 \leq x \leq a. \tag{1.20}$$

Likewise, by substituting for k from Equation (1.19) into Equation (1.17), E_n can be written as

$$E_n = \frac{n^2\pi^2\hbar^2}{2ma^2},\ n = 1, 2, 3, \ldots \tag{1.21}$$

The integer n in Equations (1.19), (1.20), and (1.21) is called a *quantum number.*

Comparing Equations (1.10) and (1.21), we can see a striking difference between the free-electron case and the electron confined in an infinite-potential well. The boundary conditions $\psi(0) = 0$ for $x = 0$ and $x = a$ allow only certain discrete energies, referred to as *energy levels*. The first four levels are shown in Figure 1.4*a*, and the corresponding wave functions and probability densities $|\psi|^2 = \psi\,\psi^*$ are depicted in Figures 1.4*b* and 1.4*c*. The wave function plots in Figure 1.4*b* are standing waves, and are consistent with the idea that an electron is reflected from the walls.

EXAMPLE 1.3. Consider an infinite-potential well of width 0.25 nm. What are the three lowest allowable energy levels for (1) an electron and (2) an 80 kg man confined to this energy well?

Solution. Equation (1.21) can be simplified to $E_n = \dfrac{n^2h^2}{8ma^2}$, where

$$n = 1, 2, 3, \quad h = 6.6 \times 10^{-34}, \quad a = 2.5 \times 10^{-10}\text{ m}$$

Electron

$$m = 9.11 \times 10^{-31}\text{ kg}$$

Therefore, $$E_1 = \frac{(6.6 \times 10^{-34})^2}{8 \times 9.11 \times 10^{-31} \times (2.5 \times 10^{-10})^2} = 9.56 \times 10^{-19}\text{ J}$$

$$= 6.07\text{ eV}$$

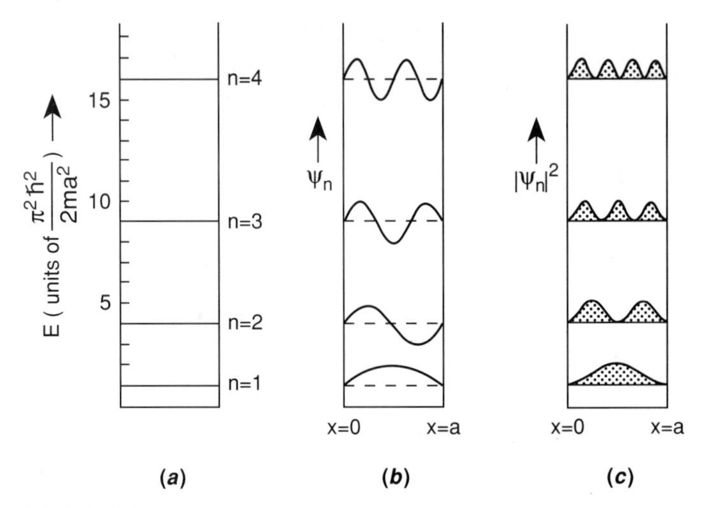

FIGURE 1.4
Particle in an infinitely deep, one-dimensional potential well. (*a*) First four allowed energy levels, (*b*) wavefunctions, and (*c*) $|\psi|^2$ associated with the first four energy levels. $|\psi|^2$ is proportional to the probability of finding the particle at a given point in the potential well.

$$E_2 = 4E_1 = 24.3 \text{ eV}$$

$$E_3 = 9E_1 = 54.6 \text{ eV}$$

The above a results are a good representation of a hydrogen atom where an electron is confined in a 0.1 nm radius (width ~0.2 nm). Consequently, we should expect ~10 eV for the ground state of a hydrogen atom.
Mass = 80 kg

Therefore,
$$E_1 = \frac{(6.6 \times 10^{-34})^2}{8 \times 80 \times (2.5 \times 10^{-10})^2} = 1.089 \times 10^{-50} \text{ J}$$

$$= 6.92 \times 10^{-32} \text{ eV}$$

E_2 and E_3 can be calculated from E_1. This example shows that the quantum mechanical effects are important only with extremely small masses.

1.2.2 Finite-Potential Barrier

Let us now examine a situation where an electron of mass m, propagating in the positive x direction, encounters a potential barrier whose potential energy V_0, that is, height of the barrier, is greater than the total energy E of the electron but is still finite. This situation is shown in Figure 1.5. To analyze this situation, regions I and II in Figure 1.5 have to be considered separately. In region I ($x < 0$) the electron is

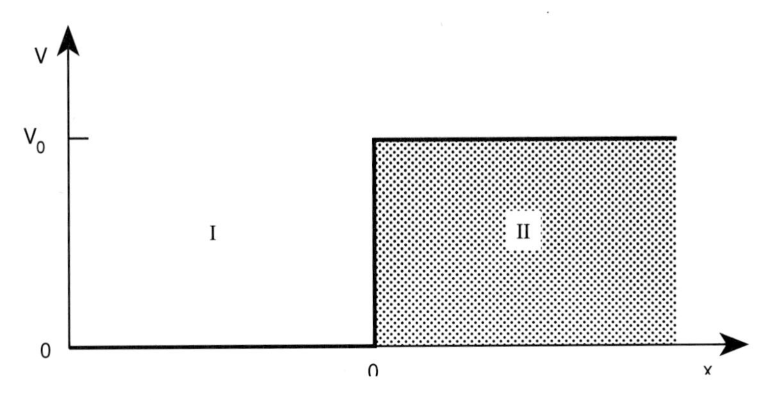

FIGURE 1.5
Schematic of a finite-potential barrier.

free, and the corresponding Schrödinger equation is

$$\frac{d^2\psi}{dx^2} + \frac{2m}{\hbar^2} E\psi_I = 0 \quad (I). \tag{1.22}$$

The Schrödinger equation for region II is

$$\frac{d^2\psi}{dx^2} + \frac{2m}{\hbar^2}(E - V_0)\psi_{II} = 0 \quad (II). \tag{1.23}$$

Again the solutions to Equations (1.22) and (1.23) are

$$\psi_I = Ae^{i\alpha x} + Be^{-i\alpha x} \quad (I), \tag{1.24}$$

where

$$\alpha = \sqrt{\frac{2mE}{\hbar^2}}, \tag{1.25}$$

and

$$\psi_{II} = Ce^{i\beta x} + De^{-i\beta x} \quad (II), \tag{1.26}$$

where

$$\beta = \sqrt{\frac{2m(E - V_0)}{\hbar^2}} \tag{1.27}$$

Since we assumed that E is less than V_0, $(E - V_0)$ is negative. Therefore, β is imaginary. To avoid this situation, we can define a new constant

$$\gamma = -i\beta. \tag{1.28}$$

Substituting $-\gamma$ for $i\beta$ in Equation (1.26), we have

$$\psi_{II} = Ce^{-\gamma x} + De^{\gamma x}. \tag{1.29}$$

We can evaluate the constants A, B, C, and D by considering the following boundary conditions:

1. When $x \to \infty$, Equation (1.29) indicates that

$$\psi_{II} = C \cdot 0 + D \cdot \infty. \tag{1.30}$$

Since the probability of finding the electron in region II is finite, this is possible only if

$$D = 0 \tag{1.31}$$

2. The wave functions ψ_I and ψ_{II} are continuous at $x = 0$. Therefore, from Equations (1.24) and (1.26), we obtain

$$A + B = C. \tag{1.32}$$

3. The slopes of the wave functions in regions I and II are continuous at $x = 0$.

$$\left.\frac{d\psi_I}{dx}\right|_{x=0} = \left.\frac{d\psi_{II}}{dx}\right|_{x=0}. \tag{1.33}$$

Taking the derivatives of ψ_I and ψ_{II} with respect to x, evaluating them at $x = 0$, and equating the results, we obtain

$$iA\alpha - iB\alpha = -\gamma C. \tag{1.34}$$

If we use Equations (1.32) and (1.34) to solve for A and B in terms of C, we obtain

$$A = \frac{C}{2}\left(1 + i\frac{\gamma}{\alpha}\right), \tag{1.35}$$

and

$$B = \frac{C}{2}\left(1 - i\frac{\gamma}{\alpha}\right). \tag{1.36}$$

Substituting for A and B from Equations (1.35) and (1.36) into Equation (1.24), ψ_I can also be expressed in terms of an amplitude C. Furthermore, substituting $D = 0$

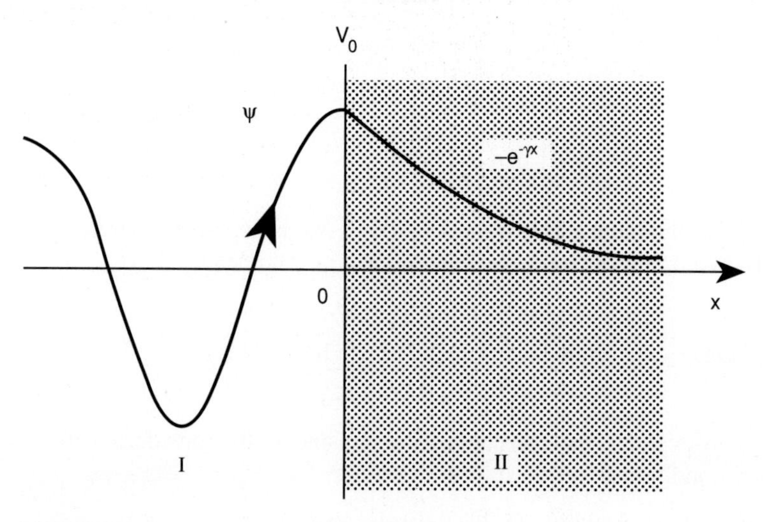

FIGURE 1.6
The wave-function ψ penetrating a potential barrier.

in Equation (1.29), we obtain

$$\psi_{\mathrm{II}} = \mathrm{C}e^{-\gamma x} \qquad (1.37)$$

The amplitude of the wave function in region II decreases exponentially, as shown in Figure 1.6. The larger the magnitude of γ—that is, the higher potential barrier V_0—the faster the wave function decays. Figure 1.6 indicates that ψ has a value to the right of the barrier, implying that the particle might be found beyond the barrier. Since its total energy is less than the barrier height V_0, the particle does not go over the barrier. Instead, the barrier is penetrated by "tunneling." This phenomenon is intimately linked with the uncertainty principle and is very important in the transport of electrons in solids, for example, in oxide/superconductor and metal/semiconductor junctions.

1.3 ORIGIN OF BAND GAPS IN A CRYSTAL

Electrons moving in a crystalline solid may be treated as particles in a three-dimensional box with a complicated interior. Without sacrificing essential physics, we consider its one-dimensional analogue. This approach greatly simplifies the mathematics.

1.3.1 Electron in a Periodic Potential of a Crystal (Kronig-Penney Model)

The potential energy function in a one-dimensional crystal may vary as shown in Figure 1.7*a*. The positively charged nuclei provide the attractive potential for the negatively charged electrons. To evaluate the movement of an electron in such a potential, Kronig and Penney (1931) simplified it into the situation depicted in Figure 1.7*b*. This coarse simplification does not take into consideration that the inner electrons are more tightly bound to the core, but it is adequate. Figure 1.7*b* shows potential wells of length a that are labeled as region I. These wells are separated by potential barriers of height V_0 and width b (region II); V_0 is assumed to be larger than the energy E of the electron. This situation is very similar to the finite potential barrier discussed in section 1.2.2 except that the arrangement repeats forever.

As before, the Schrödinger equations for regions I and II are

$$\frac{d^2\psi_{\mathrm{I}}}{dx^2} + \frac{2m}{\hbar^2}E\psi_{\mathrm{I}} = 0 \quad (\mathrm{I}) \qquad (1.38)$$

and

$$\frac{d^2\psi_{\mathrm{II}}}{dx^2} + \frac{2m}{\hbar^2}(E - V_0)\psi_{\mathrm{II}} = 0 \quad (\mathrm{II}), \qquad (1.39)$$

where

$$\alpha^2 = \frac{2m}{\hbar^2}E \qquad (1.40)$$

and

$$\beta^2 = \frac{2m(V_0 - E)}{\hbar^2} \qquad (1.41)$$

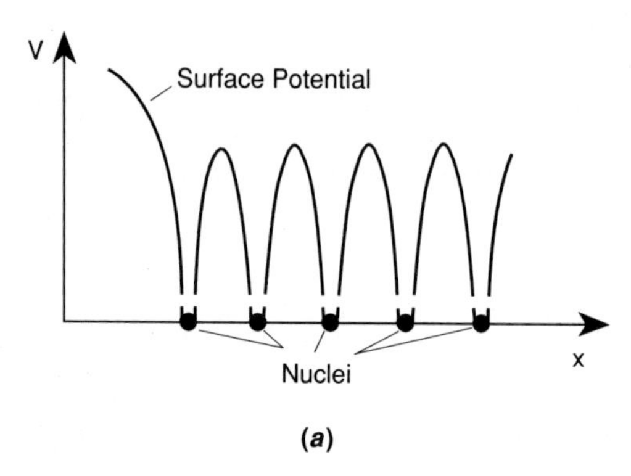

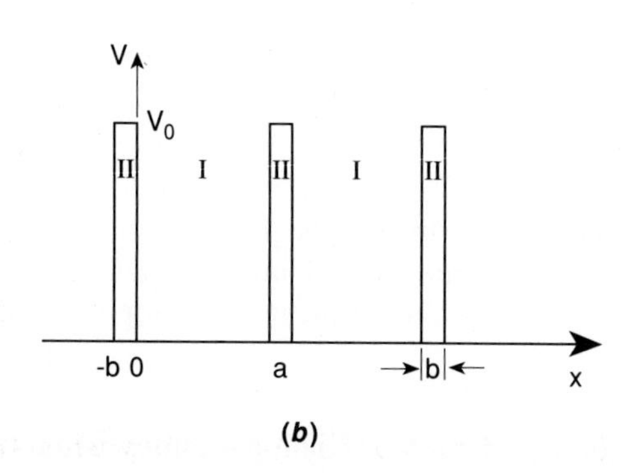

FIGURE 1.7
Schematics of (*a*) one-dimensional periodic potential distribution for a crystal and (*b*) simplified one-dimensional periodic potential distribution used in the Kronig-Penney model.

The complexity of the present case in comparison to the finite potential barrier is obvious. ψ_I, ψ_{II}, $d\psi_I/dx$, and $d\psi_{II}/dx$ not only have to be continuous at $x = 0$ in Figure 1.7*b*; they are also assumed to be continuous at $x = a + b$ and at other equivalent positions; that is, the existence of a periodic potential imposes a periodicity on wave functions in the crystal. Bloch (1930) mathematically addressed this problem and showed that in a one-dimensional periodic potential the simultaneous solution of Equations (1.38) and (1.39) has the following form:

$$\psi_{(x)} = u(x)e^{ikx}. \tag{1.42}$$

In comparison to the cases discussed in sections 1.1.3.1 and 1.2, the amplitude u(x) is no longer a constant, but changes periodically with x, which has a period equal to the lattice constant.

Calculating the value of $d^2\psi/dx^2$ from Equation (1.42) and substituting it into Equations (1.38) and (1.39), we obtain the following equations for u:

$$\frac{d^2u}{dx^2} + 2\,ik\frac{du}{dx} - (k^2 - \alpha^2)u = 0 \quad (I) \tag{1.43}$$

and

$$\frac{d^2u}{dx^2} + 2\,ik\frac{du}{dx} - (k^2 + \beta^2)u = 0 \quad (II) \tag{1.44}$$

The solutions of Equations (1.43) and (1.44) are

$$u = e^{-ikx}(Ae^{i\alpha x} + Be^{-i\alpha x}) \quad \text{(I)} \tag{1.45}$$

and

$$u = e^{-ikx}(Ce^{-\beta x} + Ee^{\beta x}) \quad \text{(II)} \tag{1.46}$$

The constants A, B, C, and D can be determined from the boundary conditions. The functions $\Psi(x)$ and $d\Psi/dx$, and therefore $u(x)$ and du/dx, pass over continuously from region I to region II at $x = 0$. These boundary conditions yield

$$A + B = C + D \tag{1.47}$$

and

$$A(i\alpha - ik) + B(-i\alpha - ik) = C(-\beta - ik) + D(\beta - ik). \tag{1.48}$$

Imposing the periodic boundary condition that $u(x)$ from Equation (1.45) at $x = 0$ must be equal to $u(x)$ from Equation (1.46) at $x = a + b$, and equivalently from Figure 1.7*b*, Equation (1.45) at $x = a$ is equal to Equation (1.46) at $x = -b$, we have

$$Ae^{(i\alpha - ik)a} + B^{(-i\alpha - ik)a} = Ce^{(ik+\beta)b} + De^{(ik-\beta)b}. \tag{1.49}$$

Finally, $\dfrac{du}{dx}$ is periodic in $(a + b)$:

$$\begin{aligned} &Ai(\alpha - k)e^{ia(\alpha - k)} - Bi(\alpha + k)e^{-ia(\alpha + k)} \\ &= -C(\beta + ik)e^{(ik+\beta)b} + D(\beta - ik)e^{(ik-\beta)b}. \end{aligned} \tag{1.50}$$

To determine the values of $u(x)$, Equations (1.47) through (1.50) have to be solved simultaneously. The solutions exist if the determinant of the coefficients A, B, C, and D vanishes. Using this criterion, the energy-restricting condition after extensive simplifications is

$$\frac{\beta^2 - \alpha^2}{2\alpha\beta} \sinh\beta b \, \sin\alpha a + \cosh\beta b \, \cos\alpha a = \cos k(a + b). \tag{1.51}$$

Assuming that the potential barriers in Figure 1.7*b* are such that b is very small and V_0 is very large, Equation (1.51) can be simplified to

$$P\frac{\sin\alpha a}{\alpha a} + \cos\alpha a = \cos ka, \tag{1.52}$$

where

$$P = \frac{maV_0 b}{\hbar^2}. \tag{1.53}$$

P represents the strength of the barrier, and from Equation (1.40) α is a function of the energy. In Figure 1.8 the left side of Equation (1.52) is plotted as a function of αa for $P = 1$, 3, and 6. Since $|\cos ka|$ cannot exceed one, Equation (1.52) is satisfied only for certain values of αa. These values correspond to the allowed energy states delineated by dashed regions. The gaps between the allowed energy states are referred to as *band gaps*. The higher the value of P, the wider the gap. Figure 1.8 shows that the points of transition between the allowed and forbidden energy levels occur when $\alpha a = \pi, 2\pi, 3\pi \ldots n\pi$. Substituting this value of αa in Equation (1.52), we find that $\cos ka = \cos(\alpha a)$; that is, $k = \alpha$. Since $\alpha a = n\pi$, $k = n\pi/a$. These k values have a physical basis as we will show later in this section.

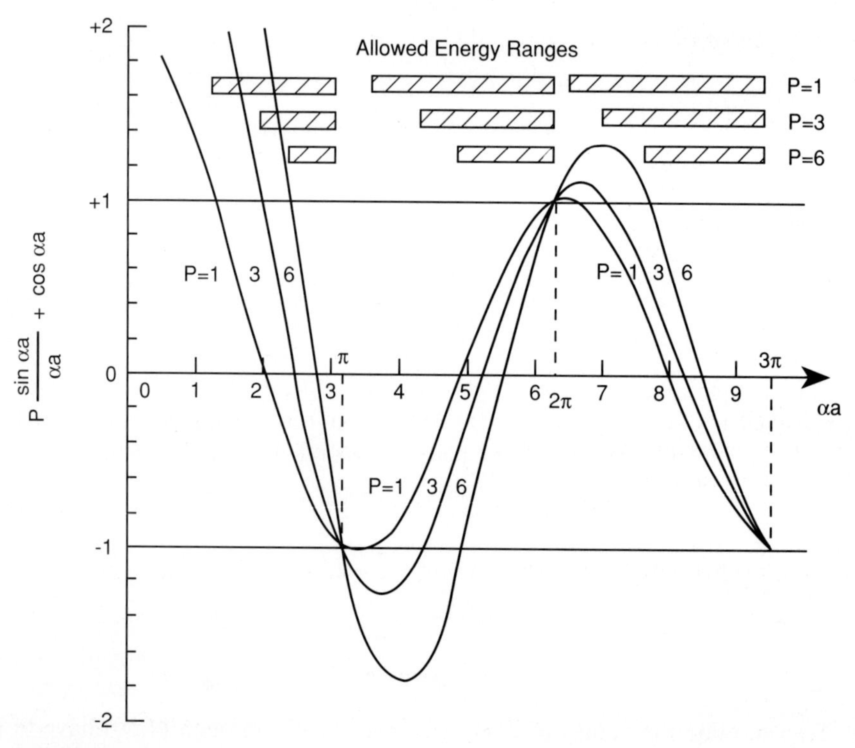

FIGURE 1.8
Allowed and forbidden bands for the Kronig-Penney approximation to the periodic series of square-well potentials for various values of the "strength of barrier" parameter P.

Let us now consider two special cases: (1) the potential-barrier strength becomes smaller and smaller and finally disappears completely, that is, P goes to zero, and (2) the potential-barrier strength is very large, that is, P approaches infinity. When P goes to zero, we obtain from Equation (1.52)

$$\cos \alpha \mathrm{a} = \cos \mathrm{ka} \cdot \tag{1.54}$$

or $\alpha = \mathrm{k}$. It follows from Equation (1.40) that

$$\mathrm{E} = \frac{\hbar^2 \mathrm{k}^2}{2\mathrm{m}},$$

which is the well-known Equation (1.12) for free electrons that we developed in section 1.1.3.1.

When $\mathrm{P} \to \infty$, $\dfrac{\sin \alpha \mathrm{a}}{\alpha \mathrm{a}} \to 0$ because the left side of Equation (1.52) must stay within the limits ± 1. This condition is possible only if $\alpha \mathrm{a} = \mathrm{n}\pi$ or

$$\alpha^2 = \frac{\mathrm{n}^2 \pi^2}{\mathrm{a}^2} \quad \text{for } \mathrm{n} = 1, 2, 3 \ldots \tag{1.55}$$

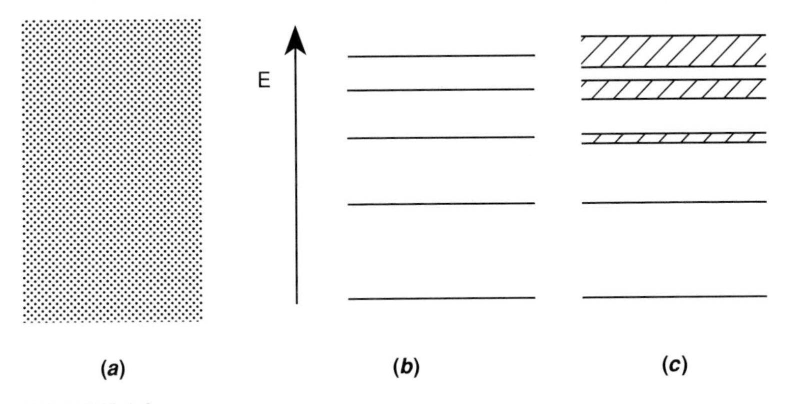

FIGURE 1.9
Allowed energy levels for (*a*) free electrons, (*b*) bound electrons, and (*c*) electrons in a periodic potential of a solid.

Substituting for α^2 from Equation (1.55) into Equation (1.40), we obtain

$$E = \frac{n^2\pi^2\hbar^2}{2ma^2},$$

which is the result for an electron confined in an infinite barrier, see Equation (1.21).

According to Bragg's law for diffraction, $n\lambda = 2d\sin\theta$, where n is the order of the diffraction, λ is the wavelength of the wave that is being diffracted, d is interplanar spacing, and θ is the diffraction angle. In a one-dimensional crystal, Bragg's law is modified to $n\lambda = 2a$ because $d \to a$ and $\theta = 90°$. Using Equation (1.13), λ can be replaced with k and we obtain $k = n\pi/a$. The implication is that the energy bands allowed according to the Kronig-Penney model terminate exactly where Bragg reflections occur in the one-dimensional lattice. Therefore, the electron states corresponding to $k = n\pi/a$ are standing waves; that is, electrons populating these states cannot propagate through the crystal.

The schematics in Figure 1.9 compare the energy levels associated with free electrons, electrons in an infinite-potential well, and electrons propagating in a one-dimensional periodic potential. The confinement of electrons in a well results in the quantization of the energy levels. In a periodic potential the constraint on the electron movement lies between the free and confined cases. As a result, some of the levels broaden and are continuous, but gaps are still present.

1.3.2 Perturbed Free-Electron Model

This section examines an alternative approach for understanding the origin of band gaps in solids. As indicated in section 1.3.1, the allowed energy bands terminate exactly where Bragg reflections occur in the one-dimensional lattice, a consequence of the periodic potential. We plan to combine this result with the

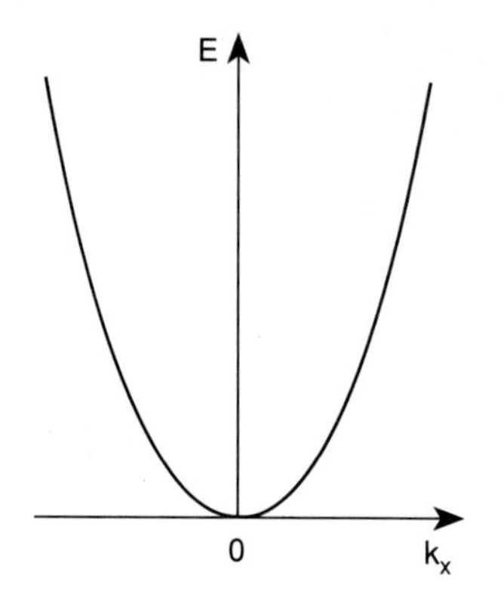

FIGURE 1.10
Electron energy E versus the wave vector k_x for free electrons.

free-electron model to understand the movement of electrons in periodic crystalline solids.

According to Equation (1.10), the energy E of a free electron varies linearly with k^2, where k is the wave vector and can take on all possible values. Figure 1.10 depicts this behavior. When we impose a periodic potential, however, we have seen from the results of the Kronig-Penney model that electron waves with $k = n\pi/a$ cannot propagate through the crystal, because they undergo Bragg reflection. For these values of k, electron waves are represented by standing waves. Let us now impose Bragg diffraction at $k = n\pi/a$ and see if energy gaps will develop in the E versus k curves of a free electron.

The imposed perturbation couples the $+n\pi/a$ state to the $-n\pi/a$ state and can be analyzed as follows. At $k = \pi/a$ two standing waves ψ_+ and ψ_- exist. These wave functions for $n = 1$ can be written as

$$\psi_+ \propto \left(e^{i\pi x/a} + e^{-i\pi x/a}\right) = 2\cos\left(\frac{\pi x}{a}\right) \tag{1.56}$$

and

$$\psi_- \propto \left(e^{i\pi x/a} - e^{-i\pi x/a}\right) = 2\sin\left(\frac{\pi x}{a}\right). \tag{1.57}$$

The spatial probability distributions associated with these two waves are $|\psi_+|^2 \propto \cos^2\left(\frac{\pi x}{a}\right)$ and $|\psi_-|^2 \propto \sin^2\left(\frac{\pi x}{a}\right)$. Our one-dimensional crystal has atoms at $x = a, 2a, 3a \ldots$ Substituting different values of x in the preceding relations, we find that $|\psi_+|^2$ has maxima in the regions of positively charged ion cores where the potential energy for electrons is lowest, whereas $|\psi_-|^2$ has maxima midway between the ion cores where the potential energy for electrons is highest. Therefore, wave functions ψ_+ and ψ_- have two different energies at $k = \pi/a$ such that an energy gap opens up equal to $V(|\psi_+|^2 - |\psi_-|^2)$, where V is the periodic potential of the lattice.

For values of k other than $n\pi/a$, we can treat electrons as free. However, the curve shown in Figure 1.10 is modified in the vicinity of $k = n\pi/a$ as illustrated in

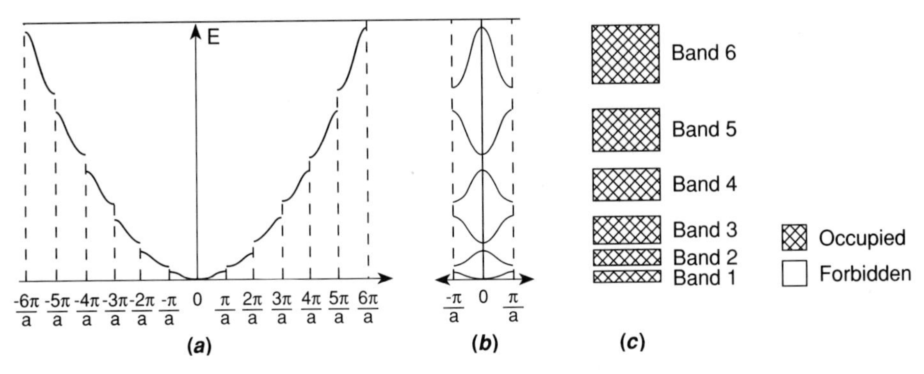

FIGURE 1.11
(*a*) The free-electron model of electrons as perturbed by a small periodic potential causing Bragg reflection conditions for $k = n\pi/a$ in the extended representation. (*b*) The reduced representation in the first Brillouin zone achieved by translating band segments in (a) by $2\pi n/a$ to bring them into the basic zone. (*c*) Separate bands are indicated.

Figure 1.11*a*. This representation is often referred to as an *extended representation* because E(k) is plotted explicitly for all values of k. Since the crystalline structure and the solution for electron waves in this structure are periodic, a displacement of k by $2\pi n/a$ does not alter the solution. It is therefore possible to display the same information as in Figure 1.11*a* using a "reduced representation" as illustrated in Figure 1.11*b*. In this case all values of E are plotted for values of k between $-\pi/a$ and $+\pi/a$. The reduced representation has been achieved by translating the individual segments of Figure 1.11(a) by multiples of $2\pi/a$ to bring them into the $-\pi/a$ and $+\pi/a$ range. Figure 1.11(c) shows the corresponding occupied and forbidden energy bands.

The band of permitted energy between $-\pi/a$ and $+\pi/a$ in Figure 1.11*c* is called the *first Brillouin zone* of this one-dimensional lattice. Values of k between $-2\pi/a$ and $-\pi/a$ and between $+\pi/a$ and $+2\pi/a$ constitute the second Brillouin zone in the extended representation. Likewise, the nth Brillouin zone corresponds to values of k between $-n\pi/a$ and $(n - 1)\pi/a$ and between $+(n - 1)\pi/a$ and $+n\pi/a$.

1.3.3 Velocity of an Electron in an Energy Band

The velocity of an electron is equal to the group velocity of an associated wave v_g, that is, velocity with which the boundaries of the wave propagate, and is given by

$$v_g = \frac{d}{dk}(2\pi\nu), \tag{1.58}$$

where ν is the frequency of the wave. We can rewrite Equation (1.58) as

$$v_g = \frac{d}{dk}\left(\frac{2\pi h\nu}{h}\right). \tag{1.59}$$

Replacing $h\nu$ by E and $h/2\pi$ by $\hbar$, we obtain

$$v_g = \frac{1}{\hbar}\frac{dE}{dk}. \tag{1.60}$$

At the edges of the permitted energies, the E-k curves in Figures 1.11*a* and 1.11*b* are flat; that is, dE/dk is equal to zero. When dE/dk is equal to zero, the electron velocity v_g is also zero. Otherwise, the electron would have a velocity at the edge of a Brillouin zone and would therefore escape into the forbidden energy gap.

1.3.4 Effective Mass of an Electron in an Energy Band

We implied in sections 1.3.1 through 1.3.3 that the mass of an electron in a solid is the same as the mass of a free electron. However, experimental measurements indicate that in some solids the mass is larger, while in others it is slightly smaller than the free-electron mass. This electron mass in the solid is referred to as the *effective mass m**. The difference between the two masses is usually attributed to interactions between the drifting electrons and the atoms in a crystal.

We now develop an expression for the effective mass. The acceleration a experienced by an electron can be obtained by taking the derivative of v_g in Equation (1.60) with respect to t:

$$a = \frac{dv_g}{dt} = \frac{1}{\hbar}\frac{d^2E}{dk^2}\frac{dk}{dt}. \tag{1.61}$$

Substituting for $k = \frac{p}{\hbar}$ from Equation (1.12) into Equation (1.61), we obtain

$$a = \frac{1}{\hbar^2}\cdot\frac{d^2E}{dk^2}\cdot\frac{dp}{dt} = \frac{1}{\hbar^2}\cdot\frac{d^2E}{dk^2}\cdot\frac{d(mv)}{dt} = \frac{1}{\hbar^2}\cdot\frac{d^2E}{dk^2}\cdot F, \tag{1.62}$$

where F is the force. In classical mechanics the force F acting on a mass m and experiencing an acceleration a is given by the following expression:

$$a = \frac{F}{m}. \tag{1.63}$$

Comparing Equations (1.62) and (1.63), the effective mass m* can be written as

$$m^* = \hbar^2\left(\frac{d^2E}{dk^2}\right)^{-1}. \tag{1.64}$$

We see from Equation (1.64) that the effective mass is inversely proportional to the curvature of an E versus k curve, that is, an electron band. The higher the magnitude of the curvature, the smaller the effective mass, which implies that for the same electric field, the velocity of an electron will be higher in that region of the crystal.

1.3.5 Holes in Solids

Previous sections considered the behavior of free electrons. In real solids the number of free electrons may exceed 10^{22} cm^{-3}. Since each energy state can be only occupied by two electrons differing in their spins, fully occupied Brillouin zones evolve. When an electric field is applied to such a solid, electrons cannot move within the solid. As a result, the current is zero. Let us now imagine a situation where some of the energy states near the zone boundary are unoccupied by electrons. The presence of unoccupied states would allow the field-induced movement of electrons within the zone, resulting in a net current. These unoccupied states are termed *holes*. In this abstraction holes are an analogue of electrons and have a positive charge.

1.4 ELECTRONS IN SOLIDS

We now apply the ideas of previous sections to the behavior of electrons in real solids. As indicated earlier the situation is complex because the number of free electrons may exceed 10^{22} cm^{-3}. This section attempts to extend the results of the preceding sections to real, three-dimensional crystals.

1.4.1 Fermi Energy and Fermi Distribution Function

Electrons within a solid occupy energy states spanning a range. The Fermi energy E_f is the highest energy that an electron can have at 0 K. A more general definition of E_f is developed later in this section. In a three-dimensional $\vec{k}$-space, a Fermi surface replaces the one-dimensional Fermi energy. We can use the following modification of Equation (1.10) to illustrate.

$$E_f = \frac{\hbar^2(k_x^2 + k_y^2 + k_z^2)}{2m}, \tag{1.65}$$

where k_x, k_y, and k_z are components of the wave vector $\vec{k}$. For the same value of k, Equation (1.65) can be satisfied for different values of k_x, k_y, and k_z. Therefore, the endpoints of all possible ks define a spherical surface. The outermost spherical surface defined by Equation (1.65) corresponding to the electrons with the highest energy is called the *Fermi surface*.

We can use statistical mechanics to calculate the distribution of the energies among a large number of particles and its change with increasing temperature. The Fermi-Dirac statistics are applicable in the present case. Consequently, the probability that a certain energy level E is occupied by an electron is given by

$$\text{Fermi function} = F(E) = \frac{1}{1 + \exp[(E - E_f)/k_B T]}, \tag{1.66}$$

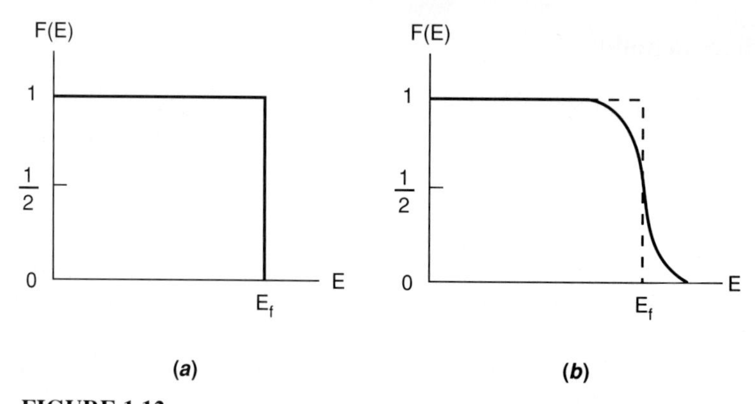

FIGURE 1.12
(*a*) Fermi-Dirac distribution at 0 K. (*b*) Fermi-Dirac distribution at temperatures above 0 K.

where E_f is the Fermi energy, k_B is the Boltzmann constant, and T is the absolute temperature. When an energy level E is occupied by electrons, $F(E) = 1$; for an empty level, $F(E) = 0$. Figure 1.12*a* shows F(E) as a function of E at 0 K. It is clear that the probability of occupation of an energy level below E_f is unity. Above absolute zero some of the electrons near E_f acquire enough energy to move into levels whose energy is higher than E_f, emptying some of the levels immediately below E_f. This situation is depicted in Figure 1.12*b*.

Now let us calculate the value of F(E) when $E = E_f$. Substituting $E = E_f$ in Equation (1.66), we obtain

$$F(E) = \frac{1}{2}. \tag{1.67}$$

That $F(E) = 1/2$ at E_f serves as the generic definition for the Fermi energy.

1.4.2 Density of States in a Band

We want to know how energy levels are distributed over a band. To address this question, we assume that free electrons are confined to a cubical potential box from which they cannot escape. The dimensions of the box are identical to those of the crystal under consideration. We assumed as an approximation in section 1.3.2 that electrons move freely within the bands. Therefore, conceptually this problem is similar to the one-dimensional potential discussed in section 1.2.1. Extending Equation (1.21) to three dimensions, we can write an energy level E_n as

$$E_n = \frac{\pi^2 \hbar^2}{2ma^2}(n_x^2 + n_y^2 + n_z^2), \tag{1.68}$$

where n_x, n_y, and n_z are principal quantum numbers associated with E_n. Treating n_x, n_y, and n_z as variables, replacing $n_x^2 + n_y^2 + n_z^2$ with R^2, and replacing E_n with E, we

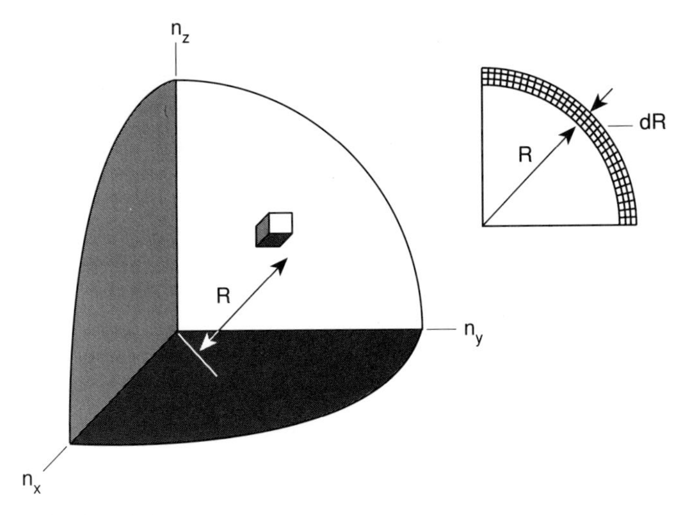

FIGURE 1.13
Energy cubes on spherical energy surface.

can rewrite Equation (1.68) as

$$R = \sqrt{\frac{2ma^2E}{\pi^2\hbar^2}}, \tag{1.69}$$

where E is the energy of a level whose quantum numbers are n_x, n_y, and n_z. Equation (1.69) describes a sphere.

In a one-dimensional case, the principal quantum number n defined a discrete energy state. However, in the present case, sets of values of n_x, n_y, and n_z define a unique energy. Therefore, the problem of finding the energy states between E and E + dE reduces to determining the number of lattice points between R and R + dR on a spherical surface. Since only positive values of n_x, n_y, and n_z are allowed, we need to consider only the positive octant of a sphere as shown in Figure 1.13.

The number of points between R and R + dR is equivalent to the volume between R and R + dR:

$$\begin{aligned}\frac{1}{8}[4\pi R^2 dR] &= \frac{4\pi}{8}\left(\frac{2ma^2E}{\pi^2\hbar^2}\right)\left(\frac{2ma^2}{\pi^2\hbar^2}\right)^{1/2}\frac{E^{-1/2}}{2}\,dE \\ &= \frac{m^{3/2}a^3E^{1/2}\,dE}{\sqrt{2}\pi^2\hbar^3} \\ &= \frac{1}{\sqrt{2}\pi^2}\left(\frac{m}{\hbar^2}\right)^{3/2}E^{1/2}a^3\,dE \\ &= \frac{1}{4\pi^2}\left(\frac{2m}{\hbar^2}\right)^{3/2}E^{1/2}\,\Omega\,dE,\end{aligned} \tag{1.70}$$

where $\Omega = a^3$ is the volume of the crystal in which the electron moves. Thus the number of states per unit energy, that is, the density of the energy states Z(E), is given by

$$Z(E) = \frac{\Omega}{4\pi^2}\left(\frac{2m}{\hbar^2}\right)^{3/2} E^{1/2}. \tag{1.71}$$

We can calculate the number of electrons per unit energy N(E) within an energy interval dE by multiplying the number of possible energy levels Z(E) and the probability for the occupation of these levels. Bear in mind that according to the exclusion principle, there are two electrons per energy level, since the spin quantum number has two states. Therefore,

$$N(E) = 2\,Z(E)\,F(E). \tag{1.72}$$

Substituting for F(E) and Z(E) in Equation (1.72), we obtain

$$N(E) = \frac{\Omega}{2\pi^2}\left(\frac{2m}{\hbar^2}\right)^{3/2} E^{1/2}\frac{1}{1+\exp[(E-E_f)/k_BT]}. \tag{1.73}$$

N(E) is called the electron population density. When E is less than E_f and $T \to 0$, $F(E) \to 1$. Therefore, $N(E) = 2Z(E)$. For $T \neq 0$ and $E \cong E_f$, F(E) smears out N(E). This smeared out region is $E_f \neq k_BT$. Only electrons in this energy range can scatter, which affects thermal and electrical conductivities.

1.4.3 Differences between Metals, Insulators, and Semiconductors

The band picture helps us to distinguish between metals, insulators, and semiconductors. Let us consider the case of a metal with one valence electron per atom. As shown in Figure 1.14*a*, the valence band is half filled because we have two energy states per atom. Upon the application of an external field, an electron can move

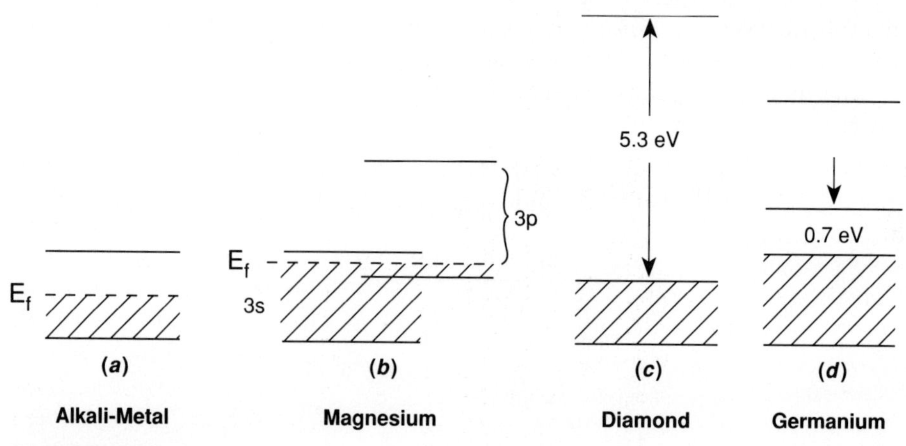

FIGURE 1.14
Schematics showing energy bands in (*a*) alkali metal, (*b*) bivalent metal, (*c*) diamond, an insulator, and (*d*) germanium, a semiconductor.

through the solid because of the availability of empty states. As a result, the solid exhibits conductivity.

In the case of bivalent metals such as Mg or Ca, the situation is slightly different. For example, in Mg the lower portion of the 3p band overlaps with the top portion of the filled 3s band. To lower the energy of the system, some of the electrons from the top of the 3s band are transferred to the bottom of the 3p band, as shown in Figure 1.14*b*. The electrons in the 3s band can occupy these empty states during their field-induced transit through the crystal, leading to high conductivity.

The bands in insulators are shown in Figure 1.14*c*. In this case the states within the valence band are fully occupied, and the valence band and the conduction band are widely separated—that is, the band gap is very large. As a result, thermal hops by electrons from the top of the valence band to the bottom of the conduction band are rare. Following Shockley, the occupied states within a valence band may be likened to a full parking garage. Cars will not be able to move around until a vacancy is created. Likewise, electrons in a filled valence band cannot undergo field-induced drift. Therefore, these materials do not conduct.

In semiconductors the energy states within the valence band are also fully occupied, as shown in Figure 1.14*d*. Therefore, field-induced conduction is not possible, and they behave as insulators at 0 K. However, as the temperature increases, electrons from the top of the valence band can move into the bottom of the conduction band. These transitions produce electrons and holes in the conduction and valence bands, respectively. When an electric field is applied, both electrons and holes can move in their respective bands, leading to electrical conduction.

1.5 SEMICONDUCTORS

We now examine some of the characteristics of semiconductors that make them unique materials.

1.5.1 Direct and Indirect Semiconductors

Equation (1.42) represents the wave function for an electron in a one-dimensional periodic potential. The amplitude u(x) is not constant and reflects the periodicity of the lattice in the direction in which the electron is moving. The existence of a periodic potential introduces energy gaps in a one-dimensional crystal. As shown in Figure 1.11*b*, the resulting situation can be represented using a reduced representation. Since the periodicity in lattices is different in various directions, the $(E, \vec{k})$ diagrams must be plotted for various crystal directions. Figure 1.15 shows general features for the $(E, \vec{k})$ diagrams for GaAs and Si. For simplicity, only two directions for $\vec{k}$ are indicated because in three-dimensions the full relationship between E and $\vec{k}$ is a complex surface.

The analysis of complicated features shown in Figure 1.15 is beyond the scope of this book. However, Figure 1.15*a* clearly shows that the band structure of GaAs

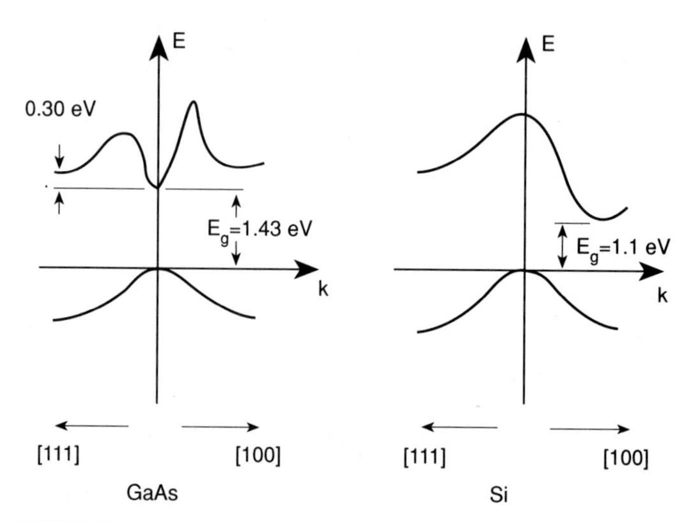

FIGURE 1.15
General features of the energy-band structures for GaAs and Si showing allowed electron energies versus electron propagation vector k.

has a minimum in the conduction band and a maximum in the valence band for the same value of $\vec{k}$, that is, at $|\vec{k}| = 0$. Thus an electron making the lowest energy transition from the conduction band to the valence band in GaAs can do so at the same $\vec{k}$ value. The energy difference is released as a photon. Such semiconductors are called *direct semiconductors* and are used in light-emitting devices. The E versus $\vec{k}$ curves for Si are shown in Figure 1.15(*b*). The minimum in the conduction band and the maximum in the valence band occur at different values of $\vec{k}$ In this case an electronic transition from the conduction band minimum to the valence band maximum involves an intermediate step. Such semiconductors are called *indirect semiconductors* and are not suitable for light emitters.

1.5.2 Charge Carriers in Semiconductors

As we briefly mentioned in section 1.4.3, at 0 K semiconductors behave as insulators. However, at higher temperatures, some of the electrons from the top of the completely filled valence band can make a transition across the energy gap and occupy states in the bottom of the conduction band. As a result, electrons and holes are created in a semiconductor. These electrons and holes are referred to as *charge carriers*.

1.5.2.1 Electrons and holes

As a result of thermal excitation, electrons and holes are created in a semiconductor. The number of electrons is equal to the number of holes, as shown in the schematic in Figure 1.16. A pair of two carrier types is called an *electron-hole pair* (EHP).

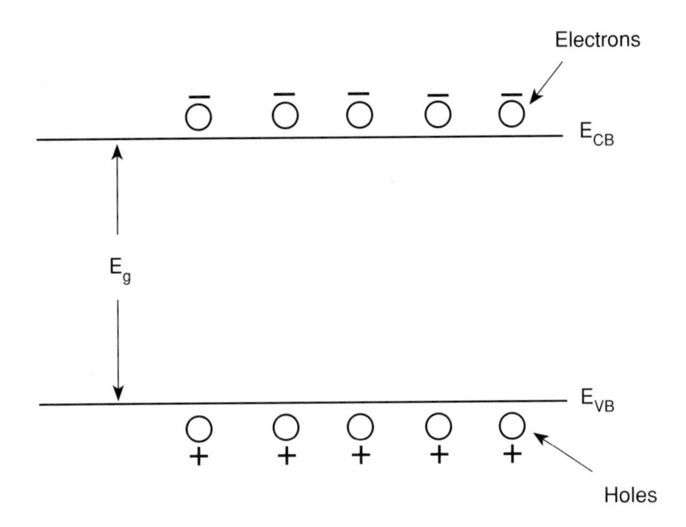

FIGURE 1.16
Schematic showing electron-hole pairs (EHP) in a semiconductor whose band gap is E_g. E_{CB} and E_{VB} denote conduction- and valence-band edges.

At a given temperature the concentration of EHPs depends on the band gap of the semiconductor: The larger the gap, the smaller the concentration of EHPs. For example, the equilibrium number of EHPs in pure Si at ambient temperature is 1.5×10^{10} cm^{-3}. In comparison to the Si atom density of more than 10^{22} cm^{-3}, the EHPs concentration is extremely small. The concentration of EHPs in Si does not increase significantly as its temperature is moderately increased. For example, EHP concentration at 350 K is $\sim 10^{15}$ cm^{-3}, implying that pure Si will have extremely high resistivity and will not be useful as a device material.

1.5.2.2 Intrinsic semiconductors

A semiconductor whose electrical conductivity is dominated by the thermally generated EHPs is called an *intrinsic semiconductor.* Since thermal excitation generates electrons and holes in pairs, the concentration of electrons (negatively charged carriers, n cm^{-3}) must be equal to the concentration of holes (positively charged carriers, p cm^{-3}). Each of these intrinsic-carrier concentrations is referred to as n_i. Therefore, for an intrinsic semiconductor

$$n = p = n_i. \tag{1.74}$$

Two processes determine the intrinsic-carrier concentration: (1) generation of carriers by thermal excitation and (2) recombination of electrons and holes due to energy considerations. In a steady state situation, the generation rate of EHPs g_i must be equal to the recombination rate of electrons and holes r_i, that is,

$$g_i = r_i. \tag{1.75}$$

Both rates are temperature dependent. However, Equation (1.75) is valid at all temperatures because the higher generation rate at a high temperature is balanced by the higher recombination rate.

1.5.2.3 Extrinsic semiconductors

The charge carriers can be introduced in a semiconductor by intentionally introducing impurities into the lattice. This approach is termed *doping*, and the impurities are called *dopants*. Doping is widely used to vary the conductivity of semiconductors. Either extra electrons or extra holes can be introduced into a crystal by doping. The respective dopants are called *donors* and *acceptors*. If the excess carriers are electrons, the doped semiconductor is called *n-type* (for negative). If the excess carriers are holes, the doped semiconductor is called *p-type* (for positive). Both types of doped semiconductors are grouped together under the term *extrinsic*.

The band diagrams of Figure 1.17 show the behavior of dopants. Figure 1.17*a* shows a situation where the energy level of a donor impurity E_d is very close to the conduction-band edge, E_{CB}. If the temperature is very low and the impurity

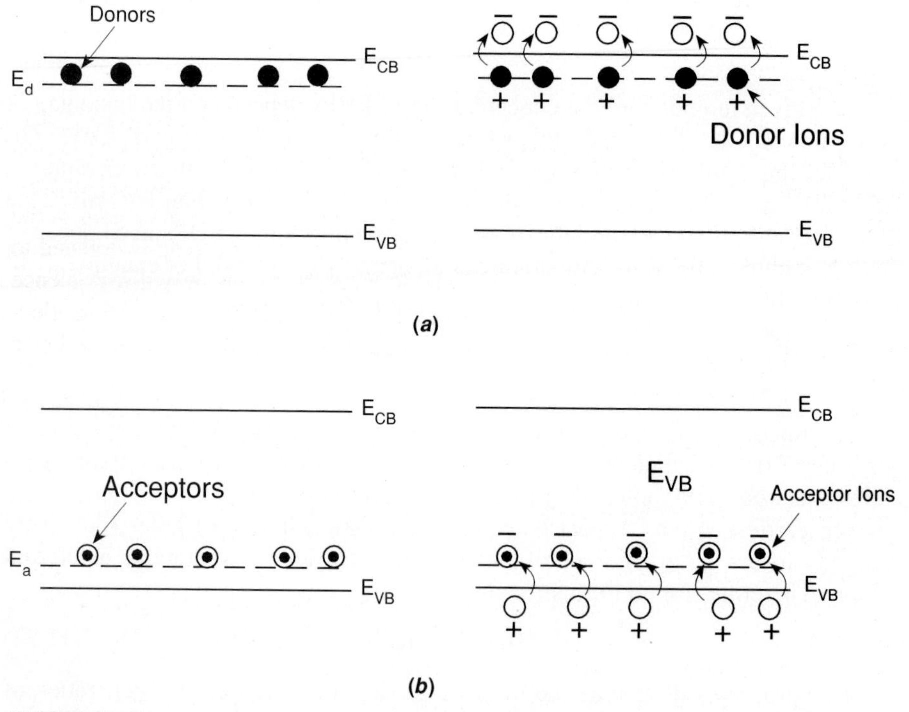

FIGURE 1.17
Band diagrams illustrating the behaviors of donor and acceptor impurities in a semiconductor. E_{CB} and E_{VB} denote conduction- and valence-band edges, whereas E_d and E_a indicate the positions of donor and acceptor levels within the gap.

atoms are not ionized, the situation shown on the left side would result. However, if the thermal energy exceeds the small ionization energy of the donor atoms, electrons will be removed from the donor atoms. The resulting electrons will move into the conduction band as shown on the right side. The concentration of these electrons depends on the concentration of the donor impurity. Therefore, the carrier concentration can be varied in a controlled way by varying the concentration of donors.

The situation regarding acceptors can be analyzed in a similar fashion. In this case the acceptor level E_a due to an impurity is close to the valence-band edge E_{VB}. At very low temperatures, electrons from the filled states in the valence band cannot access the impurity level as shown on the left side of Figure 1.17*b*. However, at higher temperatures this transition happens, leading to the formation of holes in the valence band, as shown on the right side of Figure 1.17*b*.

In addition to doping-induced carriers, thermally generated EHPs are always present in an extrinsic semiconductor. As mentioned in section 1.5.2.1, the concentration of these EHPs is extremely low, the ramification being that a small number of holes will always coexist with electrons in an n-type semiconductor, or vice versa. Electrons and holes in an n-type semiconductor are referred to as *majority* and *minority carriers*. Similarly, holes and electrons constitute majority and minority carriers in a p-type semiconductor.

In the case of elemental semiconductors, that is, Si and Ge, several impurities, such as As, Sb, and P, serve as n-type dopants, whereas p-type dopants are B, Al, Ga, In, and so on. The behavior of these impurities may be qualitatively understood in terms of bonding. Both Si and Ge are covalently bonded solids. Each Si or Ge atom has four valence electrons and is tetrahedrally coordinated to four other atoms. When a group V donor atom replaces the host lattice atom, one of the impurity electrons is not involved in the bonding. This extra electron is donated to the conduction band. On the other hand, when the impurity has only three valence electrons, it can tetrahedrally coordinate with the lattice atoms by pulling an electron from the top of the valence band. This process produces a hole in the valence band.

The doping behavior of compound semiconductors, such as GaAs, InP, CdTe, and ZnSe, is more complicated. To discern whether an impurity acts as a donor or an acceptor, we must specify the sublattice on which a particular impurity resides (see Chapter 3 for details on crystallography). For example, Si acts as a donor on the Ga sublattice, whereas on the As sublattice it acts as an acceptor. Experimentally, Si is observed to reside on either of the two sublattices of GaAs. The impurity exhibiting this type of behavior is called an *amphoteric* impurity. A number of studies have shown that S, Se, and Sn are useful n-type dopants in III-V compounds, whereas Cd, Zn, Mg, Be, and so on are useful p-type dopants.

The donor and acceptor levels in a semiconductor are generally very shallow. The column V donor levels lie approximately 0.01 eV below the conduction-band edge in Ge, and group III acceptor levels lie about 0.02–0.04 eV above the valence-band edge. In Si the usual donor and acceptor levels lie about 0.03–0.06 eV from a band edge.

1.5.3 Carrier Concentrations

This section addresses the following issues: (1) influence of doping on the Fermi level, (2) electron and hole concentrations at equilibrium, (3) effect of temperature on carrier concentration, and (4) charge compensation and space-charge neutrality.

1.5.3.1 Influence of doping on the Fermi level

To evaluate the influence of doping type and concentration on the Fermi level in intrinsic, n-type, and p-type semiconductors, we use the Fermi distribution function F(E) from Equation (1.66). F(E) is equal to 1/2 at the Fermi level, and at temperatures greater than 0 K, F(E) is symmetrical about E_f. In an intrinsic semiconductor the concentrations of electrons and holes are equal. Therefore, E_f must lie at the center of the gap because of symmetry. The electron probability tail of F(E) extending into the conduction band in Figure 1.18*a* is symmetrical with respect to the hole probability tail [1 − F(E)] in the valence band. The Fermi distribution function has values within the band gap, but no energy states are available; that is, Z(E) = 0. Therefore, no electron occupancy results from F(E) in this range.

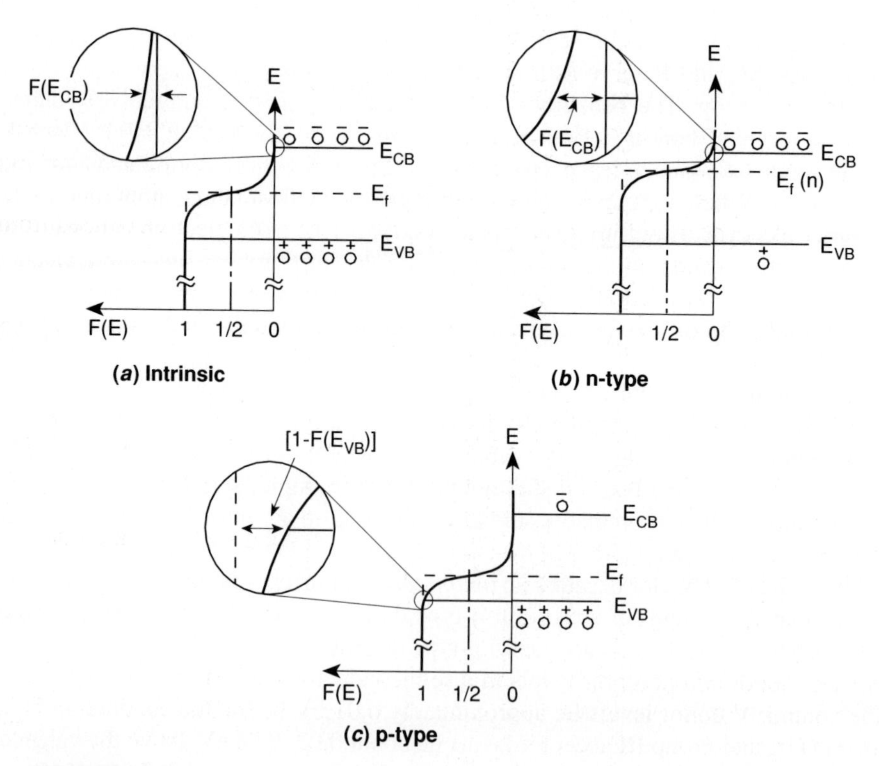

FIGURE 1.18
The Fermi distribution function applied to semiconductors: (*a*) intrinsic material, (*b*) n-type material, and (*c*) p-type material. After Streetman (1980).

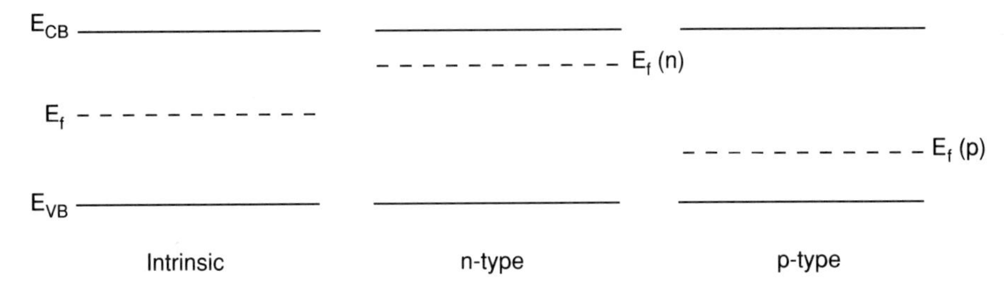

FIGURE 1.19
Schematics showing positions of the Fermi level E_f in intrinsic, n-type, and p-type semiconductors.

In an n-type material the concentration of electrons in the conduction band is considerably larger than the concentration of holes in the valence band. For F(E) to remain symmetrical about E_f, F(E) must lie above its intrinsic position as shown in Figure 1.18*b*. Therefore, as a result of n-type doping, E_f shifts toward the conduction-band edge. The higher the carrier concentration, the greater the shift. Similarly, in a p-type material E_f lies closer to the valence-band edge (see Figure 1.18*c*). The trends from the preceding discussion are schematically displayed in Figure 1.19.

1.5.3.2 Concentrations of electrons and holes at equilibrium

The concentrations of majority carriers can be estimated from doping levels, but the concentration of minority carriers is more difficult to compute. Therefore, we need to develop expressions that can adequately describe these situations.

The concentration of electrons in the conduction band can be calculated from the following expression:

$$n_0 = \int_{E_{CB}}^{\infty} N(E)\, dE, \tag{1.76}$$

where N(E) is defined by Equation (1.73). The subscript 0 indicates equilibrium conditions. The result of the integration in Equation (1.76) is the same as that obtained by representing all the distributed electron states in the conduction band by an effective density of states N_c located at the conduction-band edge E_{CB}. Therefore, the concentration of electrons in the conduction band is simply a product of the effective density of states and the probability of occupancy at E_{CB}:

$$n_0 = N_c F(E_{CB}), \tag{1.77}$$

where $N_c = 2(2\pi m_n^* k_B T/h^2)^{3/2}$ and m_n^* is the electron effective mass. Equation (1.77) assumes that E_f is at least several $k_B T$ below E_{CB}. Under this condition $F(E_{CB})$ can be simplified to

$$F(E_{CB}) = \frac{1}{1 + e^{(E_{CB}-E_f)/k_B T}} \cong e^{-(E_{CB}-E_f)/k_B T}. \tag{1.78}$$

Substituting for $F(E_{CB})$ from Equation (1.78) into Equation (1.79), we obtain

$$n_0 = N_c e^{-(E_{CB}-E_f)/k_B T}. \qquad (1.79)$$

Using similar arguments, the concentration of holes in the valence band is

$$p_0 = N_v[1 - F(E_{VB})], \qquad (1.80)$$

where N_v is the effective density of states at the valence-band edge. The probability of finding an empty state at E_{VB} is

$$\begin{aligned} 1 - F(E_{VB}) &= 1 - \frac{1}{1 + e^{(E_{VB}-E_f)/k_B T}} \\ &\cong e^{-(E_f-E_{VB})/k_B T} \end{aligned} \qquad (1.81)$$

because we assumed that E_f is greater than E_{VB} by several $k_B T$. Substituting the value of $[1 - F(E_{VB})]$ from Equation (1.81) into Equation (1.80), we obtain

$$p_0 = N_v e^{-(E_f-E_{VB})/k_B T}, \qquad (1.82)$$

where $N_v = 2\left(\frac{2\pi m_p^* k_B T}{h^2}\right)^{3/2}$ and m_p^* is the hole effective mass.

The electron and hole concentrations given by Equations (1.79) and (1.82) are valid for doped and undoped materials, provided thermal equilibrium is maintained. In an intrinsic material, E_f is at the same level E_i in both cases. Therefore, the intrinsic electron n_i and hole p_i concentrations are

$$n_i = N_c e^{-(E_{CB}-E_i)/k_B T}, \; p_i = N_v e^{-(E_i-E_{VB})/k_B T}. \qquad (1.83)$$

The product $n_0 p_0$ for a particular material is given by

$$\begin{aligned} n_0 p_0 &= N_c N_v e^{-(E_{CB}-E_f)/k_B T} e^{-(E_f-E_{VB})/k_B T} \\ &= N_c N_v e^{-(E_{CB}-E_{VB})/k_B T} \\ &= N_c N_v e^{-E_g/k_B T}, \end{aligned} \qquad (1.84)$$

where E_g is the band gap of the material. Multiplying n_i and p_i in Equation (1.83), we obtain

$$n_i p_i = N_c N_v e^{-E_g/k_B T}. \qquad (1.85)$$

Since $n_i = p_i$, Equation (1.85) can be simplified to

$$n_i^2 = N_c N_v e^{-E_g/k_B T}. \qquad (1.86)$$

Substituting n_i^2 for $N_c N_v e^{-E_g/k_B T}$ from Equation (1.86) into Equation (1.84), we have

$$n_0 p_0 = n_i^2. \qquad (1.87)$$

For example, at room temperature the intrinsic-carrier concentrations (n_i) in Si and Ge are 1.5×10^{10} cm^{-3} and 2.5×10^{13} cm^{-3}. These values are extremely small. Therefore, undoped materials would have high resistivities.

EXAMPLE 1.4. Given that the Fermi level in a p-doped Si is 0.2 eV above the valence band edge at 300K, determine the concentrations of electrons and holes in the doped semiconductor.

Solution. Substituting for N_v from Equation (1.82) into Equation (1.81), we obtain

$$p_0 = p_i \exp(E_i - E_f)/k_B T.$$

Therefore, $p_0/p_i = \exp(E_i - E_f)/k_B T$

$$E_i - E_f = 0.55 - 0.2 = 0.35 \text{ eV}$$

$$\frac{p_0}{p_i} = \exp\frac{0.35}{8.62 \times 10^{-5} \times 300}$$

$$= 7.4 \times 10^5.$$

Since $p_i = 1.5 \times 10^{10}$ cm^{-3}, $p_0 = 1.1 \times 10^{16}$ cm^{-3}.

Since $n_0 p_0 = (n_i)^2 = (p_i)^2$, $n_0 = \dfrac{2.25 \times 10^{20}}{1.1 \times 10^{16}} = 2.05 \times 10^4$ cm^{-3}.

1.5.3.3 Temperature dependence of carrier concentrations in doped semiconductors

From Equations (1.84) and (1.86) we know that carrier concentrations depend on temperature. These dependences are qualitatively described in Figure 1.20, which shows the carrier concentration of doped Si (10^{15} donors/cm^3) as a function of $1/T$. Three distinct regions are observed. In the first region, labeled "ionization," the carrier concentration increases with temperature as more and more donor impurities are *ionized,* that is, valence electrons are removed from the impurities. In the second region, labeled "extrinsic," the carrier concentration is essentially constant.

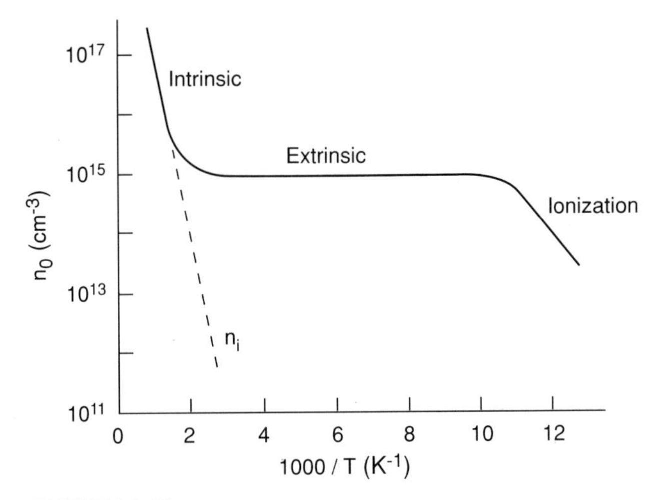

FIGURE 1.20
Carrier concentration versus inverse temperature for Si doped with 10^{15} donors/cm^3.

In the third region, labeled "intrinsic," the carrier concentration increases dramatically with temperature because of thermally generated EHPs.

Based on the preceding discussion, we expect the position of the Fermi level in a doped semiconductor to change with increasing temperature. At low temperatures where impurities are fully ionized, E_f will lie close to the conduction-band edge in an n-type material. At higher temperatures it would shift toward the middle of the gap. A similar shift in E_f toward the intrinsic level occurs in a p-type semiconductor on increasing the temperature.

1.5.3.4 Compensation and space-charge neutrality

The preceding discussion assumed that the material contained either donor or acceptor impurities. In most situations this assumption is true. However, sometimes a semiconductor may contain both donors and acceptors. This discussion assumes that the concentration of donors N_d is greater than that of acceptors N_a. The corresponding band diagram appears in Figure 1.21. Since $N_d > N_a$, the material is n-type. Therefore, the Fermi level E_f is well above the acceptor level E_a. Furthermore, the number of electrons in the conduction band will be less than the number of donors because some of the electrons may have recombined with holes in the valence band. As a result, we expect the net donor concentration to be $N_d - N_a$, instead of N_d. This reduction in carrier concentration in a doped crystal, containing both n- and p-type dopants, is called *compensation*.

The doped material is electrically neutral, so the positive charges (holes and ionized donor atoms) must be equal to the negative charges (electrons and ionized acceptor atoms):

$$p_0 + N_d^+ = n_0 + N_a^-. \tag{1.88}$$

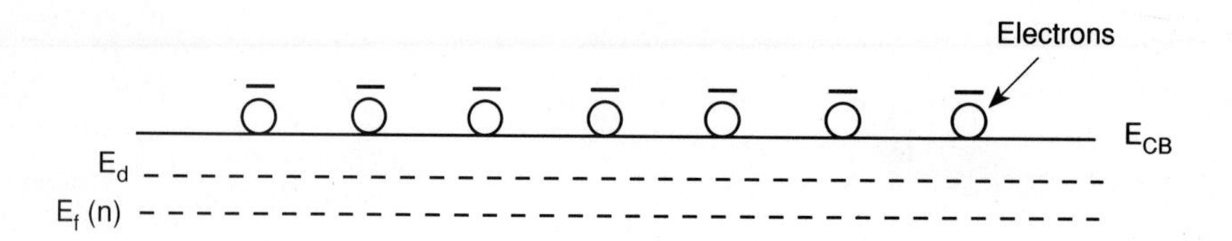

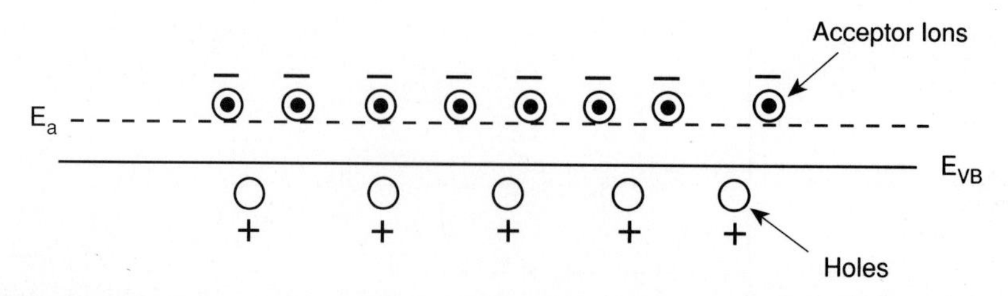

FIGURE 1.21
Schematic showing compensation in an n-type semiconductor. The concentration of holes is smaller than that of the acceptor ions because of charge compensation by electrons.

Solving for n_0, the net electron concentration in the conduction band is

$$n_0 = p_0 + N_d^+ - N_a^-. \tag{1.89}$$

If the material is n-type, that is, $n_0 \gg p_0$, then Equation (1.89) becomes

$$n_0 \cong N_d^+ - N_a^-, \tag{1.90}$$

a result consistent with the preceding qualitative discussion.

1.5.4 Excess Carriers in Semiconductors

Most semiconducting devices operate by the creation of charge carriers in excess of the thermal equilibrium value. These excess carriers can be introduced by optical excitation or electron bombardment, or as discussed in Chapter 2, they can be injected across a forward-biased p-n junction. In this section we examine the generation of carriers by optical excitation and their recombination behavior.

Let us assume that photons of a selected wavelength (energy) are directed at a semiconductor whose band gap is E_g. If the energy of the photons is greater than E_g, photons will be absorbed by the semiconductor. This absorption is due to the generation of EHPs, as shown in Figure 1.22. If the energy of the incident photons is considerably higher than E_g, an electron may be excited to a much higher level in the conduction band. Eventually, this excited electron loses energy by lattice scattering and reaches the thermal-equilibrium velocity of other conduction-band electrons. EHPs created by this absorption process are called *excess carriers*. Since these EHPs represent an unstable situation, they must eventually recombine.

The recombination of an EHP can occur in two ways: (1) direct recombination and (2) recombination involving a defect level in the band gap that is a carrier trap. The two situations are shown in Figure 1.23. Direct recombination is a fast process; the mean lifetime of EHPs is $\sim 10^{-8}$ sec or less. However, when a defect level (which may be due to an impurity) is present in the band gap, it has a strong tendency to temporarily capture (trap) electrons from the conduction band. Subsequently, the electron may be thermally reexcited to the conduction band. Finally,

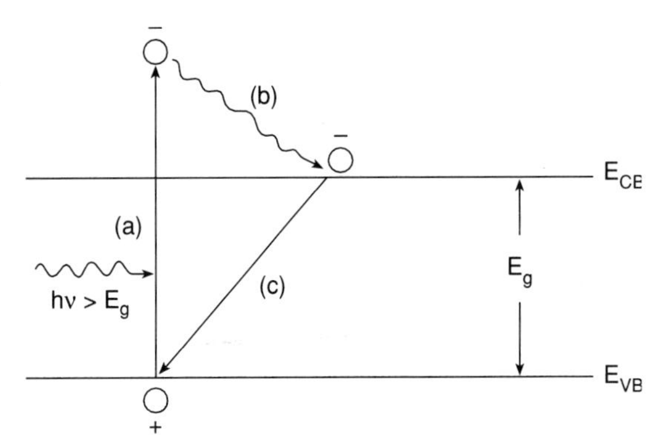

FIGURE 1.22
Schematic showing formation of an electron-hole pair by optical excitation. Energy $h\upsilon$ of the photon is greater than E_g: (*a*) an EHP is created by optical excitation; (*b*) the excited electron loses energy by lattice scattering; (*c*) the electron recombines with a hole in the valence band.

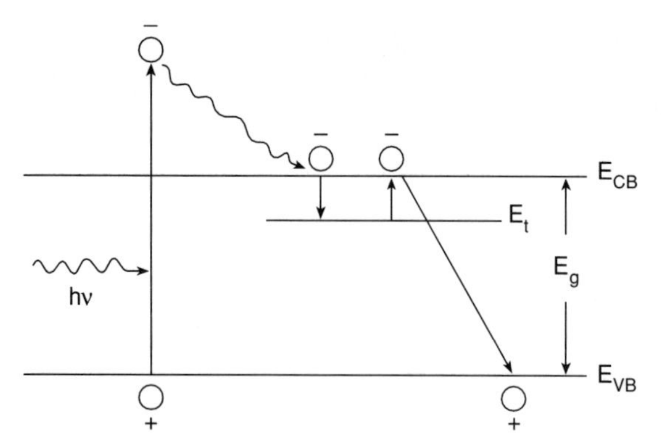

FIGURE 1.23
Schematic illustrating excitation and recombination events in a semiconductor that has an electron trap in the band gap.

direct recombination occurs as the electron recombines with a hole in the valence band. The mean lifetime of EHPs in this case could be fairly long and depends on the binding energy between the electron and the trap. The higher the energy, the longer the lifetime because thermal reexcitation of the trapped electron to the conduction band would be less probable.

Let us assume that EHPs are created at time $t = 0$ and that the initial excess electron and hole concentrations are Δn and Δp are equal. Since the electrons and holes recombine in pairs, the instantaneous concentrations of the excess carriers $\delta n(t)$ and $\delta p(t)$ are also equal. Following Streetman (1980), the rate at which EHPs recombine in a semiconductor can be written as

$$-\frac{d\delta n(t)}{dt} = \alpha_r[n_0 + \delta n(t)][p_0 + \delta p(t)] - \alpha_r n_i^2, \tag{1.91}$$

where α_r is a constant of proportionality for recombination, n_0 and p_0 are the carrier concentrations in the semiconductor, and $\alpha_r n_i^2$ represents the thermal generation rate. Equation (1.91) can be simplified to

$$-\frac{d\delta n(t)}{dt} = \alpha_r[(n_0 + p_0)\delta n(t) + \delta n^2(t)]. \tag{1.92}$$

For low-level injection we can neglect the $\delta n^2(t)$ term. Furthermore, in an extrinsic material we can usually neglect the term representing the concentration of the equilibrium minority carriers. For the sake of discussion, let us assume that $p_0 \gg n_0$. Then Equation (1.92) becomes

$$-\frac{d\delta n(t)}{dt} = \alpha_r p_0 \delta n(t). \tag{1.93}$$

The solution to Equation (1.93) has an exponential form that decays from the original excess carrier concentration Δn:

$$\delta n(t) = \Delta n\, e^{-\alpha_r p_0 t}. \tag{1.94}$$

Excess electrons in a p-type semiconductor recombine with a decay constant $\tau_n = (\alpha_r p_0)^{-1}$, called the *recombination lifetime*. Since this calculation is for the minority carriers, τ_n is referred to as the *minority-carrier lifetime*. The recombination

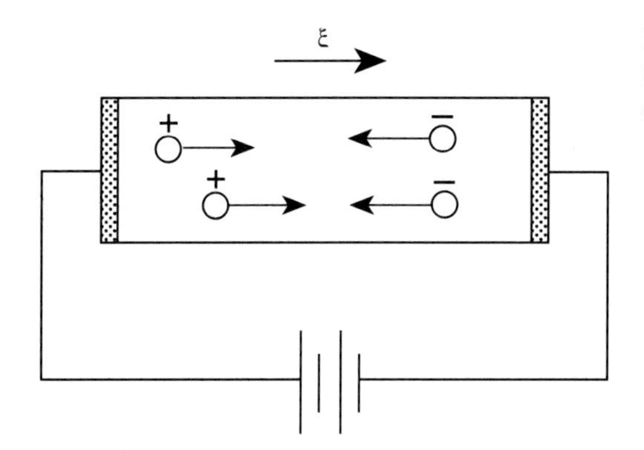

FIGURE 1.24
Schematic of carrier drift in a biased semiconductor.

of excess holes in an n-type material occurs with a decay constant $\tau_p = (\alpha_r n_0)^{-1}$. It is apparent that the minority-carrier lifetimes decrease with the increasing doping levels.

1.5.5 Carrier Drift in Electric and Magnetic Fields

Drift is defined as charged-particle motion in response to an applied electric or magnetic field. Figure 1.24 shows a piece of a semiconductor with an imposed electric field ε. As a result of the field, the force acting on the carriers tends to accelerate the positively charged holes in the direction of the electric field and the negatively charged electrons in the opposite direction. The carrier acceleration is affected by scattering from lattice atoms, impurities, other carriers, and defects. These events reduce the value of carrier acceleration.

The current is the result of carrier drift. Let us calculate this current for a p-type bar of cross-sectional area A shown in Figure 1.25. We can argue the following regarding the charge crossing the hatched plane per unit time in Figure 1.25:

Number of holes crossing the hatched plane in time $t = pv_D tA$, where p is the number of holes in the semiconductor and v_D is the drift velocity that is affected by scattering events.
Charge crossing the plane in time $t = qpv_D tA$, where q is charge on the hole.
Charge crossing the plane per unit time $= qpv_D A$.

The last quantity is clearly identical to the formal definition of current. Therefore

$$I_p(\text{drift}) = qpv_D A, \tag{1.95}$$

or

$$\text{current density} = J_p(\text{drift}) = qpv_D. \tag{1.96}$$

To relate J_p to the applied electric field, we use the following relation:

$$v_D = \mu_p \varepsilon, \tag{1.97}$$

where μ_p is the mobility of holes; its units are cm^2/V-sec. Substituting v_D from Equation (1.97) into Equation (1.96), we obtain

$$J_p = q\mu_p p\varepsilon. \tag{1.98}$$

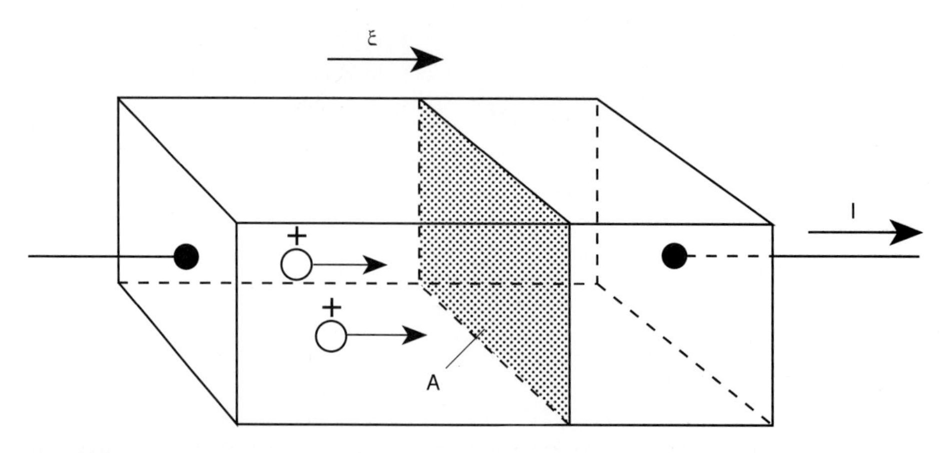

FIGURE 1.25
A biased p-type semiconductor bar of cross-sectional area A.

Likewise, we can show that

$$J_n = q\mu_n n\varepsilon, \tag{1.99}$$

where μ_n is the mobility of electrons. The total current density due to the drift of holes and electrons in an electric field ε is given by

$$J = q\varepsilon(\mu_p p + \mu_n n). \tag{1.100}$$

From Ohm's law,

$$J = \sigma\varepsilon, \tag{1.101}$$

where σ is the conductivity of the semiconductor (its units are $[\Omega\text{-cm}]^{-1}$). Substituting J from Equation (1.101) into Equations (1.98) and (1.99), we obtain

$$\sigma_p = \text{Conductivity due to the migration of holes} = q\mu_p p, \tag{1.102a}$$

or

$$\sigma_n = \text{Conductivity due to the migration of electrons} = q\mu_n n. \tag{1.102b}$$

Mobility measures the ease of carrier motion through a semiconductor. The greater the number of scattering events, the lower the mobility. Mobility is influenced by lattice (phonon) scattering and impurity scattering. Lattice scattering results from the interaction of charge carriers with lattice atoms that may be out of their ideal lattice positions due to thermal vibrations. Since the thermal agitation of atoms increases with increasing temperature, the frequency of lattice-scattering events should increase with temperature. Therefore, the carrier mobility should decrease with the increase in temperature as depicted in Figure 1.26. On the other hand, carrier scattering from ionized impurities becomes significant at low temperatures where lattice vibrations are reduced. Furthermore, cross-sections for carrier scattering from impurities are higher at low temperatures. As a result, the increase in the scattering of carriers by impurities occurs, resulting in a decrease in mobility with decreasing temperature as shown in Figure 1.26.

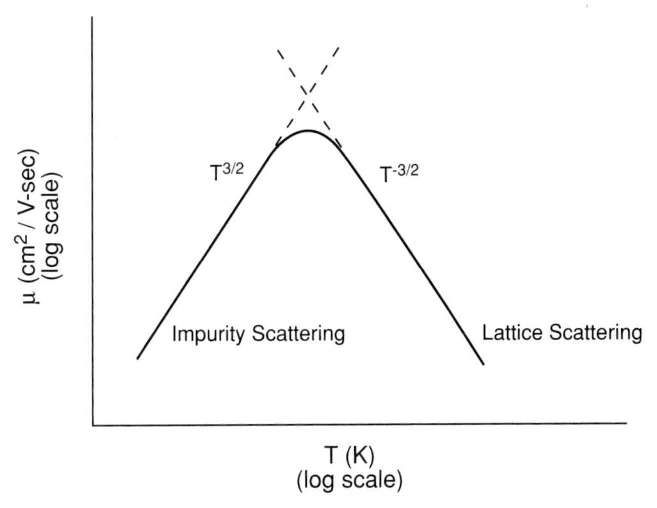

FIGURE 1.26
Influence of impurity and lattice scattering on temperature dependence of mobility.

The Hall effect measures carrier mobility and concentration. Figure 1.27 illustrates this principle. A magnetic field B_z is applied perpendicular to the current flow in a p-type bar whose length, width, and thickness are, respectively, L, w, and t. As a result of the magnetic field, the holes are deflected from the x-direction. The force on the holes in the y-direction is given by

$$F_y = q(\varepsilon_y - v_x B_z), \tag{1.103}$$

where v_x is the drift velocity in the x-direction. Equation (1.103) implies that unless an electric field ε_y is established along the y-direction, each hole will experience a

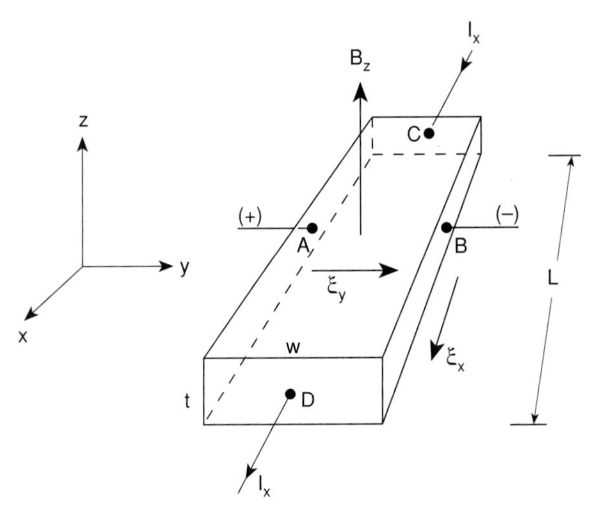

FIGURE 1.27
Schematic illustrating the Hall effect.

net force in the $-$y-direction. At first ε_y is zero. As a result, holes begin to accumulate on the long face A, which increases the value of ε_y. To maintain a steady flow of holes along the +x-direction, the electric field ε_y must balance the $v_x B_z$ product so that the net force F_y is zero. Therefore

$$\varepsilon_y = v_x B_z. \tag{1.104}$$

The establishment of the electric field ε_y is known as the *Hall effect,* and the resulting voltage $V_{AB} = \varepsilon_y w$ is called the *Hall voltage.*

Substituting v_x from Equation (1.97) into Equation (1.104) and denoting holes by p_0, we obtain

$$\varepsilon_y = \frac{J_x B_z}{q p_0}. \tag{1.105}$$

Replacing $\frac{1}{q p_0}$ with R_H, the Hall coefficient, we obtain

$$\varepsilon_y = R_H J_x B_z. \tag{1.106}$$

A measurement of the Hall voltage for a known current and magnetic field yields p_0:

$$p_0 = \frac{J_x B_z}{q \varepsilon_y} = \frac{(I_x/wt) B_z}{q(V_{AB}/w)} = \frac{I_x B_z}{q t V_{AB}}. \tag{1.107}$$

Since all the quantities on the right side of Equation (1.107) can be experimentally measured, the Hall effect measurements provide accurate values for carrier concentrations.

We can measure the sample resistance to calculate its resistivity (ρ). Equation (1.108) shows that the carrier mobility μ_p is related to the Hall coefficient.

$$\mu_p = \frac{R_H}{\rho} \tag{1.108}$$

The terms on the right side of Equation (1.108) are known. Therefore, μ_p can be computed. This approach is also applicable to n-type materials. A negative value of q is used for electrons, and both the Hall voltage and the Hall coefficient are negative.

1.6 HIGHLIGHTS OF THE CHAPTER

The highlights of this chapter are as follows:

- Free electrons exhibit particle-wave duality and can have any energy. However, when they are confined in an infinite-potential well, the energy levels become discrete and quantized. Furthermore, a finite probability exists for an electron to occur outside a finite barrier even when the electron's energy is smaller than the barrier height. This probability arises as a result of the phenomenon called *tunneling*.

- When an electron moves in a periodic potential of a solid, energy gaps develop between the allowed bands of energy states. These gaps are called *band gaps.* Semiconductors have band gaps ranging from 0.2 to 5.0 eV.
- Electrical conduction in semiconductors occurs by the field-induced migration of charge carriers called electrons and holes, whose concentrations can be tailored by doping.

REFERENCES

Bloch, F. Z. *Z. Phys.* 52, 555 (1982); 59, 208 (1930).
Kronig, D. L. and W. G. Penney. *Proc. Roy. Soc.* 130, 499 (1931).
Streetman, B. G. *Solid State Electronic Devices.* Englewood Cliffs, NJ: Prentice-Hall, 1980.

FURTHER READING

Bube, R. H. *Electrons in Solids.* Boston: Academic Press, 1988.
de Cogan, D. *Solid State Devices: A Quantum Physics Approach.* New York: Springer-Verlag, 1987.
Hummel, R. E. *Electronic Properties of Materials.* New York: Springer-Verlag, 1985.
Pierret, R. F. *Modular Series on Solid State Devices.* vol 4. Reading: Addison-Wesley, 1987.
Wilkes, P. *Solid State Theory in Metallurgy.* London: Cambridge University Press, 1973.

PROBLEMS

1.1. A material has the body-centered cubic structure at high temperature and acquires the face-centered cubic structure on cooling. Given that the lattice parameter of the body-centered cubic phase is 0.4 nm, calculate the voltage through which the electrons must be accelerated so that they can undergo Bragg diffraction at an angle of 0.1° from {111} planes of the face-centered cubic phase.

1.2. It can be shown that the energy of a particle in a three-dimensional rectangular well is given by the expression

$$E_n = \frac{h^2}{8m}\left[\frac{n_x^2}{a_x^2} + \frac{n_y^2}{a_y^2} + \frac{n_z^2}{a_z^2}\right],$$

where a_x, a_y, and a_z are dimensions of the well along the x-, y-, and z-axes. Given that the separation between {111} planes in a simple cubic crystal is 0.3 nm; the particle is an electron; and a_x, a_y and a_z are, respectively, equal to separation between the (110), $(1\bar{1}0)$ and (001) planes, calculate the energy of the ground state.

1.3. Given that the unit cell dimension for a simple cubic crystal is 0.25 nm, draw plots of

$$\frac{P \sin \alpha a}{\alpha a} + \cos \alpha a$$

for the motion of an electron in the periodic potential defined by (1) the (110) and (2) the (111) planes. Also, for each case plot E versus k for the first permitted set of energies. Assume $P = 10^{12}a$.

1.4. The behavior of an electron confined to a three-dimensional space is characterized by three independent quantum numbers. How many energy states would you expect for an electron whose principal quantum number is two and is confined to a cube? How many of these states are degenerate? Would the situation change if the electron were confined to a tetragonal box? Explain your answer.

1.5. Using the E-k diagram of a semiconductor given below, answer the following questions:

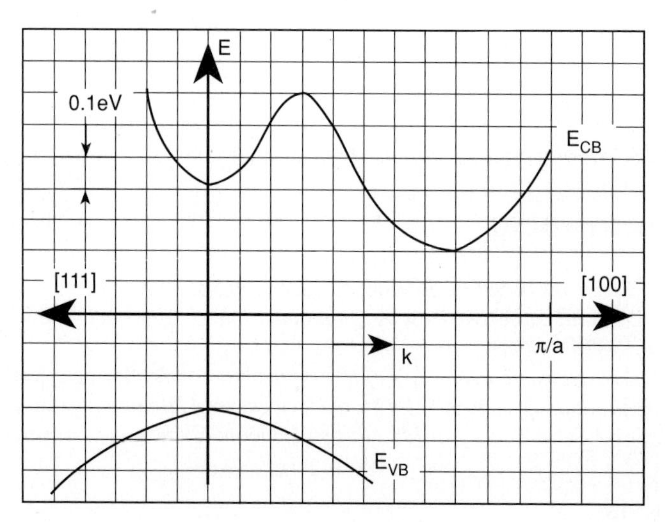

a. What is the energy gap of the semiconductor?
b. Indicate an electron in its lowest energy state in the conduction band.
c. Indicate a hole in its highest energy state in the valence band.
d. Is the gap direct or indirect?
e. Explain which is lower, the hole or electron effective mass.

1.6. The conductivity σ of an intrinsic semiconductor varies with temperature according to the relation

$$\sigma = \sigma_o \exp(-E_g/2k_B T),$$

where σ_o is a constant, E_g is the band gap, k_B is the Boltzmann constant, and T is the temperature. Given that $E_g = 1.12$ eV, calculate the temperature at which the conductivity is twice its value at 300 K.

1.7. Calculate the energy and momentum for the following situations:
(*a*) X-rays of wavelength 0.1 nm, (*b*) electrons accelerated through a potential of 300 kV, (*c*) helium ions accelerated through a potential of 2.5 MV, and (*d*) neutrons in thermal equilibrium in a nuclear reactor at a temperature of 300 K.

1.8. Given that a potential of 0.25 mV is applied to an aluminum conductor that is 100 μm long and has a square cross-sectional area with edge length of 1 μm, answer the following questions:

a. What is the electric field?
b. What is the force acting on an electron?

c. What is the magnitude of the current flowing in the conductor?
d. What is the resistance of the conductor?
e. What is the number of electrons flowing in the conductor/sec?

The density of aluminum is 2.7 gm/cm^3, its resistivity is 2.5×10^{-8} Ω and assume that only one electron/aluminum atom is involved in conduction.

1.9. Silicon is doped with phosphorus to a concentration of 4×10^{17} cm^{-3}. Assuming complete ionization of the phosphorus atoms, what are the concentrations of electrons and holes? Given that the mobilities of electrons and holes are, respectively, 700 and 200 cm^2/v-sec, what is the conductivity of the doped Si? At what temperature are the intrinsic- and extrinsic-carrier concentrations equal?

1.10. A semiconductor, whose energy gap is 1.1 eV, is doped with donors to a concentration of 10^{16} cm^{-3}. Assuming that the donor level is 0.04 eV below the conduction-band edge, sketch ln σ versus $1/T$ for the doped semiconductor and label different regions. (σ refers to the conductivity.)

1.11. Given that the conductivity of an n-type semiconductor at 300 K is 0.01 (Ω-cm)$^{-1}$ and that the mobility of electrons is 1100 cm^2/v-sec, calculate the concentrations of electrons and holes.

1.12. A Si sample is doped with phosphorus to 5×10^{16} cm^{-3}. What is the equilibrium concentration of holes at 300 K? Where is the Fermi level in the doped crystal relative to that in the intrinsic material?

1.13. The mobilities of electrons and holes in intrinsic Si are, respectively, 1300 and 500 cm^2/v-sec. Calculate the intrinsic-carrier concentration if the conductivity at 300 K is 6400 (Ω-cm)$^{-1}$. What fraction of this conductivity is due to electrons? If there are four valence electrons/Si atom and if there are eight Si atoms in a cubic unit cell of 0.543 nm, what fraction of Si atoms contribute their valence electrons to the conduction band?

1.14. Given that in an n-type Si the Fermi level is 0.2 eV below the conduction-band edge at 300 K, calculate the equilibrium concentrations of electrons and holes in the doped Si.

CHAPTER 2

Principles of Semiconducting Devices

This chapter introduces principles of some of the semiconducting devices at a basic level. We first consider p–n and metal-semiconductor junctions and then discuss the behavior of solar cells, light-emitting diodes, bipolar transistors, and field-effect transistors. We build on the conceptual framework of semiconductors developed in Chapter 1. The objective of this chapter is to provide the nonspecialist with an adequate knowledge of devices. This approach should enable the reader to understand the influence of growth- and processing-induced defects on the device behavior brought out in Chapters 5 to 10.

2.1 p–n AND METAL–SEMICONDUCTOR JUNCTIONS

This section discusses two types of junctions: p–n and metal–semiconductor. A p–n junction is formed when two pieces of a semiconductor, differing in their conductivity type, are brought together. We can produce metal–semiconductor junctions by depositing metals on clean surfaces of either n- or p-type semiconductors. We discuss the behavior of the two types of junctions without and with an applied voltage, that is, external bias.

2.1.1 p–n Junctions

We indicated in Chapter 1 that p- and n-type semiconductors differ in their majority- and minority-carrier populations. We want to understand how this population is modified at the junction when it is at equilibrium and how the carrier flow occurs across the junction with bias.

2.1.1.1 Conceptual development

Figure 2.1*a* shows two pieces of a semiconductor that are doped p- and n-type. This figure shows that the two materials contain different concentrations of the two types of charge carriers, that is, electrons and holes. Imagine that the pieces are now brought together to form a p–n junction as depicted in Figure 2.1*b*. The p-type semiconductor has many mobile holes and ionized acceptors and few electrons. The n-type material has many mobile electrons and ionized donors and few holes. Where the two types of materials meet, the carriers diffuse across the junction because of the large concentration gradients in the carrier concentrations. Holes diffuse from the p-side into the n-side, and electrons diffuse from the n–side into the p–side. This diffusion process leaves some uncompensated donors and acceptors in the vicinity of the junction as depicted in Figure 2.1*b*. As a result, an electric field ε is set up across the junction. This field creates drift components of current from n–side to p–side and from p–side to n–side that oppose the diffusion currents (Figure 2.1*c*).

At equilibrium no net current flows across the junction, and the drift current must cancel the diffusion current. In fact, the currents must cancel separately for each type of carrier because there can be no net build up of holes or electrons on either side of the junction. In symbols:

$$J_n \text{ (drift)} + J_n \text{ (diffusion)} = 0 \tag{2.1}$$

$$J_p \text{ (drift)} + J_p \text{ (diffusion)} = 0 \tag{2.2}$$

The electric field ε builds up until the net current is zero.

As shown in Figure 2.1*b*, the electric field exists in some region W about the junction. This region is called the *depletion region* because W is almost depleted of carriers in comparison to the rest of the crystal. The thickness of the depletion region depends on the doping level; the higher the doping level, the narrower the region. This condition occurs because the areal density of the acceptor or donor ions within W is high. Therefore, the electric field necessary to prevent the diffusion of carriers is created by the smaller thickness of the depletion region.

Assuming the electric field is zero in the neutral regions outside W, there is a constant potential V_n in the neutral n-side and a constant potential V_p in the neutral p–side. The difference between V_n and V_p is a potential difference V_0, the contact potential. The contact potential across W is a built-in potential barrier. V_0 is an equilibrium quantity, so no net current can result from it.

To develop an appreciation for the magnitudes of various parameters discussed above, let us consider an abrupt p–n junction in silicon with $N_a = 10^{19}$ cm^{-3} and $N_d = 10^{16}$ cm^{-3}. Following Sze (1985), we can show that the computed values for the electric field, depletion layer width, and built-in voltage for this diode are 5×10^4 V/cm, 0.34 μm, and 0.87 V, respectively. Except for the electric field, the other two quantities are relatively small.

A p–n junction is associated with two types of capacitance: (1) the junction capacitance due to dipole in the depletion layer, and (2) the charge-storage capacitance arising from the lagging behind of voltage as current changes. As the name implies, this capacitance is due to the charge-storage effects. The junction capacitance

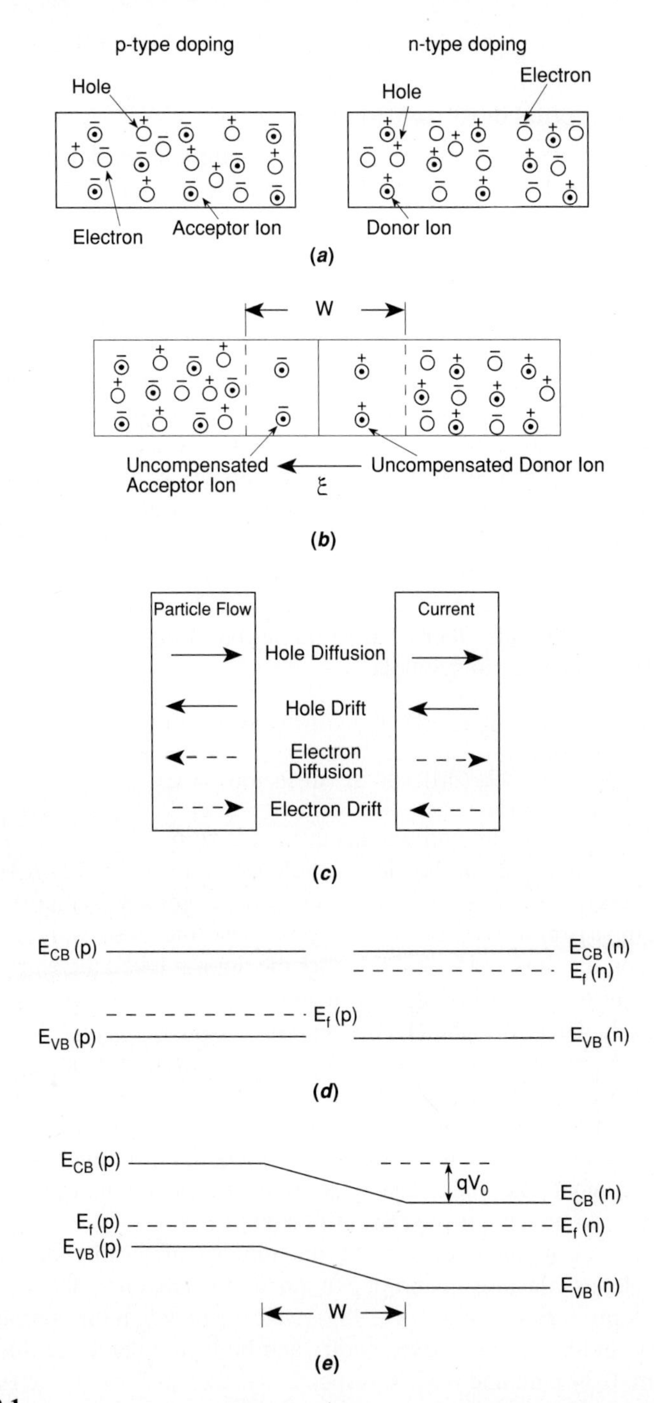

FIGURE 2.1
Schematics showing the behavior of a p–n junction at equilibrium: (*a*) isolated, neutral pieces of p- and n-type semiconductors, (*b*) junction showing the depletion region W and the resulting electric field ε, (*c*) directions of the four components of particle flow within the depletion region and the resulting current directions, and (*d*) and (*e*) the band diagrams corresponding to the situations shown in (*a*) and (*b*). The contact potential V_0 is shown in (*e*).

dominates under reverse-bias conditions, whereas the charge-storage capacitance plays a more important role under forward bias.

The junction capacitance arises from the charge distribution in the depletion region. The uncompensated acceptor ions on the p-side provide a negative charge, and the ionized donor ions on the n-side result in an equal positive charge. The resulting dipole is associated with a capacitance (C_j) that is given by

$$C_j = \frac{\in A}{W}, \tag{2.3}$$

where $\in$ is the permittivity of the semiconductor, A is the junction area, and W is the junction width.

Let us assume that a $p^+ - n$ junction is forward biased and exhibits a steady current I. The capacitance due to small changes in the stored charge is given by

$$C_S = \frac{qI\tau_p}{k_B T}, \tag{2.4}$$

where τ_p is the recombination lifetime for holes. An analogous equation holds for a p–n^+ junction where τ_p is replaced by τ_n.

The band diagrams shown in Figures 2.1*d* and 2.1*e* offer additional insights into the junction behavior. Figure 2.1*d* shows the bands when the p- and n-portions are separated from each other. The Fermi level E_f is close to the valence-band edge E_{VB} in the p-type material, whereas it is closer to the conduction-band edge E_{CB} in the n-type semiconductor. When the two portions are brought together, E_f in the two regions must line up; otherwise, there would be a net current. This situation is shown in Figure 2.1*e*. Furthermore, the contact potential separates the bands as depicted in Figure 2.1*e*; the valence and conduction bands on the p-side of the junction are higher than the corresponding bands on the n-side by the amount qV_0, where q is the charge of the carrier.

Now let us examine the situation when the p–n junction in Figure 2.1*b* is under either forward bias or reverse bias. We will assume that the applied voltage appears only across the junction, not in the neutral n- and p-regions. This assumption is reasonable because in most p–n junction devices the neutral regions are fairly thin. We will take the bias to be positive (forward bias) when the bias is positive on the p-side relative to the n-side.

On biasing, several changes occur at the junction that are schematically depicted in Figures 2.2 and 2.3. When the junction is forward biased V_f, the contact potential of the junction is lowered from the equilibrium value of V_0 to the smaller value $V_0 - V_f$. This change occurs because the forward bias raises the electrostatic potential on the p-side relative to the n-side. For a reverse bias ($V = V_r$) the opposite occurs; the electrostatic potential of the p-side is reduced relative to the n-side, and the potential barrier at the junction becomes $V_0 + V_r$.

The changes also occur in the electric field within the transition region of a p–n junction on biasing and can be assessed from the changes in the potential barrier. The electric field within W decreases with forward bias because the applied electric field opposes the built-in field. On the other hand, the electric field increases with reverse bias, since the applied and built-in fields are in the same direction.

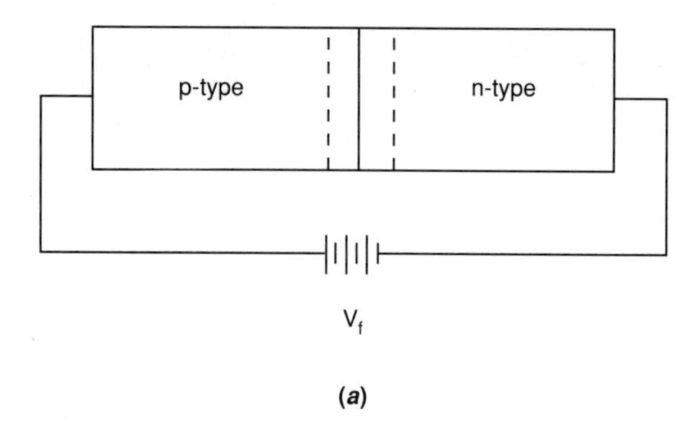

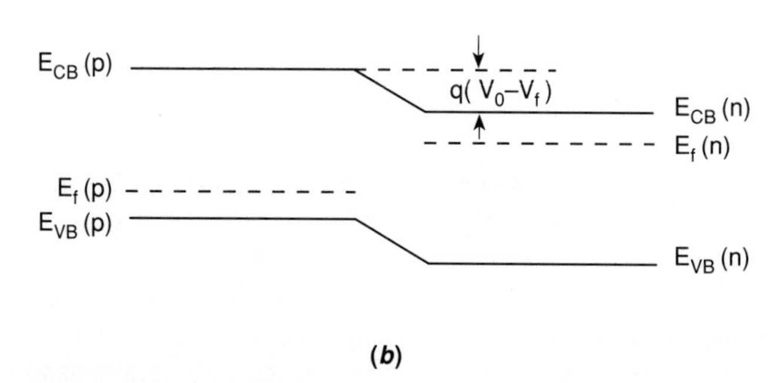

FIGURE 2.2
Schematics showing the behavior of a p–n junction under forward bias: (*a*) forward-biased p–n junction (the forward bias is V_f) and (*b*) the band diagram for a forward-biased junction.

These changes in the electric fields produce corresponding changes in the depletion region width W because the applied bias opposes the built-in field. With increasing bias the built-in field is progressively neutralized. Therefore, we expect W to decrease with forward bias.

As depicted in Figure 2.1*e*, the height of the electron energy barrier is the product of the electronic charge q and the contact potential V_0. Since V_0 is reduced to $V_0 - V_f$ on the application of a forward bias V_f, the height of the barrier is reduced to $q(V_0 - V_f)$; that is, the separations between E_{CB}(n), E_{CB}(p) and E_{VB}(n), E_{VB}(p) are reduced on forward biasing. However, the separations between the respective band edges are increased on reverse biasing. These situations are depicted in Figures 2.2*b* and 2.3*b*.

As discussed in section 1.5.3.1 of Chapter 1, the Fermi level E_f in doped crystals is determined by the dopant type and its concentration. In n-type materials E_f lies close to E_{CB}, whereas it is contiguous to E_{VB} in p-type semiconductors. Conceptually, E_f is rigidly connected to E_{CB}(n) or E_{VB}(p) in doped crystals. Since biasing does not produce any change in the carrier concentration and its type, the

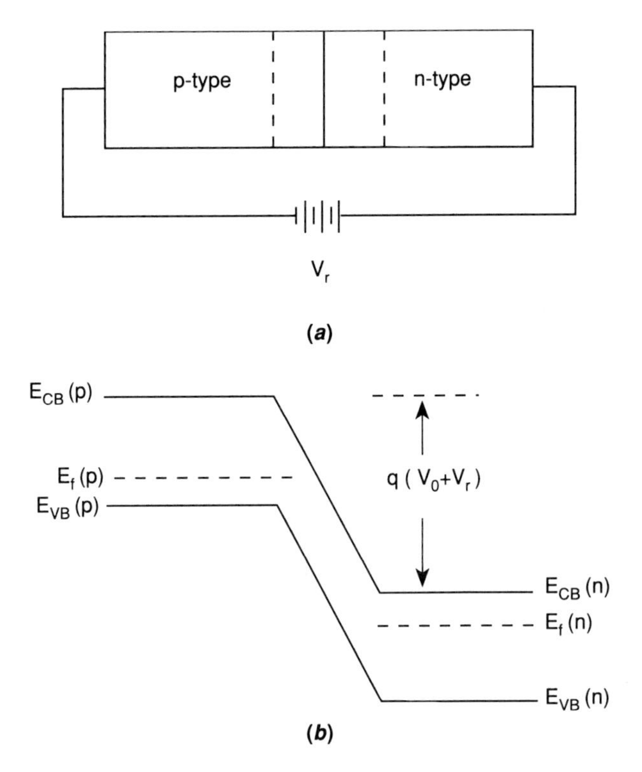

FIGURE 2.3
Schematics illustrating the behavior of a reverse-biased p–n junction: (*a*) reverse-biased p–n junction (the reverse bias is V_r) and (*b*) the band diagram for a reverse-biased junction.

movement of E_f reflects the movement of bands on the application of forward and reverse bias, as shown in Figures 2.2*b* and 2.3*b*. We can argue that in both cases in Figures 2.2*b* and 2.3*b* the Fermi levels in the two neutral regions are separated by the applied voltage.

Now let us examine the effect of bias on the flow of carriers across the junction. On forward biasing the potential barrier is lowered from V_0 to $V_0 - V_f$. As a result, many more electrons in the conduction band on the n-side have sufficient energy to diffuse from n into p; that is, J_n (diffusion) increases. Similarly, more holes can diffuse from p into n because of the reduced barrier. Therefore, J_p (diffusion) is increased on forward bias. Under reverse bias the barrier becomes so large that neither electrons on the n-side nor holes on the p-side have sufficient energy to overcome it. Therefore, the diffusion currents, J_n (diffusion) and J_p (diffusion), are negligible under reverse bias.

The drift currents, that is, J_n (drift) and J_p (drift), are relatively insensitive to the bias because the concentrations of minority carriers are extremely small (see section 1.5.2.3 of Chapter 1). Even though the bias may affect the speed of the carriers, the overall drift currents are not changed because they depend on the carrier concentration. It is therefore a good approximation that J_n (drift) and J_p (drift) at the junction are independent of the bias.

The carriers responsible for the drift currents are thermally generated EHPs. Two distinct situations may arise: EHPs are generated either in the transition region or in the neutral regions. In the depletion region the carrier density is quite small, so the carriers can drift across the region without recombining. However, this condition is not true in the neutral regions. In this case only carriers that are within a diffusion length from the depletion region drift across without recombination, the *diffusion length* being the average distance that a carrier travels without recombining.

Taking the total current crossing the junction as the sum of the diffusion and drift components, the preceding ideas on the junction behavior under different biases can be synthesized into a current I versus voltage V curve shown in Figure 2.4. In this figure the positive direction for the current is taken from p to n, and the applied voltage is positive when the positive terminal of the battery is connected to p and the negative terminal to n.

Let us now examine the behavior of a junction under increasing reverse bias. The voltage-independent generation current, referred to as the *reverse-saturation current,* is observed until a critical bias is reached. At this critical voltage V_B, the reverse breakdown occurs and the reverse current in the diode increases sharply, as illustrated in Figure 2.5. Relatively large currents can flow with little further increase in voltage. We see that a p–n junction behaves like a switch: Current flows readily in the forward direction of the diode, but almost no current flows in the reverse direction.

We can explain reverse breakdown in terms of two mechanisms: (1) Zener breakdown and (2) avalanche breakdown. The Zener breakdown occurs in a heavily doped junction, and its origin is schematically shown in Figure 2.6. Figure 2.6*a* shows the junction at equilibrium. When the junction is reverse biased, the n-side conduction band may appear opposite the p-side valence band at relatively low voltages (see Figure 2.6*b*). This arrangement may align the large number of empty

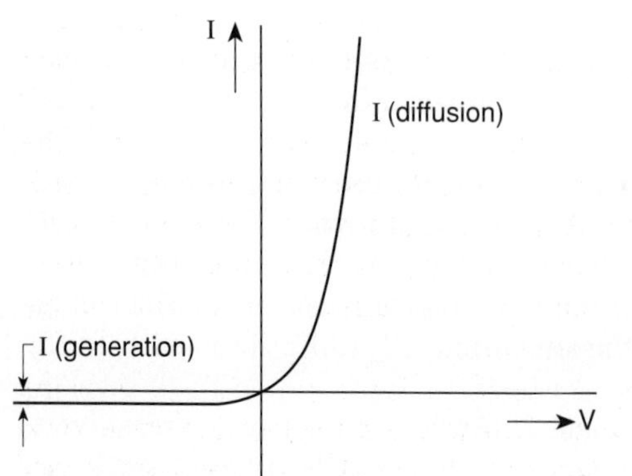

FIGURE 2.4
I-V characteristics of a p–n junction.

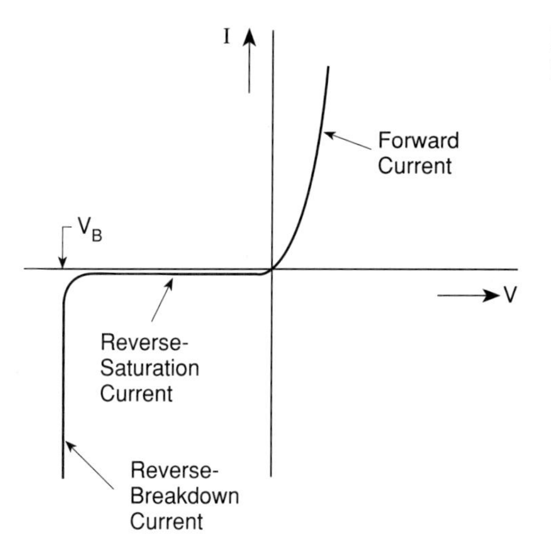

FIGURE 2.5
Reverse breakdown in a p–n junction.

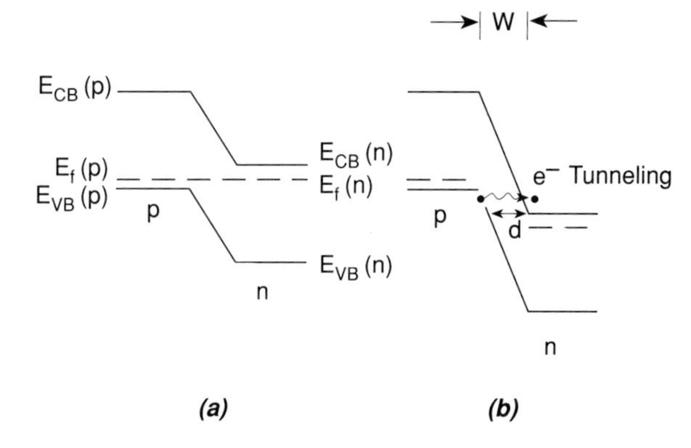

FIGURE 2.6
Schematics illustrating the Zener breakdown of a reverse-biased p–n junction: (*a*) heavily doped junction at equilibrium and (*b*) electron tunneling from p to n under reverse bias.

states in the n-side conduction band opposite the many filled states in the p-side valence band. If the barrier separating the two bands is narrow, electrons may tunnel as discussed in section 1.2.2. Tunneling of electrons from the p-side to the n-side constitutes a reverse current from n to p.

Conceptually, we can think of the Zener effect in terms of field ionization of the host lattice atoms at the junction. When a heavily doped junction is reverse biased, a large electric field may be produced within the depletion region. At a critical field strength, the electrons involved in covalent bonding may be separated from each other and accelerated to the n-side of the junction. The electric field required for this type of ionization is fairly high ($\sim 10^6$ V/cm).

When the junction is lightly doped, electron tunneling is negligible because the tunneling barrier is very wide (see section 1.22). Therefore, the junction breakdown cannot occur via the Zener mechanism. Instead, the breakdown involves impact

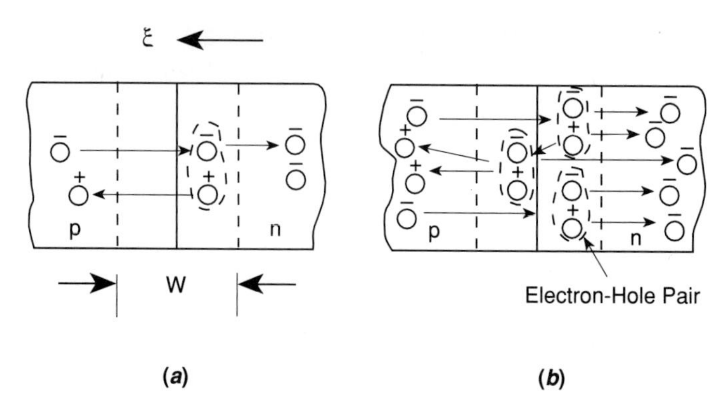

FIGURE 2.7
Schematics illustrating the generation of EHPs by impact ionization: (*a*) a single ionizing collision with a host lattice atom by an incoming electron and (*b*) multiplication of carriers due to multiple collisions.

ionization of host lattice atoms by the energetic carriers. If the electric field ε in the depletion region in Figure 2.7*a* is large, an electron coming from the p-side may acquire enough kinetic energy to cause an ionizing collision with a host lattice atom on the n-side. This collision will produce an EHP. If the carriers constituting the pair are separated by the field of the junction and the resulting carriers in turn have enough kinetic energy to produce ionizing collisions, then carriers can multiply very rapidly as shown in Figure 2.7*b*. This process is called *avalanche multiplication* and is responsible for the breakdown of lightly doped, reverse-biased p–n junctions.

The following criteria can be used to distinguish between the Zener and avalanche breakdowns: (1) the I-V characteristic in the breakdown region is soft for the Zener breakdown and hard in the case of avalanche breakdown and (2) the breakdown voltage V_B for the Zener breakdown decreases with increasing temperature, whereas it increases with temperature for avalanche breakdown.

Figure 2.8 shows the calculated critical fields at breakdown as a function of the background doping for Si and gallium arsenide one-sided abrupt junctions where the depletion layer in one of the semiconductors is very thin (Sze 1985). Also indicated is the critical field for tunneling. Consistent with the preceding discussion, tunneling occurs only in semiconductors having high doping concentrations.

To sensitize the reader to the order of magnitudes of forward and reverse currents, turn-on voltage, and breakdown voltage, we chose an example of a p–n junction in Si. When the forward bias is increased from 0 to 0.6 V, the forward current increases from 10^{-11} to 10^{-4} A. The turn-on voltage is $\sim$0.3 V. On the other hand, the reverse current increases from $\sim 10^{-12}$ to 10^{-8} A with the increase in the reverse-bias voltage from 10^{-3} to 10 V, the breakdown occurring around 10 V.

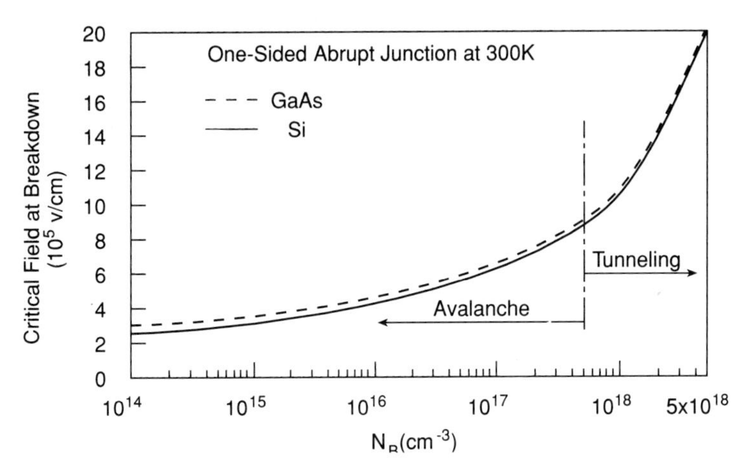

FIGURE 2.8
Critical field at breakdown versus background doping for one-sided abrupt junctions in Si and GaAs.

2.1.1.2 Quantitative treatment of I-V characteristics

We now develop an expression that describes the I-V characteristics of a p–n junction shown in Figure 2.1*b*. We carry this out in two steps. We first obtain a quantitative relationship between the contact potential V_o and the carrier concentrations on each side of the junction at zero bias. We then inject carriers across the junction by biasing it and evaluate the resulting carrier concentrations.

To obtain a relationship between V_o and the carrier concentrations, we apply the equilibrium conditions given by Equations (2.1) and (2.2). Let us first consider the case of holes. We can rewrite Equation (2.2) as

$$J_p(x) = q\Big[\underbrace{\mu_p p(x)\varepsilon(x)}_{\text{drift}} - \underbrace{D_p \frac{dp(x)}{dx}}_{\text{diffusion}}\Big], \tag{2.5}$$

where $J_p(x)$ is the current due to the migration of holes from p to n, q is the charge of the holes, μ_p is the hole mobility, $\varepsilon(x)$ represents the variation in the built-in field with distance x, D_p is the diffusion coefficient of holes, and $\frac{dp(x)}{dx}$ is the change in the hole concentration with distance. In Equation (2.5) the x-direction is arbitrarily taken from p to n. Equation (2.5) can be rearranged as

$$\frac{\mu_p \varepsilon(x)}{D_p} = \frac{1}{p(x)}\frac{dp(x)}{dx}. \tag{2.6}$$

Using the Einstein relation $\frac{-q}{k_B T} = \frac{\mu_p}{D_p}$ and writing ε as the gradient in the potential, that is, $\frac{dV}{dx}$, we can express Equation (2.6) as

$$\frac{-q}{k_B T}\frac{dV}{dx} = \frac{1}{p(x)}\frac{dp(x)}{dx}, \tag{2.7}$$

where k_B is the Boltzmann constant. If the potentials and the hole concentrations in the neutral regions on either side of the depletion region are V_p, V_n, and p_p, p_n, respectively, and assuming a one-dimensional geometry, we can integrate Equation (2.7) between the limits V_p and V_n and p_p and p_n and obtain

$$-\frac{q}{k_B T}(V_n - V_p) = \ln\left(\frac{p_n}{p_p}\right). \tag{2.8}$$

Since the potential difference $(V_n - V_p)$ is the contact potential V_o, we can rewrite Equation (2.8) as

$$V_o = \frac{k_B T}{q}\ln\left(\frac{p_p}{p_n}\right). \tag{2.9}$$

If we assume that the step junction is made up of materials with N_a acceptors/cm^3 on the p-side and N_d donors/cm^3 on the n-side, we can write Equation (2.9) as

$$V_o = \frac{k_B T}{q}\ln\left(\frac{N_a N_d}{n_i^2}\right), \tag{2.10}$$

where n_i is the intrinsic-carrier concentration because from Equation (1.87) $n_i^2 = N_d\, p_n$.

Another useful form of Equation (2.9) is

$$\frac{p_p}{p_n} = e^{qV_o/k_B T}. \tag{2.11}$$

Using the condition $p_p\, n_p = p_n\, n_n = n_i^2$, we can develop from Equation (2.11) a relationship between the electron concentrations on either side of the junction:

$$\frac{n_n}{n_p} = e^{qV_o/k_B T}. \tag{2.12}$$

From the conceptual treatment, we expect the minority-carrier concentration on each side of a p–n junction to vary with the bias because of changes in the diffusion of carriers across the junction. If V is the applied bias, we can show that Equation (2.11) becomes

$$\frac{p(-x_{p_o})}{p(x_{n_o})} = e^{q(V_o - V)/k_B T}. \tag{2.13}$$

Positions $-x_{p_o}$ and x_{n_o} are defined in Figure 2.9 that shows the minority-carrier distributions on the two sides of the depletion region of a forward-biased junction. V is positive for forward bias, whereas it is negative for reverse bias. Assuming that the diffusion of carriers across the junction has negligible effects on the

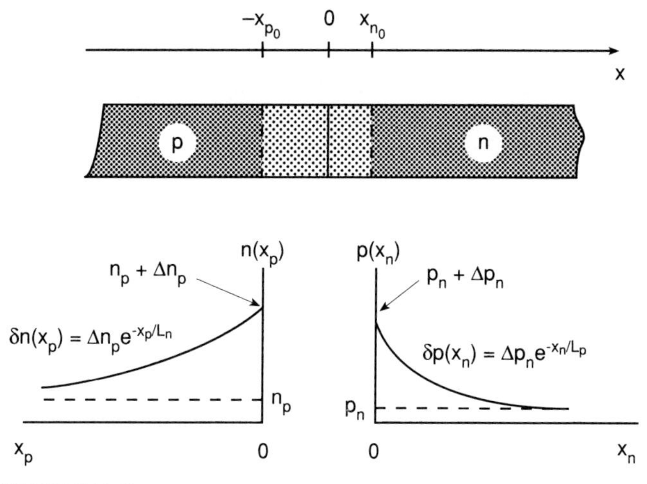

FIGURE 2.9
Minority-carrier distributions on the two sides of the depletion region for a forward-biased p–n junction. X_n and X_p are measured from the edges of the depletion region.

majority-carrier concentrations, that is, $p(-x_{p_o}) = p_p$, we can divide Equation (2.9) by Equation (2.11) and obtain

$$\frac{p(x_{n_o})}{p_n} = e^{qV/k_BT}. \tag{2.14}$$

Equation (2.14) clearly shows that with forward bias the hole concentration at the edge of the depletion region on the n-side is greatly increased over the equilibrium value. Conversely, the hole concentration under reverse bias is reduced below the equilibrium value. The exponential increase of the hole concentration at x_{n_o} in Figure 2.9 under forward bias represents an example of minority-carrier injection. We can similarly show that a forward bias V results in a steady-state injection of excess electrons into the region. These results are schematically shown in Figure 2.9.

We can estimate the excess hole concentration Δp_n at the edge of the depletion region x_{n_0} by subtracting the equilibrium hole concentration p_n from Equation (2.12) and obtain

$$\Delta p_n = p(x_{n_o}) - p_n = p_n\left(e^{qV/k_BT} - 1\right). \tag{2.15}$$

Similarly, for excess electrons on the p-side, we get

$$\Delta n_p = n(-x_{p_o}) - n_p = n_p\left(e^{qV/k_BT} - 1\right). \tag{2.16}$$

These excess carriers diffuse away from the junction because of the gradients in their concentrations. During their transit the excess minority carriers combine with the respective majority carriers. As a result, their concentrations progressively decrease, which is depicted in Figure 2.9. These curves represent exponential solutions to the diffusion equations covered in Chapter 8. The solutions are valid when the hole and electron diffusion lengths L_p and L_n are small in comparison to the lengths of the n- and p-regions. Designating the distances measured in the x-direction from x_{n_o} in the n material and in the $-$x-direction from $-x_{p_o}$in the p material as x_n and x_p, we obtain the distribution of excess carriers δp and δn:

$$\delta p(x_n) = \Delta p_n\, e^{-x_n/L_p} = p_n\left(e^{qV/k_BT} - 1\right)e^{-x_n/L_p} \tag{2.17a}$$

$$\delta n(x_p) = \Delta n_p\, e^{-x_p/L_n} = n_p\left(e^{qV/k_BT} - 1\right)e^{-x_p/L_n}. \tag{2.17b}$$

The diffusion current at any point x_n due to the migration of holes in the n material can be written as

$$I_p(x_n) = -qAD_p\frac{d\delta p(x_n)}{dx_n}, \tag{2.18}$$

where A is the cross-sectional area of the junction and D_p is the diffusion coefficient of holes. Using Equation (2.17*a*), we can write Equation (2.18) as

$$I_p(x_n) = qA\frac{D_p}{L_p}\,\Delta p_n\, e^{-x_n/L_p} = qA\,\frac{D_p}{L_p}\,\delta p(x_n). \tag{2.19}$$

By evaluating Equation (2.19) at x_{n_o}, we can obtain the hole current injected into the n material at the junction:

$$I_p(x_n = x_{n_o}) = \frac{qAD_p}{L_p}\,\Delta p_n = \frac{qAD_p}{L_p}\,p_n\left(e^{qV/k_BT} - 1\right). \tag{2.20}$$

Similarly, we can write an expression for the electron current injected into the p material at the junction:

$$I_n(x_p = -x_{p_o}) = \frac{qAD_n}{L_n}\,\Delta n_p = \frac{qAD_n}{L_n}\,n_p\left(e^{qV/k_BT} - 1\right). \tag{2.21}$$

Assuming that the diode is ideal and there is no carrier recombination in the depletion region, we can obtain the total current by simply adding I_p and I_n:

$$\begin{aligned} I &= qA\left(\frac{D_p}{L_n}\,p_n + \frac{D_n}{L_p}\,n_p\right)\left(e^{qV/k_BT} - 1\right) \\ &= I_o\left(e^{qV/k_BT} - 1\right), \end{aligned} \tag{2.22}$$

where $I_o = qA(D_p/L_n p_n + D_n/L_p n_p)$. Equation (2.22) is called the *diode equation* and gives the current through a p–n junction under forward bias $V > 0$ and reverse bias $V < 0$. If the reverse bias V_r is greater than a few k_BT/q, then $e^{-qV_r/k_BT} \to 0$. In this case then $I \to I_o$, which is the reverse-saturation current shown in Figure 2.5.

In developing Equation (2.22), we assumed that recombination and thermal generation of carriers occur primarily in the neutral regions. This assumption is not completely correct unless the width of the depletion region W is very small in comparison to the carrier diffusion lengths L_n and L_p. We can account for recombination within the depletion by modifying Equation (2.22) as follows:

$$I = I_o \left(e^{qV/\eta k_B T} - 1 \right), \tag{2.23}$$

where η is called the *ideality factor*. Its values vary between one and two depending on the material and temperature.

EXAMPLE 2.1. Assuming that a p–n junction breaks down by the avalanche mechanism, estimate the breakdown voltage for the following conditions: (1) the junction is one-sided with $N_a = 10^{16}$ cm^{-3}, (2) the saturated drift velocity V_s is 10^5 m/sec, (3) the mobility of the semiconductor μ is 0.1 m^2/v-sec, (4) its band gap E_g is 1.1 eV, and (5) its permittivity $\in$ is 1.05×10^{-12} F/cm.

Solution. Assume that the electron reaches its saturation velocity V_s in the high-field region. The mean free path between collisions L is given by the following equation:

$$L = \frac{m V_s \mu}{q},$$

where m is the mass of the electron and q is its charge. For avalanche breakdown to occur, the carrier must acquire enough energy from the field while traversing the distance L between collisions to create an electron-hole pair. In other words, the work done by the field must equal the band-gap energy of the semiconductor, that is,

$$q\varepsilon_B L = E_g,$$

where ε_B is the field at which breakdown occurs.
Substituting appropriate values of various parameters in the preceding equations, ε_B is estimated to be 1.93×10^5 V/cm. According to Sze (1985), the breakdown voltage V_B is given by the following equation:

$$V_B = \frac{\in \varepsilon_B^2}{2qN_a}.$$

By substituting the given and calculated values for different parameters, V_B can be estimated.

$$V_B = \frac{1.05 \times 10^{-12} \times (1.93)^2 \times 10^{10}}{2 \times 1.6 \times 10^{-19} 10^{16}} \text{ volts.}$$

$$V_B = 12.2 \text{ volts.}$$

2.1.2 Metal–Semiconductor Junctions

Many useful transport properties of p–n junctions discussed in the previous section can also be achieved by depositing a metal on a semiconductor, that is, by forming a metal–semiconductor junction. We have two types of these junctions: (1) rectifying and (2) nonrectifying. We now discuss their electrical behavior.

2.1.2.1 Rectifying junctions (Schottky barrier diodes)

Let us form a junction between a metal and an n-type semiconductor whose band diagrams are shown in Figure 2.10*a*. In this figure, $q\phi_M$ and $q\phi_S$ refer to the work functions of the metal and the semiconductor, respectively, and represent the energy required to remove electrons at the respective Fermi levels to the vacuum level. We assume that ϕ_M is greater than ϕ_S so that the Fermi level in the metal is at a lower position than that in the semiconductor. In addition, $q\chi$ is the electron affinity of the semiconductor and is the energy difference between the conduction-band edge E_{CB} and the vacuum level.

When the junction is formed, electrons from the conduction-band states in the semiconductor flow into the metal because they have higher energy. This flow occurs until the Fermi levels are aligned as shown in Figure 2.10*b*. The electron flow from the semiconductor to the metal leaves behind positively charged donor ions in a thickness W in the semiconductor; that is, a depletion region of thickness W is formed. Its ramification is that near the surface region the Fermi level must move deeper into the semiconductor, leading to upward bending of the energy bands (see Figure 2.10*b*). Additionally, because of the electron transfer from the semiconductor to the metal, the negative charge develops on the metal side and is contained within an atomic distance from the surface. The presence of these two types of charges on either side of the junction establishes an electric field that is directed from the semiconductor to the metal.

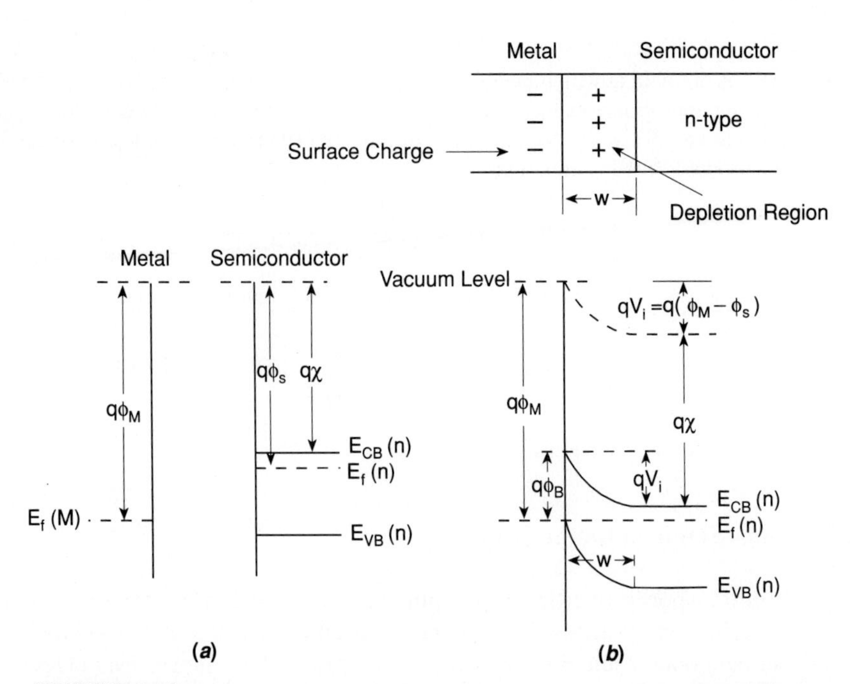

FIGURE 2.10
Energy-band diagrams for a metal/n-type semiconductor contact with $\phi_M > \phi_S$: (*a*) two materials isolated from each other and (*b*) at thermal equilibrium after the contact is made.

The equilibrium contact potential of the junction V_o prevents further flow of electrons from the semiconductor into the metal. It represents the difference between the work function potentials ϕ_M and ϕ_S. The barrier height $q\phi_B$ for the injection of electrons from the metal into the semiconductor is given by

$$q\phi_B = q\phi_M - q\chi. \tag{2.24}$$

This type of metal–semiconductor junction is called a *Schottky barrier.*

A Schottky barrier is also formed when a metal is deposited on a p-type semiconductor such that $\phi_M < \phi_S$. This condition is schematically shown in Figure 2.11. To align the Fermi levels in this case, electrons must flow from the metal to the semiconductor, resulting in a positive surface charge in the metal and a negative charge on the semiconductor. The negative charge exists within a depletion region W in which acceptor ions are left uncompensated by the holes. The potential barrier V_0 retarding the diffusion of holes from the semiconductor to the metal is $\phi_S - \phi_M$.

The two other cases of ideal metal–semiconductor junctions are when $\phi_M < \phi_S$ for n-type and $\phi_M > \phi_S$ for p-type semiconductors. Both result in nonrectifying junctions. We discuss these cases in section 2.1.2.2.

We can deduce from Equation (2.24) that for a given semiconductor ($q\chi =$ constant), the Schottky barrier heights $q\phi_B$ for different metals should be different because of the changes in the metal work functions $q\phi_M$. This ideal situation is not always realized. We find very often that the Schottky barrier heights for various metals on a particular semiconductor are the same. This behavior is attributed to the presence of a large number of interface states in the band gap of the surface region

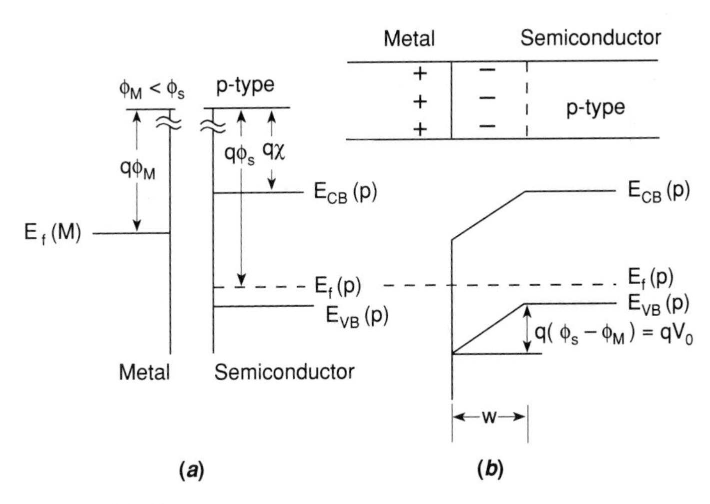

FIGURE 2.11
Energy-band diagrams of a metal/p-type semiconductor contact with $\phi_M < \phi_S$: (*a*) two materials isolated from each other and (*b*) at thermal equilibrium after the contact is made.

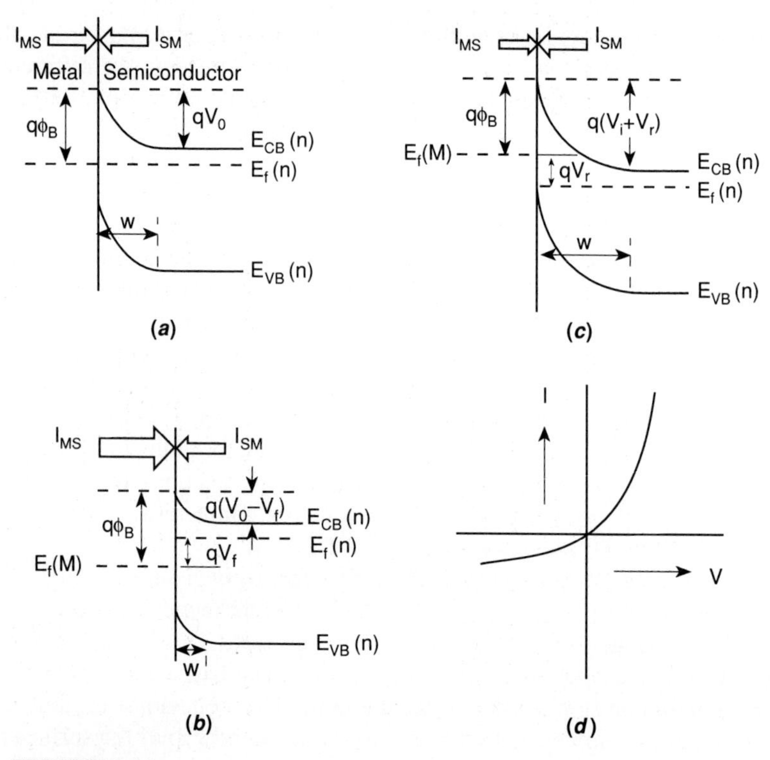

FIGURE 2.12
Energy-band diagram of a rectifying metal/n-type semiconductor contact: (*a*) at thermal equilibrium, (*b*) after forward bias, (*c*) after reverse bias and (*d*) I-V characteristics of the junction.

of the semiconductor; these states arise from the surface dangling bonds and impurities. As a result, the addition or removal of electrons from the semiconductor does not alter the position of the Fermi level at the surface and the Fermi level is said to be "pinned." Its ramification is that the position of E_{CB} at the surface in Figure 2.10*b* does not change with the metal, resulting in a constant barrier height.

That the contact in Figure 2.10 is rectifying in nature can be demonstrated using Figure 2.12. At equilibrium, electrons that have energies greater than qV_o are thermionically emitted over the barrier into the metal. Alternatively, if the surface region of the semiconductor is heavily doped so that the depletion layer becomes very thin, electrons can then tunnel into the metal through the barrier. These electrons produce a current I_{MS} that is directed from the metal to the semiconductor. Since at equilibrium the net current is zero, this current is exactly balanced by an equal and opposite current I_{SM}. When the semiconductor is biased negative with respect to the metal by voltage V_f, that is, forward bias, the barrier to the flow of electrons from the semiconductor to the metal decreases from qV_0 to $q\ (V_0 - V_f)$ as illustrated in Figure 2.12*b*. So more electrons flow from the semiconductor to the metal, and I_{MS} increases above the equilibrium value. However, I_{SM} does not change because $q\phi_B$ remains almost unaltered on biasing. Therefore, a net flow of current from the metal to the semiconductor occurs. Application of a reverse-bias V_r reduces the electron flow from the semiconductor to the metal because the barrier increases to $q(V_o + V_r)$. As a result, I_{MS} is reduced below its equilibrium value, whereas I_{SM} remains almost unchanged (see Figure 2.12*c*). Thus a small current flows. The composite I-V characteristics of a Schottky barrier diode are shown in Figure 2.12*d*. A comparison of Figure 2.12 with Figures 2.2, 2.3, and 2.4 shows the similarities between the behaviors and the current-voltage characteristics of Schottky barrier diodes and p–n junctions. Additionally, the diode equation (2.22) can describe the behavior of the Schottky barrier diodes.

EXAMPLE 2.2. Equate the equilibrium currents flowing from a metal to a semiconductor, and vice versa, to show that the Fermi energies are equal regardless of the electronic density of states.

Solution. Consider a system consisting of a metal M in perfect contact with a semiconductor S. The density of energy states in the metal and semiconductor are $\rho_M(E)$ and $\rho_S(E)$, respectively. The number of electrons filling the energy levels in the metal is $n_M(E) = \rho_M(E)F_M(E)$, where F(E) is the Fermi function. Similarly, the number of electrons filling the energy levels of the semiconductor is $n_S(E) = \rho_S(E)F_S(E)$. The number of vacant energy levels in the metal is $V_M(E) = \rho_M(E)\{1 - F_M(E)\}$. Similarly, the number of vacant energy levels in the semiconductor is $V_S(E) = \rho_S(E)\{1 - F_S(E)\}$.
At equilibrium no net transfer of electrons takes place from the semiconductor to the metal. Hence the probability of transfer of electrons from metal to semiconductor is $n_M(E)V_S(E)$, and from semiconductor to metal is $n_S(E)V_M(E)$. Therefore,

$$n_M(E)V_S(E) = n_S(E)V_M(E),$$

$$\rho_m(E)F_m(E)\rho_s(E)\{1 - F_S(E)\} = \rho_S(E)F_S(E)F_S(E)\rho_m(E)\{1 - F_m(E)\},$$

$$\rho_M(E)\rho_S(E)F_M(E) = \rho_M(E)\rho_S(E)F_S(E).$$

Further simplifying the above equation, we have

$$F_M(E) = F_S(E)$$

This relationship implies that the Fermi functions at a given energy level are identical in both the metal and the semiconductor. In other words, the Fermi level is the same in the metal and semiconductor in equilibrium.

2.1.2.2 Nonrectifying junctions (ohmic contacts)

An *ohmic contact* is a metal–semiconductor contact that has a negligible contact resistance relative to the bulk or series resistance of the semiconductor. As indicated earlier, this ohmic behavior can be achieved when metal–semiconductor junctions satisfy the following requirements: $\phi_M < \phi_S$ for n-type and $\phi_M > \phi_S$ for p-type semiconductors. These situations are shown in Figure 2.13. When $\phi_M < \phi_S$ and the semiconductor is n-type (Figure 2.13*a*), the Fermi levels are aligned by the transfer of electrons from the metal to the semiconductor. This alignment raises the semiconductor electron energies relative to that in the metal at

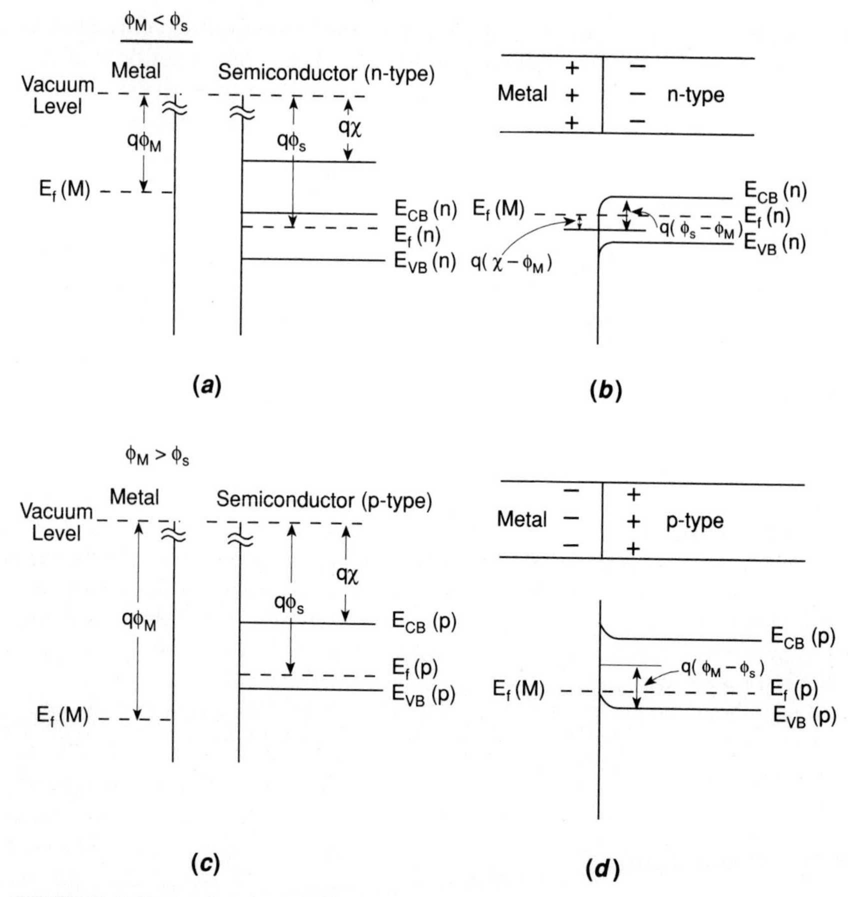

FIGURE 2.13
Band diagram for ohmic metal–semiconductor contact: (*a*) $\phi_M < \phi_S$ for an n-type semiconductor, (*b*) the corresponding equilibrium band diagram for the junction, (*c*) $\phi_M > \phi_S$ for a p-type semiconductor, and (*d*) the corresponding band diagram for the junction.

equilibrium as depicted in Figure 2.13*b*. The barrier to the flow of electrons from the metal to the semiconductor is small and can be easily overcome by the application of a small voltage. Similarly, when $\phi_M > \phi_S$ and the semiconductor is p-type, the hole flow across the junction can occur easily (see Figures 2.13*c* and 2.13*d*). Furthermore, unlike the case of the rectifying contacts discussed in section 2.1.2.1, the depletion regions do not form in the two cases, because the alignment of the Fermi levels requires the accumulation of majority carriers in the semiconductor.

Various metallizations have been explored as Schottky contacts to silicon and III-V semiconductors, and they are covered in Chapter 10. Furthermore, several authors have attempted to tailor ohmic contacts by altering chemistry at the metal/semiconductor junctions, and this approach is proving useful for producing ohmic contacts for a variety of semiconductors.

2.2 BASICS OF SELECT DEVICES

To familiarize the nonspecialist with the basics of some of the semiconducting devices, we have chosen four examples: (1) solar cells, (2) light-emitting diodes, (3) bipolar transistors, and (4) field-effect transistors. For more complete coverage of the principles and operation of devices, the reader is encouraged to refer to the textbooks listed at the end of this chapter.

2.2.1 Solar Cells

A solar cell converts solar energy into electrical energy. It is a simple device consisting of a p–n junction that is illuminated by solar radiation. To understand its operation, let us first evaluate changes in the behavior of a biased p–n junction discussed in section 2.1.1 when it is illuminated by photons whose energy $h\nu$ is greater than the band gap of the semiconductor.

We discussed in section 1.5.4 that when the energy of the incident illumination is greater than the band gap of a semiconductor, additional EHPs are produced and they complement the thermally generated EHPs. The rate of optically generated carriers g_{op} depends on the intensity of the radiation. In the case of a p–n junction, these excess carriers add to the saturation current discussed in section 2.1.1. The number of holes created per second within the diffusion length of the depletion region on the n-side is $A\,L_p g_{op}$, where A is the area of the junction and L_p is the diffusion length of holes on the n-side. Likewise, $AL_n g_{op}$ electrons are generated per second within the diffusion length of the depletion region on the p-side, where L_n is the diffusion length of electrons on the p-side. The current I_{op} due to the collection of these optically generated carriers by the junction is given by

$$I_{op} = qAg_{op}(L_p + L_n), \tag{2.25}$$

where q is charge of the carrier. Since the current given by Equation (2.25) is directed from n to p, it reduces the total current from p to n under forward bias. As a result, Equation (2.22), the diode equation, is modified on illumination:

$$I = qA\left[\frac{L_p}{\tau_p}p_n + \frac{L_n}{\tau_n}n_p\right]\left(\exp\frac{qV}{k_B T} - 1\right) - q\,Ag_{op}(L_p + L_n), \tag{2.26}$$

where τ_p is the lifetime of a hole on the n-side, τ_n is the lifetime of an electron on the p-side, p_n is the concentration of holes on the n-side, and n_p is the concentration of electrons on the p-side. Equation (2.26) assumes that the number of optically generated carriers within the depletion region is small in comparison to those produced

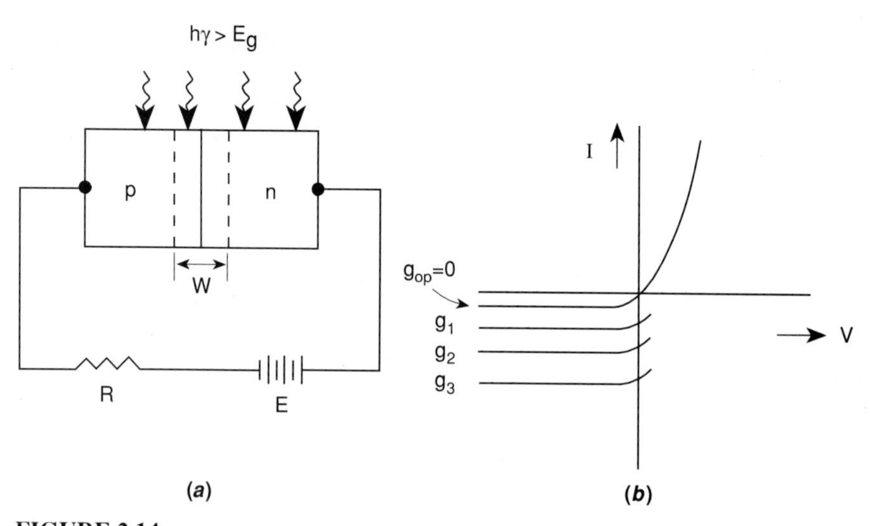

FIGURE 2.14
Optical generation of carriers in a p–n junction: (*a*) absorption of light by the junction and (*b*) I-V characteristics of an illuminated junction.

in the n- and p-sides. This assumption is reasonable because the width of the depletion region is very small (see section 2.1.1).

When we short-circuit the device, that is, $V = 0$, the first term on the right side of Equation (2.26) vanishes, resulting in a short-circuit current I_{sc} from n to p. The magnitude of this current is equal to I_{op}. Thus the I-V characteristics of an illuminated p–n junction at $V = 0$ cross the I-axis at negative values that are proportional to g_{op}, as shown in Figure 2.14. In the figure g_1, g_2, and g_3 are different generation rates that, as indicated earlier, depend on the intensity of the illumination. Let us now consider the case of an open circuit across the device, that is, $I = 0$. The open circuit voltage V_{oc} is given by

$$V_{oc} = \frac{k_B T}{q} \ln\left[\frac{(L_p + L_n)g_{op}}{(L_p/\tau_p)p_n + (L_n/\tau_n)n_p} + 1\right]. \tag{2.27}$$

The appearance of a forward voltage across an illuminated junction is known as the *photovoltaic effect.* This effect is used in solar cells to deliver power to an external circuit.

The band diagrams of junctions at equilibrium without and with illumination are compared in Figure 2.15. On illumination the contact potential V_0 is reduced by V_{oc}, whose maximum value cannot exceed V_0 because it is determined by potentials within the neutral n- and p-regions under consideration (see section 2.1.1).

Figure 2.16*a* shows the structure of a basic n–p junction solar cell. The front ohmic contact appears as narrow stripes in Figure 2.16*b*. These narrow contacts reduce the series resistance without substantially interfering with the incoming light.

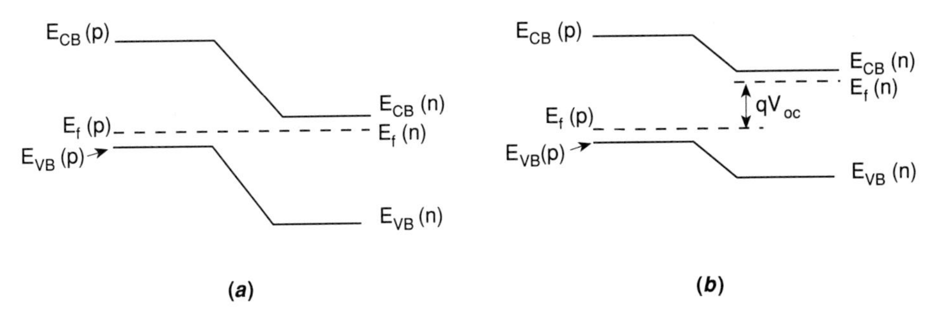

FIGURE 2.15
Effects of illumination on the open circuit voltage of a junction: (*a*) junction at equilibrium and (*b*) appearance of a voltage V_{oc} with illumination.

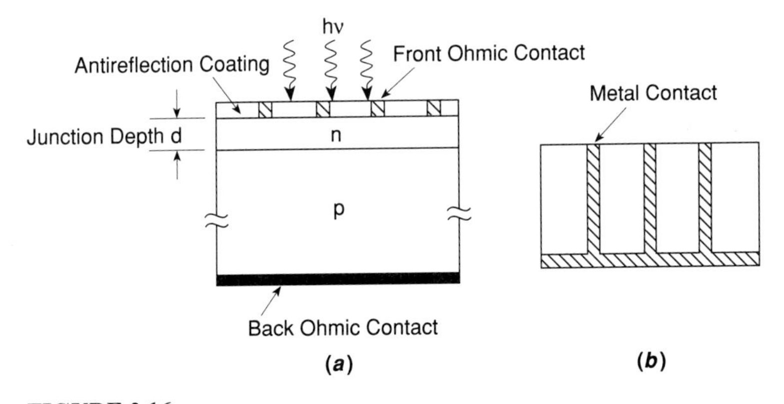

FIGURE 2.16
(*a*) Schematic of an n–p junction solar cell; front and back ohmic contacts and an antireflection coating are shown. (*b*) Top view showing metal contact stripes.

A fully metallized p-surface constitutes the back ohmic contact. The front surface is also coated with appropriate materials to reduce the reflection of the incident light and to decrease the carrier recombination at the surface.

Equation (2.25) shows that I_{op} can be increased by increasing A, g_{op}, L_n, and L_p. Therefore, to optimally absorb the incident energy, we need to design a solar cell with a large area junction that is located near the surface of the device. Additionally, the junction depth d in Figure 2.16*a* must be less than the hole diffusion length in the n-region. This situation allows holes generated near the surface to diffuse to the junction before recombining; a similar consideration applies to the p-region. We also require L_n and L_p to be as large as possible; therefore, the n- and p-regions should be lightly doped. However, to obtain a large value of V_{oc} in Figure 2.15*b*, the Fermi levels in the n- and p-regions should be close to their respective band edges; that is, the two regions should be heavily doped. Therefore, the

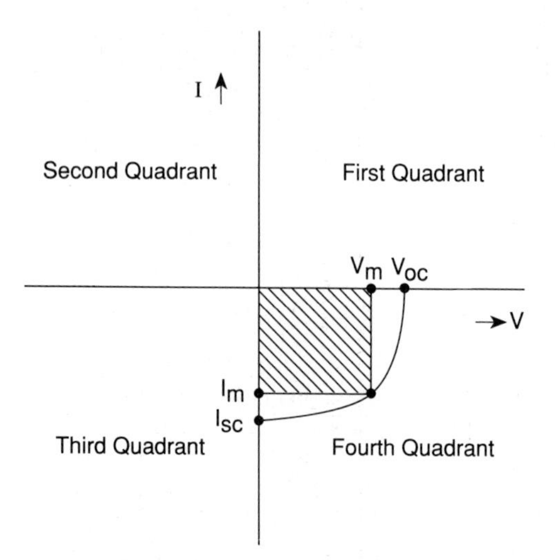

FIGURE 2.17
I-V characteristics of an illuminated solar cell. The shaded area represents the maximum power rectangle.

material requirements for increasing L_n and L_p conflict with those that produce large values of V_{oc}.

We show I-V characteristics of a solar cell in the fourth quadrant of the I-V plot in Figure 2.17. For a given light source that has a constant illumination level, V_{oc} and I_{sc} determine the cell properties. The maximum power is delivered to a load by a solar cell when the product VI_r is a maximum, where I_r is the reverse current. If we denote the respective values of V and I_r as V_m and I_m, the maximum delivered power is given by the shaded area in Figure 2.17. It is apparent that $V_m I_m$ is less than $V_{oc} I_{sc}$. The ratio $V_m I_m / V_{oc} I_{sc}$ is called the *fill factor* and is a figure of merit for solar cell design.

2.2.2 Light-Emitting Diodes

Let us now consider a situation where minority carriers are injected across a forward-biased p–n junction. In this case the current in the junction is accounted for in terms of carrier recombination in the depletion region and neutral p- and n-regions. The carrier recombination in direct-band-gap semiconductors gives off light whose wavelength is directly related to the band gap. This effect is called *injection electroluminescence* and is observed only in direct-band-gap semiconductors, such as InP and GaAs (see section 1.5.1).

The injection electroluminescent can be used to prepare two types of light-emitting devices: light-emitting diodes (LEDs), which emit incoherent spontaneous radiation, and laser diodes, which emit coherent radiation. The major applications of LEDs are in visible displays, indicators, and short-haul, infrared lightwave communication systems involving fused silica fibers as a transmission medium. On the other hand, laser diodes are extensively used as light sources in long-haul lightwave communication systems, printers, and CD-ROMS. This section covers LEDs; for information on lasers, please refer to the textbooks listed at the end of the chapter.

The emission of light from a direct-gap semiconductor basically involves two processes: generation of EHPs via an excitation process and recombination in which the excited carriers give up their energy via radiative and nonradiative processes. Before examining the operation of an LED, let us first consider the efficiency of radiative recombination (η_r). If τ_r and τ_{nr} are the radiative and nonradiative lifetimes, that is, the time that carriers have before recombining, then the total recombination time τ is

$$\frac{1}{\tau} = \frac{1}{\tau_r} + \frac{1}{\tau_{nr}}. \tag{2.28}$$

The internal quantum efficiency for the radiative process is then defined as

$$\eta_r = \frac{\frac{1}{\tau_r}}{\frac{1}{\tau_r} + \frac{1}{\tau_{nr}}} = \frac{1}{1 + \frac{\tau_r}{\tau_{nr}}}. \tag{2.29}$$

We can see that η_r can be enhanced by either reducing τ_r or increasing τ_{nr}. In high-quality direct-band semiconductors, η_r is close to unity.

The LED is essentially a forward-biased p–n junction. Electrons and holes are injected as minority carriers across the junction, and they recombine either by radiative recombination or by nonradiative recombination. The device is designed so that the radiative recombination dominates. As discussed in section 1.5.1, the recombination occurs when an electron and a hole have the same $\vec{k}$ value, where $\vec{k}$ is the wave vector.

In a typical LED, photons are generated very close to the p–n junction. The emitted light must pass through the semiconductor to reach the surface. During their transit, photons experience three loss mechanisms: (1) they may be absorbed by the semiconductor, (2) they may be refracted due to the difference in refractive indexes of the semiconductor and the outside medium, for example, air, and (3) suitably oriented light rays may undergo total internal reflection. As a result, the external quantum efficiency of the device is lower than its internal quantum efficiency.

Various materials, such as GaP, $GaAs_{1-y}P_y$, GaInP, and $Ga_xIn_{1-x}As_yP_{1-y}$, have been used to fabricate LEDs. Figure 2.18 shows the basic structure of a homojunction LED. To avoid lattice mismatch between GaAs and GaAsP materials, a graded epitaxial layer, ~10 μm thick, is grown on a GaAs substrate. (See Chapter 6 for epitaxial growth techniques.) The composition is gradually changed from $y = 0$ to $y = 0.4$. The n- $GaAs_{1-y}P_y$ layer is then deposited. A thin p-region is produced either by the diffusion of Zn into the n-ternary layer or by the growth of a doped epitaxial layer. Only a small fraction of the light generated at the p–n junction emerges from the surface because the remainder undergoes the losses discussed earlier.

2.2.3 Bipolar Junction Transistors

Two important characteristics of p–n junctions are the injection of minority carriers across the junction under the forward bias and the variation of the depletion width W

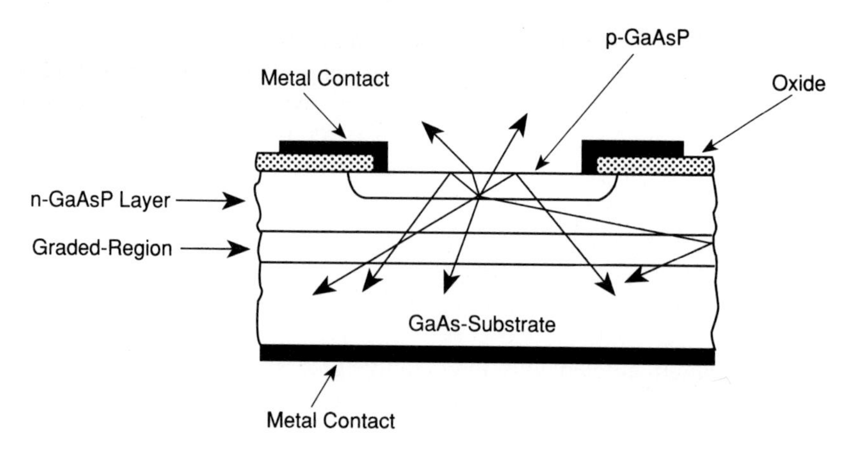

FIGURE 2.18
Schematic cross section of $GaAs_{1-y}P_y$ LED fabricated on a GaAs substrate.

with the reverse bias (see section 2.1.1). These two junction properties are used in the bipolar junction transistor (BJT) that Bardeen, Brattain, and Shockley invented in 1948 at the Bell Telephone Laboratories. BJT operates by the injection and collection of minority carriers. Since the action of both holes and electrons is important in this device, it is called a *bipolar transistor.*

BJT is a two-junction, three-terminal device in which the current through two terminals can be controlled by small changes we make in the current or voltage at the third terminal. This important feature allows us to amplify small a-c signals or to switch the device from an on state to an off state and back. Furthermore, BJTs can have two possible configurations. We use the p-n-p transistor to illustrate the principle of this device. The behavior of the n-p-n transistor can be analyzed in a similar manner.

Figure 2.19*a* shows the idealized p-n-p transistor in thermal equilibrium, that is, where all three leads are connected together or all are grounded. The emitter region in Figure 2.19*a* is more heavily doped than the collector is, while the base doping is less than the emitter doping but greater than the collector doping. Figure 2.19*b* shows the corresponding energy-band diagram. This diagram is a simple extension of the thermal equilibrium situation for the p–n junction as applied to a pair of closely coupled p^+–n and n–p junctions. Since at thermal equilibrium there is no net current, the Fermi level is a constant.

Figure 2.20 shows the corresponding situations when the transistor in Figure 2.19 is biased in the active mode. Figure 2.20*a* is a schematic of the transistor connected as an amplifier with the base lead being common to the input and output circuits. Figure 2.20*b* shows the energy-band diagram. Since the emitter–base junction is forward biased, holes are injected from the p^+ emitter into the base and electrons are injected from the n-base into the emitter. Assuming that the diode is ideal, there is no generation-recombination current in the depletion region. Then these two currents constitute the total emitter current I_E. The base–collector junction is reverse

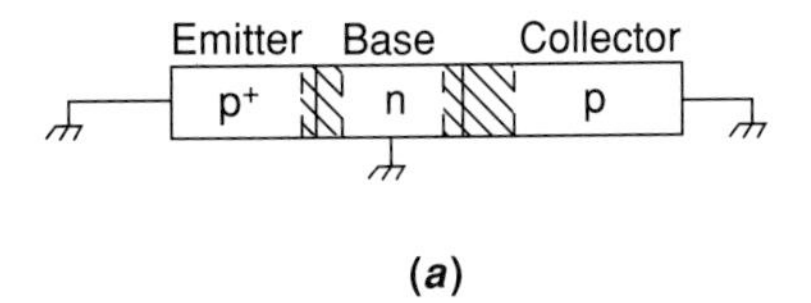

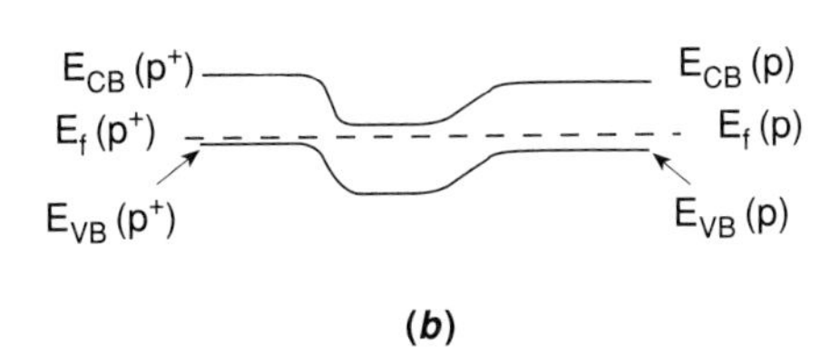

FIGURE 2.19
(*a*) A p^+–n-p bipolar transistor with all leads grounded and (*b*) energy-band diagram at thermal equilibrium.

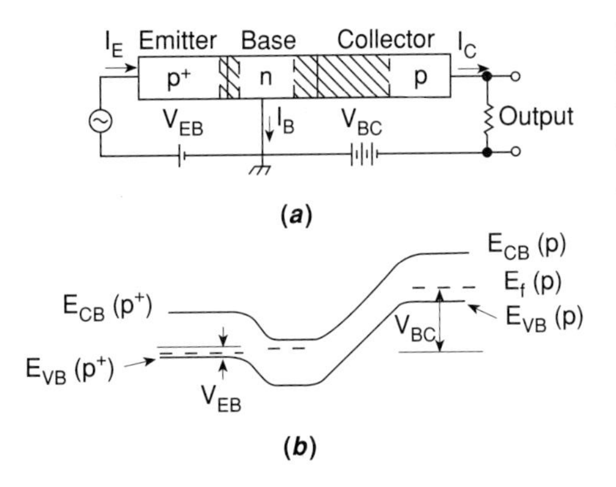

FIGURE 2.20
(*a*) The transistor shown in Figure 2.19 under the active mode of operation, and (*b*) the corresponding energy-band diagram.

biased, and a small reverse-saturation current will flow across the junction. When the base width is sufficiently narrow, the holes injected from the emitter can diffuse through the base to reach the base–collector depletion edge and be collected by the collector. If most of the injected holes can reach the collector without recombining with electrons in the base region, then the collector hole current will be very close to the emitter hole current. This situation results in a large current flow I_c in a reverse-biased collector junction, which is the transistor action. It can be realized only when the base region is thin and currents in the two junctions can interact with each other in the manner described above.

Figure 2.21 shows the various current components discussed above in an ideal p-n-p transistor operating in the active mode. We ignore the generation–recombination currents in the depletion regions. The injection of holes from the emitter into the base produces the current I_{E_P}, the largest current component in a well-designed transistor. Most of the injected holes reach the collector junction and give rise to the current I_{C_p}. The base current has three components, labeled I_{E_n}, I_{BB}, and I_{C_n}. I_{E_n} is due to the injection of electrons from the base to the emitter. I_{E_n} is not desirable and can be minimized by heavily doping the emitter. I_{BB} corresponds to electrons that must be supplied by the base to replenish electrons recombined with

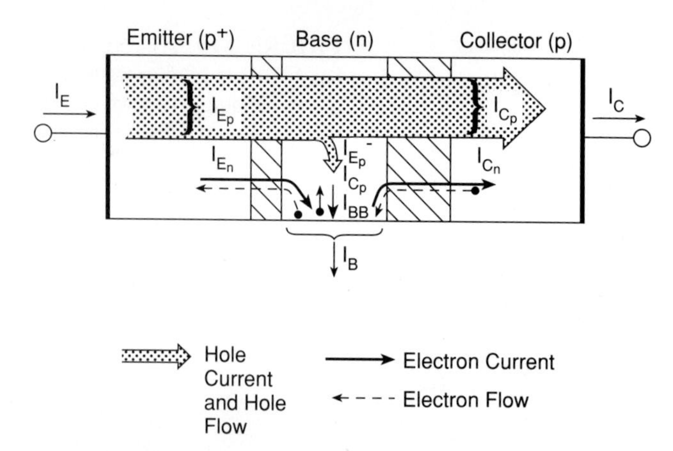

FIGURE 2.21
Various current components in a p-n-p transistor under the active mode of operation. The direction of the electron flow is opposite to the direction of the electron current flow.

the injected holes; that is, $I_{BB} = I_{E_p} - I_{C_p}$. I_{C_n} results from thermally generated electrons that are near the collector–base junction edge and drift from the collector to the base. As indicated in the figure, the current direction is opposite to that of the electron flow.

We can now express the terminal currents I_E, I_C, and I_B in terms of the various currents just described.

$$I_E = I_{E_p} + I_{E_n} \tag{2.30}$$

$$I_C = I_{C_p} + I_{C_n} \tag{2.31}$$

$$I_B = I_E - I_C = I_{E_n} + (I_{E_p} - I_{C_p}) - I_{C_n} \tag{2.32}$$

An important parameter for characterizing bipolar transistors is the common–base current gain α_o. This quantity is defined as

$$\alpha_o = \frac{I_{C_p}}{I_E}. \tag{2.33}$$

Substituting for I_E from Equation (2.30) into Equation (2.33), we have

$$\alpha_o = \frac{I_{C_p}}{I_{E_p} + I_{E_n}} = \left(\frac{I_{E_p}}{I_{E_p} + I_{E_n}}\right)\left(\frac{I_{C_p}}{I_{E_p}}\right). \tag{2.34}$$

The first term on the right side of Equation (2.34) is called the *emitter efficiency* and is labeled as γ. The second term is called the *base transport factor* α_T, which is the ratio of the hole current reaching the collector to the hole current injected from the emitter.

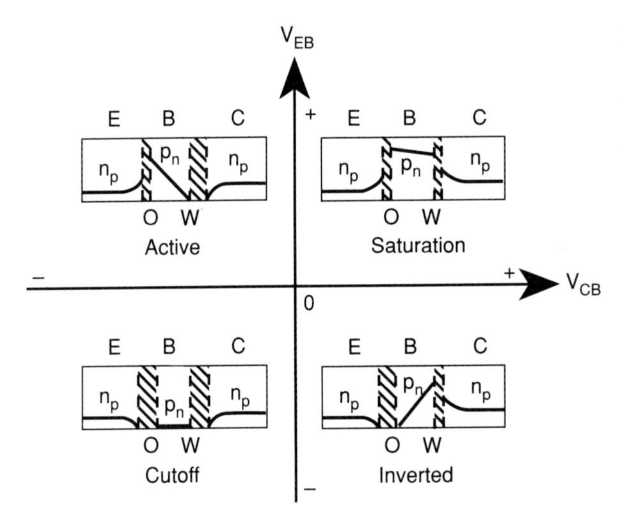

FIGURE 2.22
Junction polarities and minority-carrier distributions of a p-n-p transistor under various operational modes.

Making appropriate substitutions for the two terms in Equation (2.34), α_o can be rewritten as

$$\alpha_o = \gamma\alpha_T. \tag{2.35}$$

For a well-designed transistor, both γ and α_T approach unity, and α_o is very close to one.

Depending on the nature of the bias on the emitter–base and base–collector junctions, a bipolar transistor has four operational modes. These modes are shown in Figure 2.22 for a p-n-p transistor; the respective minority-carrier distributions are also shown. So far, we have considered only the active mode of the transistor. In this mode the emitter–base junction is forward biased, and the collector–base junction is reverse biased. In the saturation mode both junctions are forward biased. This mode corresponds to small biasing voltage and large output current; that is, the transistor is in a conducting state and acts as a closed (or on) switch. In the cutoff mode both junctions are reverse biased. There is essentially no charge stored in the base region, and the collector current approaches zero. The cutoff mode corresponds to the open or off state of the transistor as a switch. Finally, the fourth mode of operation is the inverted mode. In this case the emitter–base junction is reverse biased, and the collector–base junction is forward biased. This mode corresponds to the case where the roles of the emitter and the collector are switched with respect to those in the active mode. The current gain for the inverted mode is generally lower than that for the active mode because of poor efficiency of the collector-turned-emitter region, as it has low doping.

Figure 2.23 shows the measured output current-voltage characteristics for the common-base configuration. The various modes of operation are indicated on the figure. The collector current is nearly equal to the emitter current; that is, $\alpha_o \cong 1$ and independent of V_{BC}. This figure is in close agreement with the preceding discussion.

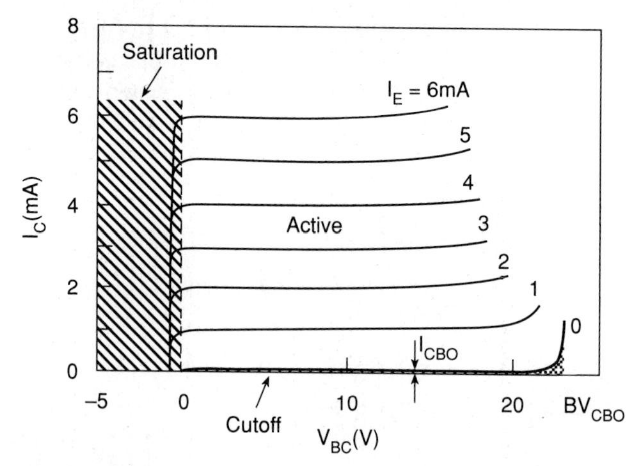

FIGURE 2.23
Output characteristics for a p-n-p transistor in the common-base configuration. I_{CBO} refers to the leakage current between the collector and the base with the emitter–base junction open.

2.2.4 Field-Effect Transistors

We now consider field-effect transistors (FETs). Like bipolar transistors, an FET is a three-terminal device in which the current between the two terminals, called the source and drain, is controlled by the application of a voltage to the third terminal, labeled as the gate. As discussed earlier, bipolars operate by the action of both holes and electrons. An FET, on the other hand, operates by the migration of only one type of carrier. Hence it is a unipolar device.

The field-effect transistors have three basic forms: (1) junction FET, labeled as JFET, (2) metal-semiconductor FET, referred to as MESFET, and (3) metal-insulator FET, labeled as MISFET. In a JFET the control (gate) voltage changes the depletion-layer width of a reverse-biased p–n junction, thus affecting carrier flow between the source and drain terminals. We discussed in section 2.1.2 that a depletion region in the semiconductor can also be produced by contact with a suitable metal. So if we replace the p–n junction in a JFET with a metal-semiconductor junction, we obtain a MESFET. Another possibility is that the metal-gate electrode in a MESFET is separated from the semiconductor by an insulator. The resulting structure is referred to as MISFET. A metal-oxide-semiconductor FET, one of the most important devices, belongs to this category and uses an oxide layer as an insulator.

2.2.4.1 Junction field-effect transistors (JFETs)

A three-dimensional view of a JFET is shown in Figure 2.24*a*. The device consists of an n-type conducting channel with two ohmic contacts, one each for the source and drain regions. When a positive potential is applied to the drain with re-

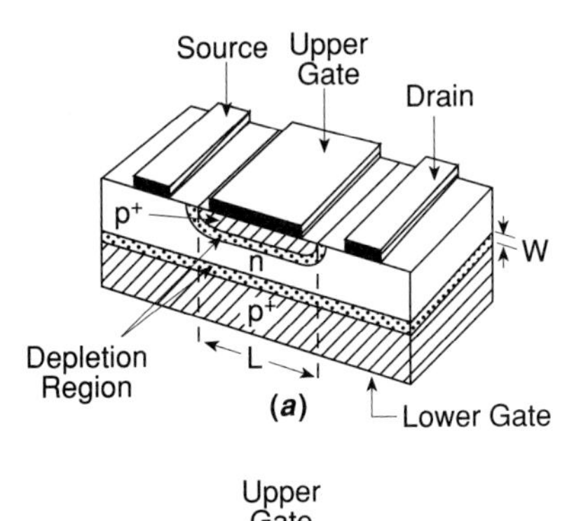

FIGURE 2.24
(*a*) Three-dimensional sketch of a JFET and (*b*) cross section of the central region of a JFET. The source is grounded, and the gate voltage and drain voltage are biased under normal operating conditions.

spect to the source, electrons flow from the source to the drain. The third electrode, the gate, forms a rectifying junction with the channel.

To analyze the behavior of this device, we will consider the symmetrical structure shown in Figure 2.24*b*, which corresponds to the central section between the dashed lines in Figure 2.24*a*. Note that the upper and lower gates are connected. The source is grounded, and the drain voltage V_D and the gate voltage V_G are measured with respect to the source. Under normal operating conditions, V_G is either zero or less than zero; and V_D is either zero or greater than zero.

When V_G is zero and V_D is small, a small drain current I_D flows in the channel as shown in Figure 2.25*a*. The magnitude of the current is given by V_D/R, where R is the channel resistance. The current varies linearly with the drain voltage. For a given drain voltage, the voltage along the channel increases from zero at the source to V_D at the drain. Thus the upper and lower gate junctions become increasingly reverse biased in going from the source to the drain. As V_D is increased, the depletion-layer width increases and the average cross-sectional area of the channel for current flow decreases. This decrease is nonuniform and is highest towards the drain. The channel resistance R also increases. As a result, I_D increases at a slower rate.

With the further increase in V_D, the depletion-layer width also increases. Eventually, at a certain voltage the two depletion regions touch each other at the drain as shown in Figure 2.25*b*. At this drain voltage the source and the drain are pinched off or completely separated by a reverse-biased depletion region. The location P in Figure 2.25*b* is called the *pinch-off point*. At this point a large drain current called the *saturation current* I_{DSAT} can flow across the depletion region. This situation is analogous to the injection of carriers in a reverse-biased depletion region as discussed in the case of bipolar transistor in the previous section.

When V_D is increased beyond the pinch-off point, the depletion region near the drain expands and the point P moves toward the source as depicted in Figure 2.25*c*.

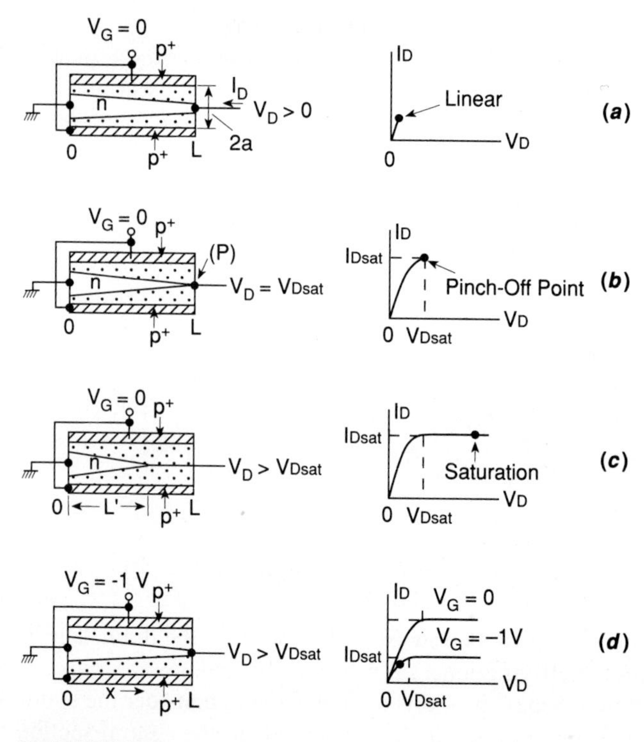

FIGURE 2.25
Variation of depletion-layer width and output characteristics of a JFET under various biasing conditions: (*a*) $V_G = 0$ and small V_D, (*b*) $V_G = 0$ and at pinch-off, (*c*) $V_G = 0$ and post-pinch-off ($V_D > V_{DSAT}$), and (*d*) $V_G = -1$ V and small V_D.

However, the voltage at point P remains V_{DSAT}. Thus the number of electrons per unit time arriving from the source to point P remains the same because the potential drop in the channel from source to point P remains unaltered. Therefore, for drain voltages larger than V_{DSAT}, the current remains essentially at the value I_{DSAT} and is independent of V_D (see Figure 2.25*c*).

When a voltage is applied to the gate of a reverse-biased p^+–n junction, the depletion-layer width increases. The ramification of this increase is that the effects of the gate voltage would complement those of V_D. As a result, the channel pinch-off will occur at a lower value of V_D. The saturation current will also be lower. These effects are shown in Figure 2.25*d*.

From the preceding discussion we know that we can control the current in the channel by manipulating V_D and bias on the gate.

2.2.4.2 Metal-semiconductor field-effect transistors (MESFET)

To illustrate the principle of a MESFET, we have chosen a GaAs MESFET whose cross-section is shown in Figure 2.26. The device can be fabricated by

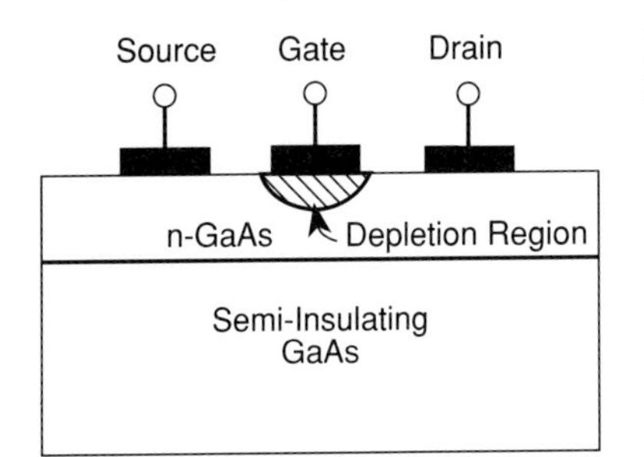

FIGURE 2.26
Schematic cross section of a GaAs MESFET.

depositing an n-type epitaxial layer on a semi-insulating GaAs substrate. The ohmic contacts to the source and drain regions are made by depositing suitable metals and alloys, followed by thermal annealing. The metal gate offers several advantages. First, a Schottky barrier contact can be formed at much lower temperatures than those required for p–n junction formation. Second, the gate length can be reduced to submicron levels.

A MESFET may be designed to operate either as a normally on (depletion mode) device or as a normally off (enhancement mode) device. In a normally on device, the thickness of the epitaxial layer is more than the zero-bias depletion region thickness of the Schottky barrier gate, and the transistor has a conducting channel at $V_G = 0$. In a normally off device, the active layer is kept thin so that the depletion layer extends across the channel at $V_G = 0$. Therefore, conduction in the channel occurs only for small positive values of V_G.

2.2.4.3 Metal-insulator-semiconductor field-effect transistors (MISFET)

The metal-insulator-semiconductor field-effect transistor (MISFET) is one of the most widely used solid state electronic devices. In this device the channel current is controlled by a voltage applied to a gate that is separated from the channel by an insulator. In the case of silicon, silicon dioxide serves an insulator, and the resulting structure is referred to as a *metal-oxide-semiconductor FET* (MOSFET). This subsection emphasizes this device to illustrate the operational principles of a MOSFET.

The two types of MOSFETs—the enhancement type and the depletion type—are shown in Figures 2.27*a* and 2.27*b*. The enhancement-mode transistor is normally off, and no current flows between the source and the drain for $V_G = 0$. A

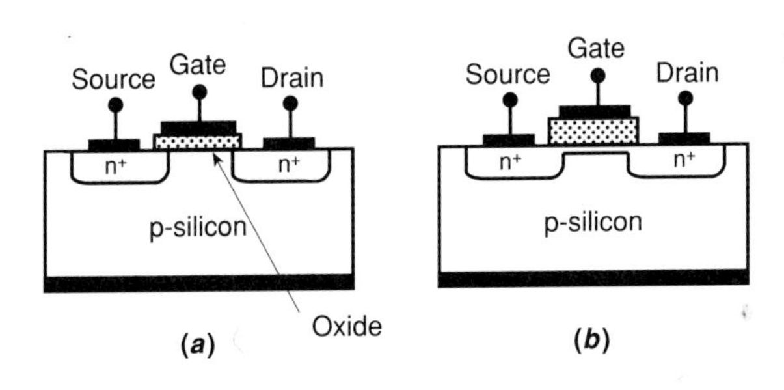

FIGURE 2.27
Schematic cross sections of n-channel MOS transistors: (*a*) enhancement-type device and (*b*) depletion-type device.

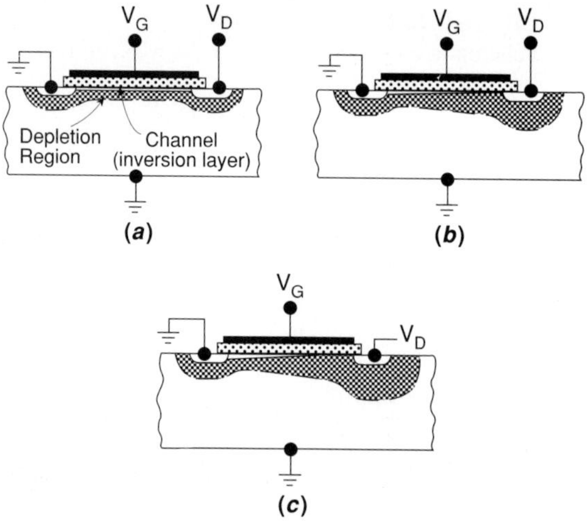

FIGURE 2.28
MOSFET cross sections showing the effect of various bias conditions on the channel thickness. (*a*) The drain voltage V_D is small, and the depletion region is uniform along the channel; (*b*) $V_D = V_{DSAT}$, and the channel is pinched off near the drain; and (*c*) V_D exceeds V_{DSAT}, and the effective channel length is reduced.

conducting channel is then induced by applying voltage of appropriate polarity to the gate. In a depletion-mode device, a conducting channel already exists, and the device is on with no bias applied to the gate. The channel is depleted of mobile carriers by a gate voltage opposite to the drain voltage.

Let us now consider the operation of the enhancement-mode device shown in Figure 2.27*a*. When a voltage is applied to the gate, the metal-oxide-semiconductor composite behaves as a capacitor. If V_G is positive, then negative charges are induced in the semiconductor underneath the gate. At a threshold gate voltage called V_{TH}, a thin layer of the semiconductor below the gate changes from being p-type to n-type; that is, an inversion layer has been formed as shown in Figure 2.28*a*. When a small positive voltage V_D is applied to the drain, the channel behaves as a resistor and electrons flow from the source to the drain.

As V_D is increased, the potential drop across the channel reduces the voltage between the gate and the inversion layer near the drain. As a result, the electron concentration in the channel near the drain decreases, which causes an increase in the channel resistance. If V_D is increased further, the gate voltage across the oxide near the drain continues to decrease until it falls below the value that is required to maintain an inversion layer. Then the channel near the drain gets pinched off, and the drain is isolated from the conducting channel by a depletion region; see Figure 2.28*b*. The drain current will then saturate to a constant value. The value of V_D at which saturation occurs is termed V_{DSAT}.

An increase in V_D above V_{DSAT} moves the pinch-off point P toward the source, and the effective channel length slightly decreases as shown in Figure 2.28*c*. However, the potential at P remains at V_{DSAT}, and the additional voltage drops across the depletion region. Thus the drain current will not be altered much when V_D is raised above V_{DSAT}.

If the gate bias is increased to a higher value, the inversion layer charge also increases. Therefore, the channel resistance decreases, causing a larger drain current for a given value of V_D. The pinch-off does not occur for a higher value of V_D.

The operating principle of a depletion-type MOSFET is basically similar to that of the enhancement-type MOSFET and can be analyzed in a similar fashion.

EXAMPLE 2.3. Find the dc resistance of a MOS transistor with length 2 μm, width 4 μm, oxide thickness 40 nm, $V_{DS} = 2.5$ V, $V_{GS} = 5$ V, and the threshold voltage of 1 V. Assume the mobility of carriers to be 600 cm^2/v-sec.

Solution. The drain to source current of a MOS transistor in the linear region is given by

$$I_{DS} = \frac{\mu \varepsilon_{ox} W_{TR}}{L_{TR} t_{ox}} \left[(V_{GS} - V_{TH}) V_{DS} - \frac{V_{DS}^2}{2} \right],$$

where μ is mobility of carriers, t_{ox} is oxide thickness, ε_{ox} is permittivity of oxide, W_{TR} is width of transistor, L_{TR} is length of transistor, V_{GS} is the gate-source voltage, V_{TH} is threshold voltage, and V_{DS} is the drain-to-source voltage. The dc resistance of the transistor is R_{DS} and is

$$R_{DS} = \frac{V_{DS}}{I_{DS}} = \frac{L_{TR} t_{ox}}{\mu \varepsilon_{ox} W_{TR} \left[(V_{GS} - V_{TH}) - \frac{V_{DS}}{2} \right]}.$$

For $L_{TR} = 2\ \mu$m, $W_{TR} = 4\ \mu$m, $t_{ox} = 40$ nm, $\mu = 600$ cm^2/v.sec, $V_{DS} = 2.5$ V, $V_{GS} = 5$ V, and $V_{TH} = 1$ V, we obtain $R_{DS} = 3.5$ kΩ.

EXAMPLE 2.4. Calculate the limiting frequencies under which an ac signal can be followed using a MOS transistor.

Solution. Let the distance between the source and drain be L. A voltage difference ΔV is suddenly established between the center of the channel and the source and drain.

The transverse field acting on the inversion layer is $\dfrac{2\Delta V}{L}$.

The drift velocity of the carriers is $2\mu \dfrac{\Delta V}{L}$, where μ is the carrier mobility.

Hence the time for a carrier to drift to the center of the channel from either the source or the drain is $\dfrac{L^2}{4\mu \Delta V}$.

When $\mu = 0.045 - 0.08$ m^2 V^{-1} Sec^{-1} for electrons,
$\mu = 0.02 - 0.04$ m^2 V^{-1} Sec^{-1} for holes and $\Delta V = 10^{-2}$ V, the limiting frequencies are in the range 5 to 30 MHZ for $L = 5$ to 2×10^{-6} m.

2.3 HIGHLIGHTS OF THE CHAPTER

The highlights of this chapter are as follows:

- A region depleted of carriers exists on either side of a p–n junction and consists of uncompensated ionized donors and acceptors. The presence of uncompensated impurities results in a built-in field. Diffusion and drift of carriers contribute to the junction current. At equilibrium the drift and diffusion currents for each carrier type add up to zero. Under the forward bias the minority carriers are injected across the junction, resulting in an increased junction current. However, when the junction is reverse biased, the junction current is very small and is independent of the bias. The current-voltage characteristics are rectifying in nature. When the reverse bias exceeds a critical value, the junction breaks down.

- Certain metal–semiconductor junctions also exhibit the rectifying behavior. In this case the depletion region exists only in the semiconductor. Nonrectifying metal–semiconductor junctions are also possible; they exhibit the linear current-voltage characteristics and are used as ohmic contacts in devices.
- When a p–n junction is illuminated with radiation whose energy is greater than the band gap of the semiconductor, a forward voltage develops across the junction. This *photovoltaic effect* is used in solar cells to deliver power to an external circuit.
- A light-emitting diode is a forward-biased p–n junction. Additional electrons and holes are injected as minority carriers across the junction. When the excess carriers recombine radiatively, light is given off.
- A bipolar junction transistor is a two-junction, three-terminal device and consists of emitter, base, and collector regions. It has two possible configurations: emitter-base-collector (p-n-p) or emitter-base-collector (n-p-n); the base region is very narrow. Depending on the nature of the biases across the emitter–base and base–collector junctions, four operational modes are possible.
- The field-effect transistors (FET) have three basic forms: junction FET, metal-semiconductor FET, and metal-oxide-semiconductor FET. The current in the channel between the source and drain regions is controlled by changing the width of the depletion regions in JFET and MESFET, whereas the carrier type is capacitively changed in the channel in MOSFETs.

REFERENCE

Sze, S. M. *Semiconductor Devices: Physics and Technology.* New York: John Wiley, 1985.

FURTHER READING

Singh, J. *Semiconductor Devices: An Introduction.* New York: McGraw-Hill, 1994.
Streetman, B. G. *Solid State Electronic Devices.* Englewood Cliffs, N.J.: Prentice-Hall, 1980.
Tayagi, M. S. *Introduction to Semiconductor Materials and Devices.* New York: John Wiley, 1991.

PROBLEMS

2.1. Compute the contact potential when an n-type silicon with donor concentration of $10^{18}/cm^3$ forms an abrupt junction with a p-type silicon with acceptor concentration $10^{16}/cm^3$. Assuming that the temperature is 300 K, calculate the width of the depletion layer. For a reverse bias of 1 V, obtain the maximum electric field in the depletion layer.

2.2. Consider an n-type silicon with concentration of majority carriers given by $10^{21}/m^3$. The silicon is inhomogeneous and has a hole concentration of $10^9\ m^{-3}$ at one end of silicon, which is 1 μm long. What current is associated with the diffusion of holes? The diffusion coefficient of holes is $1.1 \times 10^{-3}\ m^2\ s^{-1}$ at 300 K.

2.3. What is the capacitance of an MOS structure with a silicon dioxide thickness of 100 Å, a metal on one side, and a heavily doped semiconductor on the other side? If SiO_2 dielectric can withstand $10^9\ Vm^{-1}$, what voltage can you apply to the MOS capacitor? The relative permitivity of SiO_2 is four.

2.4. Consider a piece of silicon that is 2 mm $\times$ 2 mm $\times$ 0.5 mm. It is illuminated by a 1 mW pulse of light of wavelength 0.35 μm for 5 μs. If the initial resistivity of silicon is 0.5 Ω m, what are its resistivities immediately after the light pulse and 100 μ secs after the light pulse?

2.5. Calculate the built-in voltage for the following case using the depletion-layer approximation: a p–n junction of a semiconductor with an energy gap 1.1 eV. The difference between Fermi energy and the valence-band edge in p-type semiconductor is 0.15 eV. The difference between Fermi energy and the conduction-band edge in n-type semiconductor is 0.15 eV.

2.6. Consider a lightly doped silicon at 300 K. Assuming the electron and hole mobilities are 1400 and 500 cm^2/volt-sec, calculate the diffusion constants of electrons and holes.

2.7. A p–n junction is formed by joining two semiconductors, each doped to $10^{16}/cm^3$. If the acceptor and donor energy levels are 0.17 eV from the nearest band edges, what is the barrier height? What is the carrier concentration on either side of the potential barrier? Calculate the width of the depletion region.

2.8. A p–n junction is formed between an n-type semiconductor ($N_d = 10^{18}/cm^3$) and a p-type semiconductor ($N_a = 10^{16}/cm^3$). If the p–n junction is 10 μm thick and has an area $10^{-6}\ cm^2$, calculate the capacitance when the applied voltage is 10 V.

2.9. Calculate the capacitance of a p–n junction formed by doping concentration of $10^{16}/cm^3$ in each type of semiconductor. If 10 volts are applied, how does the capacitance change? Assume the area of the p–n junction is $10^{-6}\ cm^2$.

2.10. Assume an electron in a semiconductor whose energy gap is 1.1 eV reaches saturation velocity of 10^7 cm/s in the electric field ε_B. What is the mean free path between collisions? If the energy acquired by the electron is sufficient to create an EHP, what is the value of ε_B? Estimate the breakdown voltage of the p–n junction for a one-sided abrupt junction with an acceptor density of $10^{16}/cm^3$.

2.11. Consider a p-type material with a dopant concentration of 10^{22} holes m^{-3}. Assume the mobility of holes is $0.044\ m^2\ V^{-1}\ Sec^{-1}$. What is the conductivity of the semiconductor? What electric field would be necessary to produce a hole current of $10^{-7}\ A\ m^{-2}$?

2.12. Assume the Schottky model of the metal–semiconductor barrier. For n-GaAs, select the metals that will result in (1) an accumulation, (2) neutral, and (3) depletion layer on contact with the semiconductor. Draw the energy-band diagrams illustrating the situation before the contact is made and after the contact is made. Compare your predictions with experimental results.

2.13. A 10 nm thick layer of GaAs is sandwiched between two layers of $Al_{0.33}Ga_{0.67}As$. Draw the energy-band diagram of this structure.

2.14. A p^+–n-p transistor fabricated in silicon has a base with doping concentration of 10^{14} cm^{-3} and thickness 80 μm. The dopant concentration of the emitter is 10^{18} cm^{-3} and of the collector is 10^{15} cm^{-3}. In the common emitter mode, if the gain is 30 and the bias between collector to base is 300 V, what is the device collector capacitance? Comment also on the recombination lifetime.

2.15. A field-effect transistor is doped with arsenic to 10^{16} atoms/cm^3 in the channel region. For a channel width of 0.1 mm and for a dielectric constant in the channel region equal to 65, what is the pinch-off voltage as a function of channel width in the range 0 to 1 mm?

CHAPTER 3

Defects

We assumed in Chapters 1 and 2 that semiconductors are defect free; that is, they do not have any interruption in their three-dimensional periodic arrangement of atoms. This ideal situation does not exist in nature: Real materials always contain defects. This chapter classifies defects according to their dimensionality, discusses their characteristics, and emphasizes their electronic properties. These ideas form a basis for understanding the origins of growth- and processing-induced defects covered in later chapters.

3.1 CRYSTAL STRUCTURE OF IMPORTANT SEMICONDUCTORS

As indicated in Chapter 1, important semiconductors can be broadly classified into two categories: (1) elemental and (2) compound. Si, diamond, and Ge are examples of elemental semiconductors, whereas GaAs, InP, GaSb, AlN, GaN, and so forth belong to the compound-semiconductor category. Compound semiconductors can be further subdivided into III-V, II-VI, I-III-VI_2, and II-IV-V_2, where I, II, III, IV, V, and VI represent the group of the constituent atoms in the periodic table. For example, GaAs is a III-V semiconductor, while CdTe and $CuInSe_2$ are, respectively, examples of II-VI and I-III-VI_2 semiconductors.

Elemental semiconductors crystallize in the diamond-cubic structure that consists of two interpenetrating face-centered cubic (FCC) unit cells. This structure is illustrated in Figure 3.1. One of the FCC units is displaced with respect to the other by $a/4 <111>$, where a is the lattice parameter of the semiconductor. Furthermore, each lattice site in the FCC unit is occupied by the same type of atom. As shown for Si in Figure 3.2, each atom is tetrahedrally coordinated. The atoms touch each other along the four $<111>$ bond directions, and the nearest neighbor distance is $\sqrt{3}a/4$.

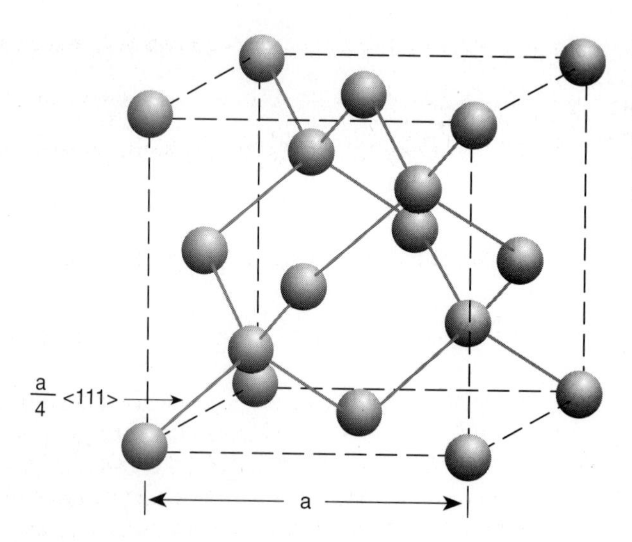

FIGURE 3.1
Arrangement of atoms in the diamond-cubic crystal. Each atom has four nearest neighbors, which are arranged at the corners of a tetrahedron.

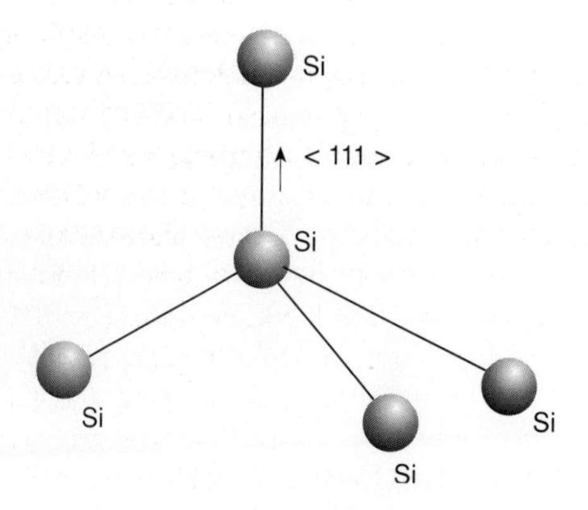

FIGURE 3.2
Schematic illustrating the tetrahedral coordination of a Si atom in silicon crystals.

Compound semiconductors, such as GaAs and CdTe, crystallize in the zinc-blende structure. Like the diamond-cubic structure, the zinc-blende structure consists of two interpenetrating FCC units (see Figure 3.3). One of the units is occupied by group III or group II atoms, whereas group V or VI atoms reside on the second unit; that is, atoms residing on each of the two units are not the same in this case. Again, each atom is tetrahedrally coordinated, and the coordination number is four. The nearest neighbors of group III atoms are group V atoms, or vice versa. This arrangement is shown in Figure 3.4 for a Ga atom in GaAs crystals. Furthermore, in assigning indexes to various directions and planes in the zinc-blende structure, we usually assume that the origin of the FCC unit on which group III atoms reside is at 0,0,0, whereas the origin of the group V unit is at 1/4, 1/4, 1/4. Ternary semiconductors, such as I-III-VI_2 and II-VI-V_2, have the chalcopyrite structure, which is shown in Figure 3.5*a*. To discern differences between the chalcopyrite, diamond-cubic, and zinc-blende structures, compare Figure 3.5*a* with Figures 3.5*b* and 3.5*c*, which show two unit cells of the diamond-cubic and zinc-blende structures.

Diamond-cubic, zinc-blende, and chalcopyrite structures are relatively loosely packed in comparison to the FCC and hexagonal close-packed structures in which

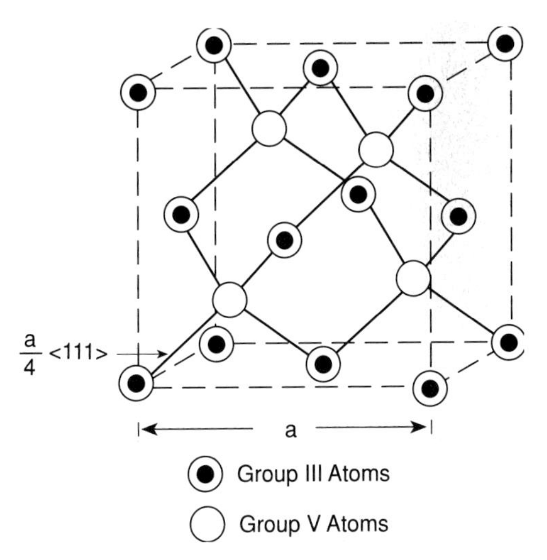

FIGURE 3.3
Arrangement of atoms in a zinc-blende structure. For example, large, open circles enclosing a solid circle could be Ga atoms, and open circles could be As atoms.

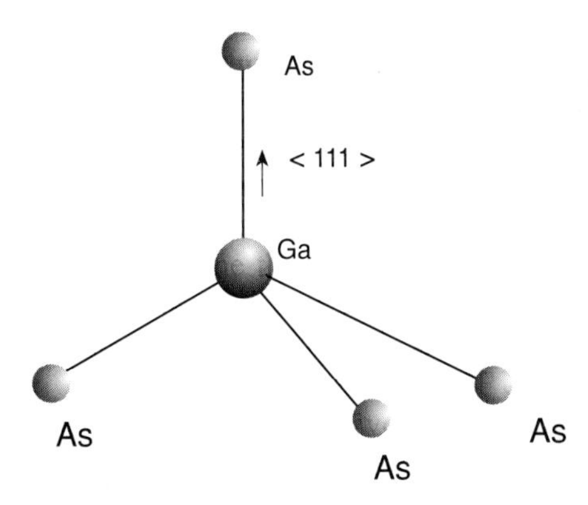

FIGURE 3.4
Schematic illustrating the tetrahedral coordination of a Ga atom in GaAs crystals; symbolically it is written as GaAs(4).

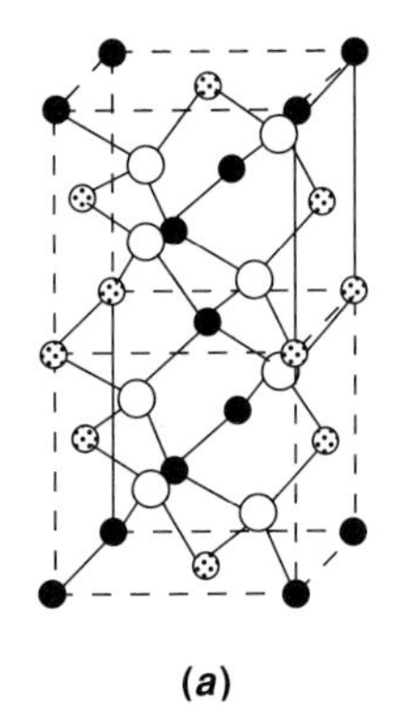

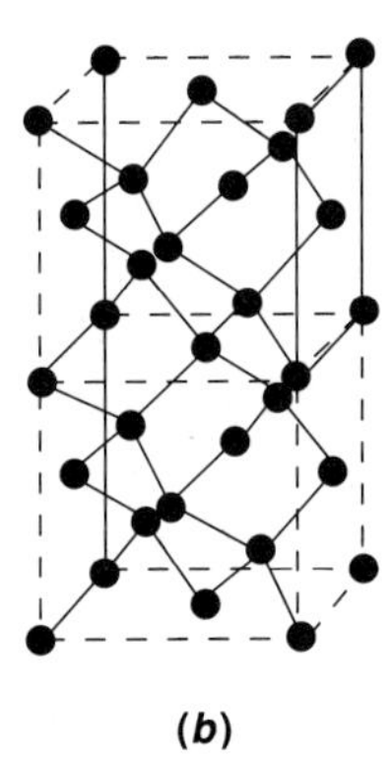

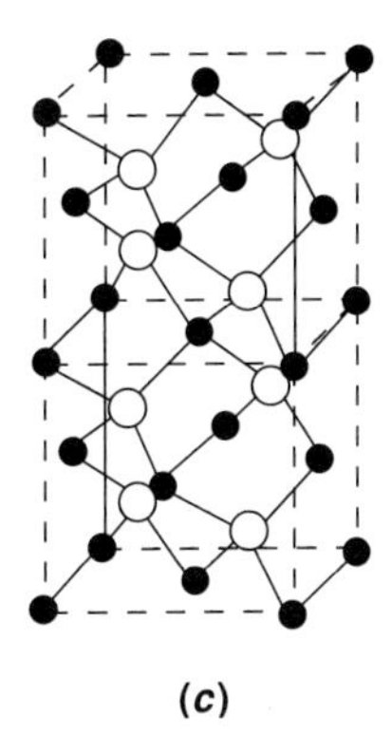

FIGURE 3.5
(*a*) Arrangement of atoms in a chalcopyrite structure. Solid, open, and dotted circles represent atoms A, B, and C of a ternary chalcopyrite semiconductor ABC_2. (*b*) and (*c*) show two units of the diamond-cubic and zinc-blende structures for comparison.

many metals and alloys crystallize. For example, based on the hard-sphere model, 34 percent of the Si lattice is occupied by atoms, whereas the packing density of an FCC crystal is about 74 percent.

3.2 STACKING ARRANGEMENT OF {111} PLANES IN DIAMOND-CUBIC AND ZINC-BLENDE STRUCTURES

We discuss in section 3.3 that dislocations glide on {111} planes of the diamond-cubic and zinc-blende structures. Stacking faults also form on these planes, so we examine the stacking arrangement of these planes. Figure 3.6 shows perspectives of the two structures along the [110] direction. Various crystallographic directions and interatomic distances are indicated on the figure. Atoms marked A, B, and C in Figure 3.6*a* are coplanar; E and F represent the [110] projection of the atoms to which atoms A and C are tetrahedrally coordinated; and D is the projection of a row of atoms that are bonded to atoms projecting along E and F. Figure 3.6*b* depicts the corresponding situation for the zinc-blende structure of a III-V compound.

If we assemble the projections shown in Figures 3.6*a* and 3.6*b* into two-dimensional arrangements, the patterns reproduced in Figures 3.7*a* and 3.7*b* will result. These patterns have the following features. First, if we examine the patterns along the $[\bar{1}11]$ direction, we find narrowly and widely separated $(\bar{1}11)$ planes, which are normal to the plane of the figure; the respective interplanar separations are

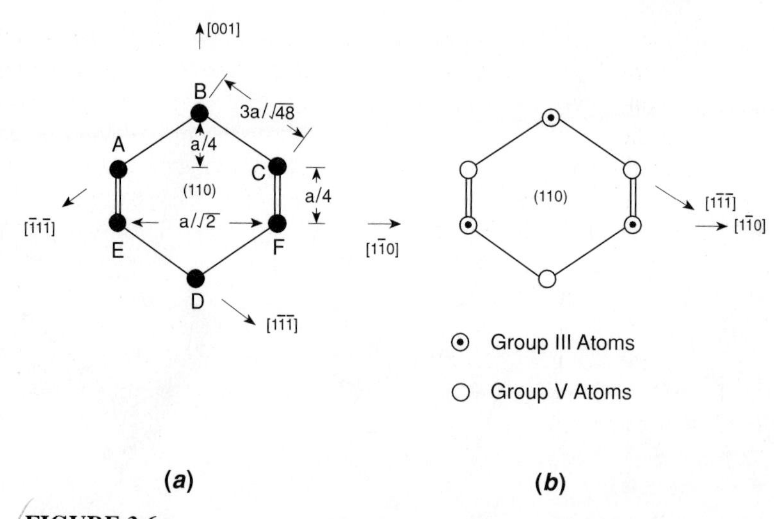

FIGURE 3.6
(*a*) Shows the [110] projection of the diamond-cubic structure; different crystallographic directions and distances are indicated in terms of the lattice parameter a. Atoms A, B, and C are coplanar, whereas E and F represent projection of atoms to which atoms A and C are tetrahedrally coordinated. (*b*) The corresponding situation in the zinc-blende structure of III-V compounds.

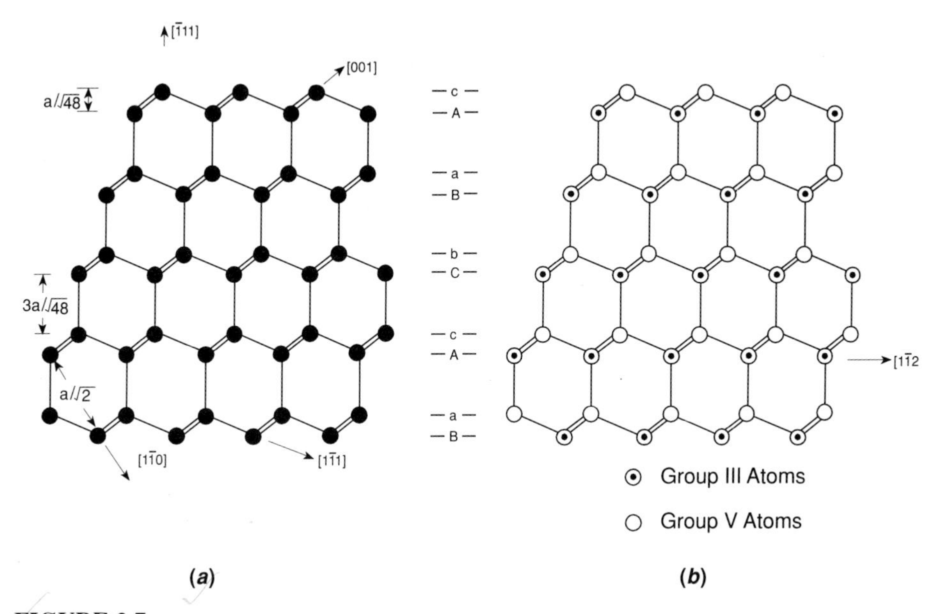

FIGURE 3.7
Assemblage of the projections shown in Figure 3.6(*a* and *b*) into two-dimensional arrangements to delineate the stacking arrangement of {111} planes: (*a*) diamond-cubic and (*b*) zinc-blende structure of a III-V compound.

$a/\sqrt{48}$ and $3a/\sqrt{48}$. Second, the stacking arrangement of the ($\bar{1}$11) planes is Aa Bb Cc Aa. . . The atoms in A and a, B and b, and C and c layers are aligned along the [$\bar{1}$11] direction; that is, when viewed along the [$\bar{1}$11] direction, atoms project on top of each other. Furthermore, these layers are widely separated. On the other hand, a and B, b and C, and c and A layers are narrowly separated. Third, in the diamond-cubic structure A, a, B, b, C and c . . . layers contain the same type of atoms, while in the zinc-blende structure the layers A, B, C . . . and a, b, c . . . contain group III and V atoms, as shown in Figure 3.7*b*. Since ($\bar{1}$11) and (1$\bar{1}\bar{1}$) planes contain the same type of atoms in the diamond-cubic structure, {$\bar{1}$11} and {1$\bar{1}\bar{1}$} planes are equivalent. Figure 3.7*b* shows that this property is not true for the zinc-blende structure: the ($\bar{1}$11) planes contain group III atoms, whereas group V atoms are located in the (1$\bar{1}\bar{1}$) planes. Therefore, {$\bar{1}$11} and {1$\bar{1}\bar{1}$} planes are nonequivalent in this structure. Current terminology refers to the {$\bar{1}$11} and {1$\bar{1}\bar{1}$} planes as $\{\bar{1}11\}_{\text{III or A}}$ and $\{1\bar{1}\bar{1}\}_{\text{V or B}}$, respectively.

3.3 STRUCTURAL CHARACTERISTICS OF DEFECTS

We can group defects into four categories in semiconductors using dimensionality as a criterion: zero-, one-, two-, and three-dimensional. The respective examples are point defects, dislocations, stacking faults, and precipitates.

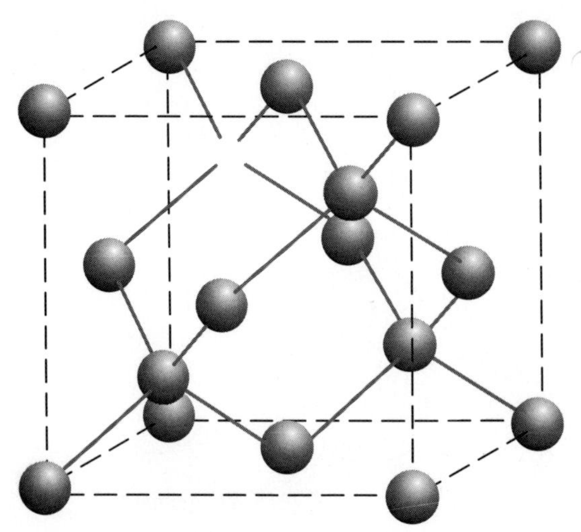

FIGURE 3.8
A vacancy in the diamond-cubic structure. Note that four unsaturated bonds, that is, dangling bonds, are present around the vacancy. The lattice must also relax around the vacant site.

3.3.1 Zero-Dimensional Defects

This category includes vacancies, interstitials, vacancy-interstitial pairs, dopant atoms intentionally added to control the conductivity of a semiconductor as discussed in Chapter 1, and impurities that are unintentionally incorporated as contaminants during material growth and processing.

Figure 3.8 shows the diamond-cubic unit cell in which one of the tetrahedrally coordinated atoms has been removed. The resulting defect is called a *Schottky vacancy.* The formation of a vacancy results in four dangling bonds that, as discussed later, impart electrical properties to these defects. The lattice must also relax around the vacancy. The energies of formation and migration for Schottky vacancies in silicon are, respectively, ~2.3 and 0.18 eV.

A Schottky interstitial is the second elementary point defect that can exist in semiconductors. Figure 3.9 shows the positions of some of the interstitial sites in the diamond-cubic structure (Rhodes 1964). An interstitial defect can be formed by inserting an atom into one of the holes in the lattice. The energies of formation of interstitials in the loosely packed diamond-cubic and zinc-blende structures are low in comparison to those in the close-packed structures because the distortions associated with the interstitials in these structures are considerably larger than those in the diamond-cubic and zinc-blende structures. A value of 1.1 eV is often associated with this type of defect in Si. Frenkel defects, that is, vacancy-interstitial pairs, can also form. An interstitial atom tends to remain in the vicinity of a vacancy so that the total distortion energy of the system is reduced.

As discussed in Chapter 1, impurities in the form of dopants are added to semiconductors to control their carrier concentration and type of conductivity. These impurities generally replace host atoms located on lattice sites. They are therefore referred to as *substitutional impurities* and belong to zero-dimensional defects. Since sizes of the substitutional impurity and the host atom are likely to be different, strains are introduced into the lattice during the impurity incorporation. The

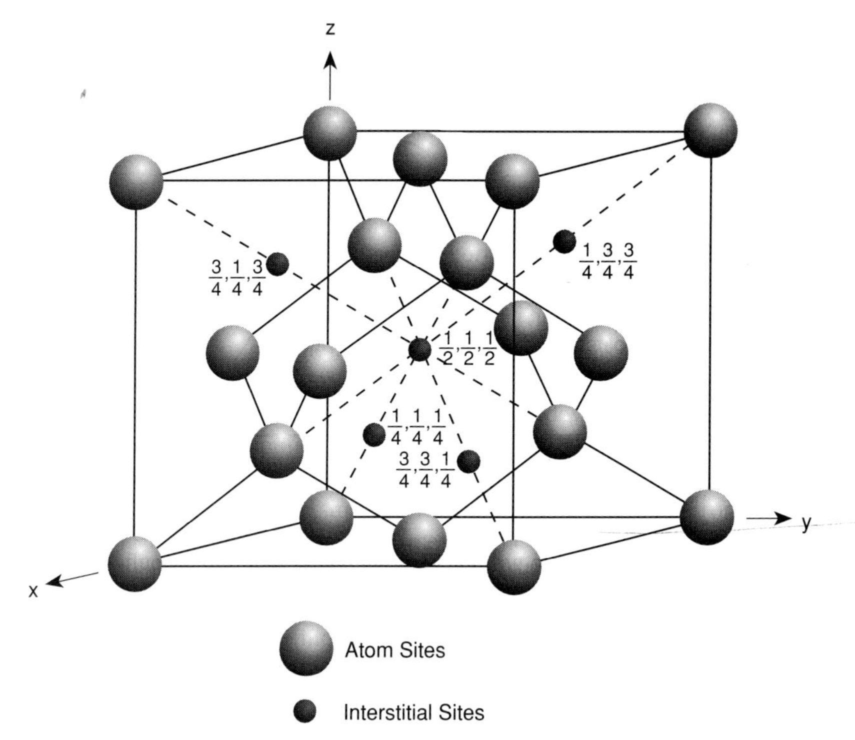

FIGURE 3.9
Some of the interstitial sites in the diamond-cubic structure.

strain issue can be addressed by referring to Table 3.1, reproduced from the pioneering work of Pauling (1960), that shows tetrahedral covalent radii of various atoms.

Before taking up the strain issue, it is pertinent to comment on how Pauling (1960) arrived at the values of tetrahedral covalent radii. Pauling based them on the following experimental findings. First, the equilibrium distances between two atoms A and B connected by a covalent bond of fixed type, that is, single, double, and so forth, in different molecules and crystals are in most cases very nearly the same. It is therefore possible to assign a constant value to the A-B bond distance for use in any molecule involving this bond. Second, the covalent bond distances are often related to one another in an additive manner; that is, the bond distance A-B is equal to the arithmetic mean of A-A and B-B distances. Third, a correction for the ionic character of the bond has to be made as follows:

$$\text{Bond distance (A-B)} = r_A + r_B - C\,(X_A - X_B), \tag{3.1}$$

where r_A and r_B are covalent radii of atoms A and B, C is correction factor due to the presence of ionic bonding mixed with covalent, and X_A and X_B are electronegativities of atoms A and B.

Since in tetrahedral semiconductors atoms touch each other along the <111> directions, the bond length is given by $(r_A + r_B)$. Using this relation, the Si-Si and

TABLE 3.1
Tetrahedral covalent radii of different atoms in nm

	Be 0.106	B 0.088	C 0.077	N 0.070	O 0.066	F 0.064
	Mg 0.14	Al 0.126	Si 0.117	P 0.110	S 0.104	Cl 0.099
Cu 0.135	Zn 0.131	Ga 0.126	Ge 0.122	As 0.118	Se 0.114	Br 0.111
Ag 0.152	Cd 0.148	In 0.144	Sn 0.140	Sb 0.136	Te 0.132	I 0.128
	Hg 0.148					

According to Pauling (1960).

Ga-As bond lengths are 0.234 and 0.244 nm, respectively. Now imagine a situation in which phosphorus atoms replace some of the Si atoms in a crystal to produce an n-type conductivity. This substitution would locally shorten the bond distance to 0.227 nm (0.117 + 0.110). As a result of the dopant incorporation, tetrahedral units having two different sizes are produced in the crystal. Therefore, lattice distortions are introduced in the doped crystal, and their magnitudes depend on the tetrahedral covalent radius of the dopant atom and its concentration.

The tetrahedral radius of the atom r_A of a host lattice can be correlated with that of a dopant impurity r_B via the following expression:

$$r_B = r_A(1 \pm \delta), \tag{3.2}$$

where δ is referred to as the *misfit factor.* The misfit factors for commonly used dopants in silicon and indium phosphide are listed in Tables 3.2 and 3.3, respectively.

The misfit factors provide a measure of the strain introduced into the crystal due to the incorporation of dopants: the larger the misfit factor, the larger the strain. We take up this issue further in Chapters 5 and 8 when we discuss the application of the strain effect to lower dislocation densities in compound semiconductors and to understand the formation of dislocation networks due to the diffusion of dopants into a host lattice.

The point-defect situations in II-VI and III-V compound semiconductors are quite complex due to the presence of two types of atoms in the structure. For example, in indium phosphide crystals, vacancies can form on the indium and phosphorus sublattices. Likewise, indium and phosphorus interstitials can exist within either of the two sublattices, resulting in four possible types of interstitials. In addition, a unique defect termed *antisite* defect can form. It refers, for example, to a phosphorus atom occupying the position of an indium atom on the indium sublattice and is labeled as P_{In}. The reverse situation is also possible. Note that the tetrahedral unit constituting an antisite defect consists of the same type of atoms; that is, in the case of P_{In}, a phosphorus atom is bonded to four other phosphorus atoms.

TABLE 3.2
Misfit factors of commonly used dopants in silicon

n-type			p-type		
P	As	Sb	B	Ga	Al
0.068	0	0.153	0.254	0.068	0.068

TABLE 3.3
Misfit factors of commonly used dopants in indium phosphide*

n-type			p-type		
S	Se	Sn	Cd	Mg	Zn
0.055	0.036	0.028	0.027	0.028	0.090

*Calculations of the misfit factors assume that S and Se atoms reside on group V sublattice, whereas Sn, Cd, Mg, and Zn atoms reside on group III sublattice.

The concentration of point defects in a semiconductor increases with increasing temperature because their presence increases both the internal energy of a crystal as well as its entropy. For example, assume that the total number of atoms in a unit volume of a crystal is N and the concentration of a Schottky defect is n_D/volume. The number of ways in which a Schottky defect can form in the crystal is given by

$$C_{n_D}^{N} = \frac{N!}{n_D!\,(N - n_D)!}. \tag{3.3}$$

The configurational entropy S associated with the above process is

$$S = k_B \ln C_{n_D}^{N}, \tag{3.4}$$

where k_B is the Boltzmann constant. The change in internal energy E of the crystal due to the formation of Schottky defects is

$$E = n_D E_D, \tag{3.5}$$

where E_D is the energy of formation of a Schottky defect. The change in free energy F at constant volume is given by

$$F = E - TS, \tag{3.6}$$

where T is the temperature of the crystal. Substituting the values of E and S from Equations (3.4) and (3.5) into Equation (3.6) we have

$$\begin{aligned} F &= n_D E_D - k_B T \ln C_{n_D}^{N} \\ &= n_D E_D - k_B T \ln \frac{N!}{n_D!\,(N - n_D)!} \\ &= n_D E_D - k_B T[\ln N! - \ln(n_D!) - \ln(N - n_D)!]. \end{aligned} \tag{3.7}$$

At equilibrium $\left(\dfrac{\partial F}{\partial n_D}\right)_{T=\text{constant}} = 0$. Differentiating F in Equation (3.7) with respect to n_D and equating the result to zero we have

$$E_D = k_B T \frac{\partial}{\partial n_D}\,[\ln N! - \ln n_D! - \ln(N - n_D)!]. \tag{3.8}$$

Applying Stirling's formula to Equation (3.8) according to which $\ln x! = x \ln x - x$, we have

$$E_D = k_B T \ln\left(\frac{N - n_D}{N}\right). \tag{3.9}$$

Equation (3.9) can be simplified further as

$$n_D = \frac{N}{1 + e^{E_D/k_B T}}. \tag{3.10}$$

Since the term $e^{E_D/k_B T}$ in Equation (3.10) is much greater than one, the equation can be reduced to

$$n_D \cong N e^{-E_D/k_B T}. \tag{3.11}$$

Equation (3.11) shows clearly that the concentration of a Schottky defect n_D increases with increasing temperature of the crystal. Furthermore, since energies of formation of Schottky defects on the two sublattices of a compound semiconductor may be different, the concentrations of defects on the two sublattices will also be different.

EXAMPLE 3.1. To illustrate the effect of temperature on the concentration of point defects, let us calculate the concentration of vacancies in silicon when its temperature is 1673 K, about 40 degrees below the melting point.

Solution. Substituting appropriate values for $E_D = 2.3$ eV, $k_B = 8.6 \times 10^{-5}$ eV/K, and $T = 1673$K in Equation (3.11), we have

$$n_D/N = e^{-15.98}$$

or

$$n_D/N = 1.5 \times 10^{-7}$$

that is, approximately one atom in 10,000,000 is missing from the solid. Now let us compare the vacancy concentration in silicon to that in aluminum when it is at 40 degrees below its melting point, that is, at 895 K, given that $E_D = 0.5$ eV. It can be shown that

$$n_D/N = 1.5 \times 10^{-3}.$$

In this case the concentration of vacancies is extremely high. The difference between the two cases implies that it is considerably more difficult to form a vacancy in silicon than in aluminum. A plausible reason could be that silicon atoms have stronger covalent bonds, whereas aluminum atoms have nondirectional, metallic bonds.

We can treat in a similar manner situations pertaining to defect complexes. However, the presence of two types of atoms in compound semiconductors makes this difficult. This issue is further compounded by the fact that the vapor pressure of group V atoms over a solid in III-V compounds is considerably higher than that of group III atoms, resulting in large differences between the vacancy concentrations on the two sublattices.

3.3.2 One-Dimensional Defects

Dislocations are one-dimensional defects and represent boundaries between slipped and unslipped regions of a crystal. In semiconductors they glide on {111} planes of the diamond-cubic and zinc-blende structures, and their Burgers vector is a/2 <110> {1$\bar{1}$1}. These slip systems are listed in Table 3.4. The 12 such systems are shown in Figure 3.10. The Peierls stress, that is, the stress required at 0 K to move a dislocation from one equilibrium position to the next, is fairly high in

TABLE 3.4
Slip systems in diamond-cubic and zinc-blende structures

Slip Plane	(111)	($\bar{1}$11)	(1$\bar{1}$1)	(11$\bar{1}$)
Slip	[1$\bar{1}$0]	[110]	[110]	[1$\bar{1}$0]
Directions	[10$\bar{1}$]	[101]	[10$\bar{1}$]	[101]
	[01$\bar{1}$]	[01$\bar{1}$]	[011]	[011]

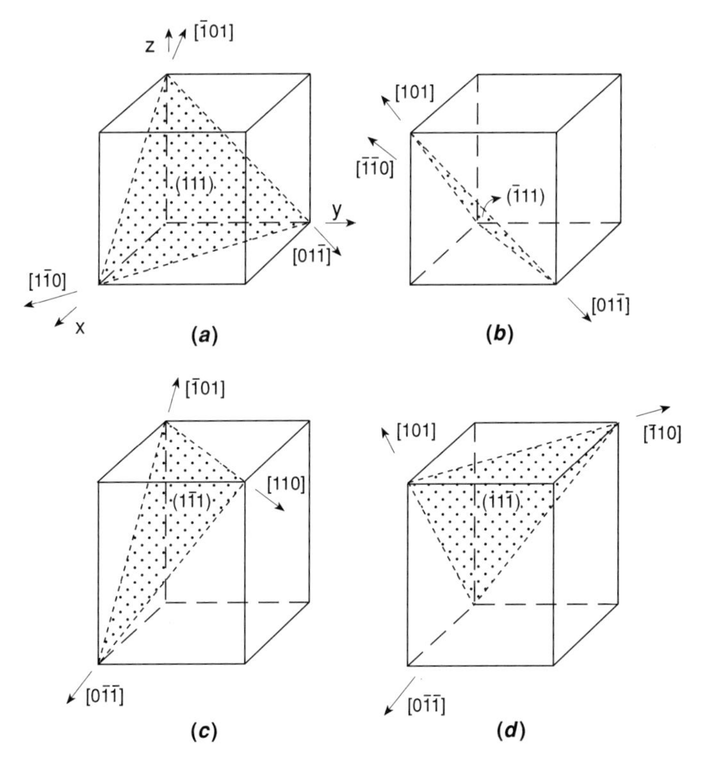

FIGURE 3.10
Schematic showing various slip systems in the diamond-cubic and zinc-blende structures.

covalently and partially ionic, covalently bonded semiconductors, and the Peierls valleys are aligned parallel to the <110> directions lying in the {111} plane (Alexander and Haasen 1968). As a result, dislocations observed in semiconductors deformed at low and moderate temperatures tend to lie along <110> directions and are of 60° type; that is, the angle between the line direction and the Burgers vector is 60° (Wessel and Alexander 1977).

Two types of a/2 <110> perfect dislocations can form in the two structures (Hornstra 1958; Hirth and Lothe 1982). Referring to Figure 3.11, a 60° dislocation in the diamond-cubic structure can be produced by removing the material enclosed by surfaces 15, 56, and 64 and subsequently welding the surfaces 15 and 64. The extra half-plane of the resulting dislocation terminates between the narrowly spaced $(\bar{1}11)$ planes, and the dislocation belongs to the "glide set." On the other hand,

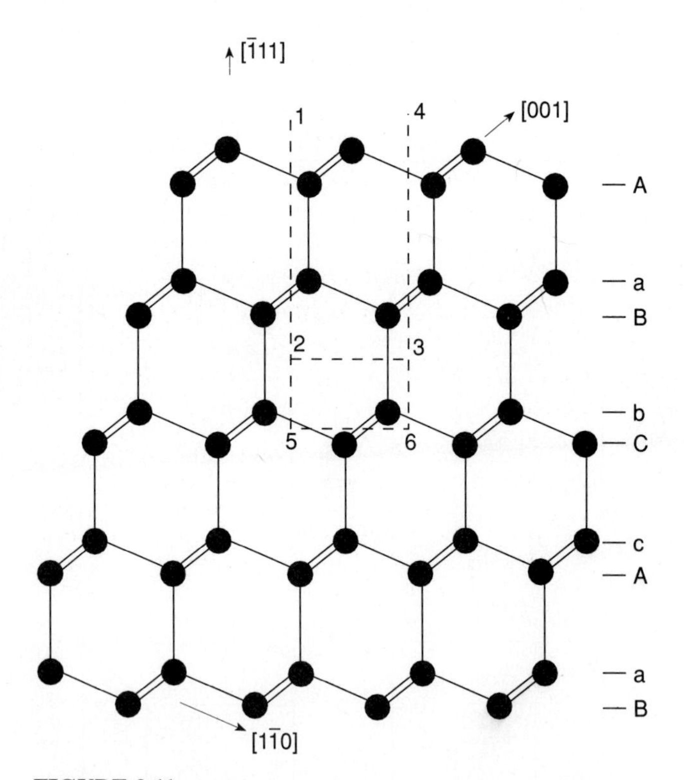

FIGURE 3.11
Diamond-cubic lattice projected onto a (110) plane. The $(\bar{1}11)$ planes are perpendicular to the plane of the paper and appear as horizontal lines. A dislocation belonging to the glide set can be formed by removing the material enclosed by surfaces 15, 56, and 64 and then welding the 15 and 64 surfaces; whereas a shuffle-set dislocation can be formed by removing the material along the 12, 23, and 34 surfaces and subsequently welding the 12 and 34 surfaces.

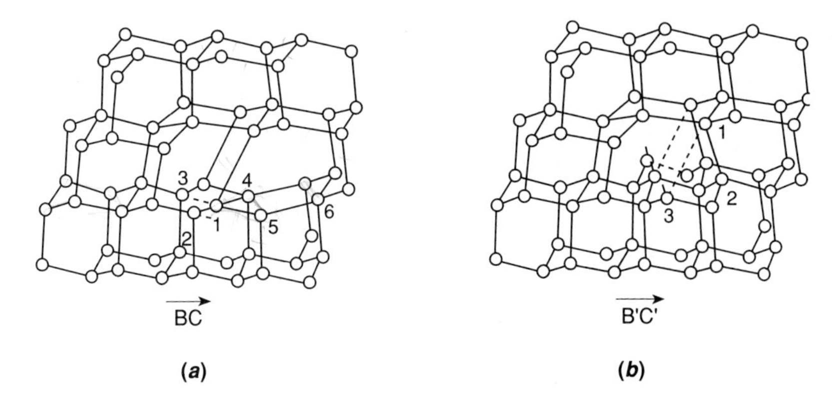

FIGURE 3.12
Possible three-dimensional projections of (*a*) a 60° glide-set dislocation and (*b*) a 60° shuffle-set dislocation. Note the presence of dangling bonds along dislocation cores in both cases. (*After Hirth and Lothe [1982].*)

when the material is removed along the 12, 23, and 34 surfaces and the surfaces 12 and 34 are welded together, the resulting imperfection is called a *shuffle set* dislocation, and its extra half-plane terminates between the widely separated $\{\bar{1}11\}$ planes.

Figure 3.12 shows three-dimensional perspectives of the 60° glide- and shuffle-set dislocations (Hirth and Lothe 1982). Their most important feature is the presence of dangling bonds along the dislocation cores. As discussed later in the chapter, these dangling bonds impart electrical activity to the dislocation and can affect carrier concentration and mobility in a semiconductor.

EXAMPLE 3.2. To discern the effect of dislocations on carrier concentration in a semiconductor, imagine that a perfect a/2 [110] edge glide-set dislocation, oriented along the $[1\bar{1}2]$ direction and gliding in $(\bar{1}11)$ the plane, is present in a 1 cm × 1 cm × 1 cm piece of intrinsic Si. The extra half-plane of such a dislocation, when viewed along the [110] direction, is depicted below.

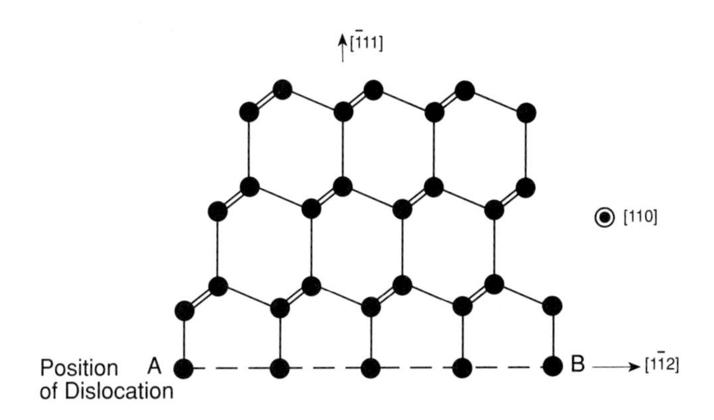

Solution. From Figure 3.6*a* we can show that the separation between the atoms along the $[1\bar{1}2]$ direction is $\sqrt{3}a/\sqrt{8}$. The number of partially bonded atoms along the dislocation core 1 cm long $= \sqrt{8}/\sqrt{3}a = 3 \times 10^7$. The number of unpaired electrons per partially bonded atom $= 3$. Therefore, the number of free electrons/(cm)^3 that can be introduced into the host lattice is 9×10^7 cm$^{-3.}$

Thus the calculated value is three orders of magnitude lower than the intrinsic EHP concentration of 1.5×10^{10} cm^{-3} in silicon at ambient temperature. Consequently, the effect of a single dislocation on carrier concentration is negligible. However, if the crystal contains a high-dislocation density, for example, 10^9 cm^{-2}, the effect could be substantial.

Since III-V semiconductors contain two types of atoms, the core of a glide-set or a shuffle-set dislocation can terminate on either a $\{\bar{1}11\}$ plane containing group III atoms or a $\{1\bar{1}\bar{1}\}$ plane containing group V atoms (Haasen 1957). Consequently, we now have two types of glide- and shuffle-set dislocations. According to the Hünfeld convention, dislocations whose cores terminate on $\{1\bar{1}\bar{1}\}_B$ and $\{\bar{1}11\}_A$ planes are termed α- and β-dislocations, respectively (Alexander et al. 1979). Using this convention, the schematic in Figure 3.13 illustrates the assignments of dislocation type. For example, dislocation 1 belongs to the shuffle set, and its core terminates on a $\{\bar{1}11\}_B$ plane. Therefore, dislocation 1 is α(s), where s refers to the shuffle set. Likewise, dislocations 2, 3, and 4 can be labeled as β(g), β(s), and α(g), where g stands for the glide set. Note that the α and β characters apply only to the edge and mixed dislocations.

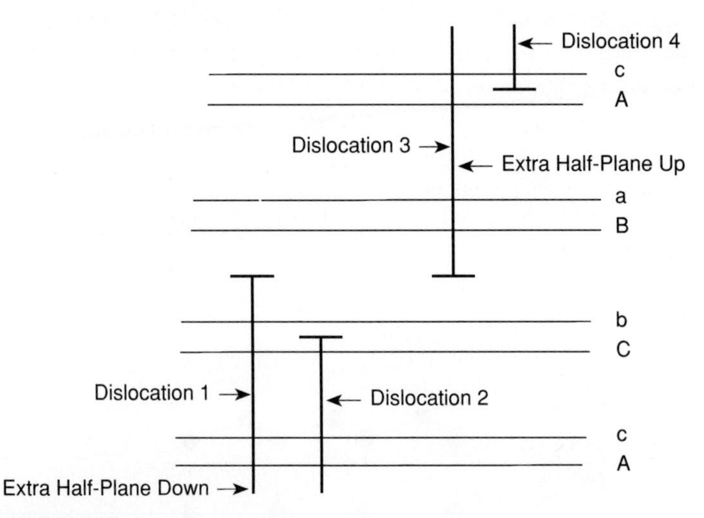

FIGURE 3.13
Schematic depicting terminations of cores of a/2 <110> perfect dislocations in the zinc-blende structure. Dislocation 1 belongs to the shuffle set, and its core terminates on a $\{\bar{1}11\}_B$ plane. According to the Hünfeld convention, dislocation 1 is α(s), where s stands for the shuffle set. Dislocations 2, 3, and 4 are, respectively, β(g), β(s), and α(g), where g refers to the glide set.

Perfect a/2 <110 > glide-set dislocations can dissociate into two a/6 <112> Shockley partials. For example, consider the case of an a/2 $[1\bar{1}0]$ glide-set dislocation moving on a (111) slip plane. This dislocation can dissociate according to the following reaction:

$$\frac{a}{2}[1\bar{1}0]_{(111)} \rightarrow \frac{a}{6}[2\bar{1}\bar{1}]_{(111)} + \frac{a}{6}[1\bar{2}1]_{(111)} \tag{3.12}$$

The schematic in Figure 3.14 illustrates this dissociation, which shows the projection of the diamond-cubic lattice on the (111) plane. Clearly, a/2 $[1\bar{1}0]$ and a/6 $[1\bar{2}1]$ plus a/6 $[2\bar{1}\bar{1}]$ displacements are equivalent, the implication of Equation (3.12). Furthermore, the subscript (111) indicates that the parent dislocation and the resulting Shockley partials can glide on the (111) plane. For this reason, the dislocation shown in Figure 3.12*a* is termed a *glide-set dislocation*.

The dislocation shown in Figure 3.12*b* has a glide plane lying between layers of the same letter index, such as B and b in Figure 3.11. This dislocation cannot dissociate directly into Shockley partials because this dissociation would place layers with different indexes on top of each other, for example, a B layer on top of an a layer. This arrangement represents a high-energy situation. In addition, as discussed by Hornstra (1958) and Hirth and Lothe (1982), the glide of the dislocation involves shuffling atoms in the core region. Hence the dislocation in Figure 3.12*b* is termed a *shuffle-set dislocation*.

The orientations of Shockley partials resulting from dissociation depend on the orientation of the perfect dislocation. The possible orientations of Shockley pairs would be 30° + 30°, 60° + 60°, and 90° + 30°, respectively, for a perfect screw, an edge, and a 60° dislocation. Hirsch (1981) has modeled the core structure of these

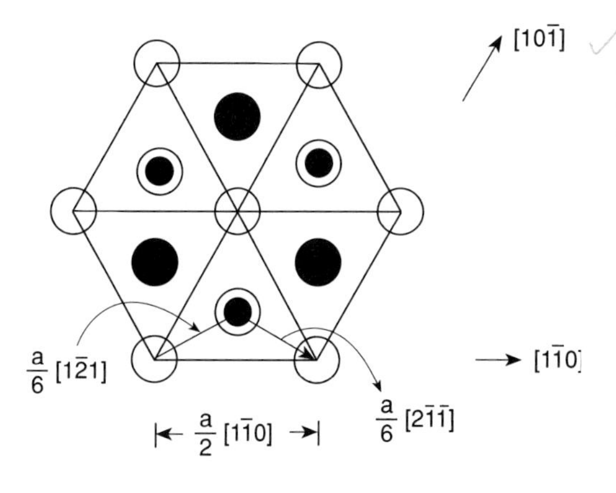

FIGURE 3.14
Schematic illustrating the projection of the diamond-cubic structure on a (111) plane.

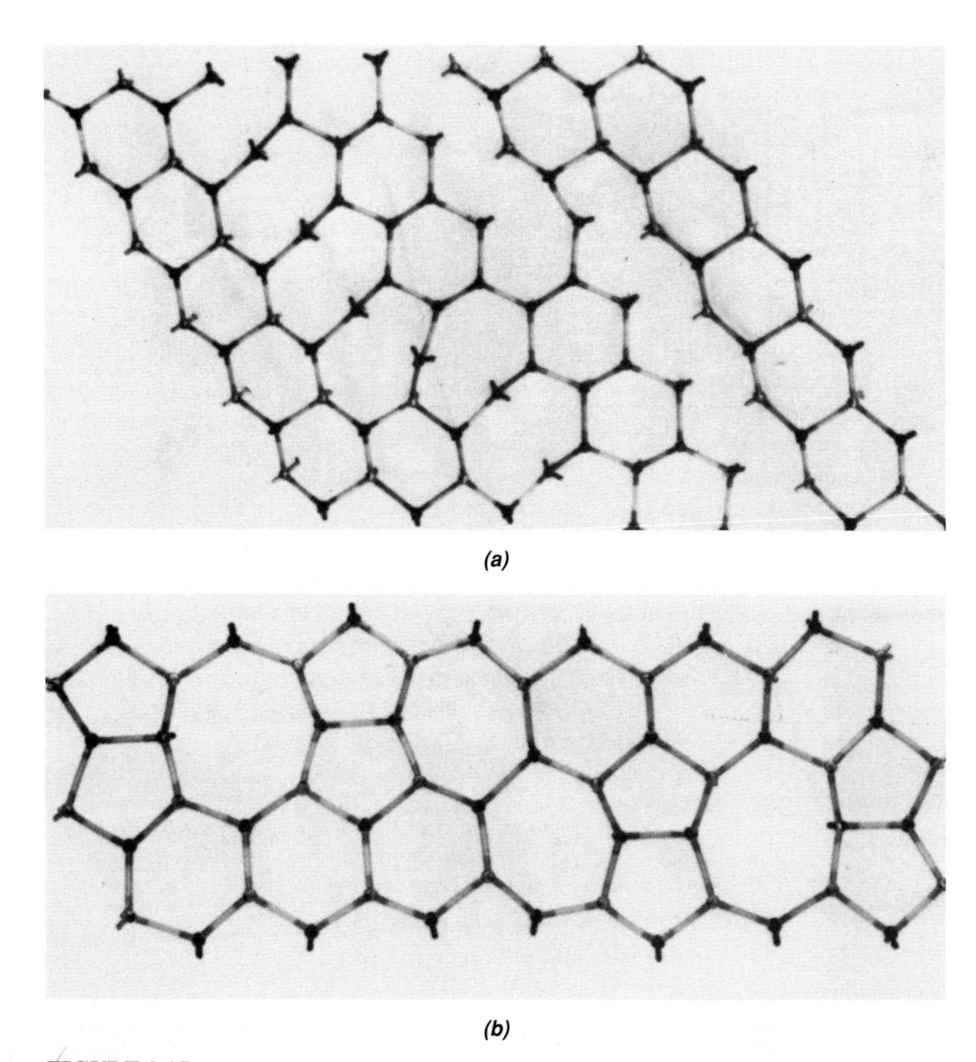

FIGURE 3.15
(*a*) Ball and stick model showing the dissociation of a 60° dislocation into a 30° Shockley partial (left) and a 90° partial (right); dislocation cores are unreconstructed. Kinks are shown on both partials. Viewed normal to the slip plane; only a layer of atoms above and below the slip plane are shown. (*b*) Reconstructed 30° partial with a kink. (*After Hirsch [1981].*)

partials. Figure 3.15*a* shows unreconstructed cores of 30° and 90° partials resulting from the dissociation of a 60° perfect dislocation. To further lower the energy, the cores can undergo reconstruction as illustrated in Figure 3.15*b* for a 30° Shockley partial.

Experimental observations indicate that dislocations in silicon, germanium, and gallium arsenide are dissociated into Shockley partials and glide in the dissociated configuration (Wessel and Alexander 1977; Gomez and Hirsch 1977; Boivin et al. 1990a, 1990b; De Cooman and Carter 1989; Kuesters et al. 1986). These results suggest strongly that glide-set dislocations are primarily involved in the deformation behavior of materials crystallizing in the diamond-cubic and zinc-blende structures.

Glide and climb can increase the length of dislocations in a material, that is, its dislocation density. If applied stresses exceed the critical resolved shear stress of a material, dislocation multiplication can occur either by the operation of a Frank-Read source (Frank and Read 1950) or by double-cross-slip of screw dislocations. The schematic in Figure 3.16 illustrates the operation of a Frank-Read source. Let us imagine that a dislocation segment AB of length L is pinned at its endpoints, Figure 3.16*a*. When a shear stress τ is applied in the direction of its Burgers vector on its glide plane, the dislocation would bow in response to this stress, Figure 3.16*b*. The critical shear stress τ_c required to bow the dislocation into a semicircular shape, Figure 3.16*c*, is given by the following relation:

$$\tau_c = \frac{Gb}{2L}, \tag{3.13}$$

where G is the shear modulus of the material. When τ_c is reached, the dislocation would begin to multiply and assume configurations shown in Figures 3.16*c* $\rightarrow$ 3.16*e*. Since dislocation segments AC and BD are opposite in nature, they will annihilate each other, resulting in the generation of a dislocation loop and regeneration of the original dislocation, Figure 3.16*e*. It is relatively easy to see that the repeated application of the above process would lead to many dislocation loops that lie on the same glide plane. Some of the dislocations in Si crystals appear to multiply by the preceding process (Dash 1956). Figure 3.16*f* shows a Frank-Reed source observed in a Si crystal.

EXAMPLE 3.3. Let us calculate the stress required to multiply a dislocation in a highly perfect crystal of Si.

Solution. Dislocation lengths are extremely short in such crystals (see Chapter 5). Since the Burgers vector of the glide dislocation is $a/2\langle 1\bar{1}0\rangle$, $|\vec{b}| = 0.543 \times \sqrt{2}/2 = 0.384$ nm. Taking L = 100 nm and G = 6.8×10^{10} Pa, we have

$$\tau_c = \frac{6.8 \times 10^{10} \times 0.384}{2 \times 100} = 1.31 \times 10^8 \text{ Pa};$$

that is, a very high stress is required to multiply dislocations in highly perfect Si crystals.

Dislocations can also multiply by the double-cross-slip process shown in Figure 3.17. Imagine a situation where a portion of a screw dislocation cross slips onto a plane ABCD, Figure 3.17*b*, and then returns to a plane that is parallel to the original glide plane, Figure 3.17*c*. This process is called *double-cross-slip* and results in the creation of dislocation lengths on the cross-slip plane. Only screw dislocations can cross slip because the dislocation line as well as its Burgers vector must lie along the line of intersection of the glide and cross-slip planes.

Before discussing the multiplication of dislocations by climb, it is important to emphasize the salient features of the climb process in elemental and compound semiconductors. Note that only mixed and edge dislocations can undergo climb.

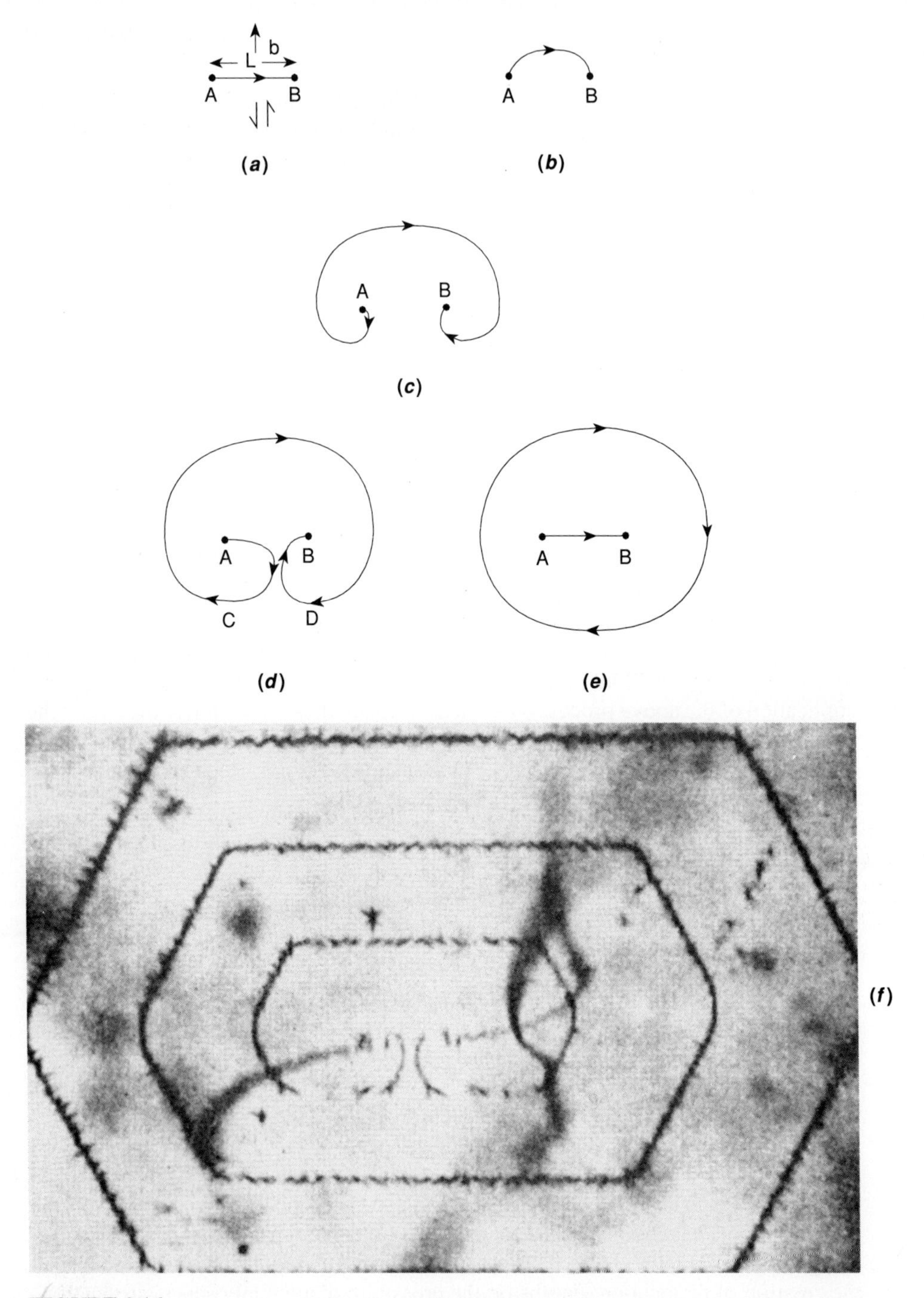

FIGURE 3.16
Schematic illustrating the multiplication of a dislocation AB according to the Frank-Read mechanism. Dislocation AB is perfect edge because the line direction is perpendicular to its Burgers vector b. Dislocation AB is regenerated in (*e*) and can in principle produce many dislocation loops that are coplanar with the mother dislocation AB. (*f*) An example of a Frank-Read source observed in a silicon crystal. (After Dash [1956].)

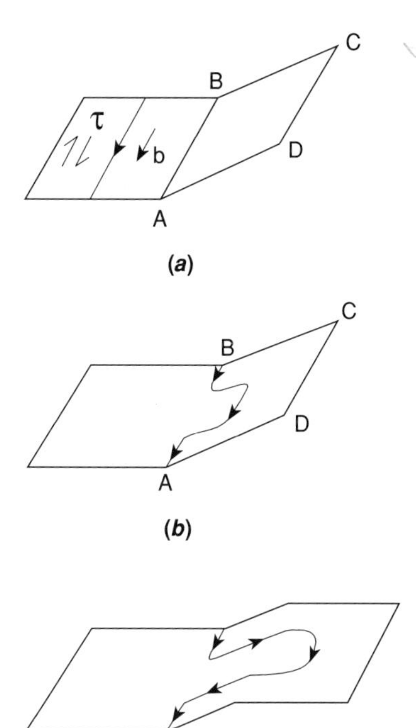

FIGURE 3.17
Schematics illustrating the multiplication of dislocations by the double-cross-slip process. (*a*) A screw dislocation with Burgers vector b gliding under the influence of an applied shear stress τ. (*b*) A portion of the dislocation has cross slipped onto the cross-slip plane. (*c*) The dislocation cross slips back onto a plane parallel to the original plane. The double-cross-slip process increases the dislocation length, that is, multiplication.

For illustrative purposes, the extra half-plane of an edge dislocation belonging to the glide set in an elemental semiconductor is shown in Figure 3.18*a*; the dislocation is oriented along $[1\bar{1}2]$ direction, its Burgers vector is a/2 [110], and it glides on a $(\bar{1}11)$ plane. Let us assume that an interstitial atom is incorporated into the dislocation core as shown in Figure 3.18*b*. As soon as this occurs, an interstitial will be created at the dislocation core with the concomitant emission of a vacancy into the adjoining lattice. This step is necessary because the stability of the diamond-cubic structure requires adding the atomic species in pairs. As a result of this process, the dislocation climbs downwards and acquires double-jogs CD and FE, Figure 3.18*b*; that is, the dislocation has multiplied. It is relatively easy to visualize that the dislocation would climb upwards if it acquires a vacancy, with simultaneous generation of a vacancy at the core and emission of an interstitial into the adjoining lattice.

The situation regarding the climb of dislocations in binary compound semiconductors is more complex because they contain two types of atoms. First, the number of elementary-point defects is twice that of the elemental semiconductors. Second, as discussed earlier α- and β-dislocations exist in these materials. Figure 3.19*a* shows the [110] extra half-plane of an α-dislocation belonging to the glide set in a III-V binary compound semiconductor. As in the case of elemental semiconductors, the glide plane of the dislocation is $(\bar{1}11)$ and its Burgers vector is a/2 [110]. Imagine a situation where a group III interstitial is absorbed at the dislocation core. This process would result in the creation of a group V interstitial at the

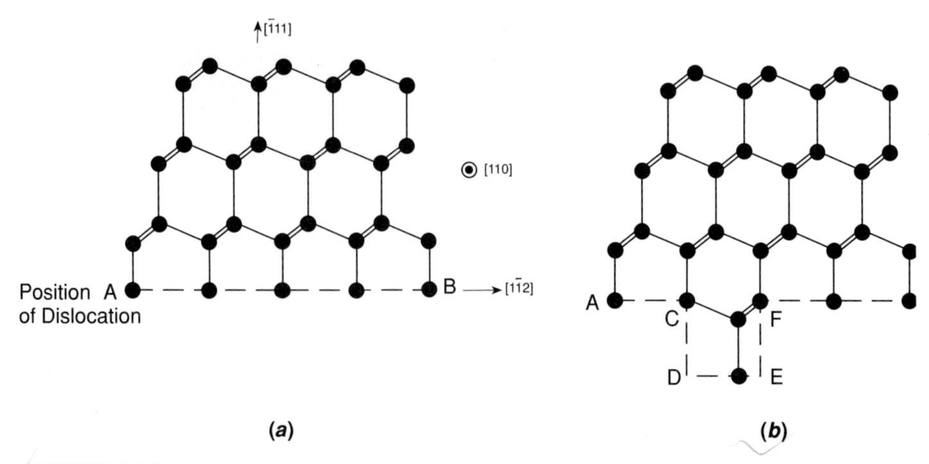

FIGURE 3.18
(*a*) Schematic showing the [110] view of the extra half-plane of an edge dislocation AB in the diamond-cubic structure; dislocation is oriented along the [1 $\bar{1}$2] direction, and its Burgers vector is a/2 [110]. (*b*) The dislocation climbs downwards by the incorporation of a pair of interstitials at the dislocation core that results in the formation of double-jogs CD and FE.

dislocation core and simultaneous emission of a group V vacancy into the adjoining lattice. As a result, the dislocation climbs downwards and acquires double-jogs, Figure 3.19*b*. These situations occur because the absorption of a group III interstitial results in an unstable situation. This instability is eliminated by the formation of a group V interstitial at the dislocation core with a concomitant emission of a group V vacancy into the contiguous region (Vardya and Mahajan 1995).

The following four distinct situations can arise during the climb of dislocations in III-V compound semiconductors:

$III_{(I)}$ absorbed at the core	+	$V_{(I)}$ created at the core	→	$V_{(V)}$ emitted into the matrix	(3.14)
$V_{(I)}$ absorbed	+	$III_{(I)}$ created	→	$III_{(V)}$ emitted into the matrix	(3.15)
$III_{(V)}$ absorbed	+	$V_{(V)}$ created	→	$V_{(I)}$ emitted into the matrix	(3.16)
$V_{(V)}$ absorbed	+	$III_{(V)}$ created	→	$III_{(I)}$ emitted into the matrix,	(3.17)

where Is and Vs refer to interstitials and vacancies, respectively. Let us now consider a situation where group V interstitials are in a considerable excess in a material that contains both α- and β-dislocations. Since the extra half of an α-dislocation terminates in a row of group V atoms, these dislocations would have very little tendency to climb. On the other hand, a β-dislocation would have a strong driving force for climb because it would eliminate the nonequilibrium concentration of

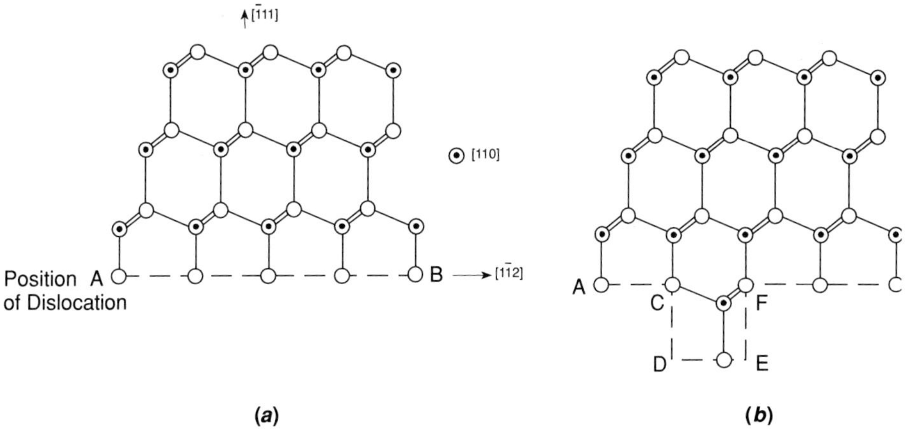

(a) (b)

FIGURE 3.19
(*a*) Schematic showing the [110] view of the extra half-plane of an edge dislocation AB in the zinc-blende structure; dislocation is oriented along the [1$\bar{1}$2] direction, and its Burgers vector is a/2 [110]. (*b*) The dislocation climbs downwards by the incorporation of a pair of interstitials at the dislocation core that results in the formation of double-jogs CD and FE.

group V interstitials present in the material. Similarly, if a material contains a large concentration of group III interstitials, α-dislocations would climb in preference to β-dislocations. Thus there is a built-in bias for the climb of α- and β-dislocations in compound semiconductors. The extent of the bias depends on the nonequilibrium concentrations of point defects existing in a material. These concentrations in turn depend on the energies of formation of various point defects and conditions prevailing during growth and processing of semiconductors.

As discussed in section 3.3.1, we can increase the equilibrium concentration of point defects in a material by raising its temperature. The nonconservative motion of jogs on a dislocation can also lead to the generation of point defects. These point defects are generated in pairs, and the pairs consist of interstitials or vacancies in the elemental semiconductors, whereas in the compound semiconductors, group V and III interstitials or group V and III vacancies make up the pairs.

EXAMPLE 3.4. Let us assess the number of point defects that are generated when the edge dislocation in Si shown in Figure 3.18*a* climbs down by a distance of $a/\sqrt{3}$ by the absorption of interstitials. Assume that the length of the dislocation is 1 cm and the crystal volume is $(1\ \text{cm})^3$.

Solution. As shown in Example 3.2, the number of atoms along the dislocation core = 3×10^7/cm.
Therefore, the number of Si interstitials required for the unit climb = $6 \times 10^7/\text{cm}^3$.
The number of interstitials supplied by the supersaturation = $3 \times 10^7/\text{cm}^3$.
The number of interstitials created at the core = $3 \times 10^7/\text{cm}^3$.

The number of vacancies created in a region contiguous to the dislocation = $3 \times 10^7/\text{cm}^3$.

A ramification of these vacancies is that interstitials and vacancies will annihilate each other, leading to a progressive elimination of the supersaturation of silicon interstitials. Subsequently, the tendency for dislocation climb will be reduced.

In the presence of supersaturation of point defects, dislocations can climb extensively and thus multiply. A situation analogous to a Frank-Read source can evolve during the climb of pinned dislocations. This situation was first discussed by Bardeen and Herring (1952). Consider a case where a prismatic dislocation ABCD, whose segment BC cannot glide in the plane, is pinned at points B and C, see Figure 3.20*a*. In the presence of the supersaturation of point defects, the dislocation BC could climb as shown in Figure 3.20*b*. With additional climb the situations depicted in Figures 3.20*c* and 3.20*d* could develop, resulting in the generation of a prismatic dislocation loop. With the repeated occurrence of this process, many dislocation loops can be generated, just as in the case of a Frank-Read source. The dislocation source shown in Figure 3.20 is, therefore, termed a *Bardeen-Herring climb source*.

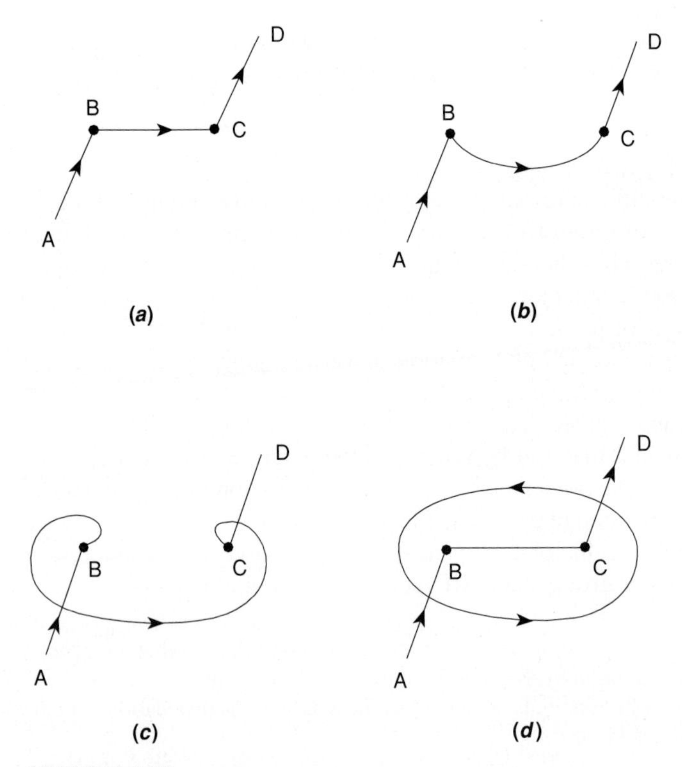

FIGURE 3.20
Schematic showing the operation of a Bardeen-Herring climb source: (*a*) A prismatic dislocation ABCD is pinned at B and C, (*b*) dislocation BC climbs by the absorption or emission of point defects, (*c*) an intermediate stage in the climb process, and (*d*) a prismatic loop has been generated with the concomitant regeneration of the original dislocation BC.

Even though Figures 3.16 and 3.20 appear similar, there are distinct differences between the two situations. First, a shear stress acting along the glide plane in the direction of the Burgers vector is necessary for the operation of a Frank-Read source, whereas the supersaturation of point defects is required for the activation of a Bardeen-Herring source. Second, the dislocation loop generated in Figure 3.16 expands by glide, whereas the loop expansion in Figure 3.20 occurs by either the absorption or the emission of point defects.

3.3.3 Two-Dimensional Defects

Stacking faults, twin boundaries, surfaces, subboundaries, and grain boundaries are typical examples of two-dimensional defects. We first consider the formation of stacking faults and twins, followed by a brief discussion of reconstruction that occurs at semiconductor surfaces and a brief introduction to the structure of subboundaries and grain boundaries.

Stacking faults and twins. Stacking faults in the diamond-cubic and zinc-blende structures form on {111} planes in two distinct ways: (1) by shear and (2) by the agglomeration of point defects. The schematic in Figure 3.21 illustrates the situations resulting from shear. As discussed in section 3.2, the stacking arrangement of {111} planes in the two structures is Aa Bb Cc Aa . . . , see Figure 3.21*a*, and its $(11\bar{1})$ projection is shown in Figure 3.14. Now imagine a situation in which an a/6 $[1\bar{2}1]$ displacement is imposed on some of the atoms in the A-a layers that move as a pair. As a result, the atoms below the doubled dashed line would move into the B-b position as shown in Figure 3.21*b*. The imposition of the a/6 $[1\bar{2}1]$ displacement on the A-a layers also causes the movements of layers below this pair as illustrated in Figure 3.21*b*. The resulting arrangement constitutes an intrinsic stacking fault. If you compare the stacking arrangement of the (111) planes in a perfect crystal with that associated with an intrinsic fault, the A-a layer pair appears to have been removed from the crystal. Actually, some of the atoms in the A-a layer have moved into the B-b position as depicted in Figure 3.22. Since faults terminating within a crystal must be bounded by partial dislocations, the position of an a/6 $[1\bar{2}1]$ Shockley dislocation that bounds the intrinsic stacking fault in Figure 3.21*b* is shown in Figure 3.22.

If we examine the layer arrangement across the C-c pair that is dashed in Figure 3.21, we find that the B-b pairs are present on either side of the C-c pair, see Figure 3.21*b*; that is, the atomic arrangement is reflected across the C-c layer. In other words, an intrinsic fault is one double-layer thick twin. In the case of the diamond-cubic structure, the "true" mirror symmetry exists across the C-c pair. This symmetry is, however, not true for the zinc-blende structure because the two sublattices are occupied by different types of atoms. Furthermore, unlike the case of a/2 $1\bar{1}0$ dislocations, the formation of an intrinsic stacking fault does not produce dangling bonds at the fault surface.

An extrinsic stacking fault can be formed by imposing a second a/6 $[1\bar{2}1]$ displacement on the C-c pair that lies below the first shear plane. The resulting layer

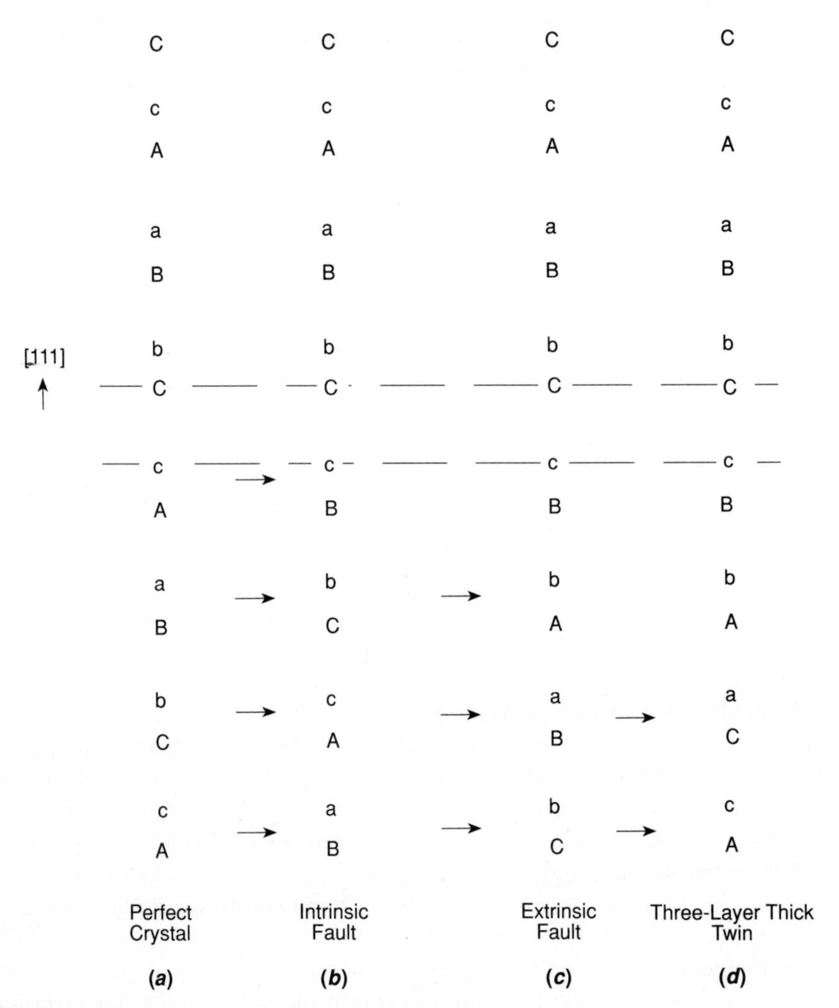

FIGURE 3.21
Schematic illustrating the formation of stacking faults and a three-layer twin in the diamond-cubic and zinc-blende structures by shearing of (111) planes: (*a*) perfect crystal, (*b*) intrinsic fault, (*c*) extrinsic fault, and (*d*) three-layer twin.

arrangement for an extrinsic fault is shown in Figure 3.21*c*. An examination of Figure 3.21*c* would reveal that the fault is a two double-layer thick twin. Likewise, a three-layer-pair thick twin can be produced as the schematic in Figure 3.21*d* shows.

Now consider a situation where vacancy discs form on the A-a planes, as shown in Figure 3.23*a*. This arrangement is unstable, and to eliminate it, the adjoining plane pairs C-c and B-b collapse around the hole, resulting in an intrinsic stacking fault as illustrated in Figure 3.23*b*. Since the displacement associated with the collapse is along $[\bar{1}\bar{1}\bar{1}]$ direction, the Burgers vector of the partial bounding the fault is $a/3\,[\bar{1}\bar{1}\bar{1}]$ and is called a *Frank partial*. As there are four different $<111>$ directions in the two

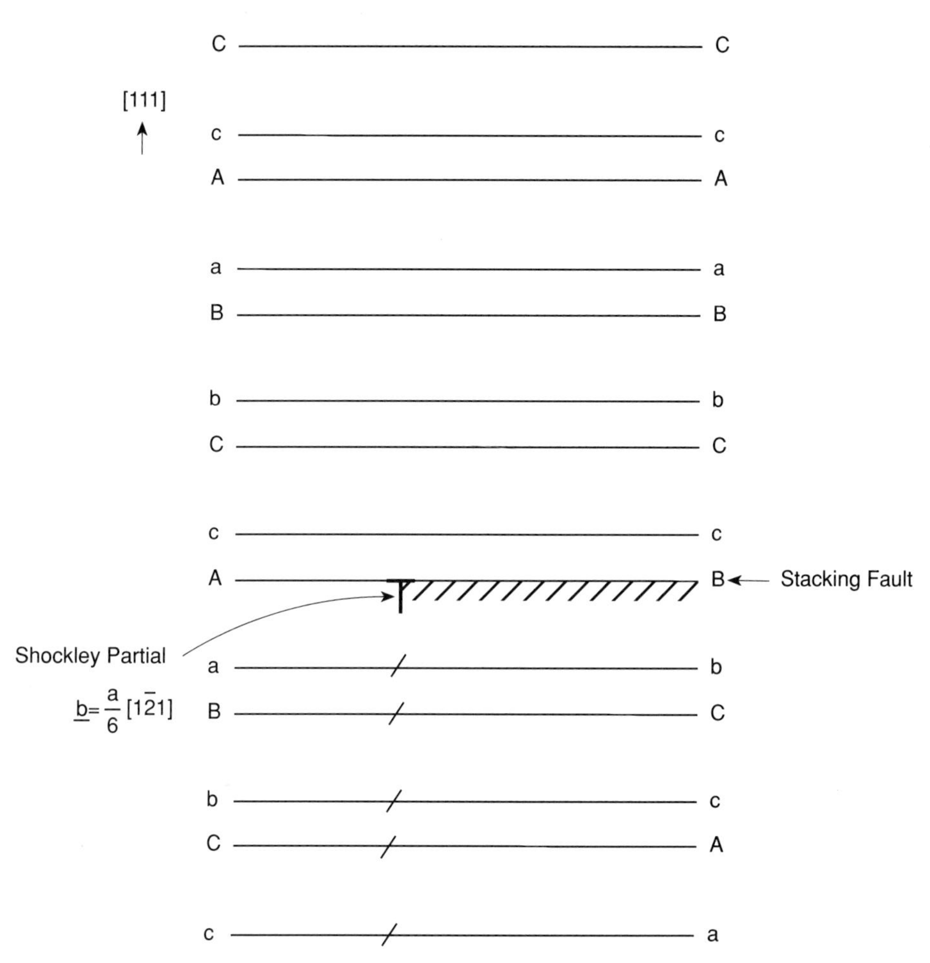

FIGURE 3.22
Schematic showing the position of the a/6 $[1\bar{2}1]$ Shockley partial that bounds an intrinsic stacking fault in the diamond-cubic and zinc-blende structures. The (111) planes are normal to the plane of the figure.

structures, four different Frank partials are possible. In addition, two vacancy discs, each consisting of group III and group V vacancies, are required to form an intrinsic stacking fault in the zinc-blende structure. For example, in InP a disc of In-vacancies must form on an A plane, and a disc of P-vacancies must precipitate on the a plane above. Furthermore, the layer arrangements in Figures 3.21*b* and 3.23*b* are identical within the faulted regions. However, the approaches used to achieve these configurations are different. In Figure 3.21*b* some of the atoms in the A-a layers have been displaced into the B-b positions, whereas they have been removed in Figure 3.23*b*.

The formation of extrinsic faults resulting from the agglomeration of interstitials can be discussed in a similar manner. The schematic in Figure 3.23*c* shows the final-layer arrangement within the faulted region and the bounding partial. This

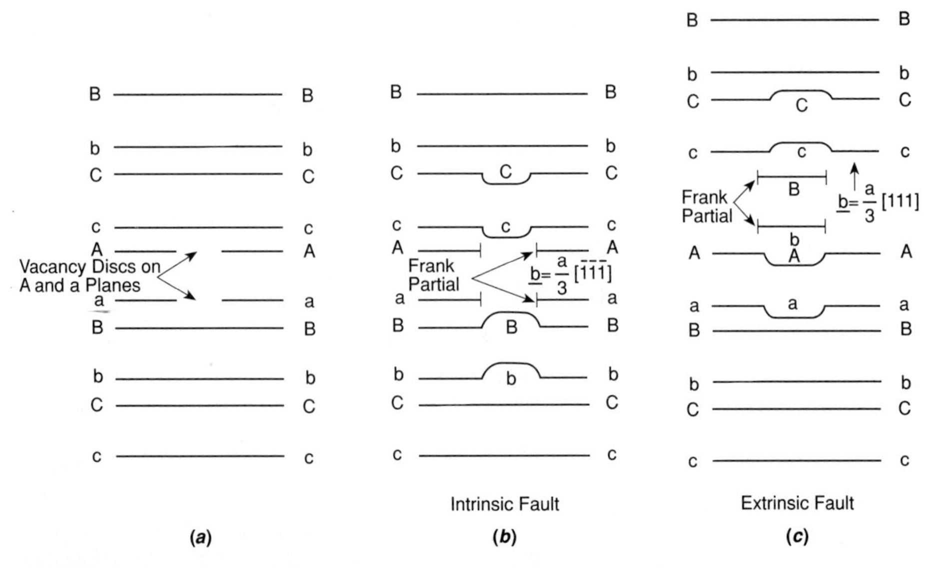

FIGURE 3.23
Schematics illustrating the formation of intrinsic and extrinsic stacking faults in the diamond-cubic and zinc-blende structures by the agglomeration of point defects on (111) planes: (*a*) layer arrangement after the formation of vacancy discs on the A-a plane pair, (*b*) intrinsic fault resulting from the collapse of the B-b and C-c pairs adjoining the vacancy discs, and (*c*) extrinsic fault resulting from the insertion of the B-b pair between the A-a and C-c plane pairs. A Frank partial bounds each fault, and the respective Burgers vectors are $a/3\,[\bar{1}\bar{1}\bar{1}]$ and $a/3\,[111]$.

arrangement is identical to that shown in Figure 3.21*c*, but it has been achieved via a different route, as discussed in the case of an intrinsic fault. Since the fault formation involves an insertion of a B-b plane-pair, the adjoining pairs would have to be displaced outwards as shown in Figure 3.23*c*. Again, a Frank partial bounds the fault, but now its Burgers vector is $a/3\,[111]$. To form such a fault in InP requires a layer each of the In and the P atoms.

EXAMPLE 3.5. Let us calculate the number of point defects required to form a (111) hexagonal-shaped extrinsic fault in GaAs; the fault edges are aligned along the $\langle 1\bar{1}0\rangle$ directions lying in the (111) plane. Assume that the edge of the fault is 1 μm long.

Solution. Surface area of the fault $= \frac{1}{2} \times 10^{-6} \times 10^{-6} \times \sin 60° \times 6$

$$= 3 \times 10^{-12} \sin 60° \,(\text{m})^2.$$

Area of one of the triangles in Figure 3.14

$$= \frac{1}{2} \times \frac{0.565 \times 10^{-9}}{\sqrt{2}} \times \frac{0.565 \times 10^{-9}}{\sqrt{2}} \sin 60°\text{C}$$

$$= \frac{(0.565)^2 \times 10^{-18}}{4} \sin 60° (\text{m})^2.$$

The "effective" number of atoms in the triangle $= \frac{1}{2}$.

$$\text{The number of atoms in the fault surface} = \frac{3 \times 10^{-12}}{(0.565)^2 \times 10^{-18}}$$

$$= 4.7 \times 10^6.$$

Since the extrinsic fault is a two-layer fault and contains Ga and As interstitials, the number of point defects required are

$$\text{Ga interstitials} = 4.7 \times 10^6$$

$$\text{As interstitials} = 4.7 \times 10^6.$$

The preceding discussion shows that the conditions required for the formation of faults by shear and for agglomeration of point defects are different. In the case of shear faults, stresses must exist in the material to move Shockley partials to expand the faulted regions. These stresses could develop during the growth of bulk crystals and epitaxial layers as discussed in Chapters 5 and 6. On the other hand, the agglomeration case requires nonequilibrium concentrations of point defects. Again, appropriate conditions could evolve during the growth of bulk crystals (Chapter 5), oxidation of silicon (Chapter 7), diffusion (Chapter 8), and ion implantation (Chapter 9).

Surfaces and steps. Substrates are required to fabricate some of the device structures discussed in Chapter 2. These substrates are obtained from bulk crystals. For certain devices, such as light-emitting devices and photodetectors, epitaxial layers are deposited on underlying substrates to tailor their device characteristics. These layers are deposited on substrates whose surface normals are slightly misoriented from specific crystallographic directions. The resulting surfaces are called *vicinal*. The vicinal surfaces can lower their energy by the formation of terraces, ledges or steps, and kinks as shown in Figure 3.24. The terrace surface normals are parallel to low-index orientations, and the ledge height is determined by the

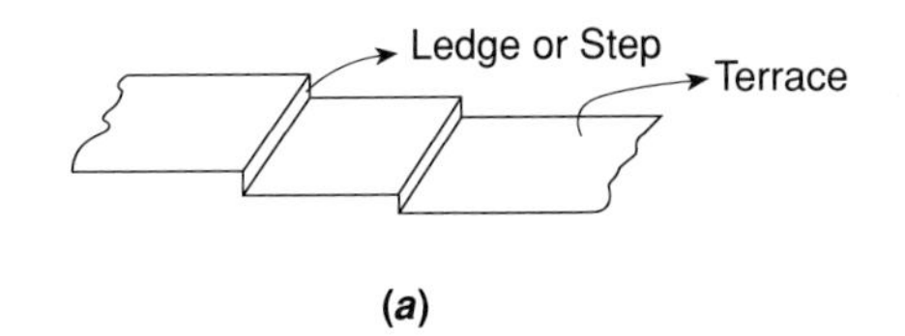

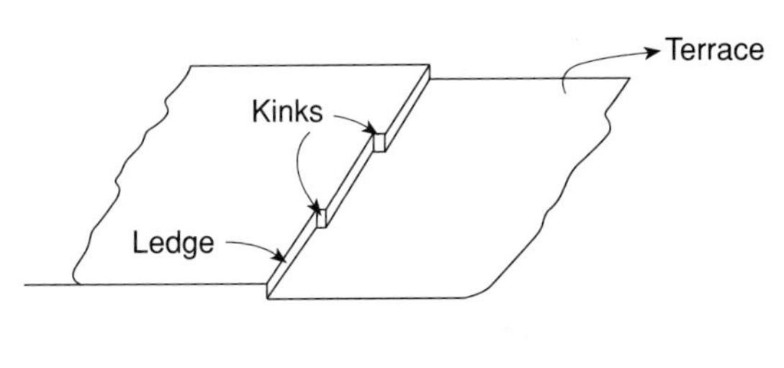

FIGURE 3.24
Schematic of vicinal surfaces showing (*a*) terraces and ledges and (*b*) terraces, ledges, and kinks.

crystallography of the surface. If the ledge height is h and the terrace length is L, then tan θ is given by

$$\tan\theta = \frac{h}{L}, \tag{3.18}$$

where θ is the angle by which the surface normal deviates from the exact orientation. Equation (3.18) indicates that for larger values of θ, the terrace length should decrease for a fixed value of h. Furthermore, by varying the orientation of the terraces from <001> to <011> to <111>, that is, the spacing between the steps and the kinks, one can synthesize any vicinal surface.

The atomic arrangement at a semiconductor surface often deviates from that existing in the bulk. This situation may occur because the surface configuration of atoms obtained by cutting a substrate from a bulk crystal does not have the lowest energy; therefore, the new atomic arrangement evolves at the surface. This phenomenon is known as *surface reconstruction.*

Figure 3.25*a* shows an unreconstructed, As-terminated (001) GaAs surface; the Ga and As atoms in the first and the second layers below the surface are also shown.

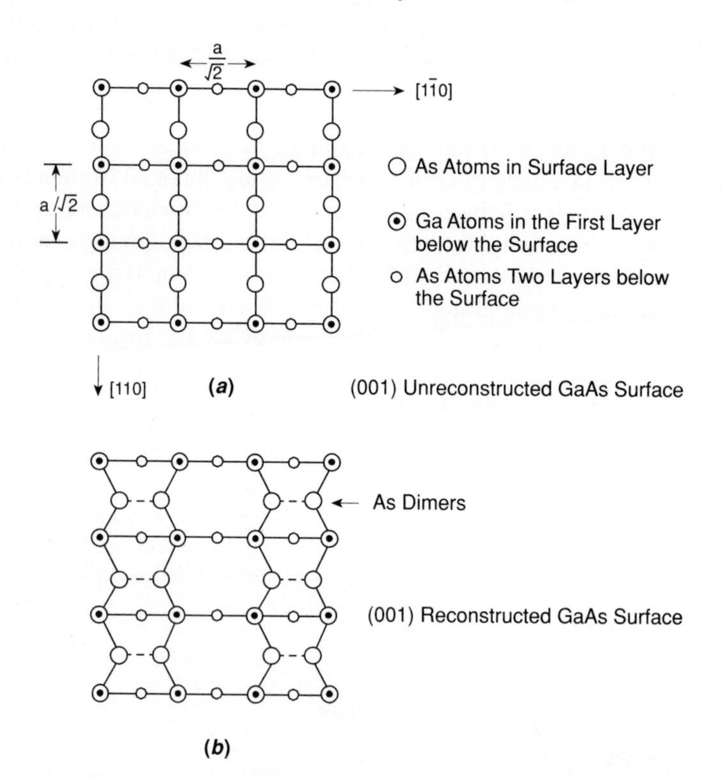

FIGURE 3.25
(*a*) (001) unreconstructed GaAs surface showing the positions of Ga and As atoms in the surface and two layers below the surface. (*b*) (001) reconstructed GaAs surface showing the formation of As dimers.

Each arsenic atom in the top layer is associated with two dangling bonds. To lower the electronic energy of the surface, the atoms in the surface undergo rearrangement, resulting in the formation of As dimers, see Figure 3.25*b*. The dimers form along the $[1\bar{1}0]$ direction because the separation between the top row and the next subsurface row is larger along this direction—the second layer of As atoms is at a distance of $a/2$ from the surface. On the other hand, reconstruction does not occur along the [110] direction as the first layer of Ga atoms below the surface is located at a depth of $a/4$ from the top As layer. If the As atoms were to form to dimers along the [110] direction, the underlying Ga atoms would have to be pushed inwards and this condition is energetically unfavorable.

Chadi (1987) has shown that the reconstructed patches are separated from each other (see Figure 3.26). Since the repeat periods along the $[1\bar{1}0]$ and the [110] directions are two and four units, respectively, the resulting reconstruction is termed (2×4). The occurrence of the (2×4) reconstruction on As-terminated (001) GaAs surfaces has been observed experimentally (Pashley et al. 1988). Furthermore, the Ga-terminated (001) GaAs surface undergoes a weakly developed (4×2) reconstruction (Chadi 1987).

As expected, (001) silicon surfaces also undergo reconstruction. The commonly observed surface reconstructions are (2×1) and (1×2), and the preponderance of

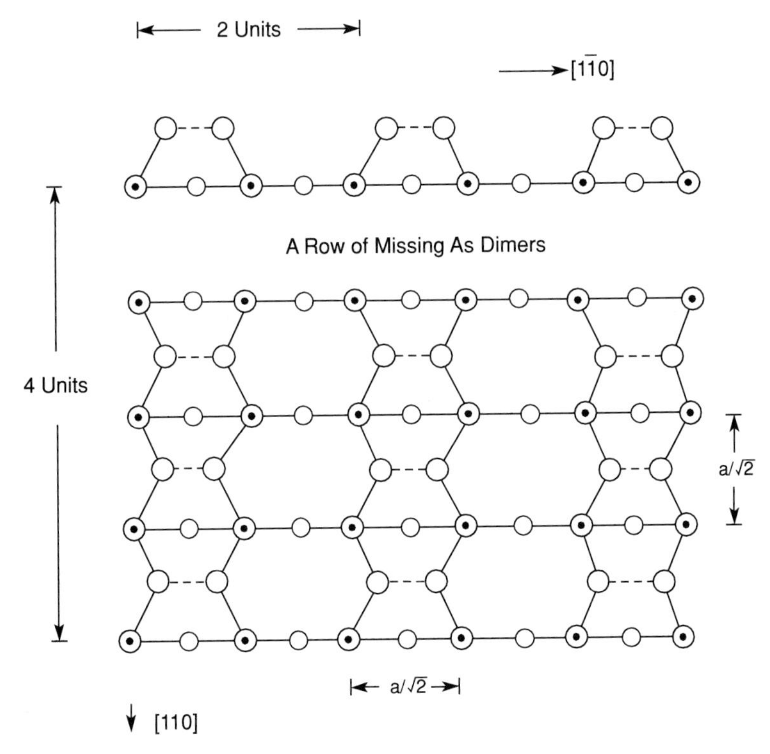

FIGURE 3.26
Schematic showing a patch of (2×4) reconstructed (001) GaAs surface.

one variant over the other can be shifted by the imposition of a stress along the surface (Men et al. 1988). Reconstruction may also vary with the orientation of the surface (Lagally et al. 1985).

Since epitaxial layers are grown on vicinal surfaces and since steps appear to play a role in the layer growth, it is instructive to consider the atomic structure of steps on typical vicinal surfaces. Figure 3.27 shows the [110] perspective of a silicon surface that has been tilted in a clockwise fashion from the exact [001] orientation; the tilt axis is [110]. Two surface configurations, ABCD and EFGH, are possible. In both cases each atom in the terrace has two dangling bonds. Furthermore, tetrahedra define the steps along the [110] direction. In the case of the ABCD surface, the row of atoms along B has two dangling bonds per atom, whereas along row C only one dangling bond/atom exists, row F has two dangling bonds/atom, and atoms are fully bonded along row G.

Figure 3.28 depicts the corresponding situation for a III-V semiconductor. It is apparent that surfaces ABCD and EFGH are not equivalent because of the noncenterosymmetric nature of the zinc-blende structure. ABCD shows a group V–terminated surface, whereas the surface EFGH contains group III atoms. In addition, group III–centered tetrahedra define steps along the [110] direction.

By examining Figures 3.7*a* and 3.7*b*, the step structure of {111} vicinal surfaces in the diamond-cubic and zinc-blende structures that have been tilted away from the exact [$\bar{1}$11] orientation around the [110] axis can be discerned. The most likely steps are of height $a/\sqrt{3}$. The step structure of surfaces having other orientations can also be analyzed in an analogous manner using the ball and stick model.

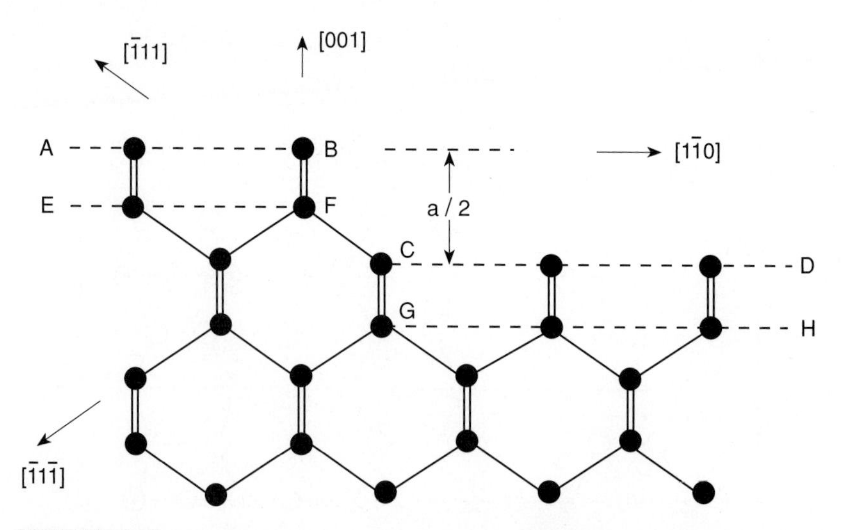

FIGURE 3.27
Schematic showing the (110) projection of a stepped (001) silicon surface. ABCD and EFGH show two possible equivalent configurations of the surface; the step height in each case is a/2.

Subgrain boundaries and grain boundaries. State-of-the-art silicon crystals are highly perfect. As a result, different regions of crystals are not misoriented from each other. However, this situation is not true in the case of III-V and II-VI compound semiconductor crystals: Misorientations exist between cellular regions present in as-grown crystals. Polycrystalline silicon is also being used as emitters and gate contacts in bipolar and metal-oxide-semiconductor transistors.

Depending on the misorientation between the adjoining crystals, the resulting interfaces are termed subgrain boundaries and grain boundaries. *Subgrain boundaries* separate regions that are misoriented by a very small amount, say 0 to 5°, whereas *grain boundaries* delineate interfaces between highly misoriented regions. An important question is, How can we describe these interfaces in terms of dislocations?

The subgrain boundaries are of three types: tilt, twist, and mixed. Consider a situation where a grain can be brought into the same orientation as another grain by a rotation around an axis. If the rotation angle is small and the rotation axis lies in the boundary, the subgrain boundary is a tilt boundary. On the other hand, if the rotation axis is perpendicular to the boundary, it is a pure twist boundary. In general a boundary is of mixed character, containing both tilt and twist components.

Read (1953) has developed a formalism to describe subgrain boundaries in terms of dislocations. He has shown that a symmetrical tilt boundary between two grains 1 and 2 and the tilt angle θ consists of edge dislocations whose Burgers vector is b. These dislocations are separated from each other by a distance D given by the following expression:

$$D \cong \frac{b}{\theta} \tag{3.19}$$

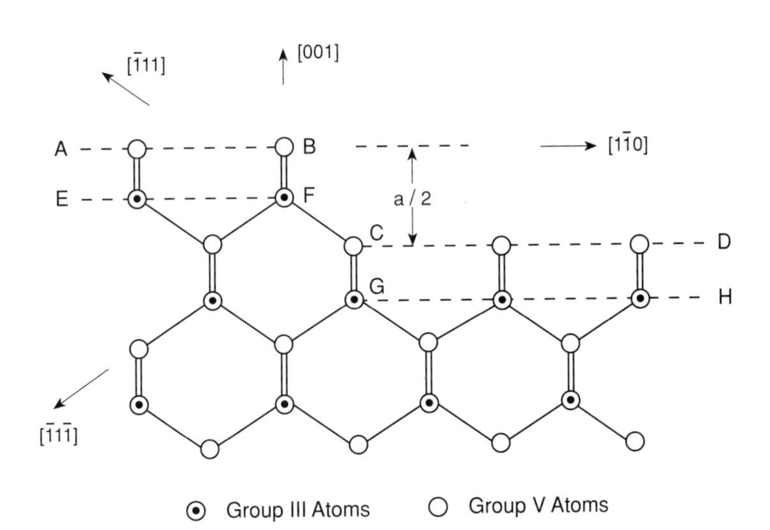

FIGURE 3.28
Schematic showing the (110) projection of a stepped (001) III-V surface. ABCD and EFGH show, respectively, group V– and group III–terminated surfaces. The step height in each case is a/2.

Equation (3.19) shows that as the misorientation θ between the two grains increases, the separation between dislocations defining a subgrain boundary decreases.

The elegant work of Vogel et al. (1953) on subboundaries in Ge crystals proved that tilt boundaries can be described in terms of dislocations. Figure 3.29 shows an example reproduced from their study. Figure 3.29*a* shows a row of etch pits observed at a subboundary between two Ge crystals, and Figure 3.29*b* shows a diagrammatic representation of the observed arrangement of dislocations, revealed by etch pits. Small-angle, pure-twist boundaries can also be interpreted in terms of networks of screw dislocations that lie in the boundary. Again, the separation between dislocations within the network varies with the angle of misorientation: The larger the angle, the smaller the separation between the screw dislocations.

The energy E per unit area of a low-angle boundary is given by an expression (Read 1953),

$$E = E_0\, \theta(A - \ln\, \theta), \tag{3.20}$$

where constant E_0 is a function of the elastic properties of a material and A is a constant that depends on the core energy of an individual dislocation. As the

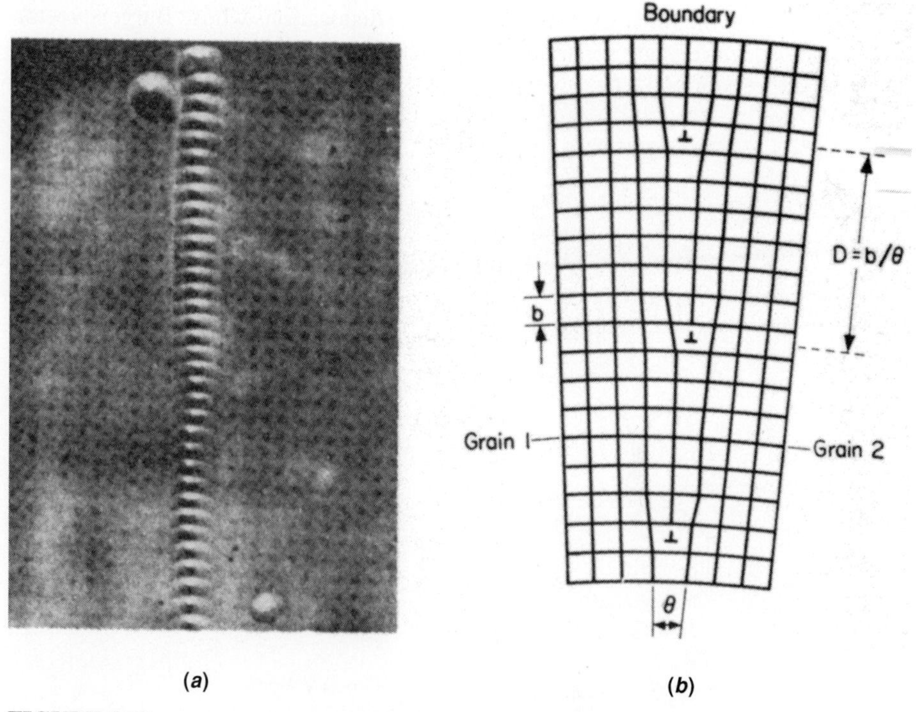

FIGURE 3.29
(*a*) A row of etch pits formed at the boundary between two Ge crystals. (*b*) Diagrammatic representation of the arrangement of dislocations revealed by the etch pits. (*After Vogel et al. [1953].*)

misorientation angle increases, the dislocation spacing decreases, leading to core overlap. As a consequence, the distortion near a dislocation no longer corresponds to that of an isolated dislocation. Under these conditions, the Read-Shockley description of grain boundaries has little physical significance. An alternative approach is based on the concepts of the coincident-site lattice of Kronberg and Wilson (1949) and the displacement-shift complete lattice of Bollman (1970). Grain boundaries are modeled in terms of arrays of dislocations belonging to the coincident-site lattice, and local deviations in misorientation between the two grains are described in terms of dislocations belonging to the displacement-shift complete lattice.

3.3.4 Three-Dimensional Defects

Precipitates and inclusions constitute three-dimensional defects. Such defects could form during the growth of doped crystals when the solid solubility limit of a dopant is exceeded and also during processing of semiconductors. Precipitates can exist in three forms: coherent, partially coherent, and noncoherent. If a precipitate has the same crystal structure and a lattice parameter similar to that of the matrix, the precipitate can form low-energy coherent interfaces with the matrix on all sides as shown in Figure 3.30*a*. This type of coherency requires the respective lattices to have orientation relationships. However, coherency can still be maintained at the precipitate–matrix interface even when the precipitate volume is slightly different from that of the matrix consumed in its formation. This situation is shown in Figure 3.30*b* when the precipitate volume is smaller than that of the matrix consumed. Generally, to produce a coherent matrix–precipitate interface, both lattices are distorted, as illustrated in Figure 3.30*b*.

Based on the interface energy considerations, it is favorable for a precipitate to be surrounded by low-energy coherent interfaces. However, this condition is

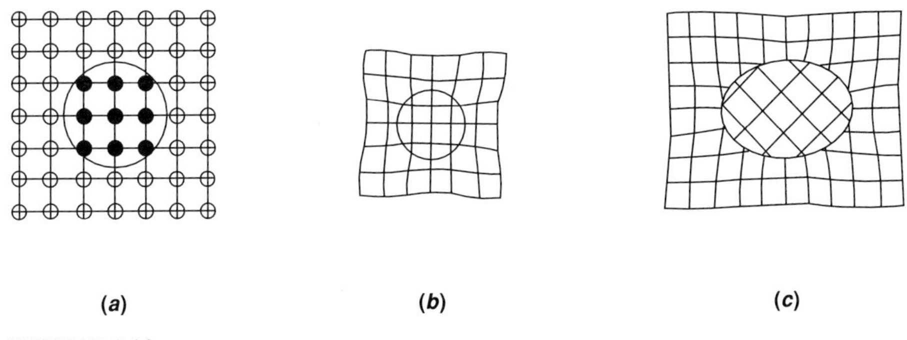

FIGURE 3.30
Schematics showing (*a*) a fully coherent precipitate associated with no strains, (*b*) a fully coherent precipitate associated with coherency strains, and (*c*) a fully noncoherent precipitate associated with misfit dislocations and strains.

not generally feasible when the precipitate and the matrix have different crystal structures. For certain situations there may be a plane that is more or less identical in each phase, and by choosing the correct orientation relationship, a low-energy coherent interface can be formed. However, along other planes of the precipitate, matching with the matrix is poor, resulting in high-energy incoherent interfaces that can be described in terms of dislocations. On the other hand, when the two phases have completely different crystal structures or when the two lattices have a random orientation relationship, the formation of coherent or partially coherent interfaces is not possible. In this case the matrix–precipitate interfaces are noncoherent, and the precipitate is said to be "incoherent." This situation is shown schematically in Figure 3.30*c*. Except in the case of perfect coherency shown in Figure 3.30*a*, the coherency strains exist at the coherent and partially coherent interfaces. Furthermore, the noncoherent interfaces are defined by interfacial dislocations.

Now let us consider a case where the volume of a precipitate is different from that of the volume of matrix consumed in its formation. An important example is the formation of SiO_2 precipitates during annealing of oxygen-containing Czochralski silicon (see Chapter 5). The volume difference can be accommodated by the formation of interstitial—or vacancy—type prismatic loops in the adjoining matrix. The nature of the loops depends on whether the precipitate volume is larger or smaller than the volume of matrix consumed, which is illustrated in Figure 3.31. As envisaged by Ashby and Johnson (1969), a segment of the dislocation loop associated with the precipitate–matrix interface can undergo repeated cross-slip, from Figure 3.31*a* → 3.31*d*, to generate a prismatic loop. The repeated application of the above process could move the extra material or its lack of, that is, the volume difference, in the form of prismatic loops.

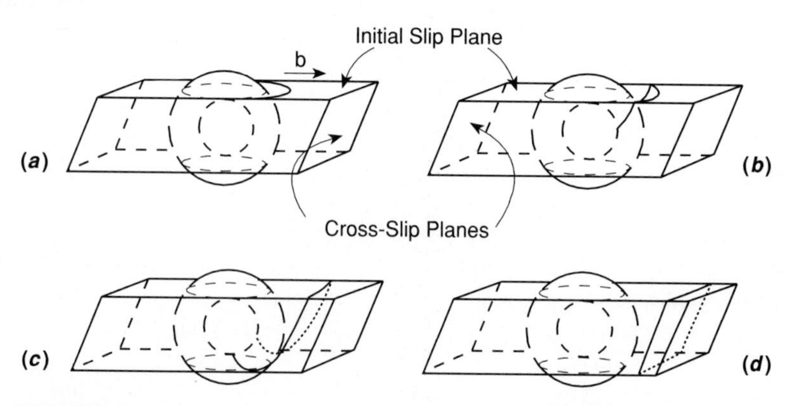

FIGURE 3.31
Schematic illustrating the generation of prismatic loops at a particle. The shear loop in (*a*) expands until its screw segments lie parallel to highly stressed cross-slip planes. Repeated cross-slip generates a prismatic loop as shown in (*a*) → (*d*). (*After Ashby and Johnson [1969].*)

3.4 ELECTRONIC PROPERTIES OF DEFECTS

The formation of defects in semiconductors creates a local electronic disturbance because of the nature of the atomic bonding. The complexity of the disturbance depends on the dimensionality of the defect and the type of semiconductor. Plausible situations that arise in elemental and compound semiconductors are discussed next.

3.4.1 Zero-Dimensional Defects

When a vacancy is formed in the diamond-cubic structure of an elemental semiconductor (refer to Figure 3.8), each atom constituting a tetrahedron has an unpaired electron. Consequently, there is a tendency for the pairing of these unpaired electrons. As a result, vacancies tend to exhibit acceptorlike behavior. In principle, a distinct energy level within the band gap should be associated with each pairing, and each energy level must occur at a progressively higher energy because of the electrostatic interactions between the captured electrons. Consequently, a vacancy in silicon can give rise to four distinct energy levels in the band gap. Experimentally, the acceptor levels at 0.11 and 0.4 eV from the conduction-band edge have been identified in silicon by a number of investigators (Gandhi 1983). In addition, a donor level at 0.35 eV from the valence-band edge has also been observed and is attributed to bond distortion that must occur in the vicinity of a vacancy.

The situation regarding interstitials can be addressed in a similar manner. An interstitial has four valence electrons that are not involved in covalent bonding with the adjoining atoms. The successive loss of these unpaired electrons to the conduction band could, in principle, result in four different donor levels within the band gap. Experimentally, a singly ionized donor level at 0.91 eV below the conduction band edge is seen in silicon. The energy levels associated with point defects in silicon are also fairly deep. They serve as centers for minority carrier recombination and thus cause a decrease in the carrier lifetime. This decrease is inversely proportional to the concentration of point defects.

Energy levels associated with point defects in compound semiconductors are poorly understood for two reasons. First, the perfection as well as purity of these materials is not as high as that of elemental semiconductors. As a result, it is difficult to discern the electronic properties of point defects. Second, because of the inherent nature of these materials, a variety of point defects and associated complexes can form, as presented in section 3.3.1. The list of energy levels of vacancies and some of the vacancy-impurity complexes in gallium arsenide in Table 3.5 highlights this complexity (Chang et al. 1971).

3.4.2 One-Dimensional Defects

The introduction of dislocations in covalently bonded semiconductors produces two important effects. First, as a result of the elastic distortions associated with a

TABLE 3.5
Energy levels associated with vacancies and vacancy-impurity complexes in GaAs

Location from valence-band edge at 5K (eV)	Type of vacancy, and possible impurity associated with it
1.49	V_{As}(Si)
1.40	V_{As}
1.37	V_{Ga}(Cu)
1.35	V_{Ga}
1.20	O_i
1.02	V_{Ga} V_{As}(Si)
0.81	V_{Ga}
0.70	V_{As}(O)
0.58	V_{Ga}

Source: Chang et al. 1971.

dislocation, band bending occurs in its vicinity. Second, dangling bonds are created along the core of the dislocation; see section 3.3.2.

Many researchers have proposed models to rationalize the electronic properties of dislocations (Read 1954a, 1954b; Schröter and Labusch 1969; Hirsch 1979). Read (1954a, 1954b) assumes that dislocation states can be represented by a one-dimensional band that is empty when the dislocation is in the neutral state. Labusch and Schröter (1980) argue that this model cannot account for the observed reduction of the hole density in p-type germanium after deformation. To rationalize this behavior, Schröter and Labusch (1969) envisage that the dislocation band is half filled in the neutral state. On the other hand, Hirsch (1979) envisages that energy bands associated with screws and 60° dislocations are split and that dislocation kinks give rise to deep donor and acceptor levels, which are located between the dislocation levels. The latter two models are depicted in Figure 3.32.

We can use the Schröter-Labusch and the Hirsch models to speculate on the electronic behavior of perfect dislocations in n- and p-type elemental semiconductors. In an n-type material, donor impurities can provide electrons to fill either the half-filled band of a dislocation, see Figure 3.32*a*, or E_{DA} in Figure 3.32*b*; that is, the dislocation behaves as an acceptor. As a result of the acquisition of electrons, the dislocation becomes negatively charged. The ramification is that the majority carriers will be repelled from the conduction-band states in a region contiguous to the dislocation. In addition, the positively charged donor impurities could be attracted to the dislocation to preserve the space-charge neutrality (Mataré 1971). This behavior leads to a space-charge region in the form of a cylinder in which positively charged donor ions surround the negatively charged dislocation. On the other hand, in a p-type material dopant atoms accept electrons

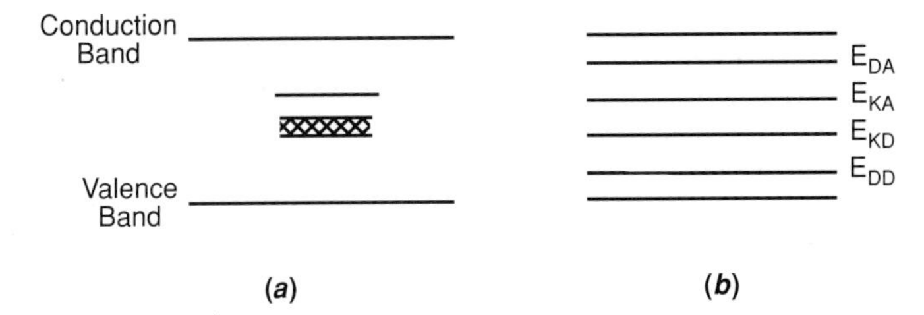

FIGURE 3.32
Schematic of dislocation band models due to (*a*) Schröter and Labusch (1969) and (*b*) Hirsch (1979, 1981). E_{DA}, E_{KA}, E_{KD} and E_{DD} refer to acceptor and donor levels due to a dislocation and a kink, respectively. D and K stand for a dislocation and a kink.

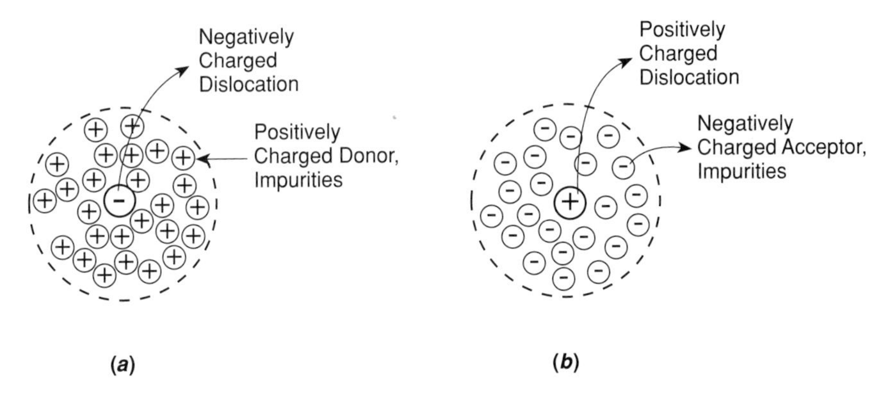

FIGURE 3.33
Schematics showing (*a*) space-charge region around a negatively charged dislocation in an n-type semiconductor and (*b*) space-charge region around a positively charged dislocation in a p-type semiconductor.

from either a partially filled band or E_{DD}. Thus dislocation behaves as a donor. The space-charge region is also formed in this case and consists of a positively charged dislocation surrounded by negatively charged acceptor ions. Figures 3.33*a* and 3.33*b* depict the resulting situations when the dislocation behaves as an acceptor and a donor.

The preceding discussion assumes that a/2 <110> perfect dislocations are not dissociated into Shockley partials and that core reconstruction to eliminate dangling bonds does not occur (Hornstra 1958). Hirsch (1981) has considered the occurrence of reconstruction at the cores of Shockley partials and finds that a 30° Shockley is likely to be reconstructed, whereas the situation is not clear for a 90° partial. This result implies that the electrical activity of dislocations dissociated into Shockley partials could vary with the orientation of the perfect dislocation. However, the dissociation does not affect the overall trend.

If one neglects the free carrier densities n and p, the radius of a space-charge cylinder R is given by the following expression (Mataré 1971):

$$R = \left[\frac{F(E)}{\pi C(N_d - N_a)}\right]^{1/2}, \quad (3.21)$$

where F(E) is the Fermi distribution function as defined in section 1.4.1 and is in the range of 0.1 at 300 K, C is the spacing between dangling bonds, and $(N_d - N_a)$ is donor minus acceptor concentration, that is, the net carrier concentration. The Fermi distribution function F(E), in turn, can be expressed in terms of the dislocation energy level E_D and the Fermi level E_f by the following equation (see section 1.4.1):

$$F(E) = \left[1 + \exp\frac{(E_D - E_f)}{k_B T}\right]^{-1}. \quad (3.22)$$

For a net carrier concentration of 10^{15} cm^{-3} and $F(E) = 0.1$, the radius of the space-charge cylinder is estimated to be ~ 1 μm in silicon.

The presence of a space-charge cylinder results in the creation of a potential barrier at the dislocation (Read 1954a) that affects the decay rate of injected carriers in dislocated crystals (Galenzer and Jordan 1969). The decay rate is found to be slower than the exponential decay predicted by the Shockley-Read model for defect-free crystals (Shockley and Read 1952; Morrison 1956). Thus the measured lifetimes in dislocated crystals can be longer than that in dislocation-free crystals. The existence of space-charge cylinders around dislocations can also reduce the conductivity of the crystal in directions both parallel and normal to the dislocation lines. Of course, this effect depends on the density of dislocations and their distribution. At low dislocation densities commonly observed in as-grown semiconducting crystals, their effect on conductivity is not significant.

A number of workers (Haasen and Schröter 1971; Labusch and Schröter 1975; Kimerling and Patel 1979; Patel and Kimerling 1981) have investigated the energy levels associated with dislocations in deformed silicon. Haasen and Schröter (1971) and Labusch and Schröter (1975) show that 60° dislocations in deformed p-type silicon introduce a band of levels at $\sim$0.34 eV above the valence-band edge. On the other hand, Grazhulis et al. (1977) find a band of levels at 0.42 and 0.67 eV above and below the valence-band and conduction-band edges in deformed n- and p-type silicon. It is difficult to fully reconcile various experimental observations with our current understanding of the dislocation structures in deformed silicon.

Jones et al. (1981) have computed energy levels of different types of dislocations in GaAs using a tight-binding Hamiltonian and a supercell containing a pair of dislocations of opposite signs. A summary of their results (see Table 3.6) shows that the position of an energy level associated with a dislocation changes with its nature. In particular, the levels due to different types of b-dislocations lie above the conduction-band edge of GaAs.

Gwinner and Labusch (1979) and Zozime and Schröter (1989) have attempted to determine experimentally the energy levels associated with dislocations in plastically deformed GaAs and InP, respectively. Their results are not conclusive because during deformation a high density of point defects was generated. To separate

TABLE 3.6
Computed energy levels due to different types of dislocations in GaAs

Dislocation type	Energy level* (eV)	Width of level (eV)
60°α(s)	0.5	0.4
60°β(s)	2.0	0.3
90°α(g)	0.2	0.5
90°β(g)	1.6	0.2
30°α(g) Shockley	1.1	0.3
30°β(g) Shockley	3.0	0.9

*Energies are measured from the valence-band edge. Levels associated with different types of β dislocations lie above the conduction-band edge of GaAs.

Source: Jones et al. 1981.

the levels due to dislocations and point defects, the levels should be measured after annealing the deformed crystals.

3.4.3 Two-Dimensional Defects

Since faulted and unfaulted regions are coherently bonded to each other, dangling bonds are not created at the fault surface. Therefore, the surfaces of intrinsic and extrinsic faults and coherent twin boundaries in elemental as well as in compound semiconductors should not be electrically active. On the other hand, partials bounding various faults should be electrically active because of the presence of dangling bonds along the dislocation core. Furthermore, the electrical activity will vary with the orientation and character of the bounding partial. Also, the occurrence of reconstruction at the core may further affect the electrical activity.

As discussed in section 3.3.3, we can treat subboundaries and grain boundaries as an assemblage of dislocations whose Burgers vector, orientation, and density depend on the crystallography of the boundary. Therefore, the electrical activities of grain boundaries can be analyzed in terms of the electrical behavior of dislocations constituting a boundary. To gain additional insight into this issue, refer to the review article by Queisser (1983).

3.4.4 Three-Dimensional Defects

In the case of semicoherent and noncoherent precipitates, dislocations are present at the matrix–precipitate interface. These dislocations can impart electrical activity to these interfaces. Thus the influence of the semicoherent and noncoherent precipitates on the electronic properties of semiconductors can also be analyzed in terms of the electronic behavior of dislocations constituting the matrix–precipitate interface.

3.5 HIGHLIGHTS OF THE CHAPTER

The highlights of this chapter are as follows:

- We can use dimensionality as a criterion to classify defects in semiconductors into four categories: zero-, one-, two-, and three-dimensional. The respective examples are point defects, dislocations, stacking faults, and precipitates.
- Equilibrium concentrations of point defects increases with increasing temperature. Dislocations and stacking faults can multiply in the presence of either stresses or point defects. Subgrain boundaries, grain boundaries, and noncoherent precipitate–matrix interfaces can be described in terms of dislocations.
- The presence of vacancies and dislocations in semiconductors produces dangling bonds. As a result, point defects, dislocations, partials bounding stacking faults, subgrain boundaries, grain boundaries, and interfaces of noncoherent precipitates are electrically active. Dangling bonds are also present on semiconducting surfaces and are responsible for surface reconstruction.

REFERENCES

Alexander, H. and P. Haasen. In *Solid State Physics,* ed. H. Ehrenreich, F. Seitz, and D. Turnbull. New York: Academic Press, 22 (1968).

Alexander, H.; P. Haasen; R. Labusch; and W. J. Schröter. *J. de Physique, Paris* 40, C6 (1979).

Ashby, M. F. and L. Johnson. *Phil. Mag.* 20, 1009 (1969).

Bardeen, J. and C. Herring. In *Imperfections in Nearly Perfect Crystals.* New York: John Wiley, 1952, p. 261.

Boivin, P.; J. Rabier; and H. Garem. *Phil. Mag.* A, 61, 619 (1990a).

Boivin, P.; J. Rabier; and H. Garem, *Phil. Mag.* A, 61, 647 (1990b).

Bollman, W., *Crystal Defects and Crystalline Interfaces.* Germany: Springer-Verlag, 1970.

Chadi, D. J. *J. Vac. Sci. Tech.* A5, 834 (1987).

Chang, L. L.; L. Esaki; and R. Tsu. *Appl. Phys. Lett.* 19, 143 (1971).

Dash, W. C. J. *Appl. Phys.* 27, 1193 (1956).

De Cooman, B. C. and C. B. Carter, *Phil. Mag. A.* 60, 245 (1989).

Frank, F. C. and W. T. Read. In *Symposium on Plastic Deformation of Crystalline Solids.* Carnegie Institute of Technology, Pittsburgh: 1950, p. 44.

Galenzer, R. H. and A. G. Jordan. *Solid State Electron.* 12, 247 (1969).

Gandhi, S. K. *VLSI Fabrication Principles.* New York: John Wiley, 1983, p. 42.

Gomez, A. M. and P. B. Hirsch. *Phil. Mag.* 36, 169 (1977).

Grazhulis, V. A.; V. V. Kveder; and Y. V. Mukhina. *Phys. Stat. Sol.* (a) 43, 407 (1977).

Gwinner, D. and R. Labusch. *J. Phys. Paris* 40, C6, 117 (1979).

Haasen, P. *Acta Met.* 5, 598 (1957).

Haasen, P. and W. Schröter. In *Fundamentals of Dislocation Theory,* ed. J. A. Simmons, R. deWit, and R. Bullough, Washington, DC. National Bureau of Standards, 1971, p. 1231.

Hirsch, P. B. *J. de Physique,* Colloque C6, Supplement 6, 40, C6, 117 (1979).

Hirsch, P. B. In *Defects in Semiconductors,* ed. J. Narayan and T. Y. Tan. *MRS Symp. Proceedings* vol. 2. Pittsburgh: Materials Research Society, 1981, p. 257.

Hirth, J. P. and J. Lothe. *Theory of Dislocations,* New York: John Wiley, 1982, p. 373.

Hornstra, J. *J. Phys. Chem. Solids* 5, 129 (1958).
Jones, R.; S. Oberg; and S. Marklund. *Phil. Mag.* B, 43, 839 (1981).
Kimerling, L. C. and J. R. Patel. *Appl. Phys. Lett.* 34, 73 (1979).
Kronberg, M. L. and F. H. Wilson. *Trans. AIME,* 185, 501 (1949).
Kuesters, K.-H.; B. C. De Cooman; and C. B. Carter. *Phil. Mag.* A, 53, 141 (1986).
Labusch, R. and W. Schröter. *Inst. Phys. Conf.* 23, 56 (1975).
Labusch, R. and W. Schröter. In *Dislocations in Solids,* ed. F. R. N. Nabarro. Amsterdam: North-Holland Publishing Company, 1980, p. 129.
Lagally, M. G.; Y. -W. Mo; R. Kariotis; B. S. Swartzentruber; and M. B. Webb. In *Kinetics of Ordering and Growth of Surfaces.* New York: Plenum Press, 1985.
Mataré, H. F. *Defect Electronics in Semiconductors.* New York: John Wiley, 1971, p. 208.
Men, F. K.; W. E. Packard; and M. B. Webb. *Phys. Rev. Lett.* 61, 2469 (1988).
Morrison, S. R. *Phys. Rev.* 104, 619 (1956).
Pashley, M. D.; K. W. Haberern; W. Friday; J. M. Woodall; and P. D. Kirchner. *Phys. Rev. Lett.* 60, 2176 (1988).
Patel, J. R. and L. C. Kimerling on "*Dislocation States in Deformed Si*" In *Defects in Semiconductors,* eds. J. Narayan and T. Y. Tan. *MRS Symp. Proceedings* vol. 2. Pittsburgh: Materials Research Society, 1981, p. 273.
Pauling, L. *The Nature of the Chemical Bond.* Ithaca: Cornell University Press, 1960, p. 246.
Queisser, H. J. on "Electrical Behavior of Dislocation and Boundaries in Semiconductors" in *Defects in Semiconductors II,* eds. S. Mahajan and J. W. Corbett. *MRS Symp. Proceedings* vol. 14, Pittsburgh: Materials Research Society, 1983, p. 323.
Read Jr., W. T. *Dislocation in Crystals,* New York: McGraw-Hill, 1953, p. 155.
Read Jr., W. T. *Phil. Mag.* 45, 775 (1954a).
Read Jr., W. T. *Phil. Mag.* 45, 1119 (1954b).
Rhodes, R. G. *Imperfections and Active Centers in Semiconductors.* New York: Pergamon-Macmillan, 1964.
Schröter, W. and R. Labusch. *Phys. Stat. Sol.* 36, 359 (1969).
Shockley, W. and W. T. Read Jr. *Phys. Rev.* 87, 835 (1952).
Vardya, R. and S. Mahajan. *Phil. Mag.* A 71, 465 (1995).
Vogel, F. L.; W. G. Pfann; H. E. Corey; and E. E. Thomas. *Phys. Rev.* 90, 489 (1953).
Wessel, K. and H. Alexander. *Phil. Mag.* 35, 1523 (1977).
Zozime, A. and W. Schröter. *Phil. Mag.* A 565 (1989).

PROBLEMS

3.1. The {111} planes in the diamond-cubic and zinc-blende structures can define an octahedron as shown below:

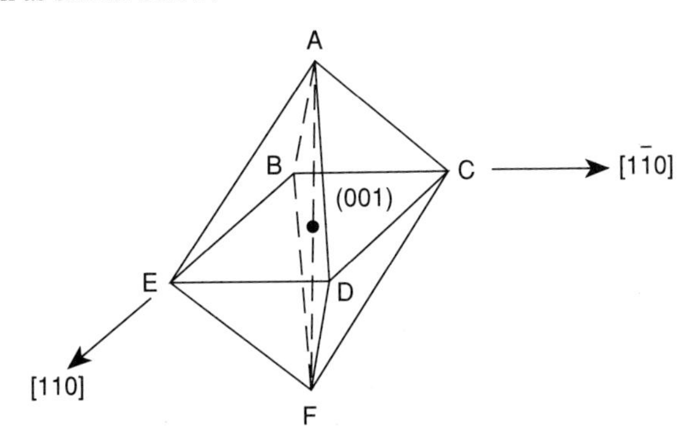

Assuming that origins of the two FCC sublattices are at 0,0,0 (group III) and 1/4, 1/4, 1/4 (group V), assign indexes to the *outer* surfaces of octahedrons in silicon and indium phosphide. Also, indicate whether the {111} plane contains indium or phophorus atoms.

3.2. Calculate the number of atoms/m^2 for $\{111\}_A$ and $\{\bar{1}\bar{1}\bar{1}\}_B$ planes in gallium arsenide, given that its lattice parameter is 0.565 nm. Also, determine the number of dangling bonds/m^2 present on these planes. Using this information, compare chemical reactivities of $\{111\}_A$ and $\{\bar{1}\bar{1}\bar{1}\}_B$ surfaces.

3.3. A hypothetical III-V compound AB crystallizes in the zinc-blende structure and has ionic and covalent bonding. Given that the covalent tetrahedral radii of A and B atoms are 0.1 and 0.3 nm, calculate the changes in bond lengths and misfit factors for the following impurities: Be, Mg, Cd, C, (p-type) and Sn and Si, Ge, S, Se, and Te. (n-type)

3.4. *a.* Compare the concentrations of vacancies and interstitials in silicon when it is heated to 1673 K, given that the respective formation energies are 2.3 and 1.1 eV.
b. If you were to reduce the vacancy concentration in (*a*), which impurities from Table 3.1 would you add to silicon and why?

3.5. Given that when a glide dislocation leaves a crystal, it produces a step on a surface whose height is equal to the resolved component of the Burgers vector along the surface normal and the step, that is, slip trace, and is oriented along the line of intersection of the glide plane with the surface, calculate step heights and slip traces for the following situations in GaAs:

a. The activated slip system is a/2 $[1\bar{1}0]$ (111) and the crystal surface is $(\bar{1}11)$.
b. The activated slip system is a/2 $[1\bar{1}0]$ (111) and the crystal surface is (123).
c. The activated slip system is a/2 $[1\bar{1}0]$ (111) and the crystal surface is (001).
Given that the lattice parameter of GaAs is 0.565 nm.

3.6. *a.* A perfect a/2 $[\bar{1}10]$ edge dislocation, oriented along the $[\bar{1}\bar{1}2]$ direction and gliding between the narrowly separated (111) planes, is introduced into a 1 cm cube of silicon crystal; the edges of the cube are aligned along the [111], $[\bar{1}10]$, and $[\bar{1}\bar{1}2]$ directions. Calculate the number of dangling bonds 1 cm^3 that are present on the dislocation.
b. If 10^{10} cm^{-2} a/2 $[\bar{1}10]$ edge dislocations were introduced in the above silicon block that is doped with boron to 10^{18} cm^{-3}, calculate the change in carrier concentration resulting from the introduction of these dislocations.

3.7. *a.* When a silicon-germanium layer is grown on a (001) silicon, a network of misfit dislocations is observed. During additional growth, portions of the misfit dislocations act as Frank-Read sources. Given that the Burgers vectors of dislocations are a/2 [110] and a/2 $[1\bar{1}0]$ and that they are aligned along the [110] and $[1\bar{1}0]$ direction, calculate the source length if the resolved component of the misfit stress at which multiplication occurs is 10^7 Pa; the lattice parameter a and shear modulus G for silicon are 0.543 nm and 10^{10} Pa, respectively.
b. Show schematically the dislocation arrangement after multiplication if the substrate and the overgrowth are viewed along the [110] direction.

3.8. *a*. A perfect a/2 $[\bar{1}10]$ edge dislocation, oriented along the $[\bar{1}\bar{1}2]$ direction, gliding between the narrowly separated (111) planes, and whose extra half-plane terminated on a $(\bar{1}\bar{1}1)_P$ plane is introduced into a 1 cm cube of an InP crystal. The cube edges are along the [111], $[\bar{1}10]$, and $[\bar{1}\bar{1}2]$ directions. Calculate the number of point defects required to climb the dislocation by a distance of $a/\sqrt{3}$ along the $[\bar{1}\bar{1}\bar{1}]$ direction. The lattice parameter (a) for InP is 0.586 nm.

b. Given that the crystal in (*a*) is supersaturated with In interstitials, calculate the local supersaturation of P vacancies produced by the climb of dislocation in (*a*). Assume that the vacancies produced by climb in (*a*) are confined within a cylinder around the dislocation whose radius is 100 nm, the energy of formation of a P vacancy is 2 eV, and the sample temperature is 873 K.

3.9. Under the influx of point defects, a Bardeen-Herring source is observed to operate in GaAs. As a result, a circular extrinsic fault, 1 mm in diameter and bounded by an a/3 <111> Frank partial, is produced on a {111} plane. Given that the lattice parameter of GaAs is 0.565 nm, calculate the number of point defects involved in the formation of the fault.

3.10. Imagine a cube of silicon that is bounded by the (111), $(1\bar{1}0)$, and $(\bar{1}\bar{1}2)$ faces. If intrinsic faults were to form on {111} planes in the center of the cube and expand until they cut the cube faces, determine the indexes of directions along which the four faults would intersect various external surfaces.

3.11. An indium-terminated vicinal surface has been prepared by tilting away from the $[\bar{1}11]$ orientation toward the $[1\bar{1}2]$ direction; the tilt axis is [110]. Show schematically the [110] projection of such a surface in InP. Also, if the tilt angle is 6°, calculate the terrace length of the vicinal surface.

3.12. Imagine a situation where a spherical precipitate of radius 1.25 μm forms in a hole of radius 1μm in silicon. Assuming that the extra material to be removed to accommodate the precipitate is transported as extrinsic, prismatic circular loops, calculate the number of loops that would form, given that the loop radius is 10 nm and the lattice parameter of silicon is 0.543 nm.

3.13. *a*. Calculate the radius of the space-charge cylinder that would form around an a/2 [110] edge dislocation, oriented along the $[1\bar{1}2]$ direction, in a silicon crystal doped with arsenic to 10^{16} cm^{-3}. Assume that the Fermi distribution function is 0.1.

b. If 10^8 cm^{-2} regularly spaced a/2 [110] edge dislocations were introduced in a block of Si, 1 cm × 1 cm × 1 cm, and space-change cylinders of the radius calculated in (*a*) were present around each dislocation, comment on the carrier mobility along the [110], $[1\bar{1}2]$, and $[\bar{1}11]$ directions.
The lattice parameter of silicon is 0.543 nm.

CHAPTER 4

Evaluation of Semiconductors

This chapter covers some of the techniques that are currently being employed for evaluating the structural, chemical, electrical, and optical characteristics of semiconductors. It emphasizes their principles as well as their limitations. The objective is to provide the reader with a background of the techniques that are used for assessing the growth- and processing-induced changes in semiconductors.

4.1 STRUCTURAL EVALUATION

The structural evaluation of semiconductors involves determining the density, distribution, and characteristics of various types of defects considered in Chapter 3. Since the semiconductors used in technology are highly perfect materials, both macroscopic and microscopic evaluations are necessary to obtain the complete information. The macroscopic characterization provides details on the density and distribution of dislocations, stacking faults, twins, grain boundaries, and large-sized precipitates ($>1\ \mu$m). On the other hand, we use microscopic evaluations to investigate, for example, the precipitation behavior of dopants, the core structure of dislocations, and the crystallography of precipitates. Etching, X-ray topography, and double-crystal diffractometry are primarily used for the macroscopic evaluation, whereas transmission electron microscopy (TEM) and high-resolution electron microscopy (HREM) are the tools for microscopic structural characterization.

4.1.1 Defect Etching

Etching of a semiconductor entails inserting it in a suitable chemical solution called an *etchant*. Since the defective regions etch differently from the matrix, topological features develop on the etched surface. These features can be discerned using optical microscopy. Etching is a powerful method for the evaluation of perfection of as-grown crystals, processed wafers, and epitaxial layers. There are several reasons for its widespread applicability: (1) high reliability, (2) high speed of testing, and (3) low cost. Etching is used primarily for investigating the density and distributions of dislocations in semiconductors. This technique is not very useful when the dislocation density exceeds $\sim 10^6 \text{cm}^{-2}$ because the etching-induced features called *etch pits* begin to overlap.

We can divide the etching of semiconductors into two categories: polishing etch and selective or defect-revealing etching (Weyher 1994). The polishing etch occurs when the diffusion of reactants and reaction products towards or away from the surface is rate controlling. The etch rate dn/dt, that is, the number of molecules removed from the surface/unit time, is given by Fick's law of diffusion:

$$\frac{dn}{dt} = \frac{DAC}{\delta}, \tag{4.1}$$

where D is the diffusion coefficient of reactants in the etch, A is the projected surface area assuming small surface irregularities in comparison to the thickness δ of the diffusion boundary layer, and C is the concentration of reactants. Under ideal conditions the diffusion-controlled etching produces mirrorlike, flat surfaces. This result occurs because diffusion, being a slow process, tends to eliminate inhomogeneities.

In selective etching the chemical activity at the surface is the rate-limiting step. This activity is, in turn, affected by the crystallographic, passivation, and electronic effects. The etch rate of such a process is kinetically controlled and can be written as

$$\frac{dn}{dt} = kA_1C, \tag{4.2}$$

where k is reaction rate constant and A_1 is the real surface area. Since k may vary locally in the presence of composition inhomogeneities and defects, dn/dt will correspondingly vary. These variations will produce a nonplanar etched surface that reflects the distribution of chemical inhomogeneities and defects.

The etching of semiconductors is electrochemical in nature. For example, chemical etching of a silicon surface is controlled by two basic reactions: oxidation of silicon followed by the dissolution of the oxide. Consequently, typical etchants contain a strong oxidizing agent and HF that dissolves the oxide. In many situations the oxidation rate at the defect, that is, dn/dt, is greater than that of the surrounding silicon, and this differential produces etch pits. Sometimes the oxidation rate of the defect is lower than that of the defect-free region, resulting in moundlike features at the defect sites.

Several material parameters affect the etch rate and the etch rate differential between the defective and defect-free regions: (1) orientation of the surface, (2) type of dopant and its concentration, and (3) composition of the etchant. Since holes are

TABLE 4.1
Etchants for defect delineation in Si

Name of etch	Composition	Remarks
Sirtl and Adler (1961)	1 part concentrated HF 1 part CrO_3 (5M) 500g/liter of solution	~3.5 μm/min etch rate; good on {111}, poor on {001}; faceted pits.
Secco-D'Aragona (1972)	2 parts concentrated HF 1 $K_2Cr_2O_7$ (0.15 M) 44g/liter of solution	~1.5 μm/min etch rate; best with ultrasonic agitation, good on all orientations, particularly suitable for {001}; noncrystallographic pits.
Schimmel (1976)	HF: HNO_3 155: 1	Applicable to p-type material.
Wright-Jenkins (1977)	2 parts concentrated HF 2 parts concentrated CH_3COOH 1 part concentrated HNO_3 1 CrO_3 (4M) 400 g/liter of solution 2 $Cu(NO_3)_2$ $3H_2O$ (0.14M) 33 g/liter of solution	~1.7 μm/min etch rate; ultrasonic agitation not required; good on all orientations; faceted pits; good shelf life.

Source: Miller and Rozgonyi (1980).

required for the oxidation of silicon, the etching of an n-type surface is accentuated in the presence of light (Brattain and Garrett 1955); this behavior is attributed to the generation of electron-hole pairs (EHPs) discussed in section 1.5.1. As expected, the effect of light on the etching of p-type silicon is small.

Table 4.1 lists the composition of etchants that are widely used for silicon. Figure 4.1, taken from the study of Miller and Rozgonyi (1980), shows some of the etch features that have been delineated in silicon using these etchants. The two etch pits on the left side of Figure 4.1 are caused by dislocations D_t, which terminate on the sample surface. Both pits follow the dislocation line down and to the left. The Secco pit is noncrystallographic, while the Wright-Jenkins pit appears to be crystallographic, possibly because the Secco etch is stronger than the Wright-Jenkins etch, thus eliminating the differences in etch rates on different planes. Decorated dislocation clusters produce shallow pits termed *saucer pits* or *S-pits*. The Sirtl etch produces a faceted S-pit, whereas S-pits delineated by the Secco etch are quite smooth. We do not fully understand the origin of S-pits. Perhaps the decoration of dislocations with precipitates relaxes their strain fields, resulting in the reduced tendency for etching. Furthermore, the difference in S-pits revealed by the two etchants can again be attributed to the difference in chemistries of the etchants. The Secco etch can be used also to delineate precipitates H, dislocation loops D_l, oxidation-induced bulk-stacking faults OSF_B, oxidation-induced surface-stacking faults OSF_S, and epitaxial stacking faults SF_{EPI} as shown in Figure 4.1.

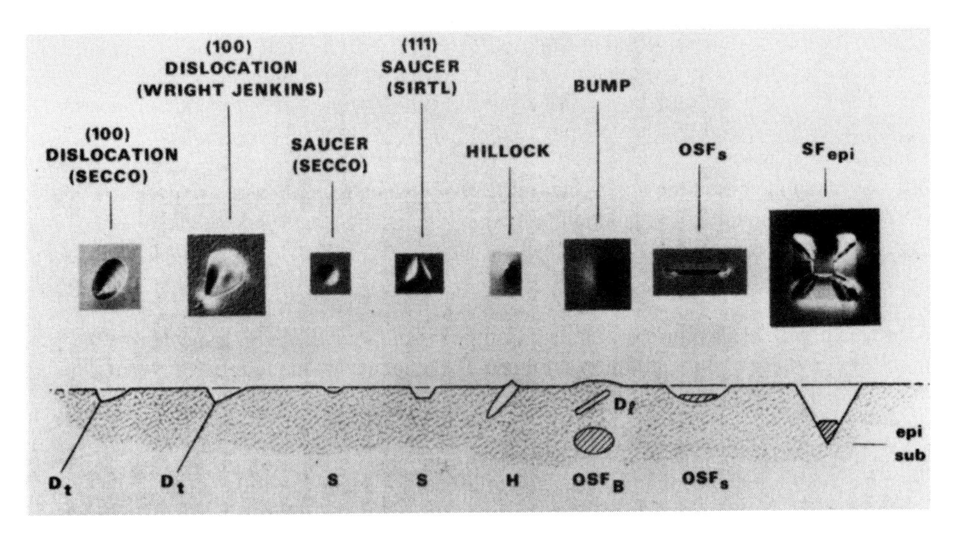

FIGURE 4.1
Etch features observed on Si surfaces using the three etchants listed in Table 4.1. (*After Miller and Rozgonyi [1980].*)

Dislocation etchants have also been developed for various compound semiconductors, and some of them are listed in Table 4.2. We show in Figures 4.2*a* and 4.2*b*, respectively, the etch pits delineated by molten KOH on the surfaces of Si- and

TABLE 4.2
Etchants for defect delineation in some compound semiconductors

Material	Chemical composition	Remarks
GaAs	Molten KOH at 300°C (Ishii et al. 1976)	Good crystallographic dislocation pits.
	1 ml HF, 2 ml H_2O 8 mg $AgNO_3$ and 1 gm CrO_3 (A-B etch) (Abrahams and Buiocchi 1965)	Dislocation lines and striations.
GaP	Behaves similarly to GaAs; above etches can be used	
InP	2 parts H_3PO_4, 1 part HBr (Huber and Linh 1975)	Good etch for {001} surfaces; ambient temperature; etching time up to 2 min.
	HNO_3 : HBr 1 : 3 (Chu et al. 1982)	Works well on {001} and $\{\bar{1}\bar{1}\bar{1}\}_B$.
CdTe	HF : H_2O_2 : deionized H_2O 30 : 20 : 21 (Nakagawa et al. 1980)	Works well on $\{\bar{1}\bar{1}\bar{1}\}_B$.

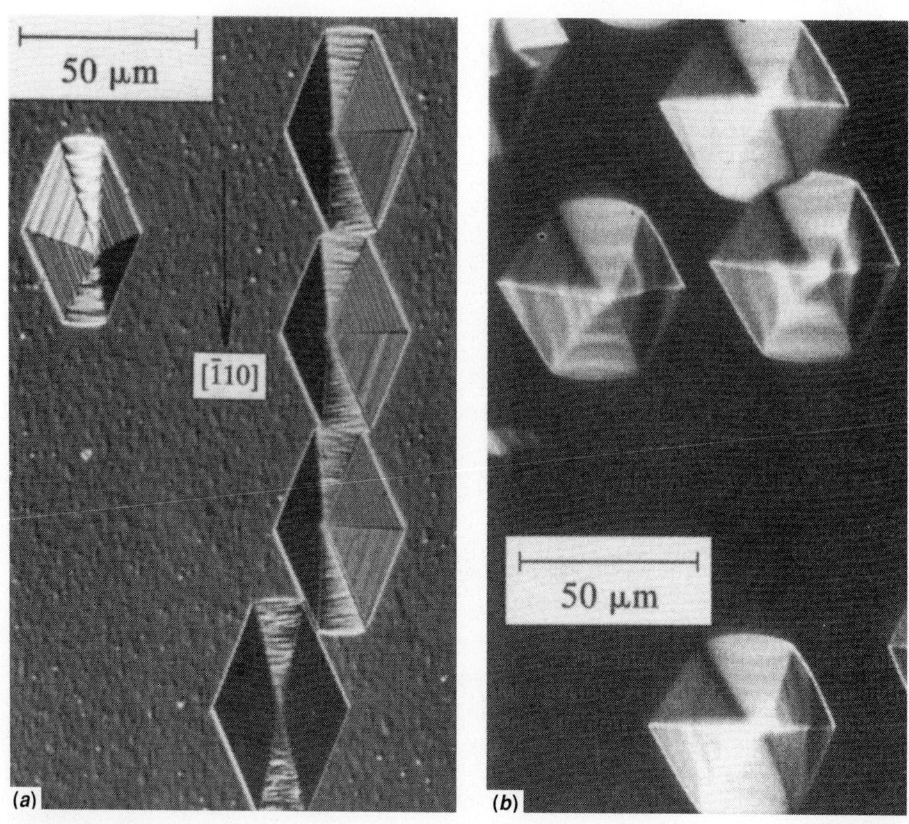

FIGURE 4.2
Micrographs illustrating the influence of sample composition on morphology of etch pits observed on the (001) surfaces of GaAs: (*a*) silicon-doped and (*b*) indium-alloyed materials (differential interference contrast image). (*After Weyher [1994].*)

In-doped GaAs crystals (Weyher 1994). The shape of the pits clearly depends on the type of the dopant in the semiconductor. The pits are elongated in the $[1\bar{1}0]$ direction in the Si-doped sample, whereas this anisotropy is absent in the In-doped specimen. Since the tetrahedral radius of In atoms is much larger than that of Ga atoms (see Chapter 3), some of the In atoms may segregate to the dislocations, resulting in a change in their etching characteristics.

As shown for Si in Figure 4.1, the composition of an etchant affects the shape of etch pits. Figures 4.3*a* and 4.3*b* show pits observed on (001) InP surfaces after etching in H_3PO_4-HBr and HBr-H_2O, respectively (Weyher 1994). After etching in H_3PO_4-HBr solution, pits are only slightly anisotropic, whereas the HBr-H_2O etch accentuates the anisotropy. Again the presence of a stronger etch, that is, H_3PO_4-HBr solution, reduces the anisotropy in the etch rates on different planes.

We indicate in section 5.4 that dopants in as-grown semiconducting crystals are not uniformly distributed. The crystals exhibit dopant-rich and dopant-poor regions

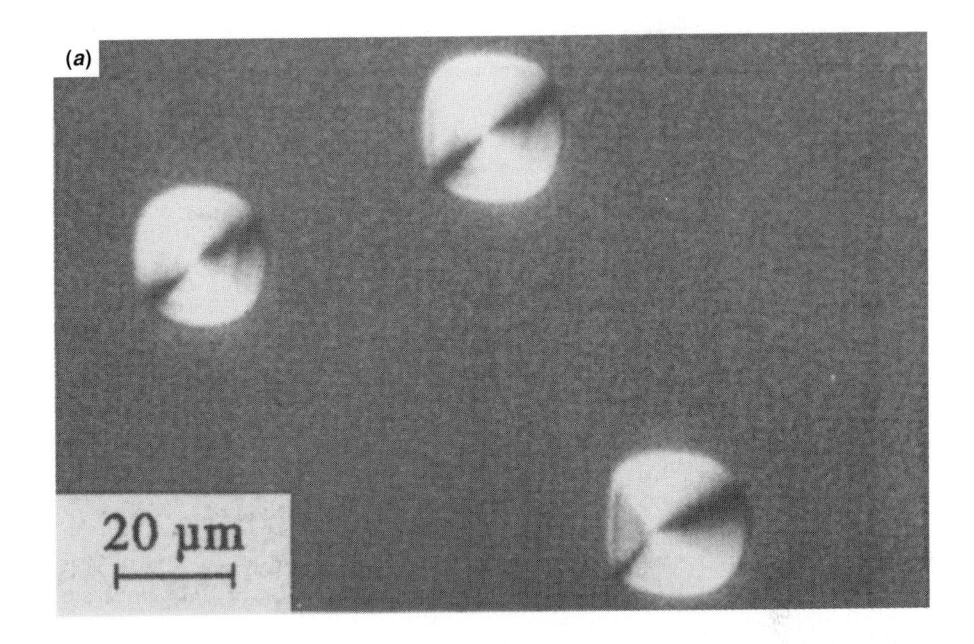

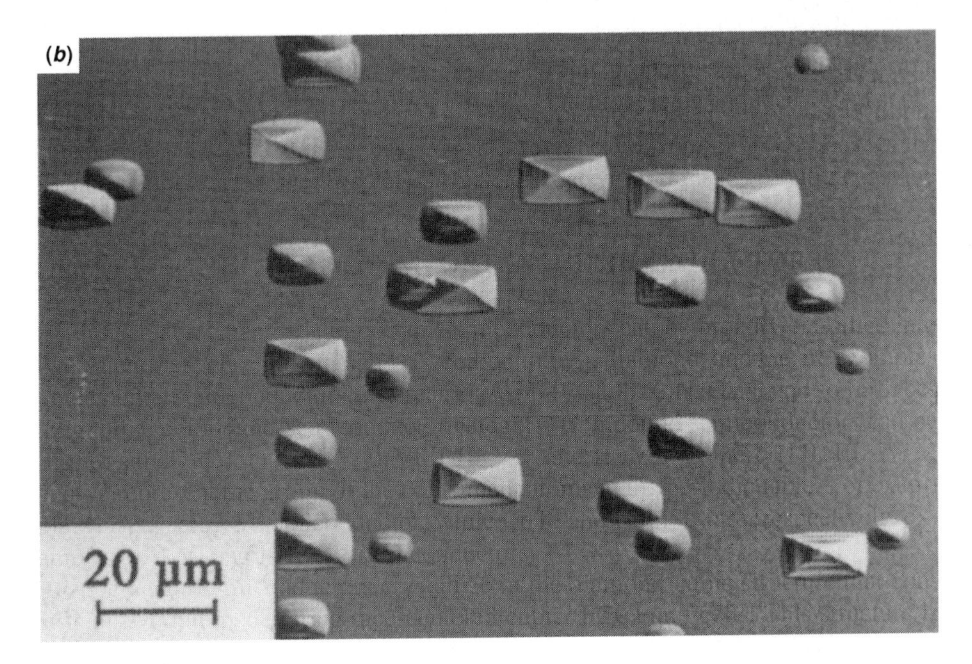

FIGURE 4.3
Changes in morphology of etch pits observed on the (001) surfaces of InP due to different compositions of the etchants: (*a*) H_3PO_4-HBr solution and (*b*) HBr-H_2O solution. (*After Weyher [1994].*)

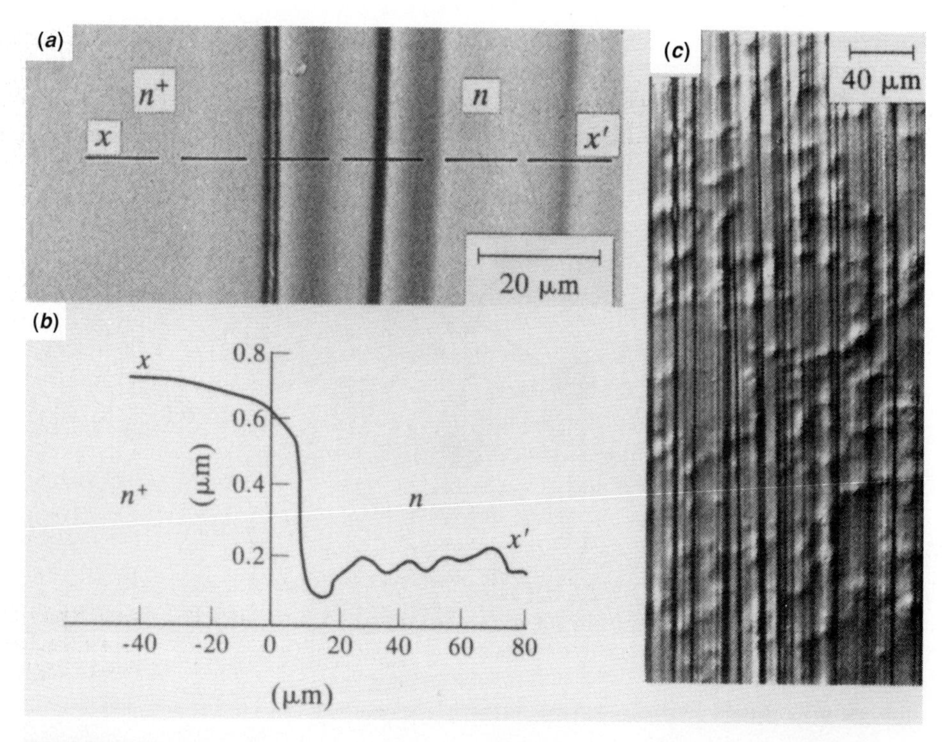

FIGURE 4.4
Remarkable alteration of the etch rate during photoetching of an n-n$^+$ area in GaAs: (*a*) differential interface contrast image of the surface, (*b*) surface profile, that is, relative etch depth, plotted along x-x′ line, and (*c*) well-resolved growth striations related to the fluctuation of deep recombinative centers revealed by photoetching on the (110) longitudinal section of an undoped GaAs crystal. (*After Weyher [1994].*)

that are referred to as *impurity striations*. Figure 4.4 shows an example of sharply defined striations in high-resistivity undoped GaAs (Weyher 1994). The striations in this case are due to the residual impurities and are revealed by etching because the impurity-rich and impurity-poor regions differ in the values of internal strains, leading to the variations in the etch rates.

EXAMPLE 4.1. (*a*) Assuming that square etch pits observed on a (001) surface of Si are bounded by {111} facets, show these pits schematically and label various directions and planes. (*b*) Imagine situations where (001) surfaces of a III-V semiconductor have been preferentially etched in two different solutions. One solution preferentially attacks $\{111\}_{\mathrm{III}}$ planes, whereas the second solution removes $\{\bar{1}\bar{1}\bar{1}\}_{\mathrm{V}}$ planes at a faster rate. Assuming that dislocation-induced etch pits are bounded by {111} planes, sketch the shape of pits in the two cases and identify various planes and label different directions. Assume that the origin of group III sublattice is at 0, 0, 0.

Solution. (*a*) Since the four {111} planes in the Si lattice are equivalent, the pit will appear as an inverted pyramid with the (001) surface as a base. Its schematic representation is given below:

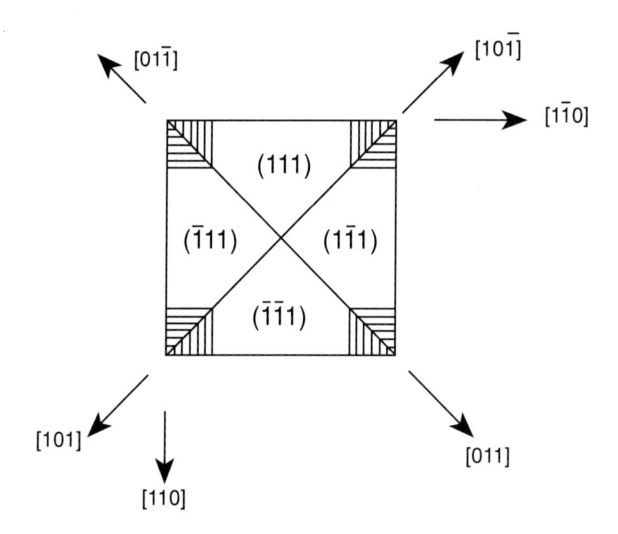

We can determine the different directions [110], [101], [011], [1$\bar{1}$0], [10$\bar{1}$], and [01$\bar{1}$] by finding the lines of intersection between various planes, that is, by taking the dot product of vectors normal to two planes involved in the intersection. For example, the line of intersection between the ($\bar{1}\bar{1}$1) and (1$\bar{1}$1) planes is [011], as indicated in the preceding figure.

(*b*) In the case of Si, {111} planes bounding a pit on a (001) surface are equivalent. However, pits on the (001) surfaces of III-V semiconductors will be bounded by $(\bar{1}11)_V$, $(1\bar{1}1)_V$ and $(111)_{III}$, $(\bar{1}\bar{1}1)_{III}$ planes. If each pair of planes is etched at a different rate, pits will delineate a rectangle instead of a square at the (001) surface. Two possible shapes are shown below:

Faster etching of $\{\bar{1}\bar{1}\bar{1}\}_V$ planes

Note that when the origin of group III sublattice is at 0, 0, 0, $\{\bar{1}\bar{1}\bar{1}\}_V$ planes intersect the (001) surface along the [110] direction.

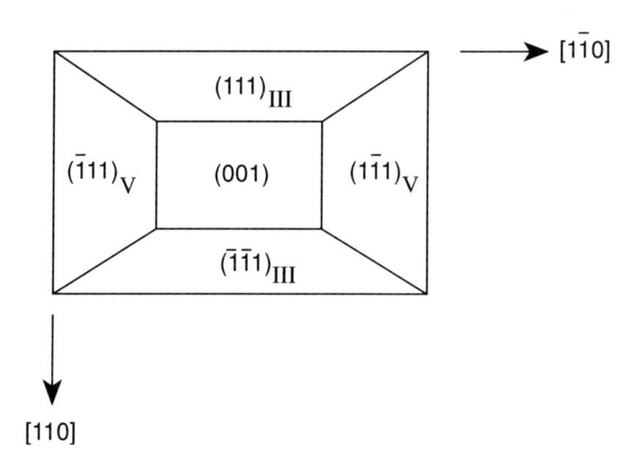

Faster etching of $\{111\}_{III}$ planes

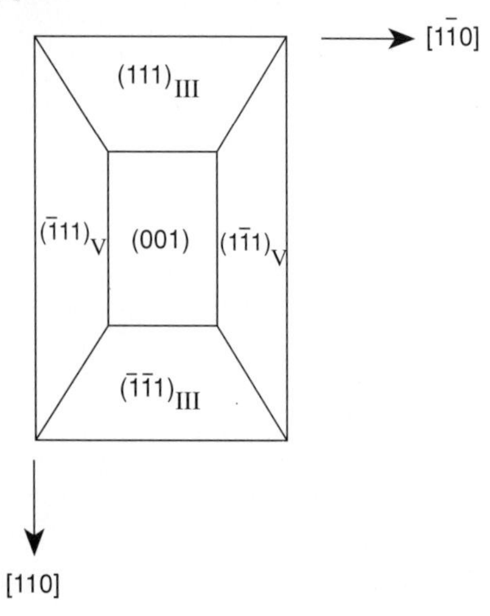

4.1.2 X-Ray Topography

As discussed in section 4.1.1, the preferential etching of semiconductors provides two-dimensional information on the density and distribution of defects, primarily the dislocations. In principle serial sectioning enables us to obtain the characteristics of three-dimensional distribution of defects, but this procedure is very time-consuming. However, we can generate this information by X-ray topography. This technique produces a two-dimensional projection of a three-dimensional distribution and can thus reveal the defect structure in the volume of the crystal without resorting to sectioning. This technique can resolve, like etch pitting, dislocation densities up to 10^6cm^{-2} and cannot reveal features smaller than 1 μm.

Figure 4.5 shows schematically the widely used arrangement for Lang transmission topography (Franzosi 1994). An X-ray beam emanating from a point source

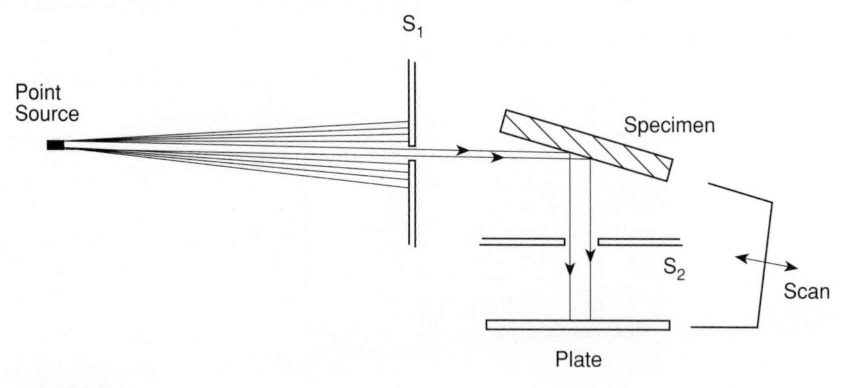

FIGURE 4.5
Schematic of a Lang camera working in the Bragg geometry. (*After Franzosi [1994].*)

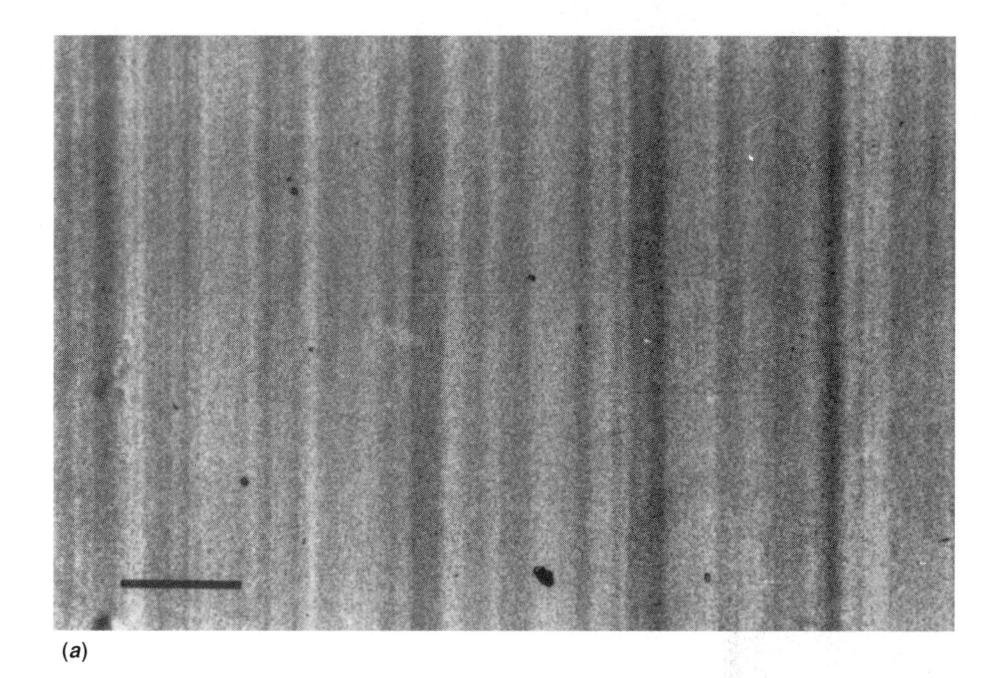

(*a*)

(*b*)

FIGURE 4.7
Growth striations observed in (110)-oriented GaAs crystals: (*a*) Si-doped crystal and (*b*) S-doped crystal. The length of the markers is 100 μm. (*After Fornari et al. [1994].*)

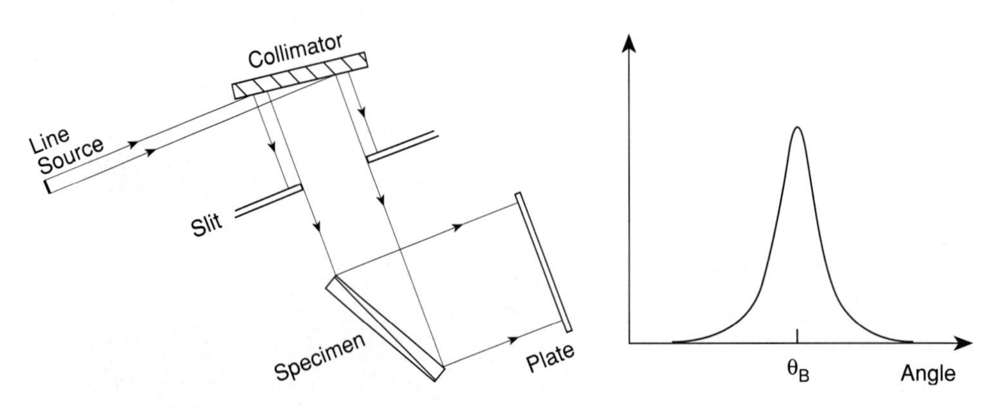

FIGURE 4.8
Schematic of a double-crystal camera. (*After Franzosi [1994].*)

FIGURE 4.9
Schematic showing a double-crystal rocking curve.

rocking curve, shown in Figure 4.9. If the crystal has defects, the rocking curve is broadened. Thus we are able to assess the perfection of crystals by examining their rocking curves.

4.1.4 Transmission Electron Microscopy (TEM)

We use TEM extensively for the microscopic evaluation of semiconductors. It is an electronic analogue of an optical microscope, but its resolution is extremely high (~0.18 nm). In this technique high-energy electrons (accelerating voltage between 125 kV to 3 MV) are focused electromagnetically and transmitted through thin samples whose thicknesses are of the order of 100 to 200 nm. Let us assume that the sample under examination is set for Bragg diffraction. The electron beam incident from the top is then diffracted from the set of vertical planes as shown in Figure 4.10*a*. As a result, two beams—direct and diffracted—emerge from the bottom of

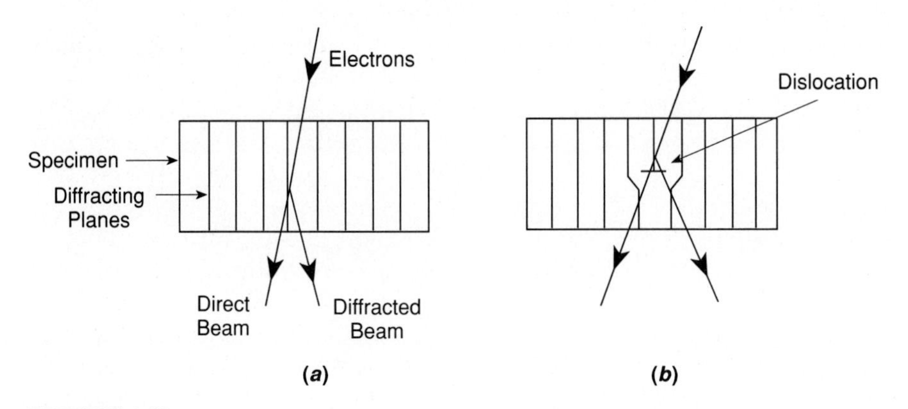

FIGURE 4.10
Schematics showing diffraction from (*a*) a perfect crystal when the Bragg condition is satisfied and (*b*) a dislocated crystal when it is slightly tilted away from the Bragg position.

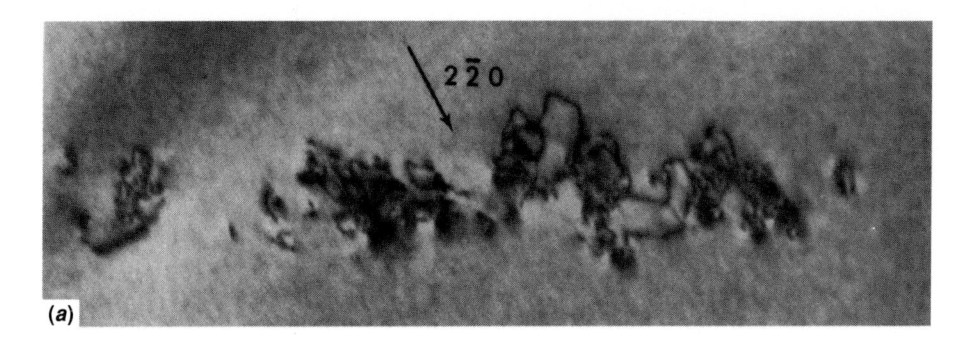

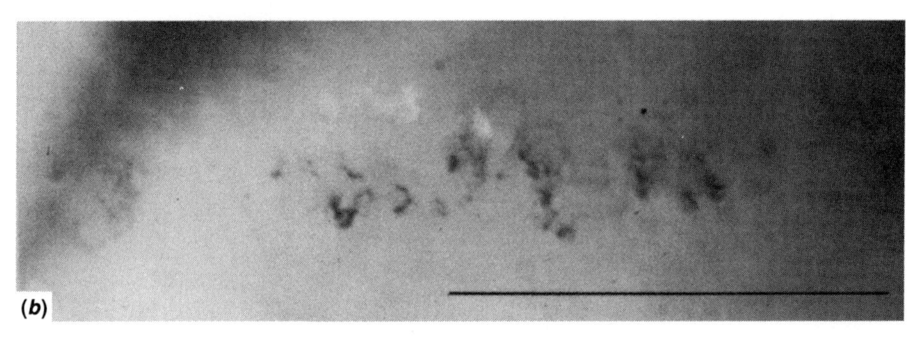

FIGURE 4.11
TEM micrographs obtained from a highly Se-doped crystal: (*a*) two-beam diffraction condition and (*b*) quasi-kinematical situation. The plane of the micrographs is ~(001), and the marker represents 1 μm. (*After Mahajan and Chin [1981].*)

the specimen. Images of the specimen can be formed by placing an aperture around either beam and are recorded on photographic plates. When the direct beam is intercepted, the image is called a *bright-field image*. On the other hand, the interception of the diffracted beam by an aperture produces a dark-field image.

We now consider a situation where a dislocation is present in a thin specimen as illustrated in Figure 4.10*b*. For the sake of discussion, we assume the dislocation to be a perfect edge. Notice that the lattice planes are bent around the dislocation core. If we tilt the specimen slightly away from the exact Bragg condition, we can produce strong diffraction from planes close to the dislocation core. This difference in the Bragg conditions results in a differential contrast between the dislocated and perfect regions of the crystal. Likewise, we can develop explanations for the origin of contrast from stacking faults, inclusions, domain boundaries, and so forth.

We have chosen two examples to illustrate the power of this technique. Figure 4.11*a* shows a dislocation substructure observed in an InP crystal doped with Se to $1 \times 10^{19}\text{cm}^{-3}$ (Mahajan and Chin 1981). When we image the same area after tilting the specimen away from the Bragg condition, we obtain an image shown in Figure 4.11*b*. On tilting, the contrast associated with the strain of the dislocation is reduced. As a result, impurities segregating to the dislocations become visible.

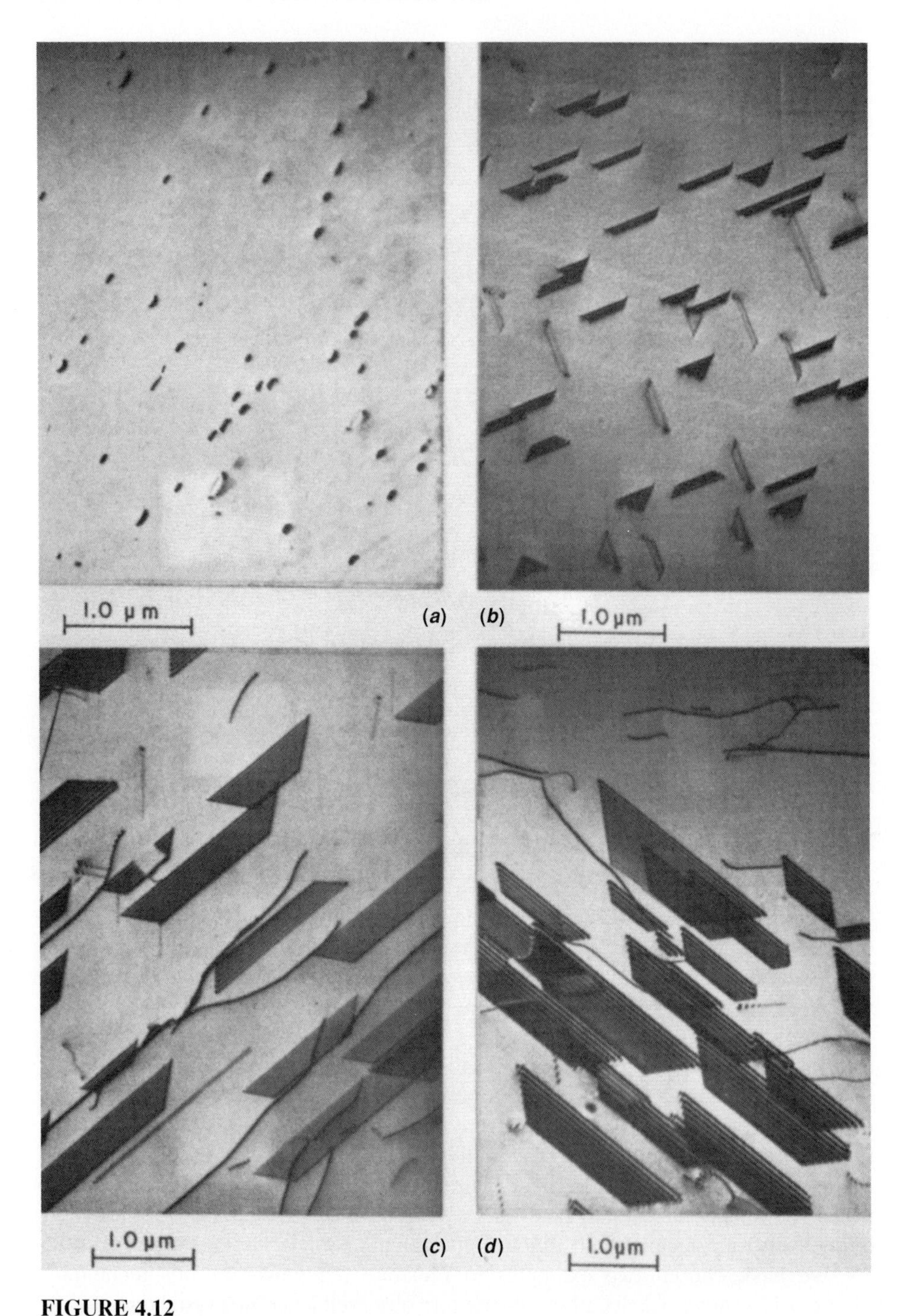

FIGURE 4.12
Series of transmission electron micrographs obtained from abraded silicon specimens oxidized for different times: (*a*) 5 min, (*b*) 10 min, (*c*) 40 min, and (*d*) 55 min at 1100°C in a steam ambient. (*After Ravi and Varker [1974].*)

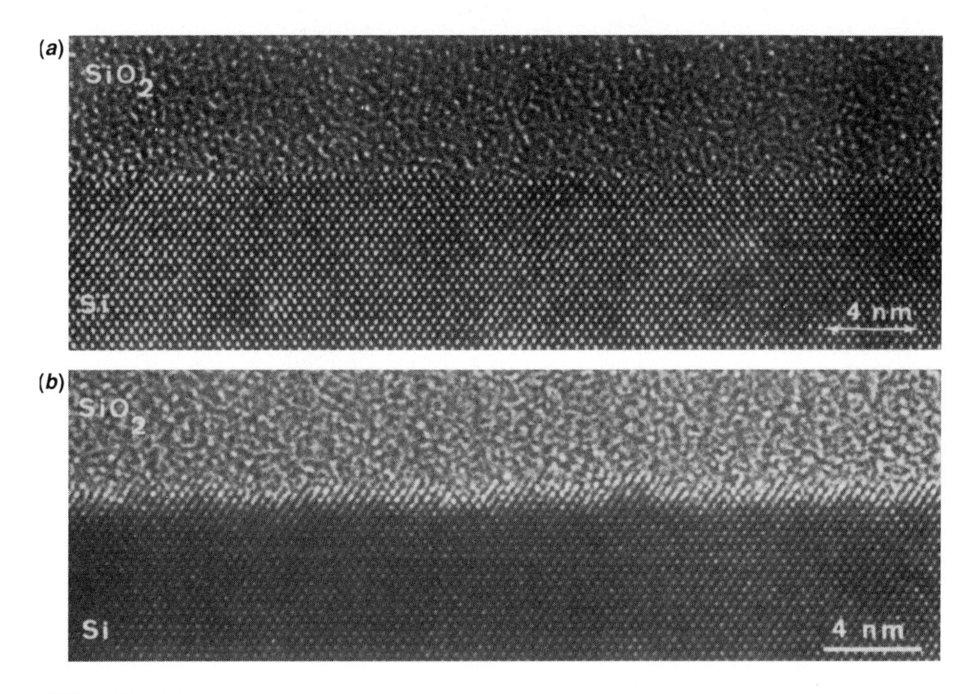

FIGURE 4.13
(*a*) HREM image of the field oxide/silicon interface and (*b*) HREM image of the chlorinated gate oxide/silicon interface. (*After Van Landuyt and Vanhellemont [1994].*)

Figures 4.12*a* through 4.12*d* show a series of transmission electron micrographs of abraded Si specimens oxidized for 5, 10, 40, and 55 min at 1100°C in a steam ambient (Ravi and Varker 1974). We see stacking faults in all the samples. These faults grow with the oxidation time. Dislocations are also observed in the samples oxidized for 40 and 55 mm. We cover the formation of these faults in Chapter 7.

Using high-resolution electron microscopy (HREM), we can image columns of atoms in a very thin sample. We can investigate, for example, interfaces, core structure of dislocations, and polytypes in SiC, by this technique. As an illustration we consider the Si/SiO_2 interfaces that the thermal oxidation of Si produces during the circuit fabrication. This interface is a key component of MOSFETs covered in section 2.2.4.3. We show an example of this interface in Figure 4.13*a* (Van Landuyt and Vanhellemont 1994). We see that the interface is extremely abrupt. If we introduce chlorine in the oxidizing ambient, the interface becomes rough, as illustrated in Figure 4.13*b*. This situation is undesirable because it produces lateral variations in the SiO_2 thickness, resulting in nonuniform capacitance of the metal-oxide-silicon capacitor.

EXAMPLE 4.2. If a dislocation in a GaAs sample is oriented along its [111] direction, determine its projection on a $(1\bar{1}1)$ electron micrograph.

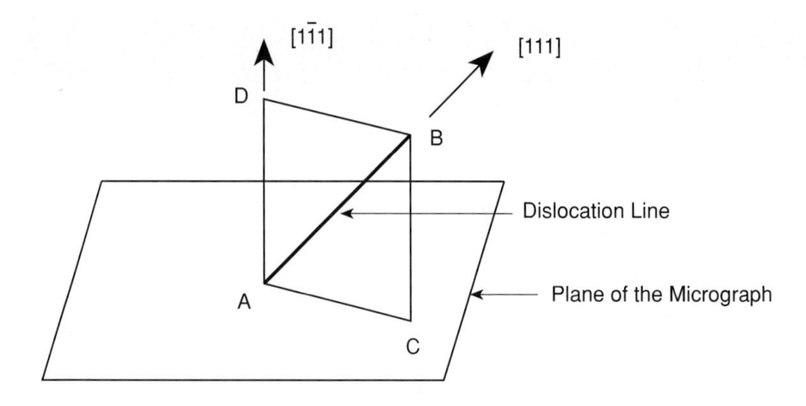

Solution. An electron micrograph represents a two-dimensional projection of a three-dimensional situation, so the given problem can be represented schematically as shown below:

AB is the dislocation line and AC is its projection on the $(1\,\bar{1}\,1)$ plane. The preceding figure shows that AC is the line of intersection of the plane ADBC and the $(1\,\bar{1}\,1)$ plane. The indexes of ADBC are $(10\,\bar{1})$ and can be determined by taking the cross-product of the $[1\,\bar{1}\,1]$ and [111] vectors. Likewise, the indexes of AC can be computed by taking the cross-product of the $[10\,\bar{1}]$ and $[1\,\bar{1}\,1]$ vectors and can be shown to be [121]. Hence the dislocation oriented along the [111] direction would project along the [121] direction on a $(1\bar{1}1)$ electron micrograph.

4.2 CHEMICAL EVALUATION

We show in Chapter 5 that the controlled amounts of impurities are intentionally added to semiconductors to increase their carrier concentrations. Impurities can also be introduced during the growth and processing of semiconductors (covered in Chapters 5 to 10). Since these impurities affect the electronic properties of a semiconductor, we need to determine their type, concentration, and location. Over the years we have developed a number of techniques for this purpose, but it is beyond the scope of this book to cover them in detail. Therefore, we discuss a few widely used techniques in this section: (1) neutron activation, (2) electron microprobe, (3) Auger electron spectroscopy, (4) secondary ion mass spectrometry, and (5) Rutherford backscattering.

4.2.1 Neutron Activation Analysis (NAA)

This technique uses neutrons to analyze the elemental composition of a material. When neutrons are absorbed by a material, nuclei of some of the elements are activated into the excited states. This activation is followed by the radioactive decay of these states. By monitoring the characteristics of the radioactive decay and neutron dose, we can identify various elements and their concentrations. In its simplest form the neutron activation process can be defined as

$$A^{x} + n^{1} \rightarrow A^{x+1}, \tag{4.3}$$

where x refers to the number of neutrons in element A (Murarka 1994). The activated species undergo decay according to the following reaction:

$$A^{x+1} \rightarrow B + \text{radiation}, \tag{4.4}$$

where B is a relatively stable product, and in the simplest case the radiation could be α, β, and γ rays. As an illustration consider the case of neutron irradiation of Ni^{62}:

$$Ni^{62} + n^{1} \rightarrow [Ni^{63}] \rightarrow Cu^{63} + \beta. \tag{4.5}$$

The activated Ni atom in the square brackets is unstable and decays to form Cu^{63} with a concomitant emission of β radiation. Furthermore, not all neutron absorptions lead to activated atoms. In these cases the irradiation may lead to the stable isotopes of the irradiated atom. For example:

$$Ni^{61} + n^{1} \rightarrow Ni^{62}. \tag{4.6}$$

The probability for reactions (4.5) and (4.6) is determined by the absorption cross-sections for these interactions. The absorption cross-sections in turn depend on the neutron energy.

In general we use thermal neutrons (energy in the range of $\sim 10^{-2}$ eV) for NAA of most of the elements in the periodic table. The sensitivity is in the range of a few parts per million (ppm) to a few parts per billion (ppb). The minimum concentrations (maximum sensitivity) in ppm of the detectable elements were computed by Meinke (1955) and are listed in Table 4.3 for a neutron flux of $10^{13}\text{cm}^{-2}\ \text{sec}^{-1}$. We can see that NAA is extremely sensitive in identifying the trace amounts of impurities.

Kruger (1971) points out several limitations of NAA: (1) its inability to distinguish between the chemical nature (bonding, ionization state) of the element under consideration; (2) the occurrence of interfering nuclear reactions; and (3) the sensitivity and accuracy limitations associated with sample conditions, irradiation conditions, postirradiation processing, and radiation measurements.

4.2.2 Electron Microprobe Analysis (EMPA)

This technique uses a focused beam of electrons to locally excite X-rays from a region as small as a few micrometers or less of a specimen. The wavelength distribution or the energy of the emitted X-rays is measured using suitable spectrometers. We can use this information to identify the impurities present.

When an electron beam having energy in the range of 20 to 30 keV strikes a sample, the majority of the beam energy is thermally dissipated. However, a very small fraction of the electrons is involved in ejecting core-shell electrons from the atoms, resulting in vacancies in the core shells. These vacancies are rapidly filled by

TABLE 4.3
Sensitivity by neutron activation analysis, calculated for a flux of 10^{13} neutrons cm^{-2} s

Element	Sensitivity	Element	Sensitivity	Element	Sensitivity
Na	0.35	Ga	2.35	Sb	0.2
Mg	30	Ge	0.2	Te	5
Al	0.05	As	0.1	I	0.1
Si	50	Si	2.5	Cs	1.5
P	1	Br	0.15	Ba	2.5
S	200	Rb	1.5	Hf	1
Cl	1.5	Sr	30	Ta	0.35
K	4	Y	0.5	W	0.15
Ca	190	Zr	15	Re	0.03
Sc	0.1	Nb	500	Os	1
V	0.05	Mo	5	Ir	0.015
Cr	10	Ru	5	Pt	5
Mn	0.03	Pd	0.25	Au	0.15
Fe	450	Ag	5.5	Hg	6.5
Co	1	Cd	2.5	Tl	30
Ni	1.5	In	0.005	La	0.1
Cu	0.35	Sn	10		

Source: Meinke (1955).

outer-shell electrons. This electronic transition is accompanied by the emission of X-rays whose energy is given by the difference in the energies of the outer and inner electronic levels. Since fewer subshells are associated with the inner electronic levels, the X-ray spectra tend to be simpler than the optical spectra that involve electronic transitions between the upper levels. Elements in a sample can, therefore, be identified on the basis of only one or two emission lines. Figure 4.14 shows a simplified energy-level diagram illustrating some of the major electronic transitions. A transition from the L shell to the K shell results in the emission of a K_α photon, whereas the M $\rightarrow$ K shell transition produces a K_β photon. In addition to the characteristic emissions indicated in Figure 4.14, the deceleration of the electrons striking the sample produces the "continuum" background radiation.

EMPA uses two complementary techniques to analyze the emitted X-rays. The wavelength of the X-rays is measured using crystal-diffraction spectrometers (CDS), whereas their energy can be determined via energy-dispersive spectrometers (EDS). The basic components of the CDS are an analyzing crystal, a detector, and readout electronics. The X-rays emitted from the sample under examination impinge on the analyzer crystal and undergo Bragg diffraction. From the knowledge of the interpla-

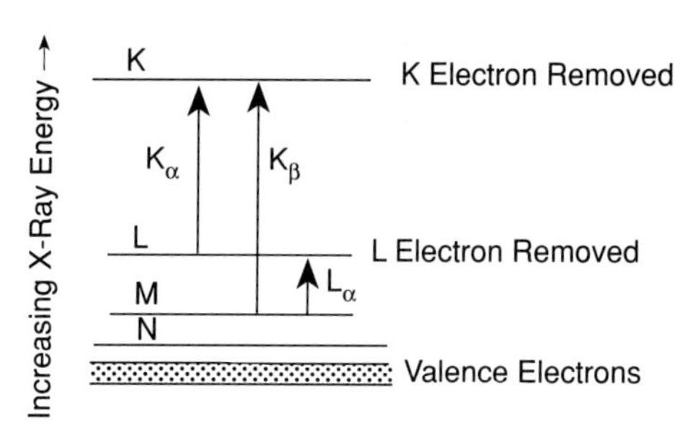

FIGURE 4.14
Simplified energy-level diagram showing some of the major electronic transitions.

nar spacing of the diffracting crystal and the Bragg angle, the wavelength associated with the emitted X-rays can be determined. Since these characteristic X-rays carry the elemental signatures, various elements in the sample may be identified.

EDS uses a reverse-biased Si detector to measure the energy of the emitted X-rays indirectly. When the X-rays strike the detector, EHPs are generated in its intrinsic region and are swept away by the applied bias. The resulting charge is converted into a voltage pulse by a low-noise preamplifier. Since the voltage of each pulse is proportional to the initial X-ray energy, the energy of the X-rays can be determined. Using this information, we can identify the elements. For most elements the detection limit with EMPA is around 1 atomic %.

State-of-the-art TEMs are equipped with EDSs. Since the beam-spot size in a TEM is very small, we can obtain the chemical information from an area as small as 20×20 (nm)^2. Figure 4.15*a* shows an application of EMPA to analyze the chemistry of different microstructural features observed in a Ag-Y-Ba-Cu alloy that is used as a precursor for the deposition of superconducting films (Vecchio 1994). Figures 4.15*b* and 4.15*c* show, respectively, X-ray spectra obtained from twinned grains and the intergranular phase indicated in Figure 4.15*a*. We see that the concentrations of Ag, Ba, C, Cu, O, and Y in the two regions are different.

4.2.3 Auger Electron Spectroscopy (AES)

AES is a powerful technique for determining the elemental composition of a few outermost atomic layers of materials. In this technique we bombard a specimen with electrons having an energy between 3 keV and 30 keV. As a result, we eject core-level electrons from atoms located within a depth of $\sim$1 μm. As we discussed in the case of EMPA in the previous section, the resulting core vacancy is then filled by an outer-level electron. The excess energy is then used to dislodge outer electrons of the atoms. These electrons are called *Auger electrons.* The schematic in Figure 4.16 illustrates this process (Grant 1994). The Auger electrons that are emitted from the few outermost layers of the specimen have kinetic energies that are characteristic of each element. By measuring these energies, we can identify all elements except hydrogen and helium in the surface layer.

The Auger spectra can be acquired in two ways: (1) the measurement of the total electron signal including Auger electrons and background and (2) the mea-

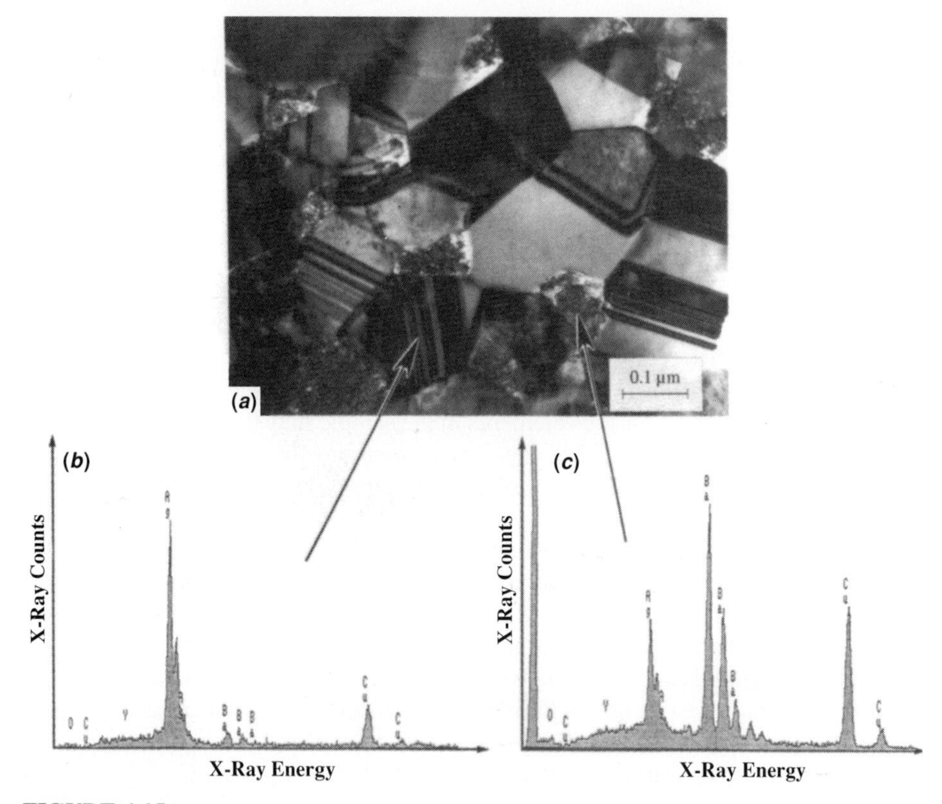

FIGURE 4.15
(*a*) TEM of a Ag-Y-Ba-Cu alloy used as a precursor for the formation of a superconducting film. (*b*) X-ray spectrum obtained from the twinned grains indicated in (*a*). (*c*) X-ray spectrum obtained from the intergranular phase. (*After Vecchio [1994].*)

surement of Auger electrons with background suppressed. By focusing the incident electron beams to small diameters, we can obtain the energy spectrum from a volume as small as $50 \times 50 \times 2\ (\text{nm})^3$. The variation in elemental composition across a surface can be mapped. This approach is referred to as *scanning Auger microscopy* (SAM). We can also determine the change in composition with depth by continuously removing layers by sputtering while monitoring the Auger signals. The detection limit for most elements with AES is between 0.1 atomic % and 1.0 atomic %. The use of electrons in SAM restricts its application to conducting materials. However, Auger electrons can also be emitted from a material surface by bombarding it with X-rays. Since X-rays cannot be focused, only large areas can be analyzed. The technique of electron spectroscopy for chemical analysis (ESCA) exploits the use of X-rays to excite Auger electrons. We can also generate information on the bonding state of the atoms on the surface.

Figure 4.17 shows stacking faults observed on the surface of a GaAlAs epitaxial layer grown by liquid phase epitaxy on a Syton-polished (001) GaAs sub-

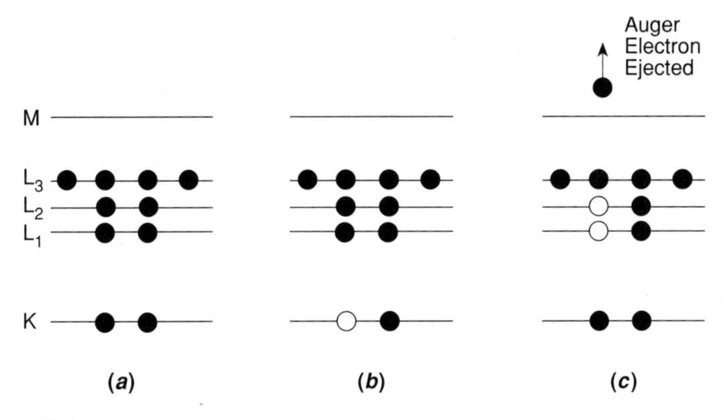

FIGURE 4.16
Schematic illustrating a K L_2L_3 Auger process: (*a*) isolated atom showing electrons present in filled K and L levels before an electron is removed from a core level, (*b*) after removal of an electron from the K level, and (*c*) following the Auger process where a K L_2L_3 Auger electron is emitted. In (*c*) one L-level electron fills the K vacancy and the other L-level electron is ejected due to the energy available on filling the K level. (*After Grant [1994]*.)

strate (Dutt et al. 1981). These faults were not seen in the layers deposited on Br-methanol polished substrates. To determine the origin of the faults, the Auger spectra were obtained from the surfaces of differently polished GaAs substrates, and these results are shown in Figure 4.18. We see that the surfaces of the Syton-polished wafers are contaminated with Ca. The presence of this contaminant disturbs growth, resulting in the stacking faults. This assessment is consistent with the fact that the faults were not observed in layers grown on Syton-polished wafers, which were sputter-cleaned prior to the deposition of the layer (see the top plot in Figure 4.18).

4.2.4 Secondary Ion Mass Spectrometry (SIMS)

Secondary ion mass spectrometry (SIMS) is used for the chemical characterization of materials. It has high sensitivity (in ppb range) to most elements in the periodic table. In this technique secondary ions are sputtered from a sample surface by bombarding it with primary ions—typically, Cs^+, O_2^+, O^-, and Ar^-—in ultrahigh vacuum. A small fraction of the sputtered atoms are ionized either positively or negatively, and they are called the secondary ions. These secondary ions are individually detected and tabulated as a function of their mass-to-charge ratio. We use this information to determine the composition of the surface.

The formation of secondary ions can be treated as a two-step process. In the first step the primary ion produces a cascade of atomic displacements in the speci-

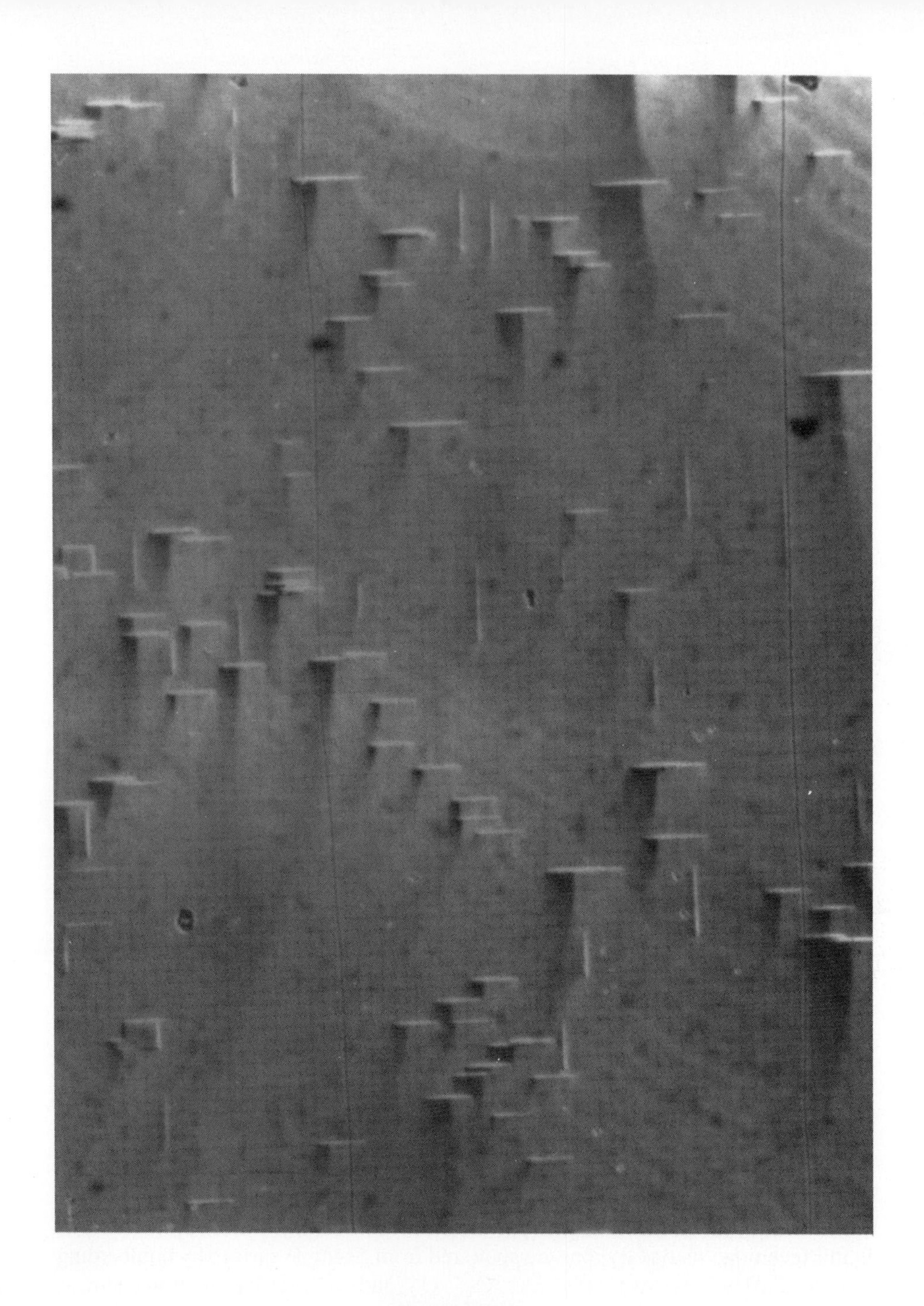

FIGURE 4.17
Stacking faults observed on the surface of (Ga,Al)As epitaxial layer grown on Ca-contaminated (001) GaAs substrates by liquid phase epitaxy. (*After Dutt et al. [1981].*)

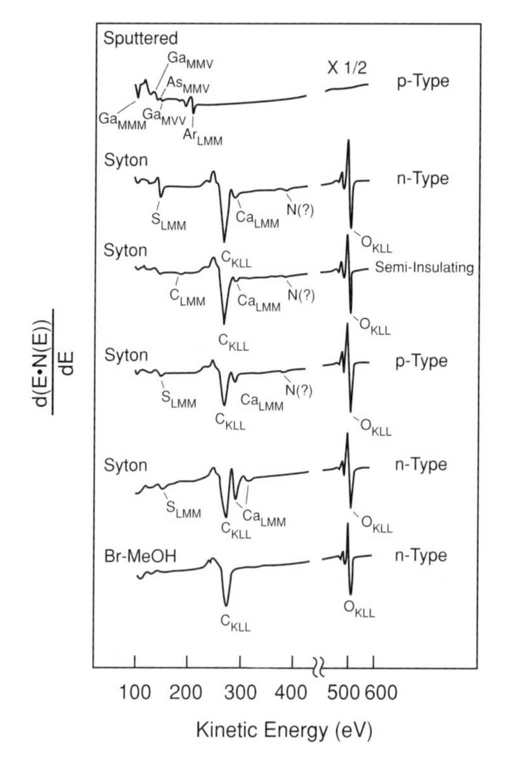

FIGURE 4.18
Electron stimulated derivative Auger spectra for four Syton-polished surfaces and one Br-methanol polished surface of GaAs. (*After Dutt et al. [1981].*)

men (see Chapter 9 for additional details), resulting in a significant mixing of the atoms in the surface region and the ejection of atoms or clusters from the surface. In the second step the ejected atoms are ionized, and the charge state of the secondary ion is determined within a very short distance from the surface. Therefore, in SIMS the depth resolution and the sensitivity are primarily determined by the sputtering and ionization steps.

We use SIMS in two modes: static and dynamic. The static mode employs a primary-ion fluence of $<10^{14}\text{cm}^{-2}$ that leaves the sample surface relatively undisturbed. Most of the secondary ions originate in the top one or two monolayers of the sample. A mass spectrum covering hundreds or thousands of atomic mass units is obtained. The dynamic mode monitors the selected secondary ion intensities as a function of the sputtering time, resulting in a concentration-versus-depth profile. The depth resolution of this technique is in the 5 to 20 nm range. This mode is widely used for the study of the electronic materials.

We use SIMS extensively for determining the concentration versus the depth profiles of dopants in semiconductors. Figure 4.19 shows a typical SIMS profile of $^{11}B^{+}$ ions, implanted into Si to a dose of $1 \times 10^{15}\text{cm}^{-2}$ at 75 keV (Schwarz 1994). We achieve a dynamic range in excess of five orders of magnitude. Using SIMS, we can detect impurity concentrations as low as 10^{16}cm^{-3}.

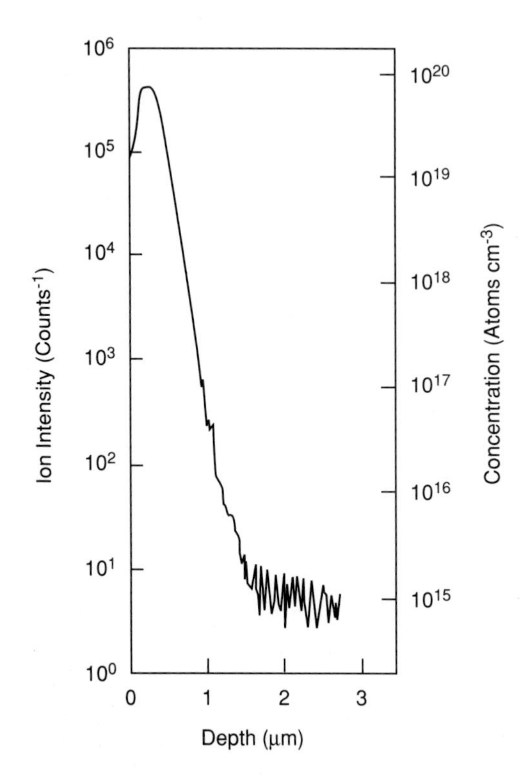

FIGURE 4.19
Boron implant SIMS depth profile in silicon. The depth scale is calibrated by measurement of the crater depth by stylus profilometry. The intensity scale is converted to boron concentration from the known dose of the implant: ^{11}B dose 1×10^{15}cm^{-2}, 75 keV. (*After Schwarz [1994].*)

4.2.5 Rutherford Backscattering (RBS)

A high-energy ion penetrating a solid can lose its energy via nuclear collisions and coulombic interactions with the electrons of the host lattice atoms (see section 9.1.1). When the ion undergoes an elastic collision in the interior, its trajectory direction changes from inward to outward. During its outward motion the ion also loses energy to the lattice atoms until it emerges from the solid. The energy loss suffered during the inward and outward motions reflects the penetration depth of the ion. The analysis technique using these energy losses is known as *Rutherford backscattering* (RBS). It is a nondestructive rapid method for the microanalysis of materials. We can use this technique to determine with high sensitivity the variations in the composition of a solid with depth. It has a depth resolution of a few tens of nanometers over a depth of several hundreds of nanometers, and we can achieve this resolution without resorting to sputtering.

The schematic in Figure 4.20 shows a typical experimental setup for RBS (Bakhru 1994). For analysis we commonly use high-energy ion beams of hydrogen (0.1 to 1 MeV) and helium (1 to 4 MeV) for three reasons. First, they can be obtained easily from small accelerators. Second, being light, these ions produce the least damage in the solid. Third, their stopping power, that is, the rate of change in their energy with the distance, has been extensively investigated. In practice, we direct a well-collimated beam of 2 MeV helium ions at the target at angle θ_1 relative to the surface normal, and a particle detector records the elastically scattered beam

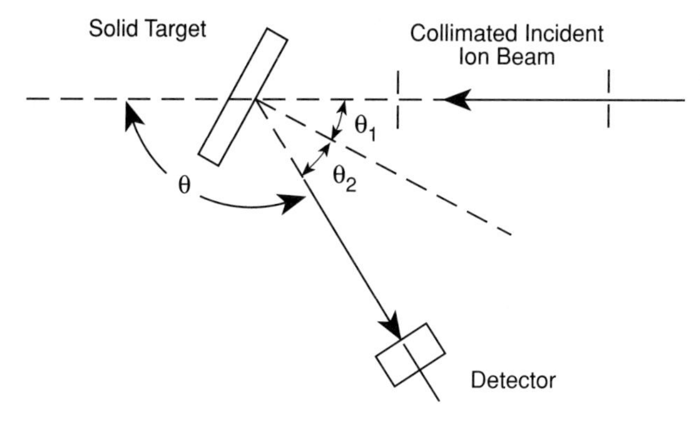

FIGURE 4.20
Schematic of the experimental geometry for the Rutherford backscattering technique. (*After Bakhru [1994].*)

emerging at angle θ_2. The helium ion beam of this energy provides a good mass resolution and is not involved in any known nuclear reactions.

In Figure 4.21*a* ions of energy E_o are incident on the surface of a thick target whose surface is covered with the light and heavy impurities. Some of the incident ions are scattered from the surface, and the scattered ions have an energy $K_M E_o$, where K_M is a kinematic factor (Bakhru 1994). If we plot the scattered energy versus the scattered yield, the ions scattered from the target atoms at the surface would

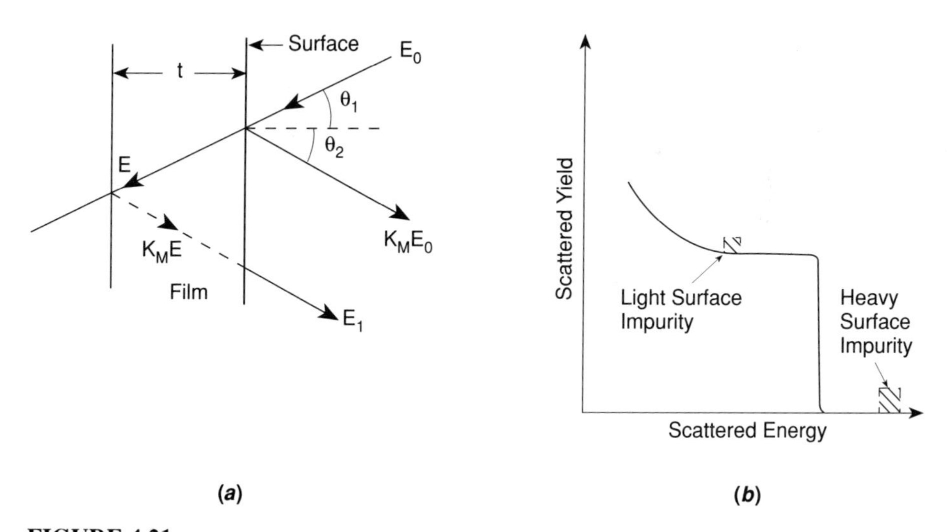

FIGURE 4.21
(*a*) Schematic illustrating the thickness dependence of energy loss of a backscattered ion; E_0 is the incident ion energy, K_ME_0 is energy of the backscattered ion from a surface atom of mass M, E is the incident ion energy at depth t, K_ME is the energy of the backscattered ion from an atom of mass M at depth t, and E_1 is the energy measured by the detector after the ion has traversed back through the film. (*b*) Schematic representation of the Rutherford backscattering yield from a thick target with surface impurities. (*After Bakhru [1994].*)

produce a steep edge on the right as shown in Figure 4.21*b*. The ions scattered from the light and heavy surface impurities would have lower and higher energies than those scattered from the surface atoms, which is shown in Figure 4.21*b*. However, most incident ions penetrate the surface and are backscattered from different depths of the sample. For example, the schematic in Figure 4.21*a* shows scattering from a depth t of the sample. The energy E_1 of the scattered beam is much smaller than the incident energy E_o because of the various losses discussed above. We can also argue that the smaller the value of t, the higher the magnitude of E_1, resulting in the Rutherford backscattering spectrum shown in Figure 4.21*b*.

4.3 ELECTRICAL AND OPTICAL EVALUATIONS

Chapter 2 covered the operation of semiconducting devices involving the motion of electrons and holes. For example, in bipolar junction transistors (see section 2.2.3), emitter, base, and collector regions have different carrier concentrations and carrier types. Another important material parameter is the minority carrier diffusion length that provides a good measure of the quality of a semiconductor. We can incorporate impurities into semiconductors during their growth and processing. If these impurities produce levels within the band gap of a semiconductor, its electronic properties are affected. This section covers the principles of some current techniques for evaluating the electrical and optical characteristics of semiconductors.

4.3.1 Mobility and Carrier Concentration

The carrier mobility in a semiconductor is an important property. This quantity represents the proportionality between the drift velocity of a carrier and the electric field for sufficiently low fields. We showed in section 1.5.5 that the conductivity σ or resistivity ρ of the semiconductors is given by Equation (1.100):

$$\sigma = \frac{1}{\rho} = qn\mu_n + qp\mu_p, \tag{4.7}$$

where q is the charge of the carrier, n and p are concentrations of n- and p-type carriers, and μ_n and μ_p are their respective mobilities. If one type of carrier dominates, then we can use the Hall effect to measure the carrier concentration and the mobility (see section 1.5.5). The van der Pauw method (1958) is used extensively for these measurements because it can be applied to samples of arbitrary shape.

The van der Pauw method requires small contacts at the circumference of a sample having uniform thickness. Since the diamond-cubic and zinc-blende crystals can be cleaved along {110} planes, the preferred geometry for a {001} Hall sample is a square with contacts at the corners as shown in Figure 4.22. $R_{AB,CD}$ is defined as the resistance obtained by dividing the voltage applied between contacts C and D by the current that enters the sample at contact A and leaves through

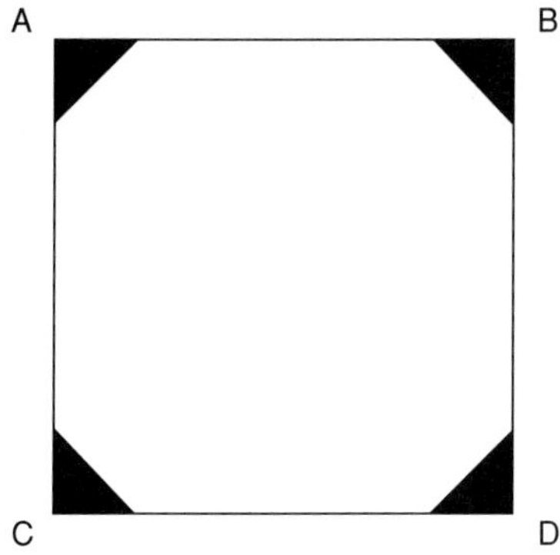

FIGURE 4.22
An electrical contact configuration for a sample prepared for Hall effect measurements using van der Pauw's method. (*After Swaminathan and Macrander [1991].*)

contact B. If $R_{BC,DA}$ can be defined in a similar fashion, then according to van der Pauw (1958), ρ is

$$\rho = \frac{\pi d}{\ln 2} \frac{(R_{AB,CD} + R_{BC,DA})}{2} f\left[\frac{R_{AB,CD}}{R_{BC,DA}}\right], \tag{4.8}$$

where d is the thickness of the sample and f is a function that is unity if $R_{AB,CD} = R_{BC,DA}$. If a magnetic field (H) is applied normal to the sample surface and if the current enters and leaves the sample at diagonally opposite contacts in Figure 4.22, then a Hall voltage ($V_H \equiv V_{AB}$ in section 1.5.5) develops between the other two contacts. As we discussed in section 1.5.5, V_H arises because the Lorentz force separates the electrons and holes, and this separation gives rise to an electric field. If j is the current entering at contact A and exiting through contact C, V_H between contacts B and D is given by

$$V_H = \frac{\mu_H H j \rho}{d}, \tag{4.9}$$

where μ_H is the Hall mobility and is closely related to the drift mobility. The Hall mobility is obtained by measuring $\Delta R_{BD,AC}$, which is the change in $R_{BD,AC}$ induced by the magnetic field. Since this measurement yields V_H/j, we can compute μ_H from the following expression:

$$\mu_H = \frac{d\, \Delta R_{BD,AC}}{H\rho}. \tag{4.10}$$

We can obtain the carrier concentration by using Equation (4.7), which simplifies to $1/\rho = qn\mu_n$ or $qp\mu_p$, depending on which type of carrier dominates.

4.3.2 Minority-Carrier Lifetime

We have several techniques for measuring the minority-carrier lifetimes in semiconductors. In all cases the basic principle is the same. It involves the generation of excess-charge carriers in a semiconductor and measuring the decay time of the generated carriers when the generating source is removed.

We generally use the photoconductive decay (PCD) technique for this purpose that involves forming ohmic contacts on a semiconductor through which we inject a constant current. We then use a high-intensity light source of short duration to generate excess carriers in the semiconductor as discussed in section 1.5.4. After the source is turned off, the excess carriers decay, and this decay process is monitored on an oscilloscope. The minority-carrier lifetime is the time required for the excess carrier pulse to decay to $1/e$ of its initial value.

When the carrier traps are present in a sample, the interpretation of lifetime measurements using PCD is difficult because the traps capture the photogenerated carriers. When the carrier source is turned off, the trapped carriers are released, giving rise to long decay times and erroneous minority-carrier lifetimes. This problem can be eliminated by saturating the traps using a steady source of light. A pulsed light source, shielded from the steady light source, is then used to generate excess carriers. The rate of decay of these carriers will represent true bulk lifetime of the material.

The excess minority carriers can be generated in several ways. A xenon flash lamp is adequate when the carrier lifetime is in excess of ~ 20 μsec. However, the measurement of lifetimes lower than 20 μsec requires light sources with faster turnoff times. Light emitting devices and Nd:YAG lasers (emission at 1.06 μm) are suitable for this purpose.

4.3.3 Deep-Level Transient Spectroscopy (DLTS)

We indicated in section 2.1.1 that due to the presence of uncompensated ionized donors and acceptors in the depletion region of a p–n junction, the junction is associated with a capacitance that is given by Equation (2.3). This capacitance is inversely proportional to the width of the depletion region. These ideas also apply to a rectifying metal–semiconductor junction. The presence of carrier traps, as well as dopant atoms, in the semiconductor will affect the capacitance of the junction. By varying the width of the depletion region, we can either fill or empty the traps. For example, as the depletion-region edge sweeps across the region containing the electron-filled traps, the traps are emptied, resulting in concomitant changes in the capacitance of the junction. By measuring the capacitance transients, we can characterize the carrier traps in a semiconductor. DLTS utilizes this approach (Lang 1974).

We can further elaborate on the principle of DLTS by referring to Figure 4.23 that shows an n^+–p junction under different bias conditions. In Figure 4.23*a* the junction is under quiescent reverse bias. As discussed in section 2.1.1, the depletion region in the n^+ semiconductor is fairly narrow because it is highly doped. However, the region is fairly wide in the p-type semiconductor. An electron trap is located at E_T and is filled in the n-region, whereas it is empty in the p-region. If we now forward-bias the junction and also subject it to an injection pulse, the situation shown in Figure 4.23*b* results. The traps in the p-region will now be occupied by electrons. If we remove the forward bias and allow the junction to return to its initial

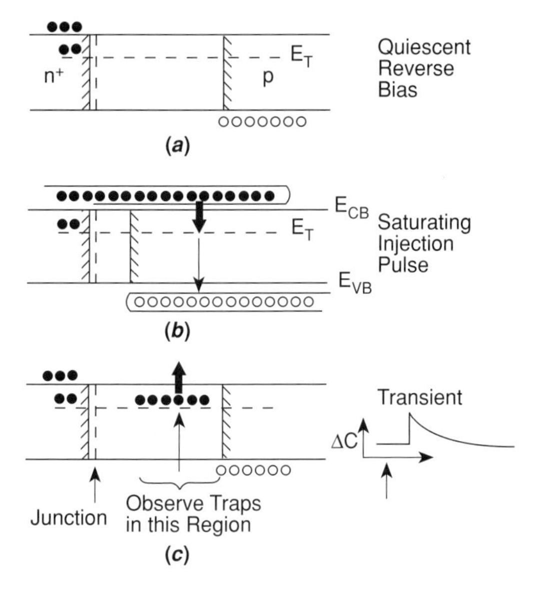

FIGURE 4.23
Injection-pulse sequence is used to produce a capacitance transient for a minority-carrier trap. The energy versus distance diagrams (with band bending omitted for simplicity) show the n^+–p junction depletion region (edges denoted by shaded lines), the capture and emission processes, and the trap occupation before, during, and after an injection pulse. (*After Lang [1974].*)

state shown in Figure 4.23*a*, electrons will be emitted from the traps located in the depletion region. This emission produces a change in the junction capacitance as a function of time, which is depicted in the lower-right corner of Figure 4.23.

Using DLTS, we can determine the activation energy of a trap, that is, its position within the band gap, its carrier-capture cross sections, and the trap concentration. We can illustrate the potential of this technique using the determination of activation energy as an example. We sample a capacitance transient at two intervals and take the DLTS signal to be the difference between the two capacitances, as illustrated in Figure 4.24. The two sampling times define an emission-rate window

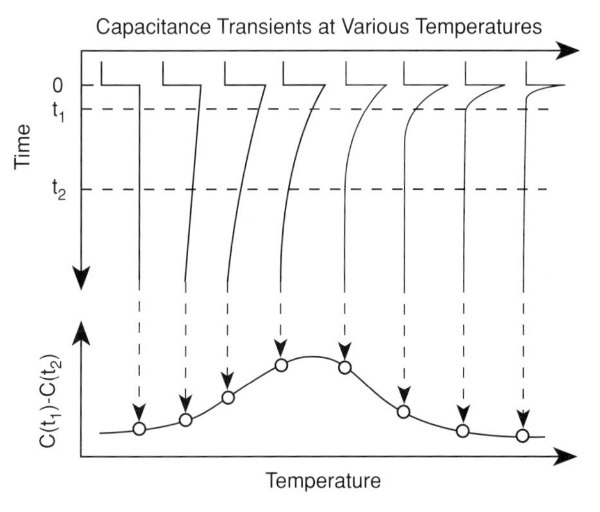

FIGURE 4.24
Schematic of a DLTS peak generated by sampling capacitance transients at two sampling points. (*After Lang [1974].*)

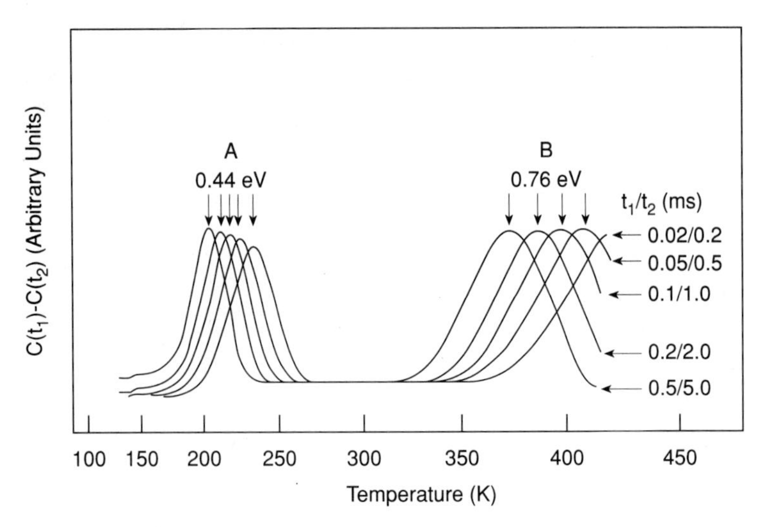

FIGURE 4.25
DLTS spectra of two hole traps in an n-GaAs layer. The transient sampling times and peaks maxima are indicated. (*After Swaminathan and Macrander [1991]*.)

over which a DLTS signal is obtained. A maximum in a DLTS signal occurs when the time constant of the capacitance transient is equal to τ_{max}, which is given by

$$\tau_{max} = \frac{t_1 - t_2}{\ln(t_1/t_2)}. \tag{4.11}$$

The activation energy can be determined by making a number of DLTS scans at different rate windows as shown in Figure 4.25. We can thus obtain data for the emission rate as a function of temperature. The slope of a plot τ_{max} as a function of the inverse of the temperature at which a DLTS peak occurs yields the activation energy.

4.3.4 Electron Beam Induced Current (EBIC)

We discussed in section 3.4 that defects in semiconductors are generally electrically active. We can assess the electrical activity of one-, two- and three-dimensional defects using the electron beam induced current (EBIC) technique, schematically shown in Figure 4.26. We carry out this type of analysis in a scanning electron microscope (SEM). The specimen is in the form of either a metal–semiconductor junction or a p–n junction. For discussion purposes, we have chosen a metal–semiconductor junction, that is, a Schottky barrier diode. A 5 to 30 keV electron beam is rastered over the surface of the sample. The high-energy electrons penetrate the specimen and irradiate a volume of the crystal that resembles a teardrop as shown in Figure 4.26. As a result of the irradiation, EHPs are created. Since the depletion layer width W is fairly small, we ignore EHPs generated in this region. However, EHPs generated in the neutral region that is within the diffusion lengths

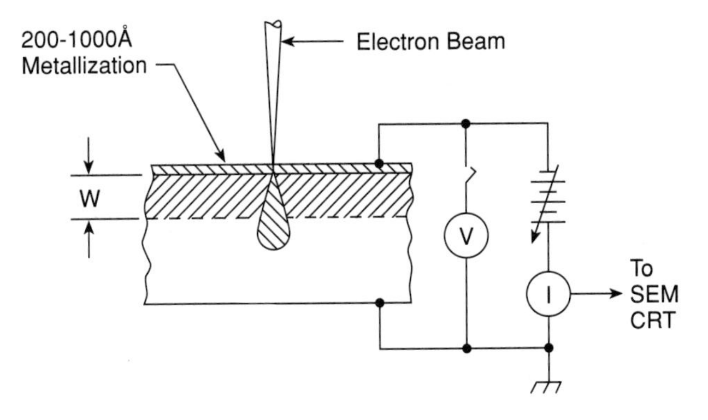

FIGURE 4.26
Schematic of the experimental arrangement employed in EBIC technique using a Schottky barrier diode. W refers to the depletion layer width.

from the edge of the depletion region are separated into electrons and holes by the electric field of the junction. These carriers migrate to the respective contacts and are collected, resulting in a current in an external circuit.

The charge collection current (I_{CC}) is given by

$$I_{CC} = \frac{1}{\alpha} I_B E_B \eta, \tag{4.11}$$

where α is the energy required to generate an EHP; I_B and E_B are, respectively, the beam current and the beam current, and η is the charge collection efficiency. Depending on whether a defect serves as recombination or generation center, η in the defective region is lower or higher than that in the surrounding matrix. This difference in η produces variations in I_{CC} as the beam is rastered across the sample, and these variations are displayed on a CRT. We thus obtain information on the distribution of electrically active defects in the sample.

Figure 4.27 shows Schottky barrier EBIC micrographs of dislocations in an n-type float-zone silicon (de Kock et al. 1977). At zero bias the dislocations exhibit a double contrast, and one of the sides appears darker than the surrounding crystal in Figure 4.27*a*. This condition implies that the dislocations behave as recombination centers; that is, the capture rate of electrons at the dislocations is much greater than the emission rate of holes. Assuming that a dislocation in silicon is associated with a half-filled band (see section 3.4), the defect can capture electrons to fill the band, leading to the decrease in I_{CC}. When the reverse bias is 1.0 V, the dislocations appear brighter than the background in Figure 4.27*b*, indicating that they are behaving as generation centers; that is, the emission rate of electrons at the dislocations is much higher than the capture rate of holes. We can also rationalize this observation in terms of the half-filled band. Perhaps when the diode is under the reverse bias, the dislocations find themselves in the depleted region. This position would require the half-filled band to be free of electrons, that is, the defect would emit electrons, leading to the local increase in I_{CC}.

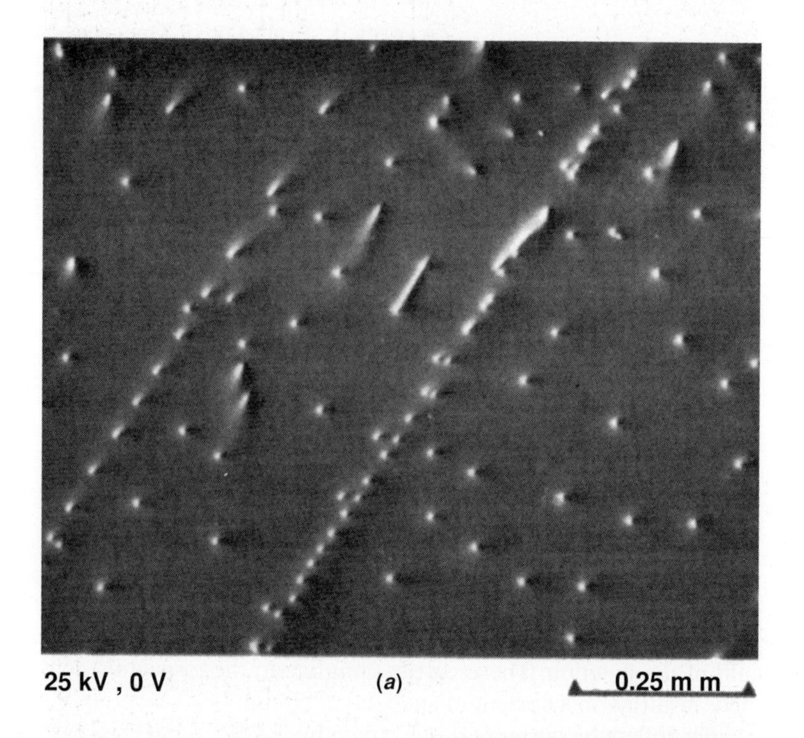

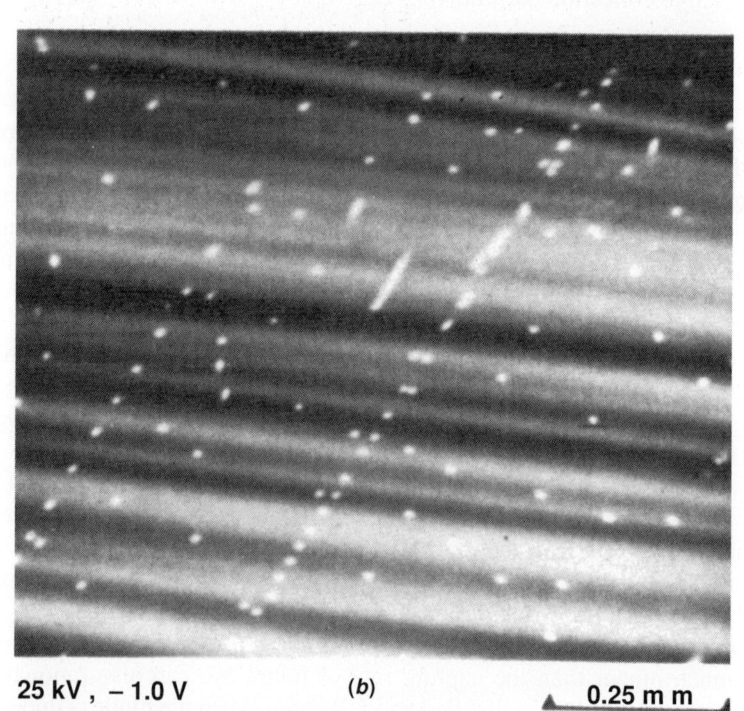

FIGURE 4.27
EBIC micrograph of dislocations in silicon. A Schottky barrier is used to provide the collecting field. (*After de Kock et al. [1977].*)

4.3.5 Cathodoluminescence (CL)

We now consider a situation where we raster the surface of a direct-band-gap semiconductor, such as InP and GaAs, with a high-energy electron beam (~10-30 keV) in an SEM in the absence of a junction. The irradiation will generate EHPs. The excess electrons will recombine with the excess holes. We discussed in section 2.2.2 the two types of recombination: radiative, and nonradiative. The radiative recombination in a direct-band-gap semiconductor produces photons. This process involving the recombination of the excess electrons with the excess holes in a direct-band-gap semiconductor to emit light is referred to as *cathodoluminescence* (CL). The emitted light is detected using a photodetector.

The radiative recombination occurs within the nearly perfect regions of the crystal, whereas the nonradiative recombination takes place at defective areas because the defects are electrically active (see section 3.4). By measuring the CL intensity across the surface, we can map the distribution of the defective areas. The lateral resolution of this technique is ~1 μm, and the information is obtained from a 3 μm thick region from the surface.

CL can also be performed in a transmission mode as schematically illustrated in Figure 4.28, and this approach has been aptly termed *transmission cathodoluminescence* (TCL) by Chin et al. (1981). TCL is accomplished by placing a detector beneath a sample. The CL radiation generated at the top surface is transmitted through the sample and is detected on the back side. As shown in Figure 4.28, both surface and volume defects decrease the CL radiation transmitted through the sample, whereas a void in the sample increases light transmission because of reduced absorption. For TCL both surfaces of the sample must be optically smooth and damage free to avoid surface artifacts in TCL images.

Figure 4.29 shows TCL images of two types of defects present in a S-doped InP sample (Chin et al. 1981). The large dark spots are due to dislocations intersecting the surface. The dislocation image is wide because its influence as a carrier

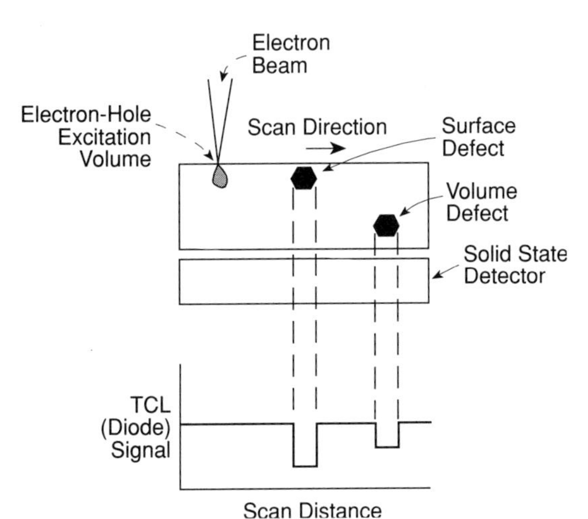

FIGURE 4.28
Schematic of TCL measurement. Both surface and volume defects are detected as a decrease in luminescent radiation by the solid state detector. The surface defect has a lower luminescing efficiency, and the volume defect shadows the detector from the luminescing surface. (*After Chin et al. [1981].*)

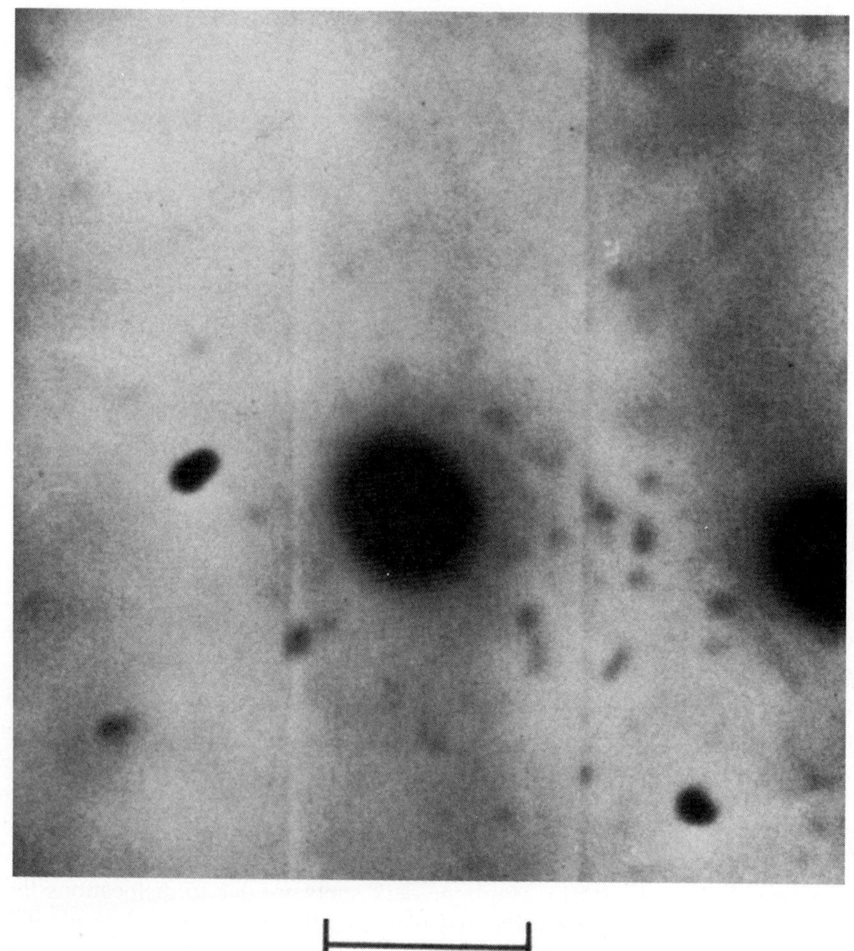

FIGURE 4.29
TCL image of heavily S-doped InP substrate large and small dark spots are dislocations and inclusions, respectively. (*After Chin et al. [1981].*)

recombination center is felt over a longer distance (Chin et al. 1981). The much smaller dark spots are inclusions located within the excitation volume.

4.3.6 Photoluminesence (PL)

We discussed in section 1.5.4 that we can create additional EHPs semiconductors by irradiating them with light whose energy is greater than the band gap of the semiconductors. The excess carriers can recombine radiatively and nonradiatively (see section 2.2.2). The radiative recombination emits light. This phenomenon of light emission is called *photoluminescence* (PL). It differs from CL in the source that generates the excess carriers.

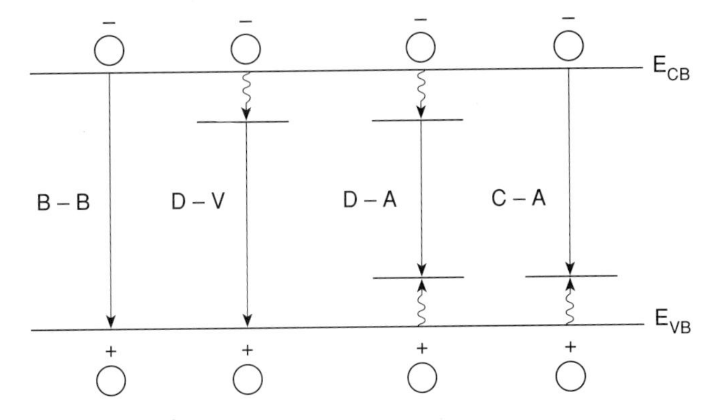

FIGURE 4.30
Schematic band diagram showing the various electron-hole recombination paths.

PL is extensively used for characterizing III-V bulk crystals and epitaxial layers. This simple and elegant technique is used both to understand the fundamentals of the recombination process and to assess the quality of the semiconductor.

Figure 1.22 showed two of the electronic processes that can occur during the optical excitation of a semiconductor when the energy of the excitation source exceeds the band gap of the material. We find that the situation is much more complicated. The excess-carrier density decays by recombination via one of the several possible paths depicted in Figure 4.30. An electron in the conduction band may recombine with a hole in the valence band, which is referred to as *band-to-band* (B-B) recombination. Recombination also occurs via shallow impurities because ionized donors and acceptors have a large capture cross section for the electrons and holes, respectively. A recombination that occurs via the donor level is called a *donor-to-valence* (D-V) recombination. Other possible recombination routes are *conduction donor-to-acceptors* (D-A) and *band-to-acceptor* (C-A).

To characterize the semiconductor, we need to measure wavelengths associated with different recombinations. For this purpose we need a light source for generating excess carriers. We then require equipment to collect, disperse, and detect the luminescene from the specimen. We should be able to vary the sample temperature. The schematic in Figure 4.31 shows a typical setup that is used for PL studies of III-V materials. Possible light sources are Ar, He-Ne, and Kr lasers. The light spot can be focused on the sample to a size of $\sim$2 μm. Calibrated neutral density (ND) filters vary the excitation intensity on the sample. A mechanical chopper operating at a frequency of 300 to 500 Hz is used with a lock-in detection of the luminescence signal. The luminescence from the sample is collected and focused onto the slit of the spectrometer for dispersion and detection.

The luminescence from electrons and holes bound to each other, that is, excitons, is observed only at low temperatures in high-purity materials (Swaminathan and Macrander 1991). As the temperature increases, the excitons break up into free carriers because the thermal energy exceeds the binding energy of the exciton.

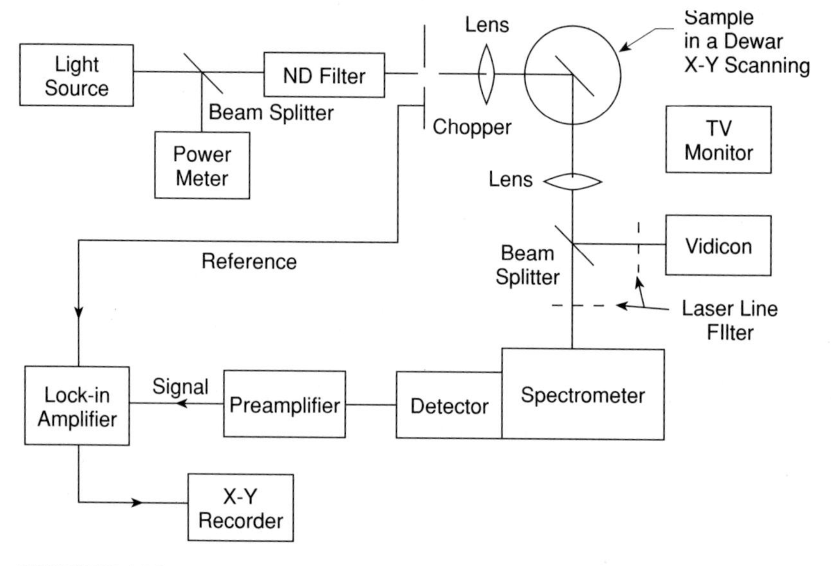

FIGURE 4.31
Experimental arrangement for PL measurements.

Likewise, if the doping increases, the excitons tend to dissociate under the local electric fields. Therefore, under these situations electrons and holes recombine via the band-to-band process discussed earlier. Since some of the electrons may not lie at the bottom of the conduction band, their recombination with holes will produce a high-energy tail in the luminescence spectrum. On the other hand, the band-to-band recombination will yield a sharp cutoff at the wavelength corresponding to the band gap of the material. These effects are illustrated in Figure 4.32, which shows a room-temperature PL spectrum of an n-type high-purity GaAs epitaxial layer (Swaminathan and Macrander 1991).

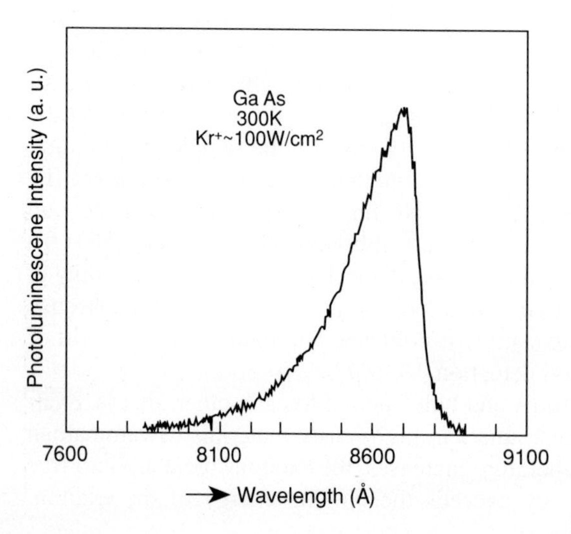

FIGURE 4.32
Room-temperature PL spectrum of an n-type high-purity GaAs sample showing the high-energy exponential tail and a sharp cutoff at the low energy side. (*After Swaminathan and Macrander [1991].*)

4.4 HIGHLIGHTS OF THE CHAPTER

The highlights of this chapter are as follows:

- We carry out three types of evaluations on semiconductors: structural, chemical, and electrical and optical.
- Defect delineation by etching, X-ray topography, and rocking curves provides information at a macroscopic level on the structural perfection of semiconductors. TEM and HREM assess the microscopic perfection of materials.
- Neutron activation analysis, electron microprobe analysis, secondary ion mass spectrometery, and Rutherford backscattering are useful techniques for discerning the chemistry of semiconductors.
- Deep-level transient spectroscopy, electron beam induced current, cathodoluminensce, and photoluminescence are used extensively for the electrical and optical characterization of semiconductors.

REFERENCES

Abrahams, M. S. and C. J. Buiocchi. *J. Appl. Phys.* 36, 2855 (1965).

Bakhru, H. In *The Encyclopedia of Advanced Materials,* "Ion Backscattering Analysis." ed. D. Bloor, R. J. Brook, M. C. Flemings, and S. Mahajan. Oxford: Pergamon Press, 1994, p. 1151.

Brattain, W. H. and C. G. Garrett. *Bell System Tech. J.* 43, 129 (1955).

Chin, A. K.; H. Temkin; and S. Mahajan. *Bell System Tech. J.* 60, 2187 (1981).

Chu, S. N. G.; C. M. Jodlauk; and A. A. Ballman. *J. Electrochem. Soc.* 129, 352 (1982).

deKock, A. J. R.; S. D. Ferris; L. C. Kimerling; and H. J. Leamy. *J. Appl. Phys.* 48, 301 (1977).

Dutt, B. V.; S. Mahajan; R. J. Roedel; G. P. Schwartz; D. C. Miller; and L. Derick. *J. Electrochem. Soc.* 128, 1573 (1981).

Fornari, R.; P. Franzosi; G. Salviati; C. Ferrari; and C. Ghezzi. *J. Crystal Growth* 69, 388 (1985).

Franzosi, P. in *Handbook on Semiconductors* "X-ray Topography." vol. 3, 2nd ed., ed. S. Mahajan. Amsterdam: North-Holland, 1994, p. 1033.

Grant, J. T. In *The Encyclopedia of Advanced Materials,* ed. D. Bloor, R. J. Brook, M. C. Flemings, and S. Mahajan. Oxford: Pergamon Press, 1994, p. 167.

Huber, A. and N. T. Linh. *J. Crystal Growth* 29, 80 (1975).

Ishii, M.; H. K. Hirano; and A. Ito. *Jap. J. Appl. Phys.* 15, 645 (1976).

Kruger, P. *Principles of Activation Analysis.* New York: Wiley, 1971.

Lang, D. V. *J. Appl. Phys.* 45, 3023 (1974).

Mahajan, S. and A. K. Chin. *J. Crystal Growth* 54, 138 (1981).

Meinke, W. W. *Science* 121, 177 (1955).

Miller, D. C. and G. A. Rozgonyi. In *Handbook on Semiconductors* "Defect characterization by Etching, Optical Microscopy and X-ray Topography." vol. 3, 1st ed. ed. S. B. Keller, Amsterdam: North-Holland, 1980, p. 218.

Morelhão, S. L.; D. L. Meier; G. T. Neugebauer; B. B. Bathey; and S. Mahajan. *MRS Proceedings,* 378, no. 29 (1995).

Murarka, S. P. In *The Encylopedia of Advanced Materials* "Neutron Activation for Chemical Analysis." ed. D. Bloor, R. J. Brook, M. C. Flemings, and S. Mahajan. Oxford: Pergamon Press, 1994, p. 1744.
Nakagawa, K.; K. Maeda; and S. Takeuchi. *J. Phys. Soc. Japan* 49, 1909 (1980).
Ravi, K. V. and C. J. Varker. *J. Appl. Phys.* 45, 263 (1974).
Schwarz, S. A. In *The Encyclopedia of Advanced Materials,* "Secondary ion Mass Spectrometry." ed. D. Bloor, R. J. Brook, M. C. Flemings, and S. Mahajan. Oxford: Pergamon Press, 1994, p. 2350.
Secco-D´Aragona, F. *J. Electrochem. Soc.* 119, 948 (1972).
Sirtl, E. and A. Z. Adler, *Metallkunde* 52, 529 (1961).
Swaminathan, V. and A. T. Macrander. *Materials Aspects of GaAs and InP Based Structures.* Englewood Cliffs: Prentice Hall, 1991, p. 304.
van der Pauw, L. J. *Philips Research Reports* 13, 1 (1958).
Van Landuyt, J. and J. Vanhellemont. In *Handbook on Semiconductors* "High Resolution Electron Microscopy for Semiconducting Materials." vol. 3, 2nd ed. ed. S. Mahajan. Amsterdam: North-Holland, 1994, p. 1109.
Vecchio, K. S. In *The Encyclopedia of Advanced Materials* "Electron Microscopy, Analytical." ed. D. Bloor, R. J. Brook, M. C. Flemings, and S. Mahajan. Oxford: Pergamon Press, 1994, p. 730.
Weyher, J. L. In *The Encyclopedia of Advanced Materials* "Compound Semiconductors: Defects Revealed by Etching." ed. D. Bloor, R. J. Brook, M. C. Flemings, and S. Mahajan. Oxford: Pergamon Press, 1994, p. 460.
Wright Jenkins, M. *J. Electrochem. Soc.* 124, 757 (1977).

PROBLEMS

4.1. ***a.*** A (111) slice of GaAs has been etched in molten KOH to delineate slip-induced dislocations. Assuming that slip occurs on the four {111} planes, schematically show the etch-pit arrangement that will be observed on the slice surface.
b. Assuming that the etch pits are bounded by {111} planes, sketch the three-dimensional shape of an etch pit in (*a*). Also, determine the line directions along which the etch pit would intersect the (111) surface.

4.2. When a lattice-mismatched layer (thickness $\cong 5\ \mu$m) is deposited on a substrate (thickness $\cong 100\ \mu$m), misfit dislocations form at the substrate-layer interface as shown:

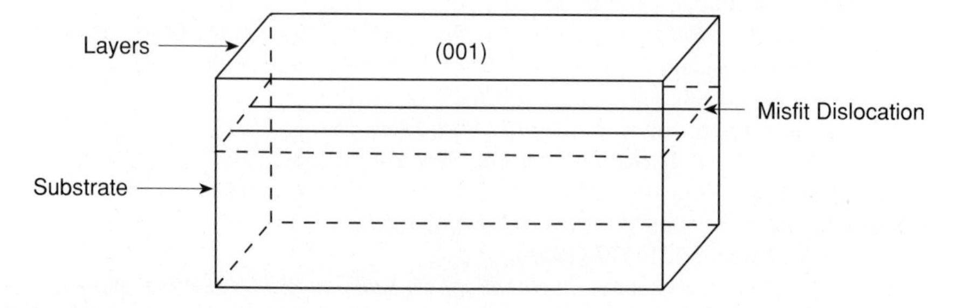

If equipment is not available to prepare samples for TEM, which technique would you use to confirm the presence of misfit dislocations? Describe in your own words the principle of the technique of your choice and its limitations.

4.3. During an electron microscopic examination of a silicon sample, a dislocation line projects along the [110] and [121] directions when the sample orientations are [001] and [111], respectively. Determine the orientation of the dislocation.

4.4. Gold-rich alloys are generally used to form ohmic contacts in InP-based devices. This is achieved by evaporating a thin film of an alloy and subsequently annealing the metal-semiconductor composite. During this anneal, gold and alloying elements migrate into the semiconductor. If the device design requires the penetration depth not to exceed 3 nm, which technique would you use to ascertain this condition? Explain the principle of the technique of your choice.

4.5. SiO_2 precipitates form during annealing of silicon containing oxygen. These precipitates are amorphous in nature and are spherical (size ~10 nm). Describe the technique that you would use to ascertain their presence, amorphous nature, and chemistry.

4.6. Carbon is a common impurity in SiGe layers grown by low-pressure, chemical-vapor deposition; its concentration is of the order of $\sim 10^{15}\text{cm}^{-3}$. Suggest the techniques that you would use to ascertain the presence of carbon and describe the principle of one of them.

4.7. Imagine a situation where you must determine the three-dimensional distribution of dislocations in as-grown SiC crystals. Assuming that the dislocation density is less than 10^5cm^{-2}, suggest the technique that you would use to address the issue and describe its principle.

4.8. A buried layer of nickel, 20 nm thick, can be introduced into silicon by ion implantation. If the layer is located at a depth of 50 nm from the surface, propose techniques to establish the presence of the nickel layer and describe the principle of one of them.

4.9. InP is a direct-band-gap semiconductor, and its band gap is 1.35 eV at 295K. Given that the material has an electron trap at 0.5 eV below the conduction-band edge and a hole trap at 0.3 eV above the valence-band edge, show schematically the photoluminescence spectra that may be obtained when this material is irradiated with a laser source whose energy is greater than the band gap.

4.10. Implanted semiconductors are generally annealed to activate implants. During this anneal, the implanted profile could change. Which technique would you use to measure this change and why? Explain the principle of the technique of your choice. Assume that the dopant concentration in the surface region is $5 \times 10^{16}\text{cm}^{-3}$.

CHAPTER 5

Growth of Bulk Crystals

The semiconducting substrates required for the fabrication of devices discussed in Chapter 2 are obtained from bulk crystals. This chapter covers the growth of these crystals. It discusses the sources of dislocations in as-grown crystals and methods to enhance their perfection. The text emphasizes the growth of doped crystals and the origins of inhomogeneous distribution of dopants. It also covers the nature of microdefects observed in silicon crystals and the effects associated with the precipitation of oxygen in Czochralski Si. This chapter forms a basis for assessing the processing-induced defects covered in Chapters 6 through 10.

5.1 PRODUCTION OF STARTING MATERIALS

5.1.1 Production of Electronic-Grade Polysilicon from Quartzite

Quartzite, a relatively pure form of sand (SiO_2), is used in the production of electronic-grade Si (EGS). The conversion process requires four steps: (1) reduction of quartzite to metallurgical-grade polysilicon (MGS), which is 98 percent pure; (2) conversion of MGS into trichlorosilane ($SiHCl_3$); (3) purification of $SiHCl_3$ by fractional distillation; and (4) conversion of $SiHCl_3$ into EGS by chemical vapor deposition.

The reduction of quartzite into MGS takes place in a submerged arc furnace where the source materials and carbon in the form of coal, coke, or wood chips are heated to over 2000°C. Under these conditions silicon dioxide is reduced to liquid Si (MGS) and carbon monoxide:

$$SiO_2 + 2C \rightarrow Si(l) + 2CO(g) \tag{5.1}$$

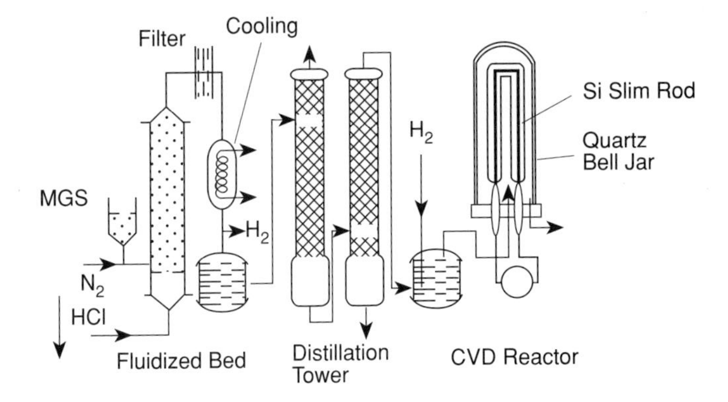

FIGURE 5.1
Schematic of fluidized bed, distillation tower, and chemical vapor deposition (CVD) reactor developed by Siemens for purifying trichlorosilane and converting it into electronic-grade polysilicon.

The liquid Si collects at the bottom of the furnace and is tapped off, and CO escapes from the furnace. At this stage MGS is 98 percent pure and contains 1 to 2 percent Fe, and Al, and other impurities such as B and P.

MGS is subsequently converted into $SiHCl_3$ in a fluidized bed reactor as shown in Figure 5.1. The conversion involves a reaction between MGS powder and anhydrous HCl:

$$Si(s) + 3HCl \rightarrow SiHCl_3 + H_2 \tag{5.2}$$

Since reaction (5.2) takes place at the silicon surface, the use of silicon powder in the reactor can enhance its kinetics. Reaction (5.2) occurs at 300°C in the presence of a catalyst. In addition to forming $SiHCl_3$, halides of Fe, Al, and B are produced in the reactor. Next $SiHCl_3$ that boils at 31.8°C is removed by fractional distillation from the halides of Fe, Al, and B. The resulting $SiHCl_3$ is highly pure.

The pure $SiHCl_3$ is converted into EGS using chemical vapor deposition. The conversion involves a reaction between high-purity hydrogen and $SiHCl_3$ as shown:

$$SiHCl_3(g) + H_2(g) \rightarrow Si(s) + 3HCl(g) \tag{5.3}$$

The deposition of the resulting Si occurs heterogeneously on a thin, heated, high-purity silicon rod referred to as a *slim rod* in Figure 5.1; the production of large-diameter rods requires several hundred hours.

5.1.2 Synthesis of Polycrystalline Compound Semiconductor Source Materials

High-purity elements from groups II, III, V, and VI are used to synthesize polycrystalline compound semiconductors. Since the vapor pressures of group V elements

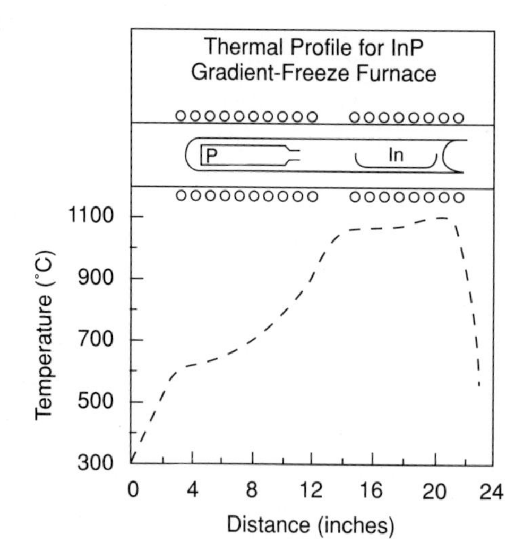

FIGURE 5.2
Schematic showing the synthesis of polycrystalline indium phosphide from its elemental sources. Evacuated quartz ampoules are used in the synthesis, and it requires an extended period.

over III-V compounds are considerably higher than those of group III species, it is very difficult to preserve the stoichiometry of the solid. Synthesizing the feed stock in a closed tube, as shown in the case of InP in Figure 5.2, can circumvent this difficulty.

The melting point of InP is 1045°C. At this temperature In and P react with each other exothermically, and the reaction can be violent. Consequently, the synthesis has to be carried out in a slow, controlled manner. To achieve this condition, In and P sources are contained in an evacuated quartz ampoule and are maintained at two different temperatures. The temperature of the In source is in the 1000 to 1080°C range, whereas the P source is kept at 600 to 800°C. Under these conditions P vapors react with liquid In to form solid InP. The temperatures for the two zones are chosen so that the vapor pressure of P over synthesized InP is lower than that over liquid P. Due to the difference in the activities of P in the two zones, P vapor will be transported to the In source region where the two can react to form polycrystalline InP:

$$\mathrm{In(l)} + \mathrm{P(g)} \rightarrow \mathrm{InP(s)} \tag{5.4}$$

The activities' difference controls the rate of the synthesis of InP and is kept low to avoid explosions. This particular synthesis requires several days to convert 500 g of In melt into InP.

The feed stock can be synthesized in situ at an accelerated rate in a high-pressure system using an injection cell (Stolté 1984). The function of the cell is to facilitate bubbling of group V elements into molten group III species. The cell is then removed prior to crystal growth. In another version the arsenic cell, mounted within the growth system, is used to inject As either continuously or periodically into a Ga melt, thereby permitting precise control of the stoichiometry of the resulting solid.

5.2 GROWTH OF BULK CRYSTALS

The EGS and polycrystalline III-V materials, synthesized in section 5.1, serve as feed stocks for the growth of bulk semiconducting crystals that are the building blocks for electronic devices. The following techniques are currently used to grow bulk crystals of elemental and compound semiconductors: (1) Czochralski (CZ), (2) float zone (FZ), (3) liquid-encapsulated Czochralski (LEC), and (4) Bridgman. Silicon crystals are grown by the first two processes, whereas the latter two techniques are used to grow compound semiconductor crystals. Silicon solar cells are finding extensive terrestrial use. As this application requires thin silicon sheets, techniques for growing silicon ribbons are treated separately in section 5.2.1.3. This growth approach avoids slicing bulk crystals into wafers.

5.2.1 Growth of Silicon Crystals

This section covers the CZ and FZ processes. The discussion emphasizes growth principles and highlights the relative advantages and disadvantages of the two processes.

5.2.1.1 Czochralski (CZ) process

The CZ process, named after its inventor Czochralski, is the main technique for the growth of bulk Si crystals. The process is relatively simple and preserves a high degree of crystal purity. The development of automatic diameter control in the late 1960s led to a rapid increase in ingot diameter and length. These trends are illustrated in Figure 5.3. As shown 200 mm diameter crystals and 75 kg charges were available in 1985. At present, the ingot diameter is $\sim$250 mm. The availability of large diameter crystals has resulted in higher device yields per wafer, thereby lowering the device cost.

A schematic of a CZ setup is shown in Figure 5.4 (Huff 1992). The molten Si is contained in a quartz crucible that is in contact with a graphite susceptor. Melting is achieved by either induction or resistive heating. If the crystal is to be doped, either the required amount of dopant is added to the melt or a doped polycrystalline charge is used.

The first step in the growth sequence is to load EGS into the crucible. Then the growth chamber is pumped down and backfilled with an inert gas, whose function is to prevent oxidation of the melt. Next a Si seed crystal of desired orientation, 5 mm in diameter and 100 to 300 mm long, is lowered into the molten Si. The seed is allowed to partially melt and is then withdrawn at a controlled rate. The melt solidifies on the seed. The seed serves as a template, and its atomic arrangement is replicated into the newly formed solid. The seed crystal and the crucible are rotated in opposite directions during the pulling process to reduce temperature nonuniformities in the growth system.

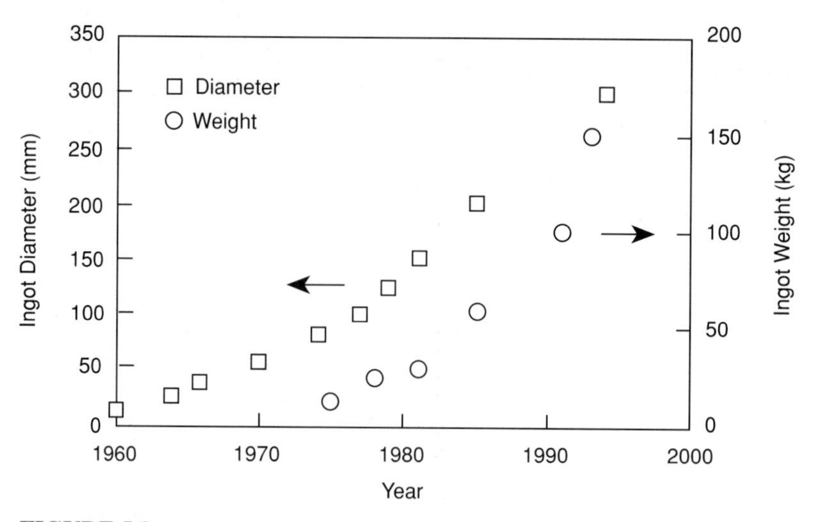

FIGURE 5.3
The increase in CZ silicon crystal diameter and charge sizes between 1960 and 1995. (*Courtesy of MEMC Electronic Mats. Inc.*)

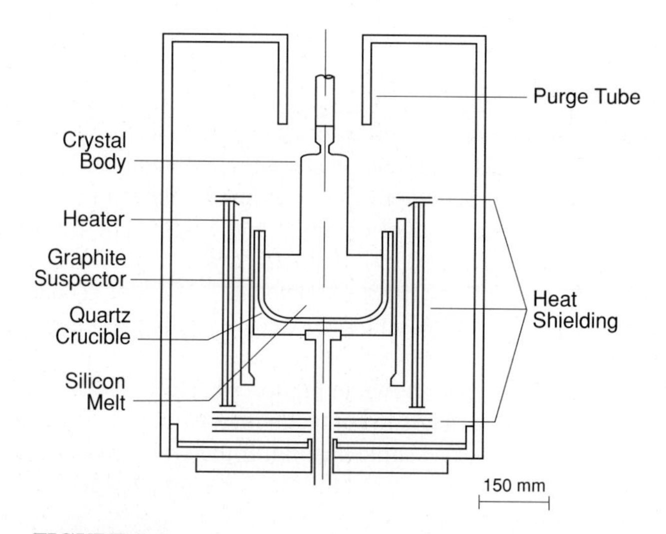

FIGURE 5.4
Schematic of the setup used to grow silicon crystals by the CZ process. (*After Huff [1992].*)

As the melt solidifies, the latent heat of fusion L is introduced into the growth system. This heat is transported from the solid-liquid interface along the growing crystal and is lost from the crystal surface due to radiation. The upper limit to the pull rate, v_{max}, is reached when the maximum heat lost by the ingot is equal to that being released by the solidifying Si. In the absence of a thermal gradient in the melt,

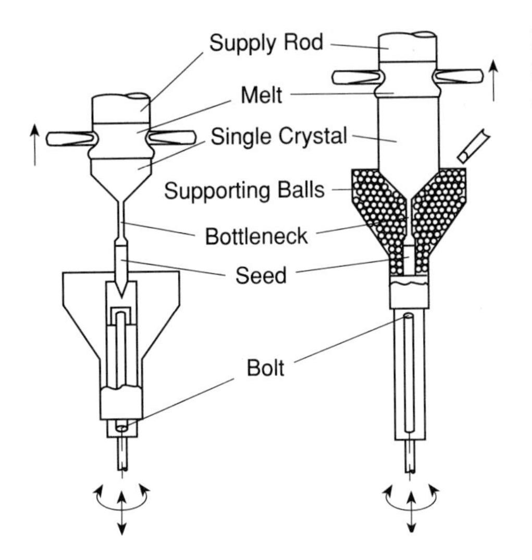

FIGURE 5.5
Schematic illustrating the basic elements of the FZ process.

v_{max} can be calculated from the relation (Rea 1981)

$$v_{max} = K_S \left(\frac{dT}{dx}\right) \Big/ (L\rho), \tag{5.5}$$

where K_S is the thermal conductivity of solid Si, (dT/dx) is the temperature gradient along the axis of the ingot, and ρ is the density of Si.

The CZ process has one significant disadvantage. Since molten Si is very reactive, it attacks the quartz crucible, converting SiO_2 into SiO:

$$SiO_2(s) + Si(l) \rightarrow 2SiO(g). \tag{5.6}$$

As SiO bubbles through the melt, it dissociates into silicon and oxygen. The released oxygen is incorporated into the melt. It is inherited by the as-grown crystals, and its concentration is relatively high, $\sim 2 \times 10^{18}$ cm^{-3} (Kaiser et al. 1956). Carbon is another common contaminant in CZ silicon, and its concentration is in the range of 10^{16} to 3×10^{17} cm^{-3} (Nozaki et al. 1970; Bean and Newman 1971). The possible sources for carbon are the starting material, the graphite susceptor, and the heating element.

5.2.1.2 Float-zone (FZ) process

The FZ process is illustrated in Figure 5.5. A polysilicon supply rod is mounted vertically in a growth chamber that is either under vacuum or filled with an inert gas. Supporting balls maintain the position of the rod. To initiate crystal growth, the rod and seed are partially melted using induction heating. The molten zone is passed along the ingot to form a "bottleneck." Its objective is to reduce the density of dislocations in the crystal that is discussed in section 5.3. After the neck formation, the molten zone is passed along the length of the rod. The controlled solidification of melt on the seed converts the polycrystalline rod into a single crystal.

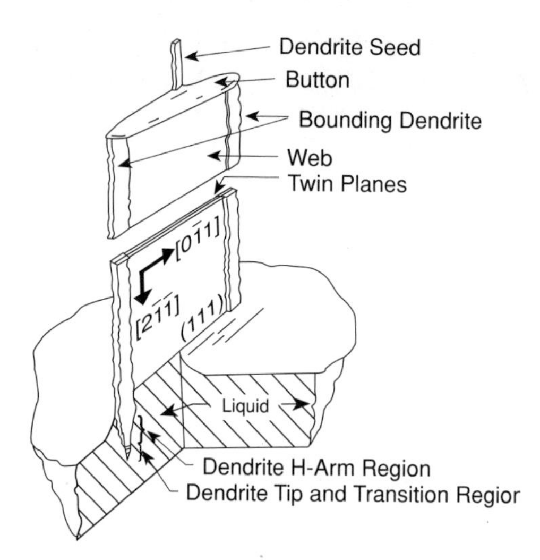

FIGURE 5.6
Schematic of dendritic-web growth showing different regions of the web and crystallographic relationships. (*After Seidensticker [1982].*)

In this process surface tension forces hold the molten zone in place. The zones larger than 125 mm in diameter appear to be unstable. Therefore, it is difficult to grow large-diameter Si crystals. Since the molten zone is not in contact with any crucible, FZ crystals have a higher purity than CZ crystals. Undoped FZ crystals can have high resistivity and are therefore the material of choice in high-voltage devices. The concentration of oxygen in FZ crystals is $\sim 10^{16}$ cm^{-3}. This concentration is considerably lower than that in CZ crystals, whereas the carbon concentrations of FZ and CZ crystals are similar.

5.2.1.3 Growth of silicon ribbons

Two techniques are used for growing silicon ribbons: dendritic-web growth (Seidensticker 1982) and edge-defined film-fed growth (EFG) (Wald 1981). In the dendritic-web growth technique, crystals are grown using a dendrite seed about 0.5 to 1.0 mm square. When the seed is dipped in a melt, heat is dissipated from the seed, resulting in a "button" that spreads laterally across the melt surface. If the melt is slightly supercooled ($\sim$2–3°C), dendrites form at the ends of the button. These features are illustrated in Figure 5.6. A ribbon can be grown by pulling the seed upwards. In principle, long ribbons can be produced by replenishing the melt.

Seeding is a crucial step in the web growth, and the schematic in Figure 5.7 shows the details. Prior to growth, a thin dendrite is etched to remove any oxide and to generate a needlelike tip. The melt temperature is adjusted so that the seed does not melt, as depicted in Figure 5.7*a*. This temperature is referred to as the *hold temperature*. With the seed in contact with the melt, the temperature is lowered by a few degrees. As the temperature drops, a button spreads laterally from the seed along the <110> directions across the melt surface as shown in Figure 5.7*b*. The shape of the button is controlled partially by the temperature distribution in the melt, partially by atomic attachment kinetics at the solid-liquid interface, and partially by heat flow along the seed. When the button reaches the correct size, it is

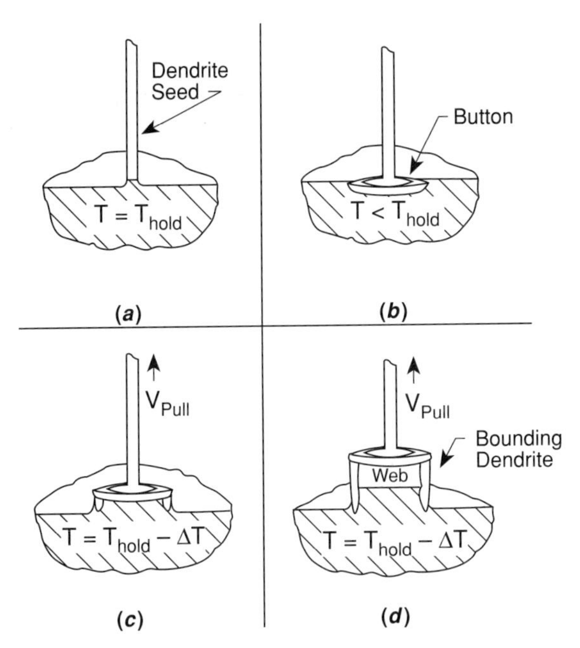

FIGURE 5.7
Schematic representation of the various stages in dendritic-web growth: (*a*) holding, (*b*) buttoning, (*c*) start of growth-dropping dendrites, and (*d*) growth.

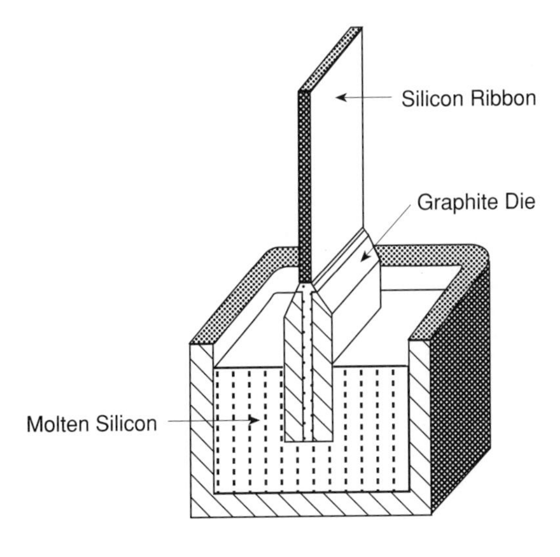

FIGURE 5.8
Schematic of "edge defined" meniscus in a ribbon die. (*After Wald [1981].*)

withdrawn from the melt. This withdrawal propagates two bounding dendrites into the melt as shown in Figure 5.7*c*. The molten Si forms a film between the dendrites. With the continued withdrawal of the button, a silicon web can be grown as shown in Figure 5.7*d*.

In contrast to the web-silicon process, a silicon film in the EFG process is shaped by a die fabricated from high-purity graphite, as shown in Figure 5.8.

Capillarity raises molten silicon in the die that shapes the melt into a film. If the shaped film is contacted with a ribbon seed and the temperature at the seed-melt interface is adjusted to cause solidification, a silicon ribbon can be grown by withdrawing the seed. Since in the EFG process molten silicon is directly in contact with a graphite die, silicon carbide particulates can form in the melt. Their subsequent incorporation into a ribbon leads to the generation of dislocations, thereby reducing the quality of the crystals.

5.2.2 Compound Semiconductors

The LEC process, a modified version of the CZ technique, is used extensively for growing III-V crystals. Its advantage is that circular substrates required for automated device processing can be produced from large diameter (~75 mm) cylindrical crystals without much loss of materials.

5.2.2.1 Liquid-encapsulated Czochralski process

The LEC technique prevents the decomposition of III-V melts due to the loss of group V species. It uses a cap of an inert liquid over the melt. This cap prevents decomposition of the melt as long as the external pressure provided by an inert gas, such as He or Ar, on the cap surface is higher than the vapor pressure of group V elements at the growth temperature.

Requirements for a suitable cap material are the following: (1) the material should float on the melt without mixing with it, (2) it should be chemically stable, (3) it should be optically transparent so that the seeding process can be monitored, and (4) it should be impervious to the out-diffusion of group V elements. Boric oxide meets these requirements. The cap material should also be free from moisture because it may be responsible for the "grappe" defects observed in InP crystals (Augustus and Stirland 1982).

Figure 5.9 shows an LEC setup used for the growth of InP crystals (Swaminathan and Macrander 1991). Comparing this figure with Figure 5.4 that shows a setup for growing CZ Si, two distinct differences are evident. First, the growth is carried out in a pressure vessel. The imposed pressure depends on the stability of the III-V compound. The more unstable the material, the higher the pressure. Second, molten boric oxide covers the melt surface.

As in the case of Si, crystals can be grown by partially melting the seed and then withdrawing it slowly from the melt contained in a boron nitride crucible. In the case of LEC, the seed as well as the melt are also rotated. The melt solidifies epitaxially on the seed, resulting in a single crystal. Since the crystal is pulled through a molten boric oxide cap, a thin layer of boric oxide adheres to the crystal surface and must be removed.

The setup shown in Figure 5.9 can also be used for *in situ* synthesis of InP (see section 5.1.2). Prior to crystal growth, phosphorus from the reservoir is bubbled through molten indium, resulting in the formation of InP. Using this approach, phos-

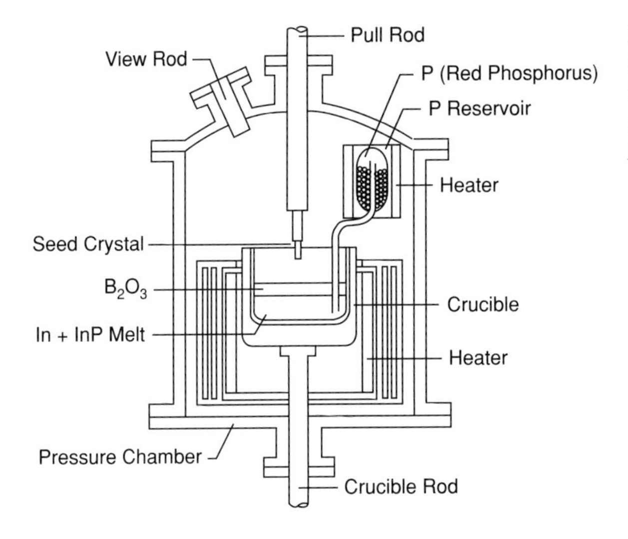

FIGURE 5.9
Schematic of an LEC setup for the growth of InP crystals. (*After Swaminathan and Macrander [1991].*)

phorus can be injected through the molten InP and In, resulting in rapid synthesis of InP.

We show in section 5.4 that dopants are inhomogeneously distributed in elemental and compound semiconductors grown by the CZ, FZ, and LEC processes. These inhomogeneities result from thermal convection in the melt, which can be reduced by applying a magnetic field during growth. The magnetic field increases the viscosity of the melt (Utech and Flemings 1966). For example, Terashima et al. (1984) decreased temperature fluctuations in GaAs melts from $\pm 18°C$ to $0.1°C$ at a magnetic field of 1.3 kGauss, resulting in reduced growth striations (see section 5.4) and reduction in dislocation densities (Kafalas 1985).

5.2.2.2 Horizontal and vertical Bridgman techniques

The schematic in Figure 5.10 shows a setup for the horizontal Bridgman technique. A polycrystalline charge and a seed crystal are placed in a long pyrolytic boron nitride boat. A portion of the seed and the charge are melted using resistive heating. The melt accesses the seed through a narrow constriction in the boat, whose function is to suppress the propagation of spurious orientations that may nucleate during seeding of the growth. The crystal growth is initiated at the seed–melt interface and is propagated along the length of the boat by displacing the imposed gradient in a controlled manner; therefore, this approach is also referred to as the *gradient-freeze technique*. The typical value of the imposed gradient is 10°C/cm that is displaced at a rate of 5 mm/sec.

The As reservoir in Figure 5.10 provides the stoichiometry control. The temperature of the As source zone is such that the activity of arsenic over the liquid arsenic in the source zone is equal to that of arsenic over the melt. Parsey et al. (1983) have examined the effects of the As reservoir temperature on the perfection of GaAs crystals, and their results are reproduced as Figure 5.11. The two activities of

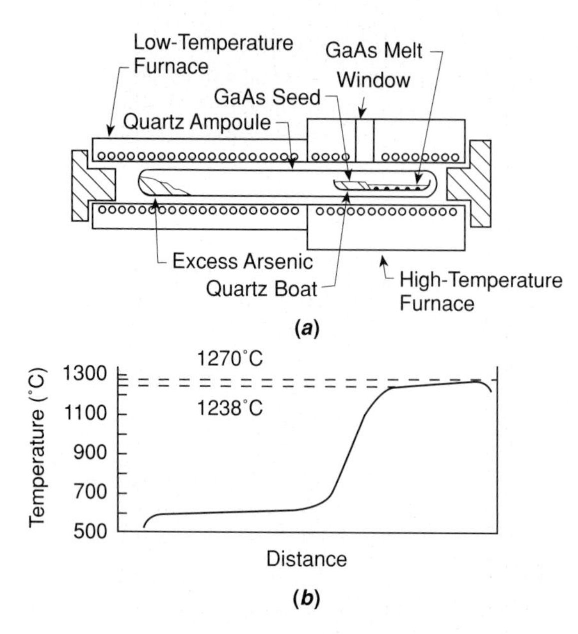

FIGURE 5.10
Schematic of the horizontal Bridgman setup used for the growth of GaAs crystals.

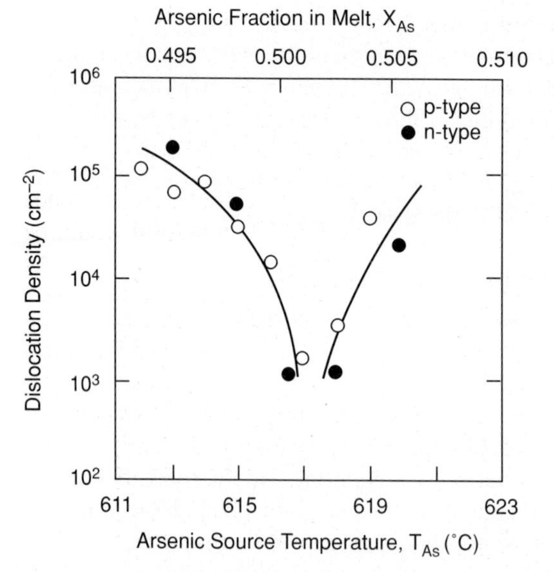

FIGURE 5.11
Variations in dislocation densities in n- and p-GaAs crystals grown under different As overpressures by the horizontal Bridgman technique. (*After Parsey et al. [1983].*)

arsenic are balanced when the source is at 614°C. The crystals grown under this condition have low dislocation densities, whereas deviations on either side of the equilibrium lead to the formation of dislocations. We discuss the origin of point-defect-induced dislocations in as-grown crystals in section 5.5.

The as-grown crystals are D shaped because the melt is flat at the top. The fabrication of (001) circular wafers requires machining, resulting in an excessive material loss in comparison to the LEC technique. On the other hand, since thermal gradients in this growth technique are lower than those in the LEC process, the Bridgman crystals have a lower dislocation density than the LEC crystals. The typical dislocation densities in D-shape GaAs crystals, 5 to 7.5 cm wide, are in the range of 3×10^3 cm^{-2} to $\sim 10^4$ cm^{-2}, whereas the dislocation densities in the LEC GaAs crystals are within the 5×10^4 cm^{-2} to $\sim 10^5$ cm^{-2} range.

In the vertical Bridgman technique, a seed is positioned at the bottom of a quartz or a pyrolytic boron nitride crucible and is covered with a polycrystalline charge. The temperature in the surrounding furnace is adjusted so as to melt the charge and a portion of the seed. Subsequently, the crucible is lowered slowly, resulting in an epitaxial solidification of the melt on the seed surface. Instead of moving the crucible, the furnace can also be moved relative to the crucible.

In the vertical gradient-freeze (VGF) technique, neither the furnace nor the crucible are moved. Instead, the imposed gradient is displaced along the length of the crucible. One of the main advantages of this technique is that the axial and radial temperature gradients are considerably lower, thereby resulting in reduced convective flow and thermal stresses. Crystals 5 to 7.5 cm in diameter and having a low dislocation density, $\sim 10^2$ to 3×10^3 cm^{-2}, can be grown. Reproducible growth of large-diameter, low-dislocation-density crystals has been demonstrated using a VGF setup (Gault et al. 1986).

5.2.2.3 Effects of existence region on stoichiometry of crystals

In binary III-V systems, stoichiometric AB compounds do not have the highest melting points, where A and B represent group III and V species, respectively. Instead, the first to solidify composition could be rich in either A or B. Retrograde solubility of A or B in the solid also exists, resulting in precipitates during the cool down. We emphasize these points by referring to Figure 5.12 that shows the existence region in the gallium-arsenic system. In this case the first to form solid would be arsenic rich. As a result of the retrograde solubility of As in GaAs, the excess As atoms should come out of solution during the cool down. These atoms could cluster together to form liquid droplets at high temperatures. Of course, the droplets would convert into As precipitates at lower temperatures. Cullis et al. (1980) have seen such precipitates on dislocations in LEC GaAs crystals (see Figure 5.13). Precipitates tend to nucleate heterogeneously on dislocations because of positive interactions between the respective elastic-strain fields.

EXAMPLE 5.1. Calculate the number of arsenic precipitates/(cm)^3 that would form in GaAs when a nonstoichiometry of 5×10^{17} cm^{-3} is eliminated. Assume that the precipitates are spherical in shape, their radius is 5 nm, and the tetrahedral radius of an arsenic atom is 0.118 nm.

Solution

$$\text{Volume of the precipitate} = \frac{4}{3}\pi(5 \times 10^{-9})^3 \text{ (nm)}^3.$$

$$\text{Volume of the arsenic atom} = \frac{4}{3}\pi(0.118 \times 10^{-9})^3 \text{ (nm)}^3.$$

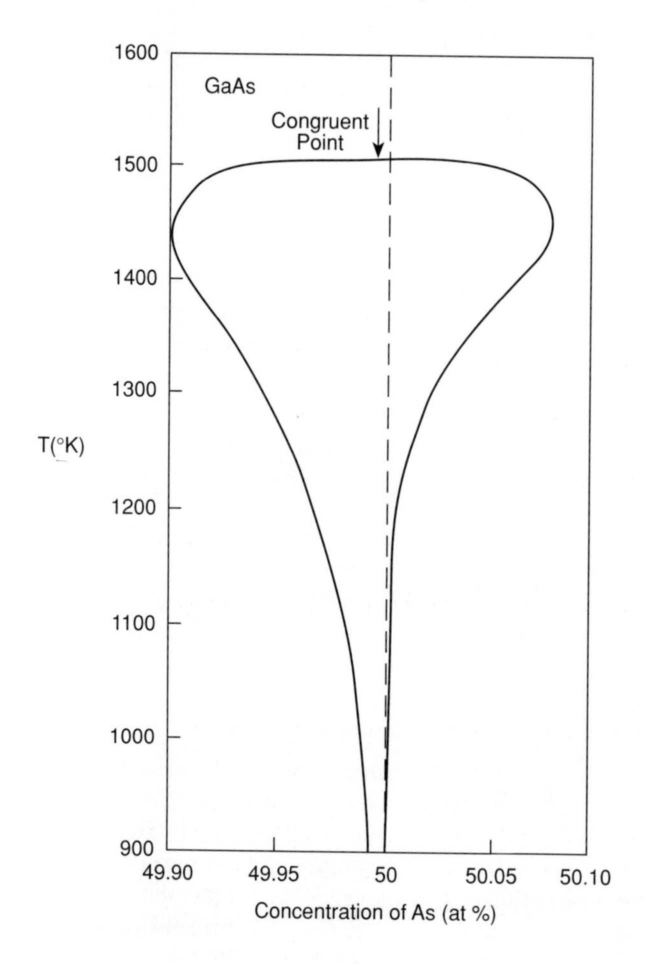

FIGURE 5.12
Existence region in the Ga-As system.

$$\text{Number of arsenic atoms in a precipitate} = \left(\frac{5}{0.118}\right)^3.$$

$$= 7.61 \times 10^4.$$

$$\text{Number of precipitates/cm}^{-3} = \frac{5 \times 10^{17}}{7.61 \times 10^4}$$

$$= 6.57 \times 10^{12}\ \text{cm}^{-3}.$$

5.2.2.4 Growth of semi-insulating crystals

We now illustrate that by achieving a suitable balance between point defects and impurity levels in III-V crystals, we can grow semi-insulating, that is, very high resistivity, crystals that are used in the fabrication of field-effect transistors. A number of workers have shown that an electron-trap level at 0.82 eV below the conduction-band edge in GaAs, commonly labeled as EL2, plays a role in semi-insulation. EL2s

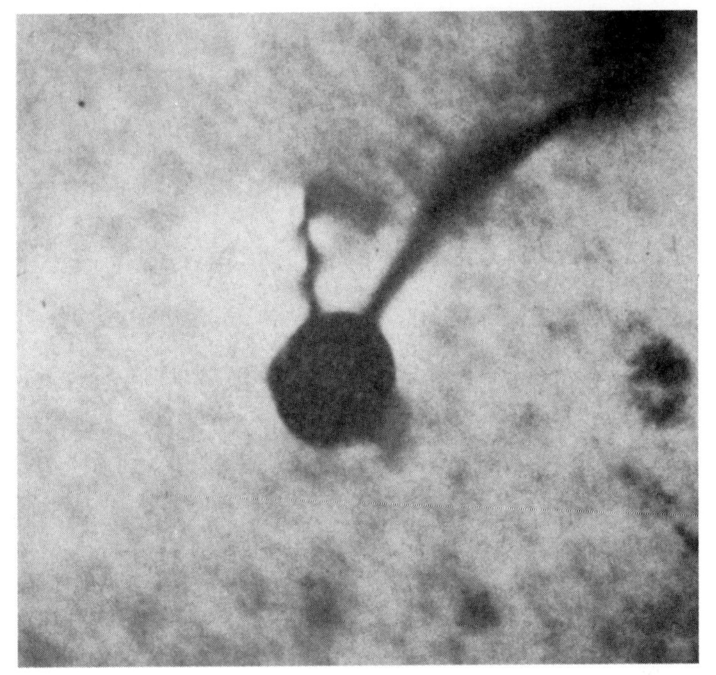

(*a*)

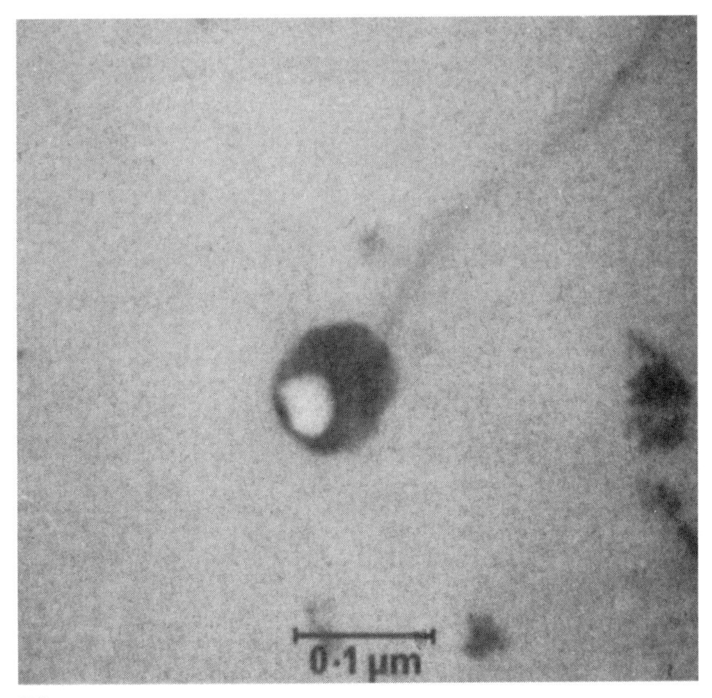

(*b*)

FIGURE 5.13
Micrograph showing As precipitates on dislocations in LEC GaAs crystals. (*After Cullis et al. [1980]*.)

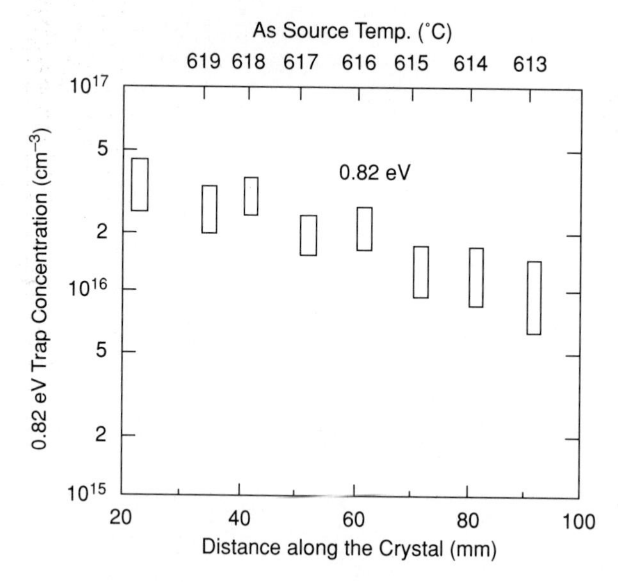

FIGURE 5.14
Dependence of concentration of EL2 level on the arsenic source temperature. (*After Lagowski et al. [1982].*)

are regarded as antisite defects in which an As atom replaces a Ga atom on the group III sublattice, that is, As_{Ga}.

Lagowski et al. (1982) have systematically investigated the influence of As source temperature, that is, stoichiometry, on the concentration of arsenic atoms on gallium sites (As_{Ga}) in crystals grown by the Bridgman technique. Figure 5.14 shows the dependence of the concentration of EL2 level on the arsenic source temperature, which in turn affects the concentration of As in the melt. At higher source temperature the melt is arsenic-rich and the concentration of EL2 is also higher. High-resistivity crystals can also be produced by doping with deep impurities. For example, semi-insulating GaAs and InP crystals have been grown by doping with Cr and Fe, respectively.

5.3 SOURCES OF DISLOCATIONS IN AS-GROWN CRYSTALS AND PERFECTION ENHANCEMENT

We showed in Chapter 3 that dislocations have deleterious effects on the electronic properties of semiconductors. These effects are also manifested in devices. Figure 5.15 shows reverse-biased I-V characteristics of p-n junctions in silicon as a function of the dislocation density (Ravi 1981). The leakage current at a given bias increases with the increasing dislocation density in the device. The breakdown voltage also tends toward a lower value with the increase in dislocation density, although the change is gradual. These effects could be due to two sources: (1) dislocations act as carrier-generation centers; and (2) dislocations may be

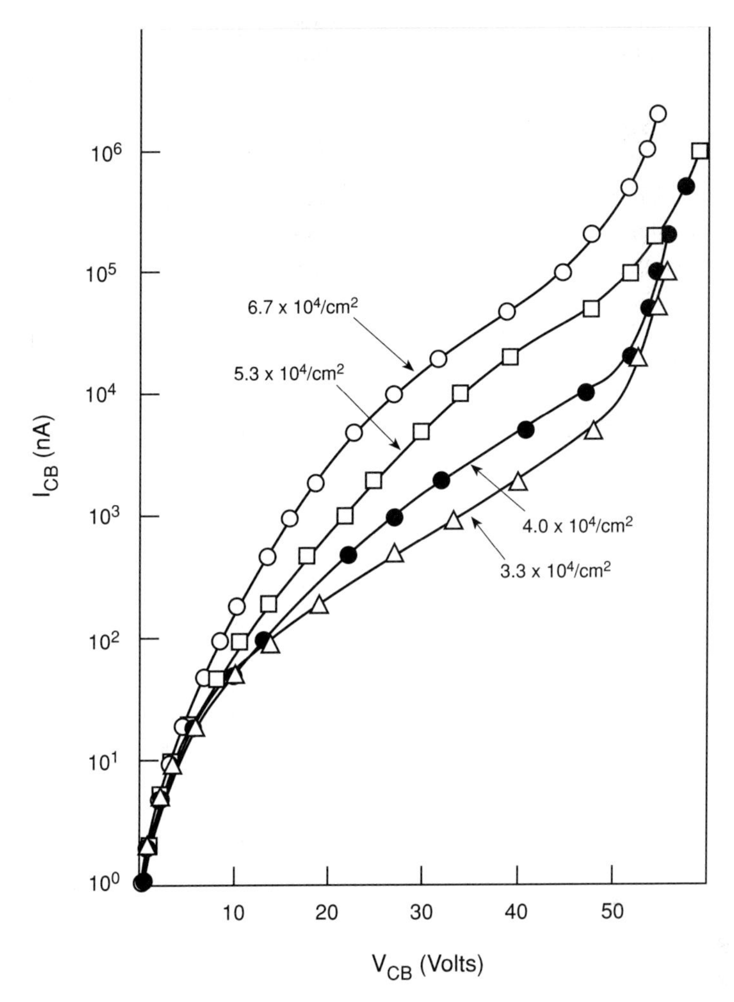

FIGURE 5.15
Reverse I-V characteristics of p-n junctions as a function of dislocation density. (*After Ravi [1981].*)

decorated with impurities, and when they thread through the junction, they may short it.

The reverse-bias I-V characteristics of InP diodes containing different dislocation densities fabricated on Fe and S-doped substrates (Beam et al. 1992) are shown in Figures 5.16*a* and 5.16*b*, respectively. The diodes were produced by the diffusion of Cd into n-type InP layers. When the dislocation density is increased from $2.2 \times 10^7\,\text{cm}^{-2}$ to $1.5 \times 10^8\,\text{cm}^{-2}$, the breakdown voltage is substantially reduced; and at a given reverse bias, the leakage current dramatically increases. Since these diodes were fabricated by diffusion, Beam et al. (1992) envisage that space-charge

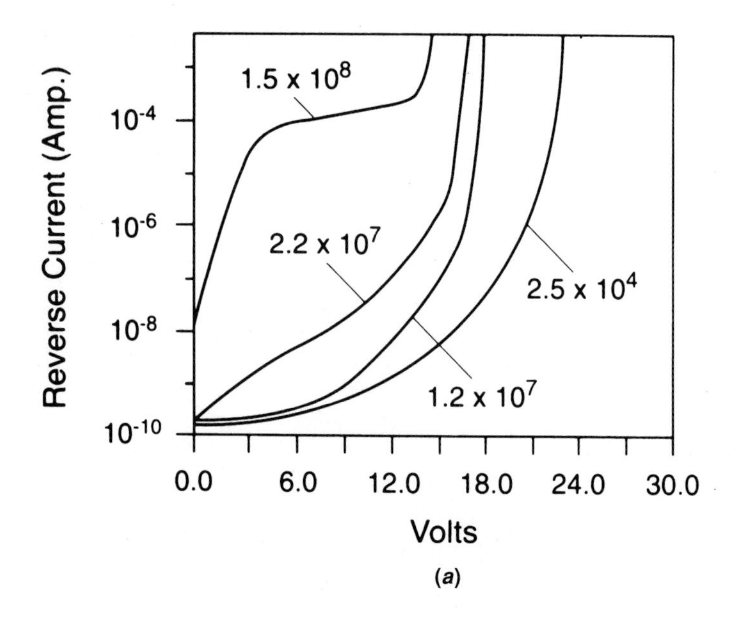

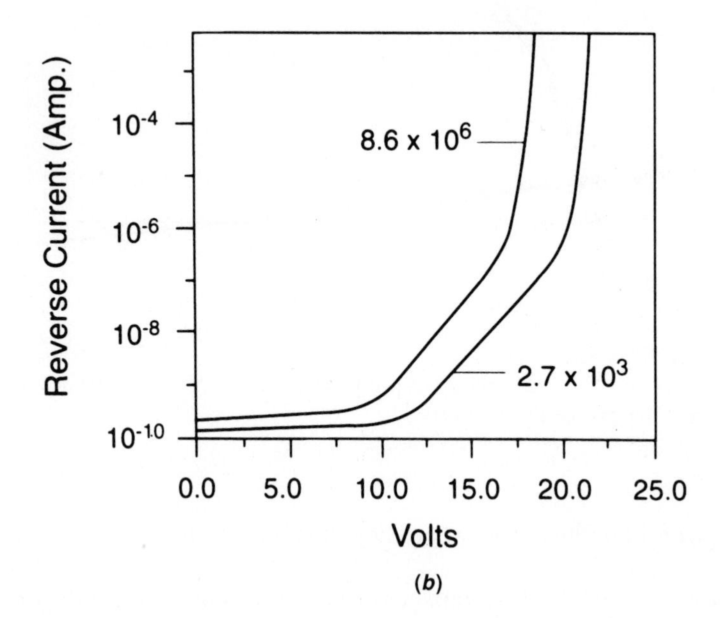

FIGURE 5.16

Reverse I-V characteristics of InP diodes containing different dislocation densities fabricated on (*a*) Fe-doped and (*b*) S-doped substrates. (*After Beam et al. [1992].*)

cylinders develop around the dislocations. At higher dislocation densities these cylinders may begin to overlap and thereby produce high-leakage currents. These two examples illustrate the need for crystals having low dislocation densities.

Dislocations in crystals could originate from three different sources: (1) dislocations present in the seed crystal could propagate into a growing crystal, (2) point defects could cluster to form dislocation loops during the cool down, and (3) under the influence of thermal-gradient-induced stresses, dislocations present in the peripheral regions of a growing crystal could propagate into the interior of the crystal.

Let us consider a situation in which dislocations of different Burgers vectors terminate at the seed surface and melt solidifies on it. Two distinct possibilities exist: (1) Burgers vectors of the dislocations are inclined to the seed surface, and (2) Burgers vectors of the dislocations are parallel to the seed surface. When the melt freezes epitaxially, dislocations in the first case will be replicated into the newly formed crystal because of the presence of growth-spirals steps at the emergence points of dislocations. In the second case Beam et al. (1990) have shown that dislocations will also be replicated because protrusions and depressions are produced at the surface where dislocations emerge. Therefore, all the dislocations present in the seed will be incorporated into the as-grown crystal. The cold seed is also thermally shocked when it is dipped into the melt, which leads to dislocation multiplication. Therefore, we need two approaches to prevent the propagation of dislocations from the seed and to reduce thermal shock.

The two current approaches for the growth of high-quality crystals are based on the original ideas of Dash (1958, 1959). The use of small-diameter seeds reduces the magnitude of the thermal stresses because the radial temperature gradient is shallower for small seeds. The total number of dislocations in a smaller seed are also less than those in a larger seed.

Dash showed that forming a neck by an initial rapid pull can prevent the propagation of dislocations from the seed into the crystal. Replicated dislocations then have a larger surface area to terminate on, as shown in Figure 5.17. Figure 5.17*a* shows three dislocations in a seed crystal. In the presence of a neck, only the central dislocation can propagate into the as-grown crystal, Figure 5.17*b*; the other two terminate at free surfaces and are eliminated. After forming the neck, the melt temperature is reduced and stabilized so that the desired ingot diameter can be produced. This procedure is referred to as the *shouldering-out step*. The desired diameter is maintained subsequently by precisely monitoring the pull rate.

The central dislocation in Figure 5.17*b* can also be eliminated if the dislocation could climb out of the crystal. To enhance the probability of climb, Dash suggested that initially a crystal should be grown at a faster rate to incorporate nonequilibrium concentrations of point defects. These point defects should facilitate climb.

Bonner (1981) concluded that to avoid the formation of twins in <111>-oriented InP crystals, the surface produced during the shouldering-out step should make an angle with the growth axis that is smaller than 19°. This conclusion is not well understood. Bonner suggested that for such a surface, {111} planes on which

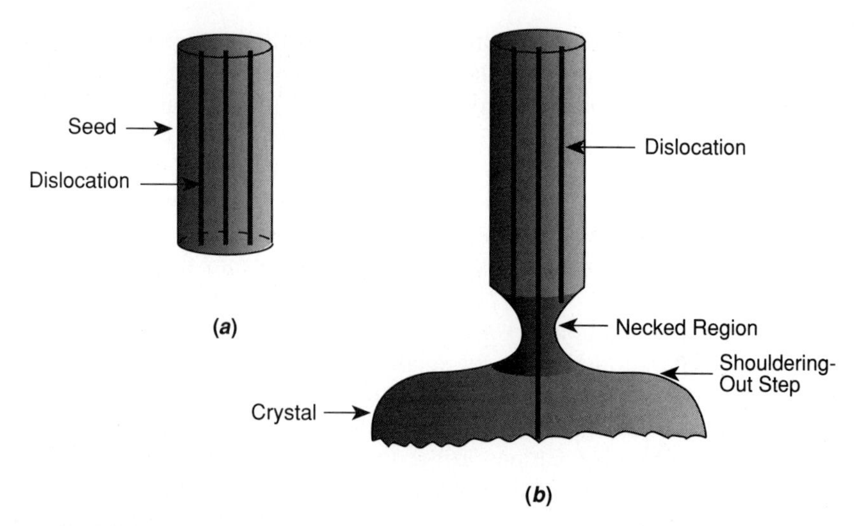

FIGURE 5.17
(*a*) Schematics showing dislocations in the seed crystal and (*b*) elimination of some of the dislocations by the formation of a necked region.

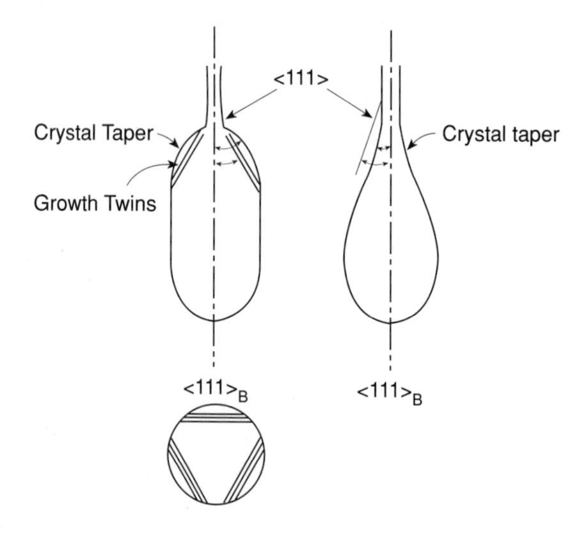

FIGURE 5.18
InP crystals having different tapers during the shouldering-out process. Note the presence of twins in crystals in which the angle between the crystal shoulder and the <111> growth axis is greater than ~19°. (*After Bonner [1981].*)

twins form are not exposed during growth. The dramatic effect of the taper angle is shown schematically for LEC InP crystals in Figure 5.18. (Bonner 1981).

The second source of dislocations is the excess concentration of point defects that is incorporated into a crystal that is grown at a very high temperature. Clustering of excess point defects into faulted dislocation loops during the cool down can eliminate supersaturation. We discuss this issue further in section 5.5. The driving force for the formation of these loops is the overall reduction in the energy of the system. For the sake of discussion, let us evaluate the energy change ΔE associated

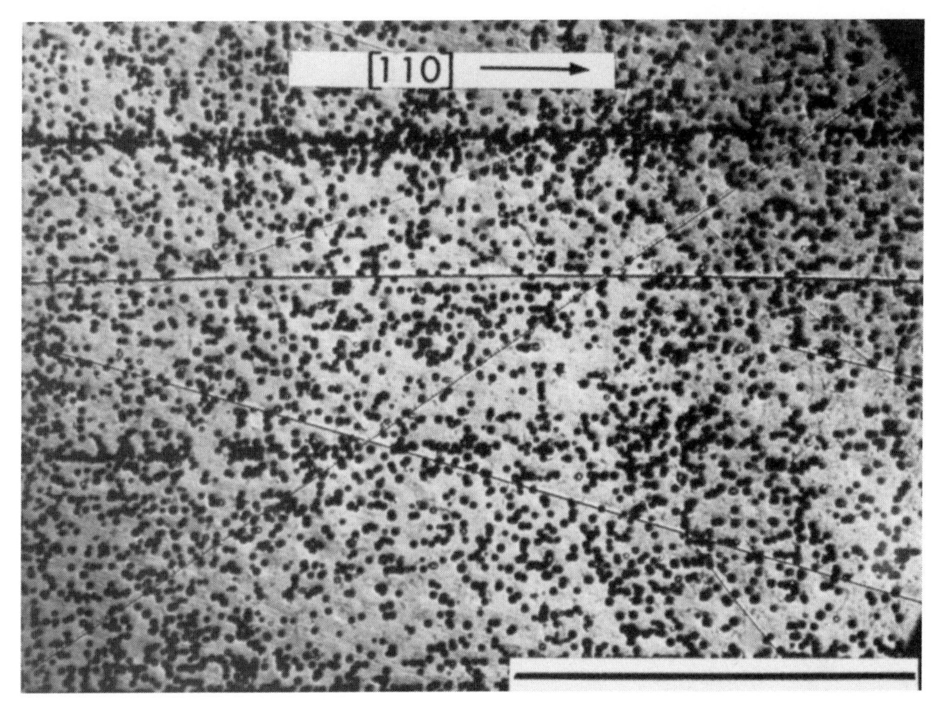

FIGURE 5.19
Typical dislocation etch-pit distribution observed on a (001) slice of highly dislocated sulphur-doped indium phosphide wafer. Marker represents 0.1 mm. (*After Mahajan [1989].*)

with clustering of n interstitials into a dislocation loop of radius R. We can show that ΔE is given by

$$\Delta E = 4\pi n\, r^2 \gamma - 2\pi R^2\, E_{SFE} - \pi RGb^2, \tag{5.7}$$

where r is the radius of the interstitial, γ is the surface energy of an interstitial, E_{SFE} is the stacking-fault energy of the crystal, G is the shear modulus, and b is the Burgers vector of the dislocation that bounds the loop. Since a large number of interstitials are involved in the formation of a loop, the $n\gamma$ term in Equation (5.7) is fairly large. Therefore, substantial savings in the internal energy of the crystal can occur if interstitials cluster together to form loops. When the term $4\pi r^2 n\gamma$ is greater than $2\pi R^2 E_{SFE} + \pi RGb^2$, point defects will cluster together to form loops. The driving force for clustering is the overall reduction in the surface energy of point defects.

Etchings of {001} and {111} substrates to reveal dislocations show characteristic etch-pit patterns. Figure 5.19 shows the etch-pit distribution observed on the (001) plane of a highly dislocated InP crystal (Mahajan 1989). The dislocation density is of the order of $2 \times 10^5\ \mathrm{cm}^{-2}$, and the etch pits appear to be aligned along the [110] and $[1\bar{1}0]$ directions. Similar etch-pit alignments have been observed in GaAs

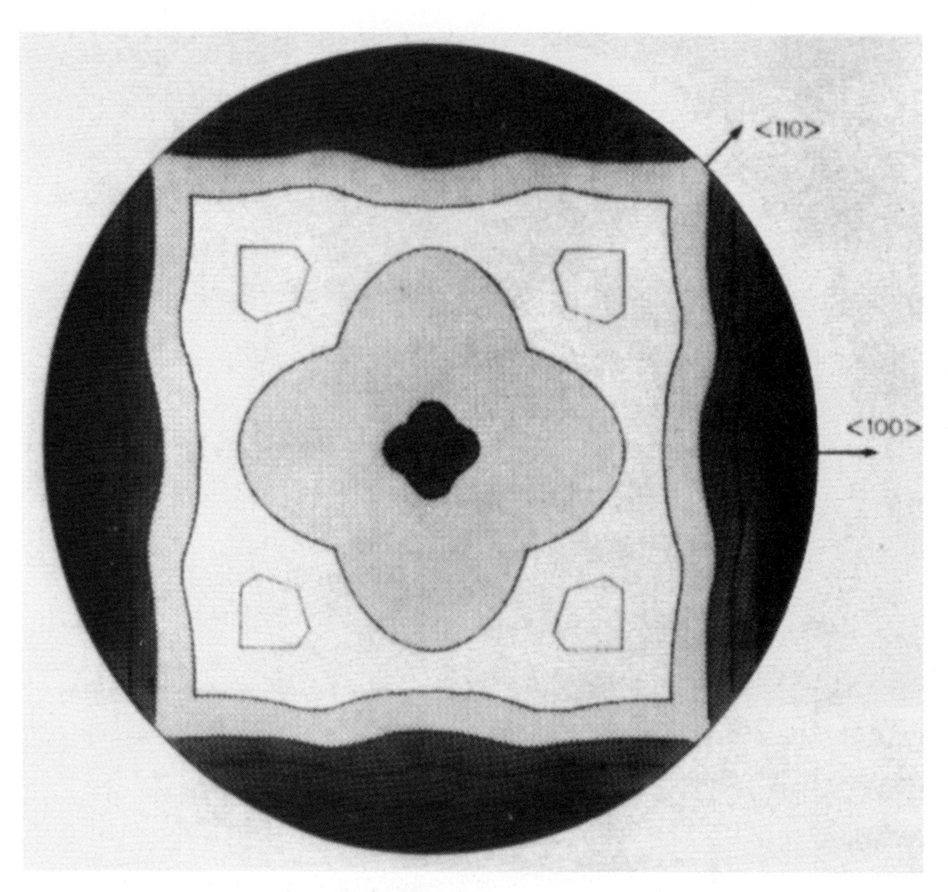

FIGURE 5.20
Dislocation-density contour lines for the top wafer of a (001) GaAs boule computed by Jordan et al. (1980).

(Jordan et al. 1980) and GaP (Nygren 1973) crystals grown by the LEC technique. The observed pit alignments imply that dislocations lie on their {111} glide planes and are slip induced.

We discussed in Chapter 3 that the slip systems in the diamond-cubic and zinc-blende structures are a/2 <1$\bar{1}$0> {111}. Assuming that slip is caused by thermal-gradient-induced stresses, Jordan et al. (1980) have computed dislocation-density contour lines for the top wafer of a <001> GaAs boule grown by the LEC technique, and their results are reproduced as Figure 5.20; the darker the area, the higher the dislocation density. The dislocation distribution exhibits fourfold symmetry. The crowding of the contours near the periphery of the crystal implies a sharp gradient in dislocation density. A photomacrograph obtained from a (001) wafer cut close to the top end of the tellurium-doped (001) GaAs boule, grown by the LEC technique, is shown in Figure 5.21 (Jordan et al. 1980). The key features of the dislocation distribution in Figure 5.21 are fourfold symmetry, maximum density at the

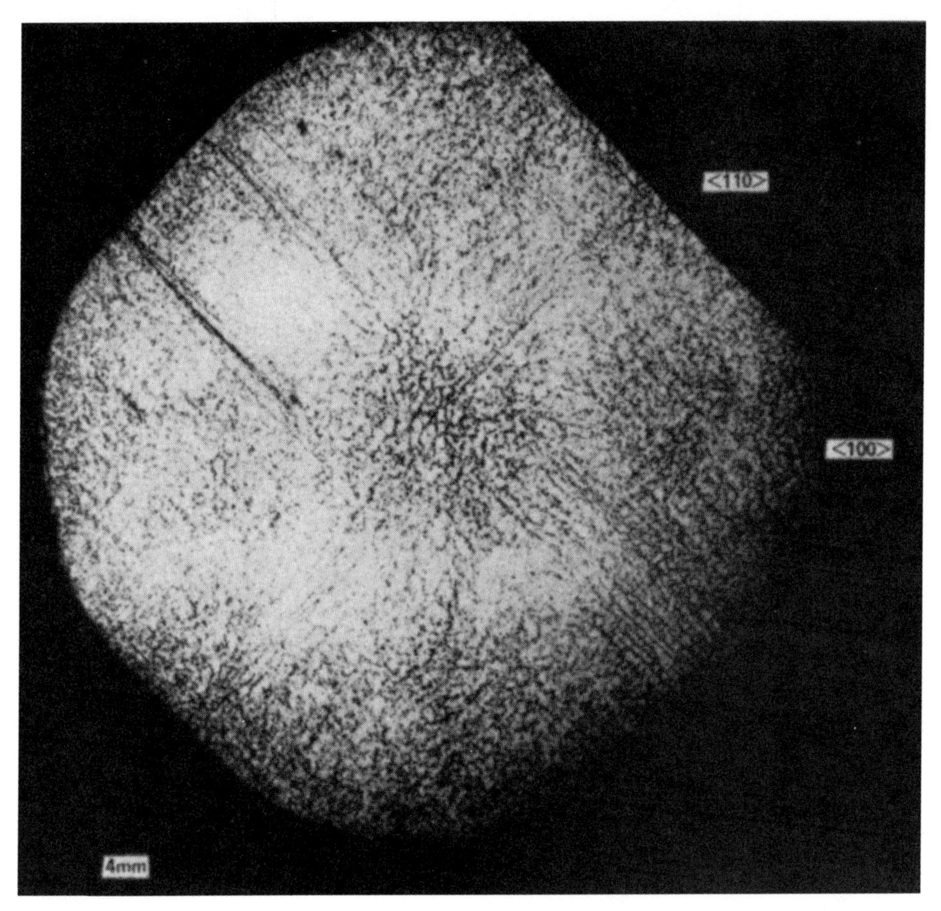

FIGURE 5.21
Macrophotograph obtained from a KOH-etched (001) GaAs wafer. The wafer was cut from the top end of a tellurium-doped GaAs boule grown by the LEC technique. (*After Jordan et al. [1980].*)

<100> edge, minimum density midway between the center and the <110> edge, and intermediate densities at the center and the <110> edge. The results in Figure 5.20 and 5.21 are nearly identical.

We have argued that the existence of thermal-gradient-induced stresses introduces dislocations in as-grown crystals. This problem is more acute in the case of compound semiconductors because they are softer than Si (Mahajan et al. 1979; Brasen and Bonner 1983; Rai et al. 1991). One approach to grow crystals with lower dislocation densities is to reduce the magnitudes of the thermoelastic stresses to levels where $a/2 \langle 1\bar{1}0 \rangle \{111\}$ slips are not activated. This stress level is very difficult to achieve because yield stresses of compound semiconductors at high temperatures are very low (Brasen and Bonner 1983; Cockayne et al. 1983; Rai et al. 1991). So the accepted approach is to strengthen the matrix by the addition of impurities (Seki et al. 1976, 1978; Cockayne et al. 1983; Rai et al. 1989, 1991). These impurities can be either isoelectronic in nature or act as dopants in the host lattice.

When an impurity is added to a tetrahedrally coordinated crystal, the bond lengths in the tetrahedra either increase or decrease, depending on the covalent tetrahedral radius of the impurity; tetrahedral covalent radii of different atoms are listed in Table 3.1. As an illustration, let us consider the substitution of some of the gallium atoms by the indium atoms on group III sublattice of GaAs. Ehrenreich and Hirth (1985) visualize that this substitution would lead to In(4) As tetrahedral units, that is, four indium atoms are bonded to an arsenic atom, within the GaAs matrix. The concentration of the units is linearly related to the concentration of In. Since the In-As bond length is greater than the Ga-As bond length, the volume of the In(4)As tetrahedral unit is larger than that of the Ga(4)As unit. The formation of In(4)As units produces distortions in the lattice. These distortions strengthen the lattice and thus impede the motion of dislocations. This strengthening reduces the probability of dislocation multiplication under the influence of thermal-gradient-induced stresses, yielding more perfect crystals.

The situation is much more complicated when an impurity can also act as a dopant in the host lattice. In addition to the size effect discussed in this section, electronic effects come into play. The ionized dopant atoms could interact with charged-point defects and thus prevent them from clustering into potential dislocation sources. The work of deKock et al. (1979) on microdefect formation in macroscopically dislocation-free CZ silicon indicates that the point defect–dopant interaction is electronic in nature and does not depend on the size of the dopant atom. A similar situation could exist in compound semiconductors.

EXAMPLE 5.2. A growing Si crystal experiences a tensile stress along the [123] direction. Determine the slip system along which the crystal will deform. Also, comment on the etch-pit patterns that appear on the $(11\bar{1})$ and (101) slices obtained from the as-grown crystal.

Solution. Deformation always occurs first on a slip system that carries the highest resolved shear stress. Assuming that the applied stress is σ, the resolved shear stress (τ) can be determined from the following expression:

$$\tau = \sigma \cos\theta \cos\phi,$$

where θ and ϕ are, respectively, the angles that the stress axis makes with the slip direction and slip plane normal. For a given σ, slip will occur on that system for which $\cos\theta$ $\cos\phi$ is the highest.

Various slip systems are listed in Chapter 3. For illustrative purposes let us calculate $\cos\theta$ $\cos\phi$ for the [101] $(\bar{1}11)$ system.

[101] $(\bar{1}11)$ Slip System

Cosine of the angle between two vectors $[u_1, v_1, w_1]$ and $[u_2, v_2, w_2]$ is given by

$$\cos\alpha = \frac{u_1u_2 + v_1v_2 + w_1w_2}{\sqrt{u_1^2 + v_1^2 + w_1^2}\sqrt{u_2^2 + v_2^2 + w_2^2}}.$$

Applying the above relation to calculate cos θ and cos ϕ, we have

$$\cos\theta = \cos[101] < [123] = \frac{1 + 0 \times 2 + 1 \times 3}{\sqrt{2}\,\sqrt{14}} = \frac{4}{\sqrt{2}\,\sqrt{14}}$$

$$\cos\phi = \cos[\bar{1}11] < [123] = \frac{-1 + 1 \times 2 + 1 \times 3}{\sqrt{3}\,\sqrt{14}} = \frac{4}{\sqrt{3}\,\sqrt{14}}.$$

$$\text{Therefore, } \cos\theta\cos\phi = \frac{16}{14\sqrt{6}}.$$

Similarily, we can calculate values of cos θ cos ϕ for other slip systems and show that when stress is applied along the [123] direction, cos θ cos ϕ is highest for the [101] $(\bar{1}11)$ system. Therefore, deformation will occur on this system during crystal growth.

$(11\bar{1})$ Slice

Due to the [101] $(\bar{1}11)$ slip, dislocations lie on the $(\bar{1}11)$ planes and will intersect the $(11\bar{1})$ surface as shown below:

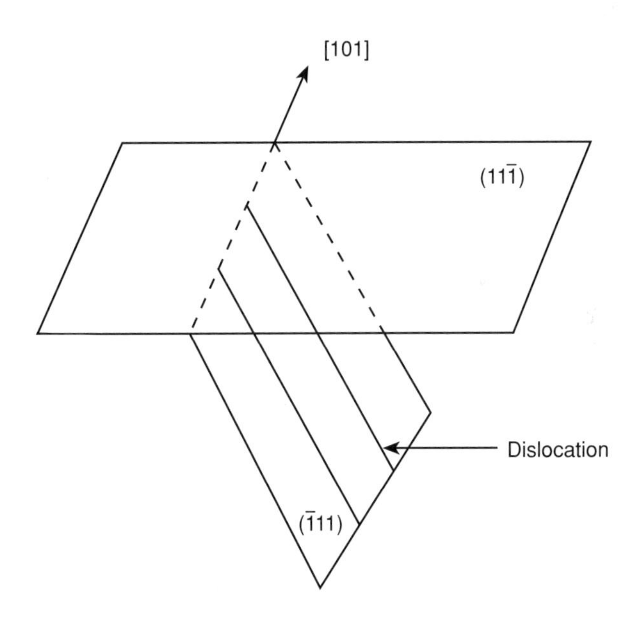

Since the Burgers vector does not have a component normal to the $(11\bar{1})$ plane, that is, the plane of the observation, etch pits may not form at the emergence points of dislocations at this surface.

(101) Slice

The activated slip plane intersects the (101) slice along the $[12\bar{1}]$ direction. Therefore, the endpoints of dislocations terminating on the (101) surface will be aligned along this direction as shown below:

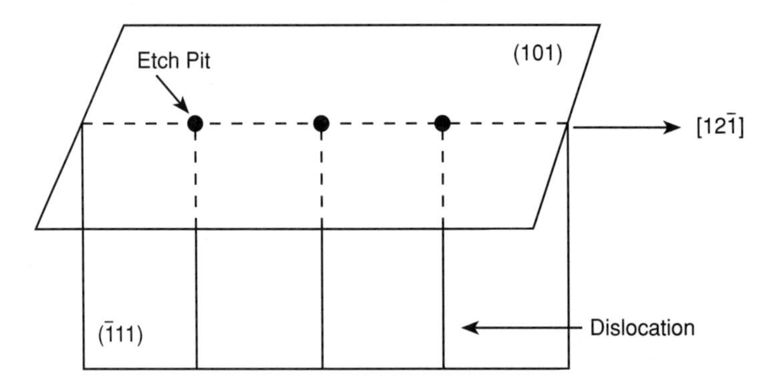

5.4 DOPING IN THE MELT

For devices we need crystals that have p- and n-type conductivities and are doped to various levels. We can achieve these characteristics by growing crystals from doped melts. However, several factors complicate the incorporation of dopants into as-grown crystals. First, fluid-flow conditions in a growth system are complicated (Brown et al. 1989). Second, generally the solubility of an impurity in a melt is different from that in a solid, causing continuous changes in the composition of the melt. Third, the externally imposed growth rate may be different from the local growth rate because of thermal conditions in the growth system. Fourth, the solid–liquid interface may not be planar and has a curvature. We discuss these issues in this section.

5.4.1 Underlying Concepts

This section introduces the concepts for understanding the partitioning of solutes between the melt and the solid during solidification. We examine how the local solidification rate affects partitioning.

5.4.1.1 Equilibrium and effective distribution coefficients of dopants

Crystals having desired conductivity as well as specific carrier concentrations can be grown from melts containing appropriate dopants. Even for very high carrier concentrations, we need to add extremely small amounts of dopants. We do so by preparing master alloys and adding them to the charge. Generally, under equilibrium conditions the concentrations of solute in the solid and the liquid are different because of different solubilities. A distribution coefficient under equilibrium conditions k_{eq} is defined as

$$k_{eq} = \frac{\text{concentration of solute in the solid}}{\text{concentration of solute in the liquid}} = \frac{C_s}{C_L} \tag{5.8}$$

Equation (5.8) assumes that the crystal is growing slowly so that the equilibrium is maintained during growth. Possible values of k_{eq} are greater than one, less than one, or equal to one. The k_{eq} values for different impurities in silicon are listed in Table 5.1 (Huff 1992). Except for oxygen, k_{eq} is less than one for all other impurities listed in the table; the highest and lowest values of k_{eq} are for B and Mo, respectively.

We envisage that the solute concentration near a solid-liquid interface may vary as shown in the schematic in Figure 5.22 for k_{eq} less than one. In Figure 5.22*a* the crystal is being grown under equilibrium conditions, and the melt is large. When the melt solidifies, the solute atoms are rejected by the solid into the melt. We assume that these solute atoms have a sufficient time to diffuse away from the solid-liquid

TABLE 5.1
Equilibrium distribution coefficients of various impurities in silicon

Elements	Group	Equilibrium distribution coefficient
Li	IA	1×10^{-2}
Cu	IB	1×10^{-4}
Ag	IB	$\sim 1 \times 10^{-6}$
Au	IB	2.5×10^{-5}
Zn	II B	1×10^{-5}
Cd	II B	1×10^{-6}
B	III B	$\sim 8 \times 10^{-1}$
Al	III B	$\sim 2 \times 10^{-3}$
Ga	III B	8×10^{-3}
In	III B	4×10^{-4}
Tl	III B	1.7×10^{-4}
C	IV B	$7 \pm 1 \times 10^{-2}$
Ge	IV B	3.3×10^{-2}
Sn	IV B	1.6×10^{-2}
N	VB	7×10^{-4}
P	VB	3.5×10^{-1}
As	VB	3×10^{-1}
Sb	VB	2.3×10^{-2}
Bi	VB	7×10^{-4}
O	VI B	1.4 ± 0.3
S	VI B	$\sim 1 \times 10^{-5}$
Cr	VI A	1.1×10^{-5}
Ti	IV A	2×10^{-6}
V	VA	4×10^{-6}
Mn	VII A	$\sim 1 \times 10^{-5}$
Fe	VIII	8×10^{-6}
Co	VIII	8×10^{-6}
Ni	VIII	3×10^{-5}
Mo	VI A	4.5×10^{-8}
Ta Pt	V A VIII	1×10^{-7}

Source: Huff 1992.

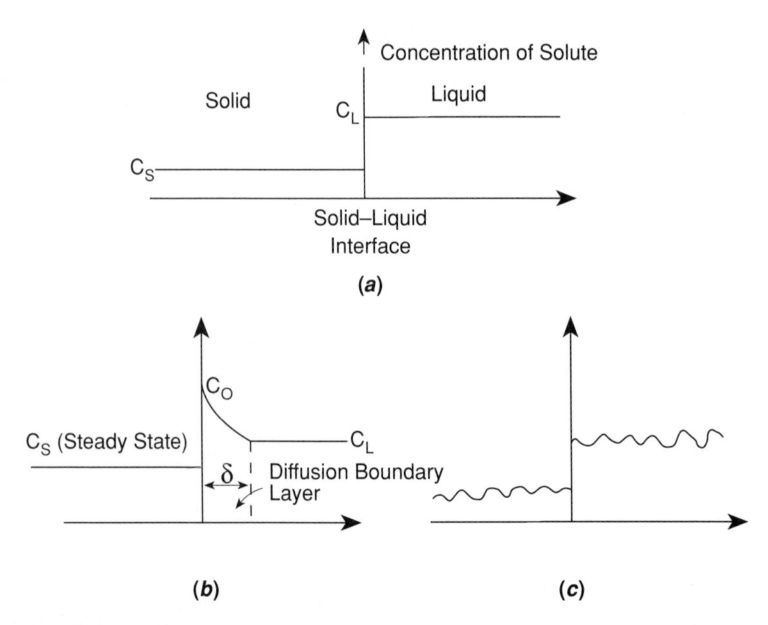

FIGURE 5.22
Possible situations prevailing at the solid-liquid interface during crystal growth: (*a*) equilibrium growth, (*b*) finite growth rate and $k_{eq} < 1$, and (*c*) periodic fluctuations.

interface. Since the melt is large, the rejection of the solute atoms into the melt does not change the solute concentration in the melt, that is, C_L.

The situation in Figure 5.22*b* prevails during the growth of a crystal under steady-state conditions at a finite growth rate. The rejected solute atoms do not have a sufficient time to diffuse away from the solid-liquid interface, with the result that their concentration builds up at the growth interface to C_O, which is greater than the concentration of solute in the liquid (C_L) in Figure 5.22*a*. Assuming that the concentration of the solute in the solid varies linearly with that in the liquid, the amount of the solute incorporated into the solid, that is, C_S, will be greater than that incorporated under the equilibrium conditions in Figure 5.22*a*. We can describe this situation using an effective distribution coefficient, k_{eff}, where

$$k_{eff} = \frac{C_S \text{ (under nonequilibrium conditions)}}{C_L}. \tag{5.9}$$

When k_{eq} is less than one, k_{eff} is always greater than k_{eq}.

We can show from fluid-flow considerations that a boundary layer of thickness δ must exist at the solid-liquid interface shown in Figure 5.22*b*. The solute atoms rejected from the solid must diffuse through this layer to reach the liquid. This diffusion occurs because of the difference in concentrations of solutes at the interface and in the melt.

The effective distribution coefficients of various impurities in GaAs and InP are listed in Tables 5.2 and 5.3, respectively (Swaminathan and Macrander 1991). For some of the impurities, the free-electron distribution is given assuming that C_S is equal to the free-electron concentration at room temperature.

TABLE 5.2
Effective distribution coefficients of impurities in GaAs

	Effective distribution coefficients	
Impurity	**LEC growth**	**Non-LEC growth**
Be		3
Cr	1.03×10^{-3}	5.7×10^{-4}
Ge	2.8×10^{-3}	1.0×10^{-2}
Si	1.85×10^{-2}	1.4×10^{-1}
Sn	5.2×10^{-3}	8.0×10^{-2}
Se	5.0×10^{-2}	3.0×10^{-1}
Te	6.8×10^{-2}	5.9×10^{-2}
	3.0×10^{-2}	2.8×10^{-2}
Fe		1.0×10^{-3}
Zn		4.0×10^{-1}
In	0.10	
Mn		0.02
Co		4.0×10^{-4}
Ni		4.0×10^{-3}
O		0.3
Sb		1.6×10^{-2}
Pb		$<1.0 \times 10^{-3}$

Source: Swaminathan and Macrander 1991.

TABLE 5.3
Effective distribution coefficients of impurities in InP

	Effective distribution coefficients	
Impurity	**LEC growth**	**Non-LEC growth**
Co	4×10^{-5}	
Cr	$(1 - 6) \times 10^{-4}$	
Fe	1.6×10^{-3}	
Mn	0.4	
Ge	2.4×10^{-2}	5×10^{-3} (liquid phase epitaxy)
S	0.5	
Se		4.2 (liquid phase epitaxy)
Sn	2.2×10^{-2}	2.0×10^{-3}(liquid phase epitaxy)
Si	0.55	30 (liquid phase epitaxy)
	0.15	
	>1.0	
Te		0.4 (liquid phase epitaxy)
Sb	2.6×10^{-2}	

Source: Swaminathan and Macrander 1991.

Figure 5.22*c* shows that C_O fluctuates periodically at the solid-liquid interface, and steady-state conditions are never achieved. As a result, the concentration of the solute in the solid varies periodically reflecting the situation at the solid-liquid interface.

Figure 5.22*b* represents the conditions normally used for the growth of crystals. The situation in Figure 5.22*a* will, of course, result in high-quality crystals, but the growth rate would be extremely low. Therefore, it is not economically feasible. On the other hand, the concentration of the solute in the crystal fluctuates in Figure 5.22*c*, and the growth rates will vary periodically, resulting in poor-quality crystals.

EXAMPLE 5.3. Certain device applications require float-zone silicon crystals, doped with As to 5×10^{17} cm^{-3}. What should be the concentration of As atoms in the melt to grow crystals having the desired carrier concentration? Assuming that the melt is 50 kg, how many grams of arsenic (atomic weight 74) should be added?

Solution. Since k_{eq} for arsenic in silicon is 0.3, the concentration of arsenic in the melt to produce a concentration of 5×10^{17} cm^{-3} is $= \dfrac{5 \times 10^{17}}{0.3}$

$$= 1.67 \times 10^{18}\ \text{cm}^{-3}.$$

The density of molten silicon is 2.53g/cm^3.

$$\text{Therefore, the volume of the melt} = \frac{50 \times 10^3}{2.53} = 1.98 \times 10^4\ (\text{cm})^3.$$

$$\text{Total number of As atoms in the melt} = 1.67 \times 10^{18} \times 1.98 \times 10^4$$

$$= 3.31 \times 10^{22}.$$

74 gms of As contains 6.023×10^{23} atoms.

$$\text{Therefore, the amount of As to add to the melt} = \frac{74 \times 331 \times 10^{22}}{6.023 \times 10^{23}}\ \text{gm}$$

$$= 4.07\ \text{gms}.$$

5.4.1.2 k_{eq} and phase diagrams

We argue that the values of k_{eq} are intimately related to the characteristics of the phase diagrams at very low solute concentrations. Figure 5.23 shows the phase diagram of semiconductor A containing an element B as a dopant. The top portion of the A-rich section of the diagram is enlarged in Figure 5.23*b*. The value of k_{eq} in this case is less than one. If we assume that the liquidus and solidus curves are essentially straight lines, then k_{eq} is independent of the composition. This assumption is reasonable because the concentrations of solutes (dopants) are extremely small. If the liquidus and solidus curves slope upwards, then the value k_{eq} will be greater than one.

The determination of k_{eq} involves careful measurement of the concentrations of the dopant in the crystal as a function of the amounts of the dopant in the liquid. Although modern characterization techniques such as secondary ion-mass spectrome-

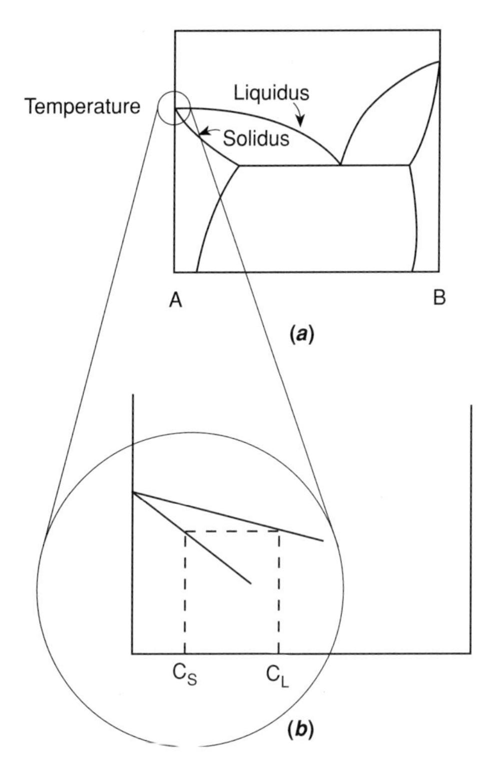

FIGURE 5.23
Interrelationship between k_{eq} and the phase diagram: (*a*) schematic phase diagram of semiconductor A and dopant B, and (*b*) enlargement of the circled region in (*a*).

try have simplified the measurement process, it is still a time-consuming determination.

5.4.1.3 Interrelationship of k_{eq}, k_{eff}, and microscopic growth rate

During crystal growth, the externally imposed pull rate differs from the microscopic growth rate at the solid-liquid interface. The latter may also vary along the interface. The higher the microscopic rate, the higher the incorporation of dopant atoms in doped crystals because the rejected atoms will have less time to diffuse away from the interface. These variations in the dopant concentration would produce nonuniform carrier concentrations in wafers. To understand this phenomenon, we need to develop a correlation between k_{eq}, k_{eff}, and the microscopic growth rate. We follow the elegant approach of Burton et al. (1953) for this purpose.

Assume that k_{eq} is less than one. The continuity equation that expresses the conservation of solute atoms in the growth system is

$$\frac{\partial C}{\partial t} = -\nabla \cdot (Cv - D\nabla C), \tag{5.10}$$

where v is the vector fluid velocity and D is the diffusion coefficient of the solute in the liquid that is independent of the concentration. This assumption is reasonable because the doping levels are low. Assume that the solid-liquid interface lies in the y-z plane at $x = o$ and that the concentration of the solute within the solid-liquid interface is fairly uniform. Under these assumptions the terms

$$v_y \frac{\partial C}{\partial y}, \; v_z \frac{\partial C}{\partial z}, \; D \frac{\partial^2 C}{\partial y^2}, \text{ and } D \frac{\partial^2 C}{\partial z^2}$$

can be neglected. Equation (5.10) is then reduced to

$$\frac{\partial C}{\partial t} = -C\nabla v - v_x \frac{\partial C}{\partial x} + D \frac{\partial^2 C}{\partial x^2}. \tag{5.11}$$

For an incompressible fluid, $\nabla v = 0$. Therefore, Equation (5.11) can be simplified to

$$\frac{\partial C}{\partial t} = -v_x \frac{\partial C}{\partial x} + D \frac{\partial^2 C}{\partial x^2}. \tag{5.12}$$

Beyond the boundary layer shown in Figure 5.22*b*, the solute concentration in the liquid approaches C_L. Within the boundary layer, the fluid velocity component normal to the growth interface approaches the crystal growth rate f, that is, the microscopic growth rate. Under these conditions

$$\begin{aligned} -v_x &= f \qquad \text{for } x < \delta \\ C &= C_L \qquad \text{for } x > \delta \, . \end{aligned} \tag{5.13}$$

f is positive for growth and is negative for melting. Replacing $-v_x$ with f in Equation (5.12), we have

$$\frac{\partial C}{\partial t} = D \frac{\partial^2 C}{\partial x^2} + f \frac{\partial C}{\partial x} \tag{5.14}$$

Under steady-state conditions, $\partial C/\partial t = 0$. Equation (5.14) then reduces to

$$D \frac{d^2 C}{dx^2} + f \frac{dC}{dx} = 0 \tag{5.15}$$

because C now depends only on x. The corresponding boundary conditions are

$$(C_0 - C_S) f + D \frac{dC}{dx} = 0 \quad \text{at } x = 0 \tag{5.16}$$

and

$$C = C_L \text{ at } x = \delta.$$

The following solution gives the concentration of solute at the interface:

$$\frac{C_0 - C_S}{C_L - C_S} = e^{\Delta}, \tag{5.17}$$

where $\Delta = \dfrac{f\delta}{D}$ and δ is the thickness of the boundary layer.

Substituting for $\frac{C_S}{C_0} = k_{eq}$ and $\frac{C_S}{C_L} = k_{eff}$ in Equation (5.17), we obtain the following relationship between k_{eff}, k_{eq}, and f:

$$k_{eff} = \frac{k_{eq}}{k_{eq} + (1 - k_{eq})\, e^{-\Delta}} \tag{5.18}$$

Equation (5.18) is referred to as the *Burton-Prim-Slichter relation.*

Let us now consider the two distinct cases that could develop during the crystal growth: *f* is very low, and *f* is very high. When *f* is very low, Δ is very small and $e^{-\Delta} \simeq 1$. Substituting this value for $e^{-\Delta}$ in Equation (5.18), we obtain

$$k_{eff} \simeq k_{eq}. \tag{5.19}$$

That is, when the growth rate is very low (Figure 5.22*a*), k_{eff} is equal to k_{eq}. In other words, the rejected solute does not pile up at the solid–liquid interface, and the solute atoms have enough time to diffuse into the melt.

When *f* is very large, $e^{-\Delta} \simeq 0$. Substituting this value for $e^{-\Delta}$ in Equation (5.18) yields

$$k_{eff} \simeq 1. \tag{5.20}$$

That is, all the solute that is present in the liquid is incorporated into the solid. The solute cannot diffuse away from the solid–liquid interface because of the high growth rate.

5.4.1.4 Solute concentration in the crystal as a function of melt fraction solidified

We assumed in section 5.4.1.3 that the volume of the melt is large in comparison to the size of the crystal that is being grown, a situation that exists during the growth of state-of-the-art CZ silicon crystals. Therefore no appreciable change in the melt composition occurs when the solute atoms are rejected from the solid into the melt during the crystal growth. Now let us consider the case where the melt volume is finite and the rejection of solute into the melt results in a continuous change in the melt composition. The growth of Bridgman crystals represents this situation.

Assume that k_{eq} is less than one, and the initial volume of the melt is unity. If the area of the solid–liquid interface is assumed to be unity, then the height of the melt is also unity. Let s be the amount of solute in the melt when the melt of thickness dx has solidified. The amount of solute incorporated into the solid is given by

$$C_S dx = -ds \tag{5.21}$$

Expression (5.21) assumes no change in volume on solidification. The amount of solute in the diffusion boundary layer is small. If the steady state prevails during growth, C_L is given by

$$C_L = \frac{s}{1 - x} = \frac{C_S}{k_{eff}}$$

or

$$C_S = k_{eff}\left(\frac{s}{1 - x}\right) \tag{5.22}$$

Substituting for C_S from Equation (5.21) into Equation (5.22), we have

$$\frac{ds}{s} = -k_{eff}\left(\frac{dx}{1-x}\right) \tag{5.23}$$

Integrating Equation (5.23) between the limits s and s_o and x and o, we have

$$\ln\left(\frac{s}{s_o}\right) = k_{eff}\ \ln(1-x), \tag{5.24}$$

where s_o is the initial amount of solute in the melt. Since the initial volume of the melt is unity, $s_o = C_M$, the initial concentration of solute in the melt. Substituting for s_o in Equation (5.24), we have

$$s = C_M\,(1-x)^{k_{eff}} \tag{5.25}$$

But $s/1 - x = C_L$ and $k_{eff} = C_S/C_L$. Substituting these values in Equation (5.25), we can write

$$C_S = k_{eff}\,C_M\,(1-x)^{k_{eff-1}} \tag{5.26}$$

Equation (5.26) indicates that when k_{eq} is less than one and a crystal is grown from a finite melt, the concentration of the solute in the first-to-freeze portion of the crystal is lower than that of the last-to-freeze section. This result implies a continuous variation in the concentration of the solute along the length of the crystal.

EXAMPLE 5.4. Assuming that k_{eff} for a dopant in silicon is 0.75, compare dopant concentrations in portions of a crystal that form on solidification of one-quarter, one-half, and three-quarters of the melt. The volume of the melt is small, and the initial concentration of the dopant is $5 \times 10^{16}\ cm^{-3}$.

Solution. Use Equation (5.26) to calculate C_S values for x = 0.25, 0.5, and 0.75.

a. Quarter of the melt

$$C_S = 0.75 \times 5 \times 10^{16}(0.75)^{0.8-1}$$

$$= 0.75 \times 5 \times \frac{1}{(0.75)^{0.2}} \times 10^{16}\ cm^{-3}$$

$$= 3.99 \times 10^{16}\ cm^{-3}$$

b. Half of the melt

$$C_S = 0.75 \times 5 \times 10^{16}\ (0.5)^{-0.2}$$

$$= 4.31 \times 10^{16}\ cm^{-3}$$

c. Three-quarters of the melt

$$C_S = 0.75 \times 5 \times 10^{16} \times\ (0.25)^{-0.2}$$

$$= 4.93 \times 10^{16}\ cm^{-3}$$

FIGURE 5.24
Dopant striations observed in a longitudinal section of a P-doped FZ silicon crystal. (*After deKock [1980].*)

5.4.2 Dopant or Impurity Striations

We argued in section 5.4.1.3 that when k_{eq} is not equal to one, dopants are not incorporated homogeneously in the crystals. These dopant inhomogeneities are manifested as dopant or impurity striations in both the nominally doped and undoped crystals. An example of the striations revealed by preferential etching of a longitudinal section of a P-doped FZ silicon is reproduced in Figure 5.24 (deKock 1980). Similar features are seen in compound semiconductors. The striations observed in a Se-doped InSb crystal are shown in Figure 5.25 (Witt and Gatos 1967). In transverse sections, that is, the sections cut normal to the growth axis, the dopant atoms are distributed in a swirl pattern. Figure 5.26 shows a typical situation seen in selenium-doped InSb crystals grown by the CZ process (Morizane et al. 1966). Residual impurities could be responsible for the striations observed in nominally undoped crystals. The striations are also seen in rotated as well as in unrotated crystals (Witt and Gatos 1967). The striations observed in the two types of InSb crystals are compared in Figure 5.27 (Witt and Gatos 1967). The striations in the rotated crystals, Figure 5.27*a,* are periodic and regularly spaced, whereas they are aperiodic in the unrotated crystals, Figure 5.27*b*.

We can rationalize the origin of impurity striations in terms of the Burton-Prim-Slichter approach (1953). As discussed earlier, during the solidification of a melt,

FIGURE 5.25
Rotational striations in the off-core-on-core transition region of a Se-doped InSb crystal grown 9° off the <111> direction. (*After Witt and Gatos [1967].*)

the concentration of a dopant or an impurity in the crystal is determined by k_{eff}. For the sake of discussion, assume that k_{eq} is less than one. For a finite growth rate, the dopant concentration builds up at the solid–liquid interface during the solidification. Consequently, the concentration of dopant in as-grown crystals could fluctuate from region to region because of convection in the melt, asymmetry in the thermal profile of the growth system, growth rate, and so forth.

Burton et al. (1953) have analyzed the striation phenomenon using Equation (5.18). In general, the microscopic growth rate f differs from the imposed growth rate and is affected by the local growth conditions. If a portion of the crystal freezes at a rate different from that of the contiguous region, the concentrations of the dopant in the two regions would differ; that is, the striations would develop. Morizane et al. (1967) developed the following relation that relates f and the crystal rotation rate R:

$$f = v_0\,(1 - \alpha_0 \cos 2\pi \mathrm{RT}). \qquad (5.27)$$

with

$$\alpha_0 = \frac{2\pi \Delta \mathrm{TR}}{v_0 \mathrm{G}},$$

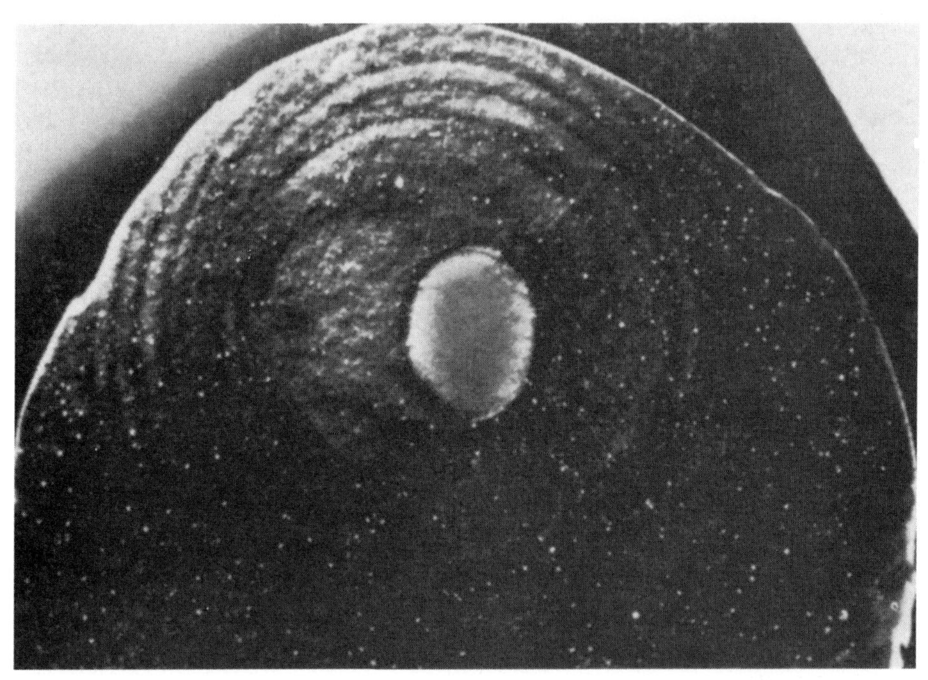

FIGURE 5.26
Dopant striations observed in a transverse section of a Se-doped InSb crystal grown without rotation. The part of the surface without swirls corresponds to the (111) facet. (*After Morizane et al. [1966].*)

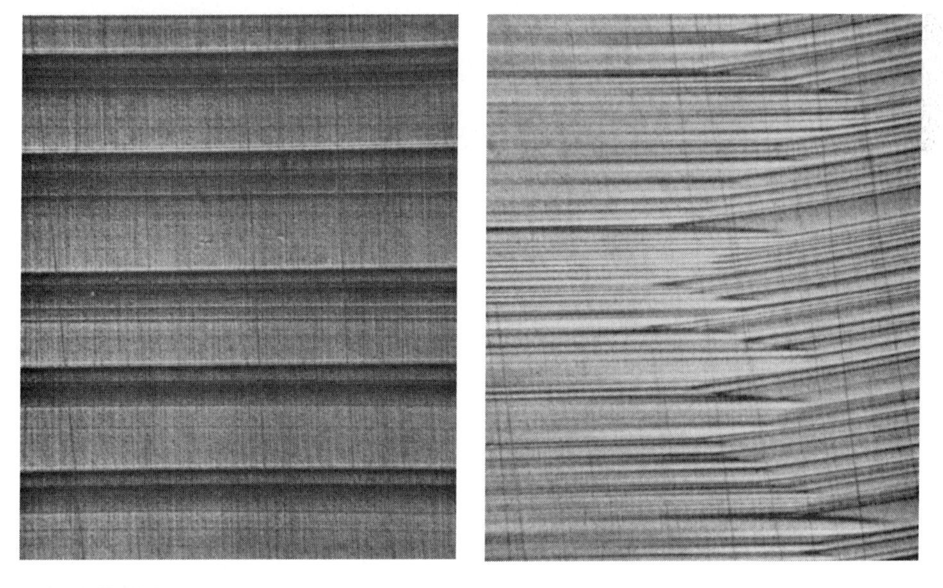

FIGURE 5.27
Dopant striations observed in (*a*) rotated and (*b*) unrotated InSb crystals. (*After Witt and Gatos [1967].*)

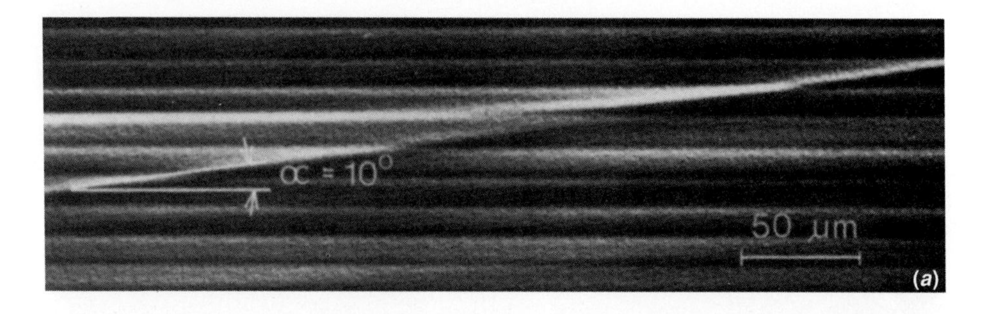

FIGURE 5.28
Examples of type II striations observed in a CZ silicon crystal. The angle α between type II and type I striations (horizontal lines) decreases from (*a*) $\rightarrow$ (*c*). (*After Bauser and Rozgonyi [1982].*)

where ΔT is the temperature variation that a given point at the solid–liquid interface is subjected to during one crystal rotation, G is the temperature gradient in the melt contiguous to the growth interface, and v_0 is the pulling rate. For $\alpha_0 > 1$, a condition that can easily be satisfied at low values of v_0 and for a large thermal asymmetry, remelting occurs during each crystal rotation. For curved interfaces this periodic remelting causes large variations in *f*, which results in large variations in the incorporation of the dopant (Morizane et al. 1967).

The solid–liquid interface can have several different shapes: planar, convex into the melt, or concave. Witt and Gatos (1967) show that, in the case of InSb,

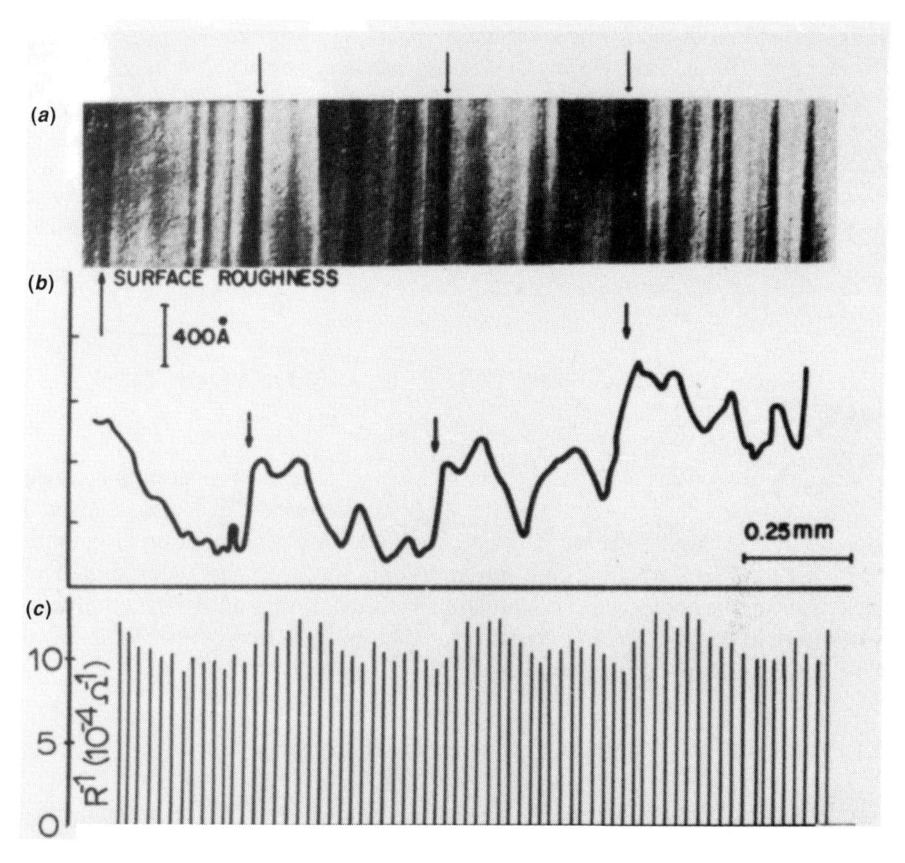

FIGURE 5.29
The correspondence between (*a*) etching rates, (*b*) topology of the etched surface, and (*c*) lateral variations in resistivity with the presence of dopant striations in P-doped FZ silicon crystal. (*After deKock et al. [1977].*)

when the central portion of the solid-liquid interface is planar, the crystal acquires a facet that does not extend to its periphery. As a consequence, striations are no longer horizontal in regions away from the core region and reflect the shape of the solid-liquid interface.

Bauser and Rozgonyi (1982) investigated dopant striations in CZ Si crystals by chemical etching. They refer to the striations discussed above as type I. In addition to type I striations, they observe feathery features termed type II striations. An example reproduced from their work is shown in Figure 5.28. The horizontal lines in Figure 5.28 are type I striations, whereas type II striations make different angles α with type I striations. Bauser and Rozgonyi suggest that type II striations are caused by a nonplanar solid-liquid interface that can be described in terms of steps. Since the orientations of the risers and terraces of the steps may be different, the segregation characteristics of the dopant to the two regions would differ, leading to type II striations.

The striations are dopant-rich and dopant-poor regions in the crystal. Therefore, local changes in electrical resistivity are expected to occur on the same lateral scale as the striations. Figure 5.29*a* shows dopant striations, revealed by preferential etching, in a longitudinal slice obtained from a phosphorus-doped Si crystal (deKock et al. 1977). The two regions etch differently because the magnitudes of the dopant-induced strains are different. The difference in the etch rate of the two regions is consistent with a Talystep profile of the surface shown in Figure 5.29*b*. A spreading resistance profile from the respective regions is reproduced in Figure 5.29*c*. Assuming that the etching rate is higher in the dopant-rich regions because of higher strains, then the observed lateral variations in resistivity can be correlated with the dopant striations.

5.5 MICRODEFECTS IN MACROSCOPICALLY DISLOCATION-FREE SILICON CRYSTALS

We indicated in sections 5.2 and 5.3 that the crystal growth from the melt is carried out under nonequilibrium conditions at high temperatures. We also showed in Chapter 3 that the higher the temperature, the higher the concentration of point defects in a solid. Therefore, high concentrations of point defects are incorporated into a crystal during the cooldown. This situation is unstable. We address the elimination of instability in this section and demonstrate that this process occurs by the formation of defect clusters referred to as *microdefects*.

5.5.1 Nature of Microdefects and Their Distribution

Microdefects have been observed in FZ and CZ silicon (Abe and Maruyama 1967; deKock 1973, 1980) and do not produce etch pits similar to those of dislocations covered in Chapter 4. Figure 5.30 shows noncrystallographic etch pits on the surface of an etched Si wafer containing microdefects (Ravi and Varker 1974). The authors argue that those nondescript etch pits are due to microdefects. The crystals that do not show etch pits corresponding to dislocations are termed *macroscopically dislocation free*.

deKock (1973) identified two types of microdefects, termed *A* and *B swirl defects,* in as-grown FZ silicon crystals. Figure 5.31 shows the distribution of A and B swirl defects revealed by X-ray topography of an FZ silicon slice. In the peripheral region, about 2 mm wide, A-type defects are absent, but B swirls are still present. The average concentrations of A and B swirls are in the range of 10^6 to 10^7 cm^{-3} and 10^7 to 10^8 cm^{-3}, respectively (deKock 1973).

Microdefects are also observed in CZ silicon crystals (deKock 1976, 1977; deKock et al. 1979; deKock and van de Wijgert 1980). In Figure 5.32, the distributions of A- and B-type swirls in FZ and CZ crystals are compared. The distributions in the two types of crystals are very similar.

Let us now discuss the microscopic nature of the swirl defects. Transmission electron microscopy studies have shown that A swirls are prismatic dislocation loops, interstitial in character, the Burgers vector is a/2 <110>, and lie on {111} and {110} planes (Föll and Kolbesen 1975; Föll et al. 1977). A typical example of an A swirl is shown in Figure 5.33. B swirls have not been unambiguously identified.

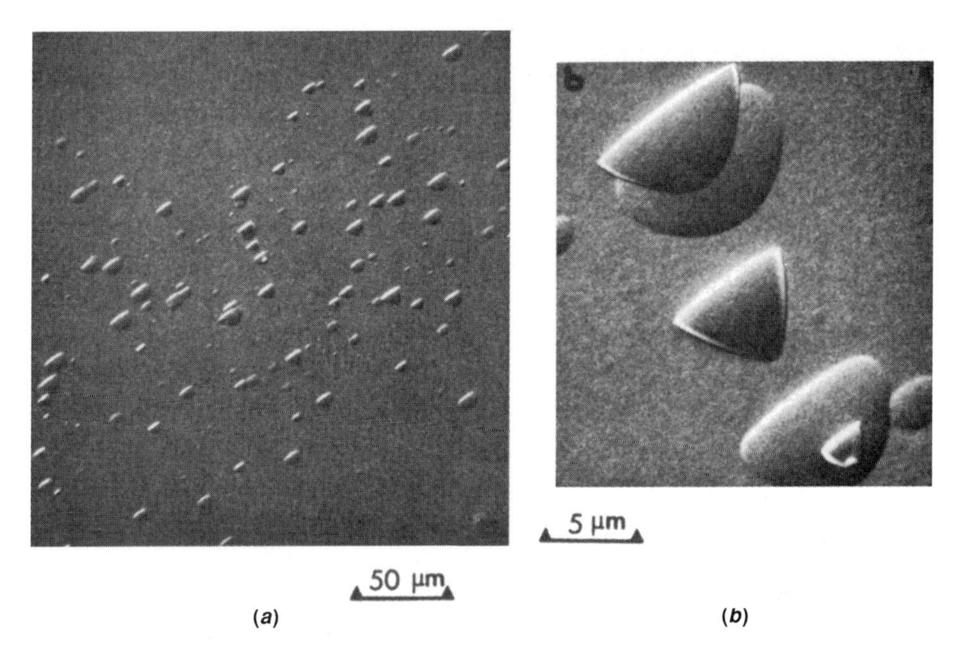

FIGURE 5.30
Shallow or empty etch pits corresponding to the microdefects observed in FZ silicon: (*a*) low-magnification micrograph and (*b*) high-magnification micrograph.
(*After Ravi and Varker [1973].*)

We now consider the evolution of A and B swirls. According to Föll and Kolbesen (1975) and Föll et al. (1977), self-interstitials are the dominant defect in silicon at temperatures above 900°C. The concentration of vacancies is assumed to be negligible. During growth the self-interstitials are incorporated into the solid, their concentration being close to the equilibrium value near the melting point. Supersaturation of the interstitials occurs during the nonequilibrium cooldown, leading to their agglomeration into spongy structures that represent B swirls. Impurities such as carbon act as heterogeneous nucleation centers because of the strain considerations. As the crystal cools further, B swirls grow by incorporating more self-interstitials. When they reach a certain size, B swirls collapse into extrinsic, faulted loops bounded by a/3 <111> Frank partials (see Chapter 3). These stacking faults grow by the absorption of Si interstitials at the bounding fault partials. When the fault grows, the energy of the defect increases. At a certain critical size, the fault can be eliminated, and the defect is converted into a perfect dislocation loop. This conversion occurs by the nucleation of two Shockley partials, one on each plane of the Frank loop. Dislocation reactions governing this reaction are as follows:

$$\frac{a}{3}\langle 111\rangle + \frac{a}{6}\langle 2\bar{1}\bar{1}\rangle \rightarrow \frac{a}{6}\langle 411\rangle$$
$$\frac{a}{6}\langle 411\rangle + \frac{a}{6}\langle \bar{1}2\bar{1}\rangle \rightarrow \frac{a}{2}\langle 110\rangle\ . \tag{5.28}$$

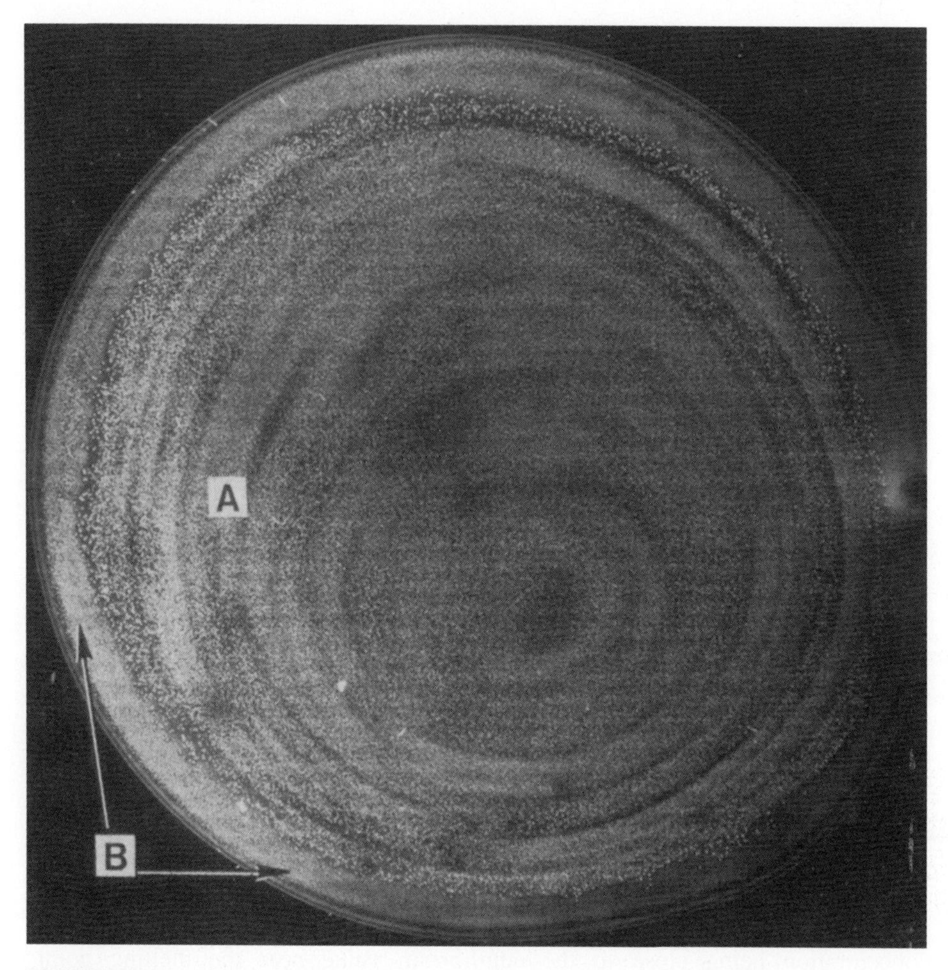

FIGURE 5.31
X-ray transmission topograph of a copper-decorated transverse section taken from dislocation-free FZ Si crystal grown at a rate of 3 mm/min in argon. A and B refer to A- and B-type swirl defects. (*After deKock [1980].*)

The Petroff-deKock (1975, 1976) model is very similar to that of Föll and Kolbesen (1975) and Föll et al. (1977). Petroff and deKock assume the incorporation of nonequilibrium concentrations of thermal-point defects into the as-grown crystal. Their assumption is reasonable because the temperature distribution in real-growth systems is not completely cylindrically symmetrical, the crystal rotation introduces temperature variations at the growth interface that could cause remelting, and the crystal growth occurs under nonequilibrium conditions.

Now let us apply the conceptual framework on swirls to some of the experimental results. In longitudinal sections of FZ crystals, swirls are distributed in a striated pattern as shown in Figure 5.34. This pattern resembles impurity striations.

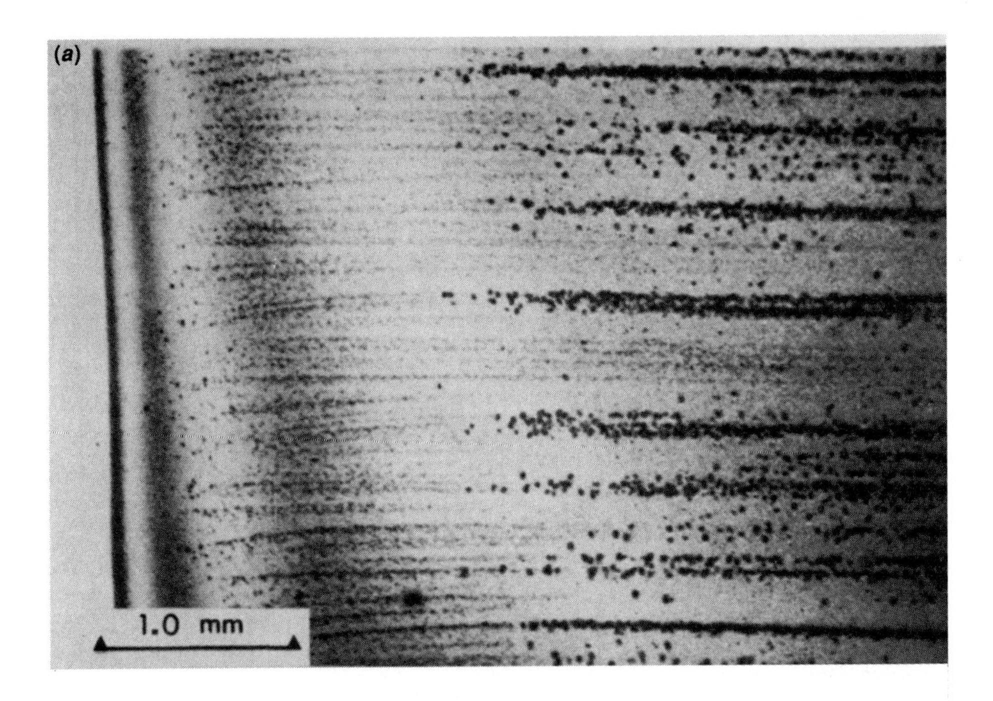

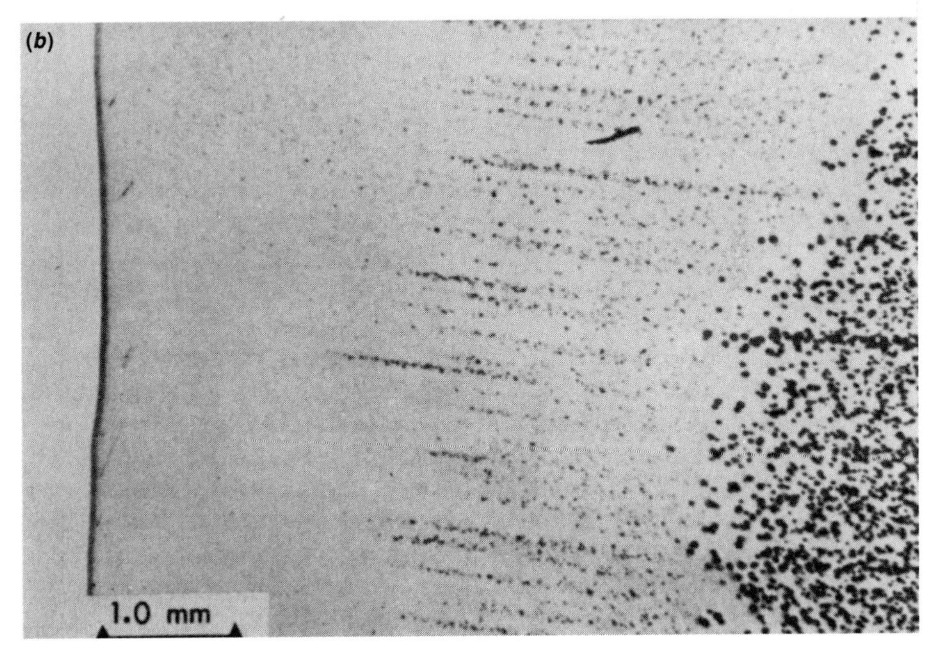

FIGURE 5.32
A and B swirl defects observed in the peripheral regions of dislocation-free silicon crystals by X-ray topography of copper-decorated samples. (*a*) Float-zone silicon and (*b*) Czochralski silicon. (*After deKock [1980].*)

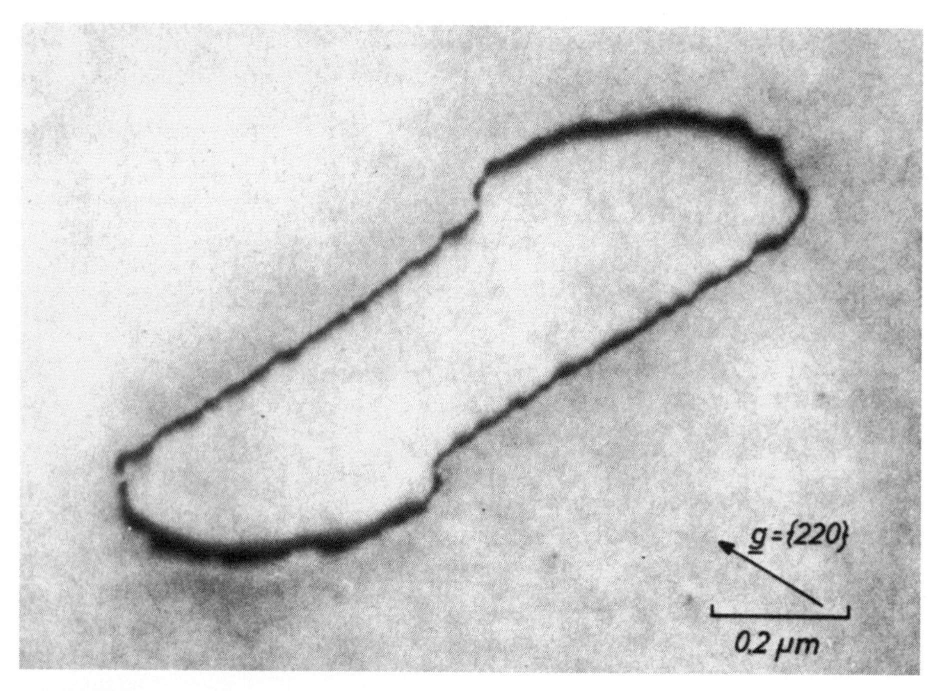

FIGURE 5.33
Micrograph showing detailed nature of an A-swirl defect. (*After deKock [1980].*)

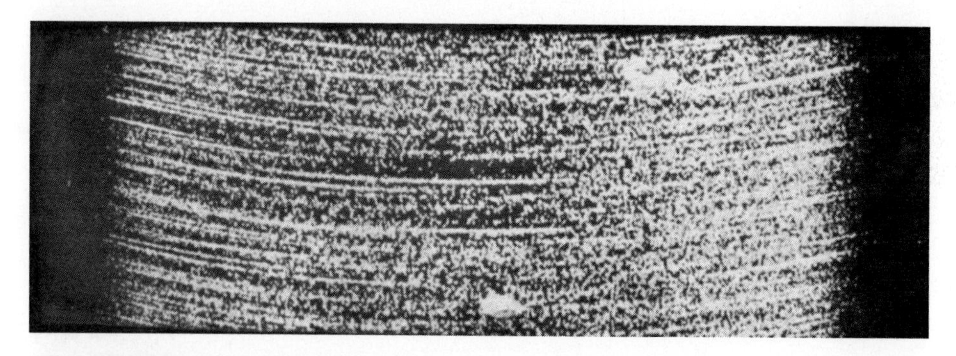

FIGURE 5.34
X-ray topograph of longitudinal section of a lithium-decorated, dislocation-free FZ Si crystal. (*After deKock [1980].*)

This result is understandable if impurities serve as heterogeneous nucleation centers for the swirls. We also know that the formation of swirls can be suppressed by imposing either low pulling rates (Roksnoer et al. 1976) or high pulling rates (deKock 1973, 1980). These effects are illustrated in Figure 5.35. When the growth rate is low, the supersaturation of point defects may not occur because they may have enough time to migrate to various sinks. However, at higher growth rates the concentrations of point defects will exceed the equilibrium values, but they may not have enough time to cluster to form microdefects.

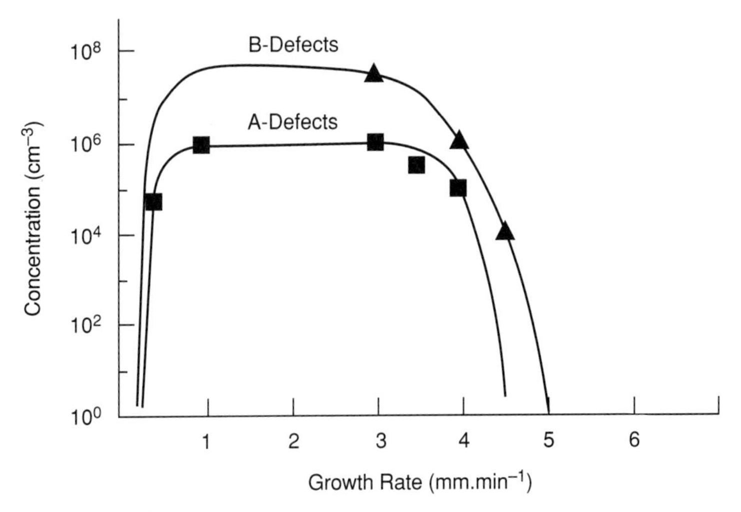

FIGURE 5.35
Illustrates the influence growth rate on the concentrations of A- and B-swirl defects in FZ Si crystals grown in argon atmosphere. Defect elimination occurs for growth rates exceeding 5 mm/min and below 0.2 mm/min. (*After deKock [1980].*)

We have shown above that A swirls are dislocation loops. Therefore, they are likely to affect device characteristics of p-n junctions. Figure 5.36 shows the influence of microdefects on the reverse I-V characteristics of silicon diodes (Varker and Ravi 1973). Figure 5.36*a* shows an optical micrograph of a split wafer. The array of diodes is on the right half of the wafer, and the left half is chemically etched to reveal the distribution of the microdefects in the wafer. Figure 5.36*b* shows the variations in leakage current at a reverse voltage of 25V for a row of diodes that cuts across the swirl pattern. The leakage currents of the individual diodes vary between < 10 and 10^5 nA. The excess leakage currents tend to correlate well with the presence of swirl defects in the diode.

EXAMPLE 5.5. Given that $5 \times 10^{17}\ cm^{-3}$ excess interstitials are incorporated into a silicon crystal, calculate the number of hexagonal stacking faults that would form from the coalescence of these interstitials. Assume that faults lie on a {111} plane, are bounded by <110> directions, and the fault edge is 5 nm. The lattice parameter of Si = 0.543 nm.

Solution. First we need to calculate the number of interstitials involved in the formation of a single stacking fault. To do this, examine the planar arrangement shown in Figure 3.15. The side of the hexagon is $a/\sqrt{2}$, where a is the lattice parameter, and the effective number of atoms in the hexagon = 3.

$$\text{Area of the hexagon} = \frac{1}{2} \cdot \frac{a}{\sqrt{2}} \frac{a}{\sqrt{2}} \sin 60^\circ$$

$$= \frac{a^2}{4} \sin 60^\circ$$

$$= \frac{(0.543)^2 \times \sqrt{3}}{2 \times 2 \times 2} (\text{nm})^2.$$

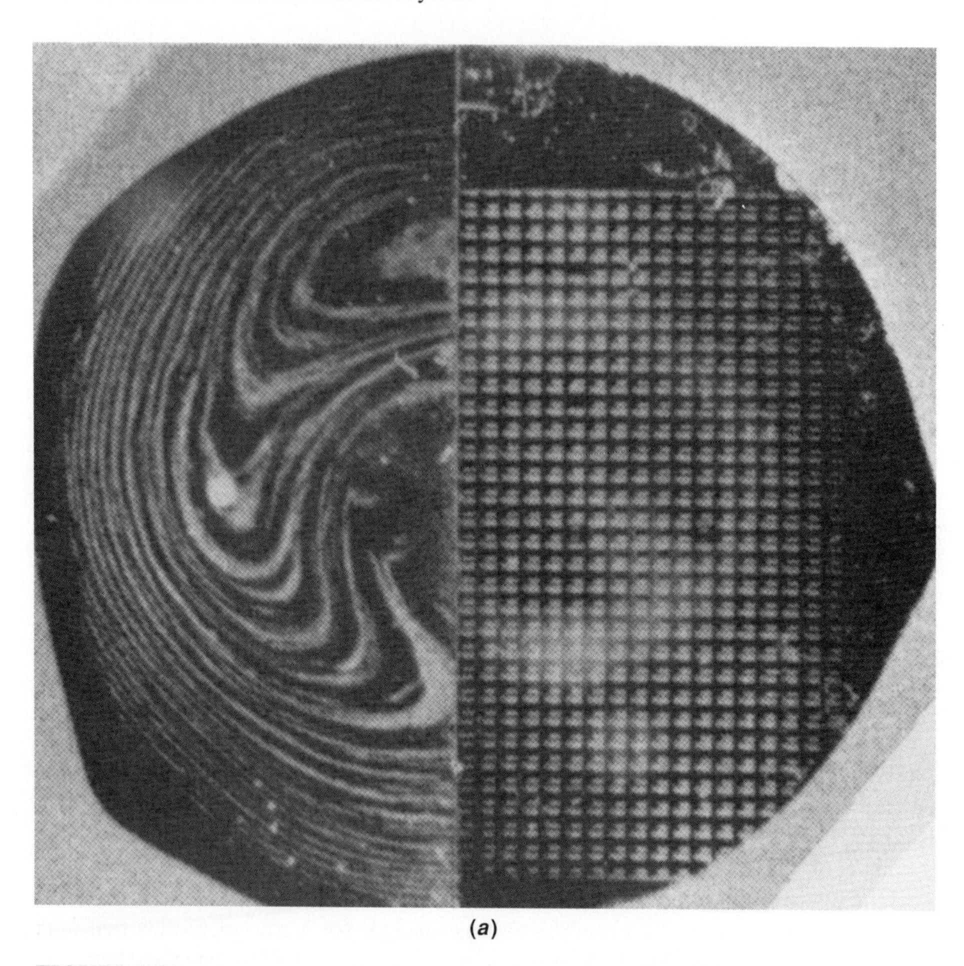

(*a*)

FIGURE 5.36*a*
Optical micrograph of a split wafer showing swirls on a preferentially etched Si wafer. (*After Varker and Ravi [1973].*)

$$\text{Area of the fault} = \frac{1}{2} \cdot 5 \times 5 \sin 60^\circ \ (\text{nm})^2.$$

$$\text{Number of atoms in each layer of the fault} = \frac{2 \times 3 \times 25}{(0.543)^2}$$

$$= \sim 508.$$

Since the fault is two layers thick, the number of atoms involved in the fault formation $= \sim 1{,}016$.

$$\text{Number of faults} = \frac{5 \times 10^{17}}{1{,}016}$$

$$\simeq 4.9 \times 10^{14} \ (\text{cm})^{-3}.$$

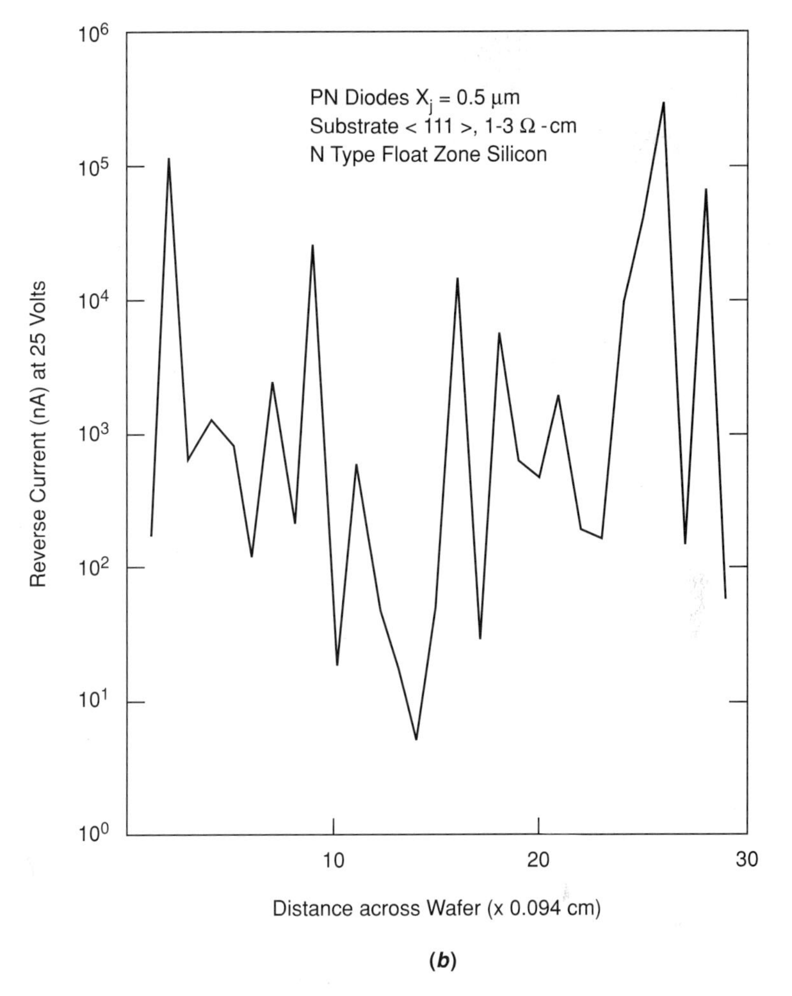

FIGURE 5.36*b*
typical diode reverse-current distribution in a column cutting across the swirl pattern shown in Fig. 5.36(*a*). (*After Varker and Ravi [1973].*)

5.6 OXYGEN IN SILICON

We discussed in section 5.2.1 that the growth of CZ silicon crystals from melts contained in quartz crucibles results in the incorporation of $\sim 2 \times 10^{18}$ cm^{-3} oxygen. This amount represents the solubility limit of oxygen at the growth temperature and exceeds the solubility limits at lower temperatures where silicon devices are processed. Given enough mobility, the dissolved oxygen comes out of solution. We consider in this section its ramification on the electrical, structural, and mechanical characteristics of CZ silicon.

5.6.1 Basic Properties of Oxygen in Silicon

Infrared spectroscopy is widely used to measure the interstitial oxygen concentration in silicon wafers. The presence of Si-O-Si bonds gives rise to a stretching (antisymmetric) mode at 1107 cm^{-1} and a symmetric mode at 515 cm^{-1} (Stavola 1984). Experimentally, the absorption coefficient, α_o, of a wafer at 1107 cm^{-1} is measured using infrared spectroscopy. The interstitial oxygen concentration, C_{OI}, is calculated from

$$C_{OI} = (3.14 \pm 0.09) \times 10^{17} \alpha_o, \tag{5.29}$$

with C_{OI} in atoms cm^{-3} and α_o in cm^{-1}. The ideal wafer thickness χ that gives the best accuracy for α_0 is given by $\chi = 1/\alpha_o$. Therefore, a 2 mm thick slice is recommended for measurement of C_{OI} in CZ silicon. Cooling the samples to liquid nitrogen or helium temperature can further enhance the sensitivity of the infrared measurements.

Results from infrared spectroscopy studies suggest that silicon and oxygen are bonded in an Si_2O configuration shown in Figure 5.37 (Stavola and Snyder 1983). In this arrangement the oxygen atom is in an interstitial position (Kaiser et al. 1956), is very slightly displaced from a <111> direction, and the angle between the Si-O bonds is 162° (Bosomworth et al. 1970). The Si-O bond length is 0.161 nm (Pajot and Cales 1986). The dissolved oxygen is also electrically inactive.

A range of values for the equilibrium distribution coefficient of oxygen in silicon appears in Table 5.4. The large variation in the reported values indicates that it may be difficult to achieve equilibrium between molten silicon and oxygen. One of the difficulties in determining the distribution coefficient is the continuous supply of oxygen from the container walls and the evaporation of silicon monoxide from the

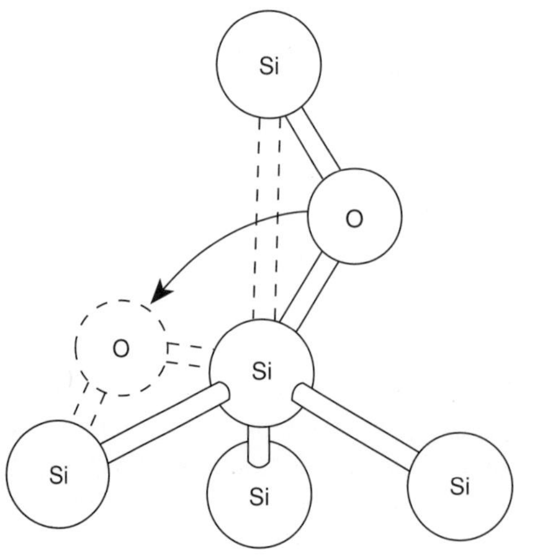

FIGURE 5.37
Position of interstitial oxygen in the silicon lattice. (*After Stavola and Snyder [1983].*)

TABLE 5.4
Distribution coefficient of oxygen in silicon

k_{eq}	Method
1.48	Growth rate—IR
1.25 ± 0.17	CPAA*
$1 + 0.25$	FTIR*
1.0 ± 0.1	Growth rate
1	Growth rate—IR, SIMS*, SRP*
$0.36 - 0.9$	Calculated
0.5	Vacuum fusion
0.3	Growth rate—IR
0.25	Growth rate

Source: Bender and Vanhellemont 1994.

FTIR = Fourier transform infrared spectroscopy, CPAA = charged particle activation analysis, SIMS = secondary ion-mass spectrometry, SRP = spreading resistance probe.

melt. Harada et al. (1985) have suppressed these effects by encapsulating the melt and report the distribution coefficient to be unity. Lin and Hill (1980) and Lin and Stavola (1985) have grown small-diameter silicon crystals at a constant growth rate under nearly equilibrium conditions and find the distribution coefficient to be in the range of 0.25 to 0.3.

As indicated earlier the solubility of oxygen in molten silicon is $2.75 \pm 0.15 \times 10^{18}$ cm^{-3} (Yatsurugi et al. 1973). The temperature dependence of the solubility has been investigated. The best fit to the various data can be expressed by the following expression (Mikkelsen 1986):

$$C_{OI} = 9 \times 10^{22} \exp\left(\frac{-1.52}{k_B T}\right) \text{ atoms cm}^{-3} \tag{5.30}$$

The diffusivity of oxygen in silicon has been measured mainly in the 700 to 1200°C range (Takano and Maki 1973; Gass et al. 1980). The agreement between the two sets of results is excellent.

The measurements of Mikkelsen (1982a, 1982b) and Stavola et al. (1983) on the diffusivity of oxygen cover a wider temperature range. The best fit to their data indicates the following temperature dependence of the oxygen diffusivity (Mikkelsen 1986):

$$D_o = 0.13 \exp\left(\frac{-2.53}{k_B T}\right) \text{ cm}^2 \text{ sec}^{-1} \tag{5.31}$$

The interstitial oxygen diffuses by hopping between different interstitial sites (Stavola et al. 1983). Results indicate that an interstitial hop mechanism dominates oxygen diffusion above 700°C.

5.6.2 Influence of Annealing on Electrical, Structural, and Mechanical Characteristics

We know from Equation (5.30) that the solubility of oxygen in CZ silicon decreases with the decreasing temperatures. If as-grown crystals are annealed at temperatures where oxygen atoms have a sufficient mobility, the extra atoms would tend to come out of solution. This decrease can occur in several ways. Near the surface regions, oxygen atoms may be lost to the ambient. However, in the interior they may form clusters and precipitates. We now consider the effects of clustering and precipitation on the electrical and structural characteristics and mechanical properties of CZ silicon.

5.6.2.1 Electrical characteristics

When CZ silicon is annealed around 450°C, oxygen-related thermal donors (TDs) are produced. Their concentration is of the order of 10^{16} cm^{-3}, and the associated energy levels are fairly shallow. Hall measurements on aged samples indicate two energy levels at 0.13 and 0.06 eV below the conduction-band edge (Gaworzewski and Schmaiz 1979), whereas energy levels at 0.15 and 0.07 eV below the conduction-band edge are observed by deep-level transient spectroscopy (Kimerling and Benton 1981; Stavola et al. 1983). The presence of p-type dopants, such as boron, gallium, and aluminium also increases the rate of TD formation and maximum concentration (Fuller and Doleiden 1958). This behavior is consistent with the donor character of TDs.

Several researchers have attempted to model the structure of TDs and to develop a mechanism to explain the associated electrical activity (Oehrlein and Corbett 1983; Stavola and Snyder 1983; Ourmazd et al. 1984; Stavola and Lee 1986). A consensus has emerged that TDs consist of silicon-oxygen clusters, but neither the number of oxygen atoms involved nor their spatial distribution is clear. As an illustration, consider the model of Oehrlein and Corbett (1983), which is an extension of the model proposed by Kaiser et al. (1958). It is assumed that Si_3O_2, Si_4O_3, and Si_5O_4 clusters are present after the 450°C heat treatment. Since the volume of this cluster is larger than the volume of silicon consumed in the cluster formation, the clusters are compressed by the adjoining silicon. The compression imposes a stress on the cluster, and Si-O-Si bond angles are severely distorted. The distortion may cause the 2p lone-pair orbitals of two oxygen atoms to interact, which is illustrated in Figure 5.38. The interaction between the lone-pair orbitals results in bonding and antibonding orbitals. If the energy of the antibonding orbital is sufficiently above the Si conduction-band edge, this state could be doubly ionized, a result consistent with the observed behavior of TD.

5.6.2.2 Structural characteristics

The silicon and oxygen atoms begin to form clusters on aging. These clusters may have volumes larger than those of silicon consumed in their formation. They grow with continued aging and may even change their nature. In this section we use these simple ideas to explain the observed structural characteristics of annealed CZ silicon.

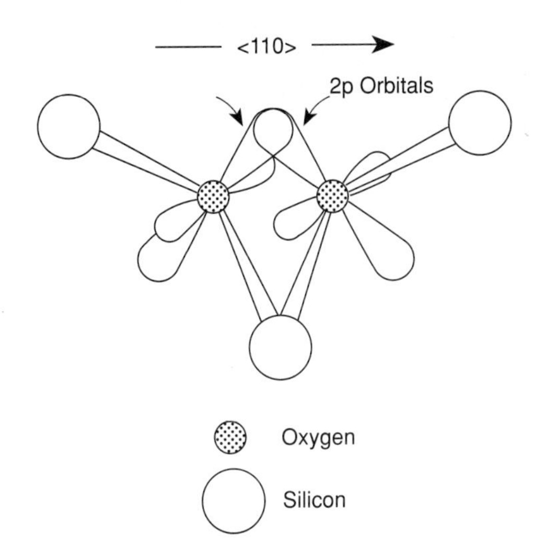

FIGURE 5.38
Schematic of the overlapping of the 2p lone-pair orbitals of two oxygen atoms in a distorted Si_mO_n-complex, where m and n are integers. (*After Oehrlein and Corbett [1983].*)

The distribution of oxygen varies from crystal to crystal. Therefore, for the precipitation studies the crystals are given an oxygen-dispersal anneal around 1300°C prior to aging at appropriate temperatures. We use two types of annealing procedures: single- and double-step anneals. The difference between the two approaches is that in the case of the double-step anneals, crystals are preannealed at low temperatures, followed by a high-temperature anneal (Bourret 1986). Since the driving force for clustering or precipitation is higher during aging at low temperatures, a large number of nucleation sites are produced where precipitation could occur during the subsequent high-temperature anneals. The structural features produced during the two types of anneals are nearly identical except that their densities and the kinetics of the precipitation may be different. For the present section we are more interested in the highlights of the structural features. Therefore, we plan to cover together the structural ramifications of the two procedures.

The principal feature observed after the single-step anneal is SiO_x precipitates (Templehoff et al. 1977). They have a platelike habit, lie on {001} planes, and their edges are parallel to <110> directions (Bender 1984a). The thickness of the precipitates does not exceed 3 to 4 nm for anneals up to 900°C (Wada et al. 1980; Bender 1984a). In addition to the platelike precipitates, octahedral-shaped precipitates are seen after the double-step anneal (Bender 1984a). An example of a truncated octahedral precipitate, viewed from various crystallographic directions, is shown in Figure 5.39*a*, whereas its high-resolution image is reproduced as Figure 5.39*b*. The platelike precipitates in the double-step annealed samples are associated with prismatically punched out dislocations as shown in Figure 5.40(*a*) to (*f*) under different imaging conditions. The composition of the precipitates changes from being SiO at low temperatures to SiO_x at high temperatures, where x is close to 2.

If we assume that the platelike precipitates are associated with interfacial dislocations, then we can explain the formation of prismatic loops observed in association with the platelike precipitates in terms of the model developed by Ashby and Johnson (1969) as covered in Chapter 3. These interfacial dislocations undergo

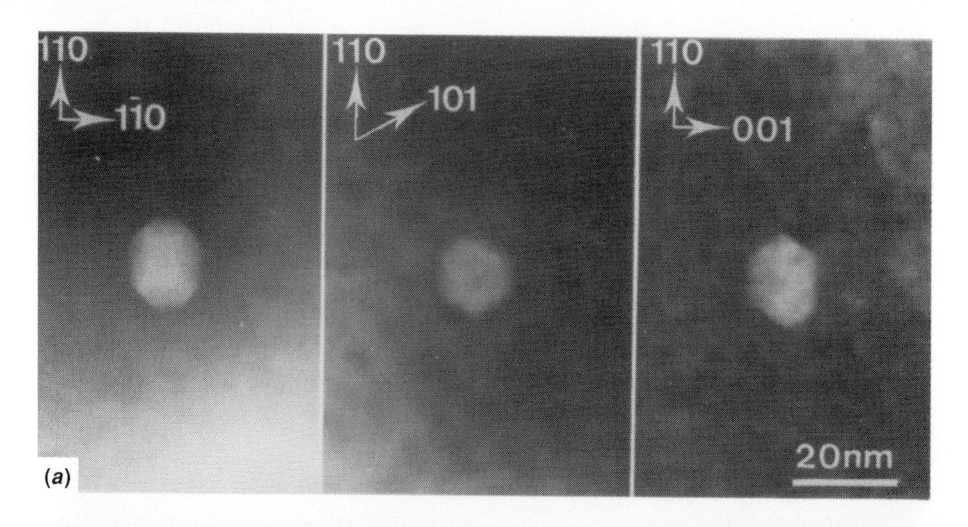

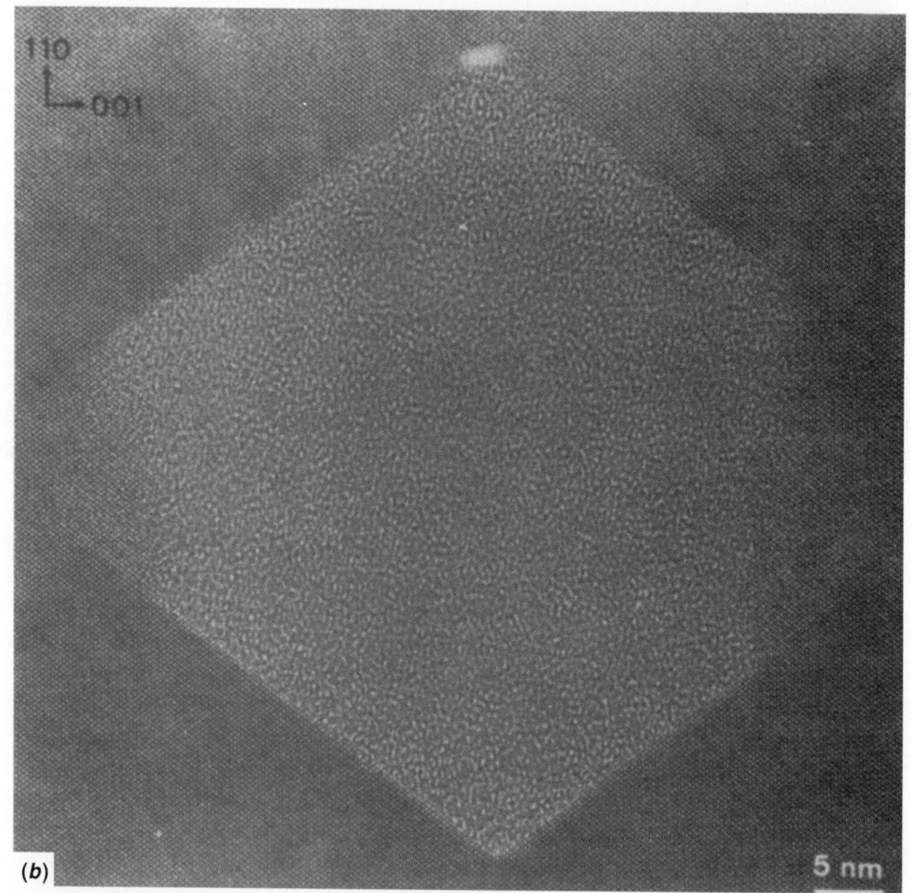

FIGURE 5.39
Truncated octahedral precipitates seen in CZ silicon after a two-step annealing (850°C for 15 hr + 1500°C for 15 hr): (*a*) same precipitate for different sample orientations; (*b*) high-resolution image showing the amorphous nature of the precipitate. (*After Bender [1984a].*)

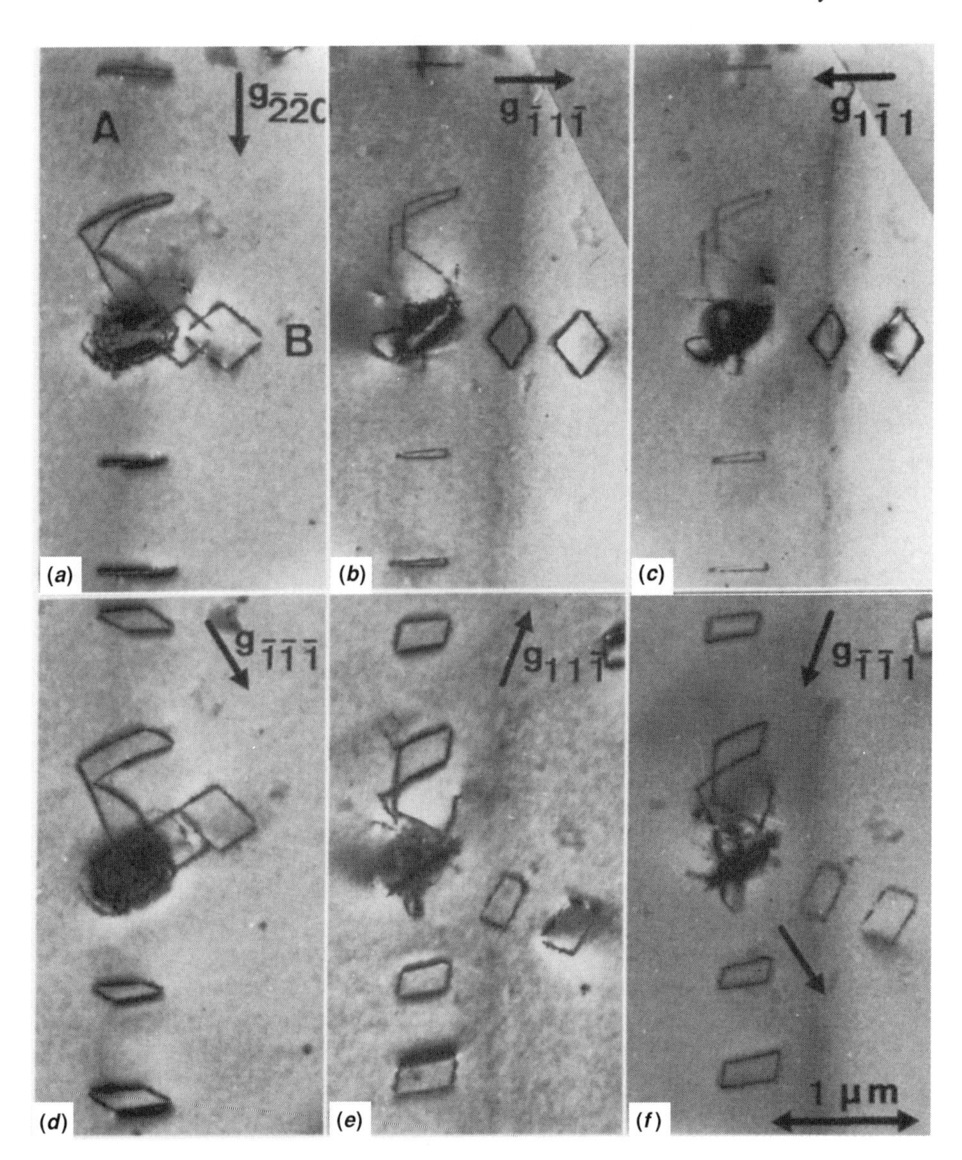

FIGURE 5.40
a → f. Electron micrographs showing the contrast behavior of prismatically punched-out loops, observed in association with platelike precipitates in CZ silicon annealed at 1050°C for 20 hr, for different reflections. (*After Bender and Vanhellemont [1994].*)

glide and climb to form prismatic loops. Under the influence of the precipitate-induced misfit stress, the loop glides away from the precipitate on the glide cylinder containing the loop segments. If we assume that this process can occur repeatedly, then we can understand the arrangement of loops shown in Figure 5.40(a) to (f).

Another notable feature of the doubly annealed samples is extrinsic stacking faults, lying on {111} planes and bounded by a/3 <111> Frank partials. An example

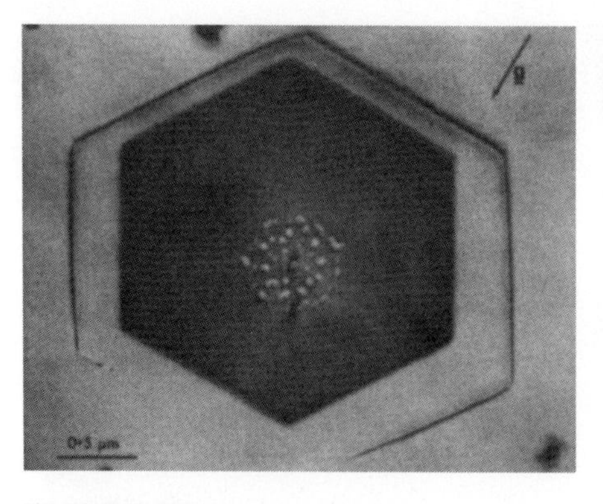
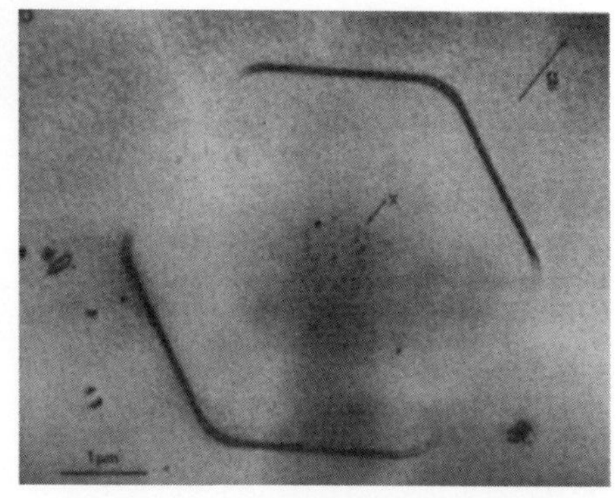

FIGURE 5.41
An example of an extrinsic stacking fault observed in CZ Si after annealing. Note the presence of precipitates in the center of the stacking fault. (*After Maher et al. [1976].*)

of a stacking fault, reproduced from the study of Maher et al. (1976), is shown in Figure 5.41. The fault surface is decorated with precipitates.

We now attempt to rationalize the formation of stacking faults in aged CZ silicon. Since the volume of the SiO_x precipitate is considerably larger than the volume of silicon consumed in its formation, silicon atoms must be emitted as interstitials into the adjoining lattice. Their concentration can be estimated using the approach developed by Bender and Vanhellemont (1988). The total number of interstitials C_I^g is given by

$$C_I^g = nw\Delta C_{OI}, \tag{5.32}$$

where n is the number of interstitials consumed per oxygen atom in the formation of precipitates assuming full-stress release, ΔC_{OI} is the initial interstitial oxygen supersaturation, and w is the fraction of the volume expansion that is accommodated by the emission of silicon atoms into the adjoining lattice and is given by

$$w = 1 - \frac{\Omega_{S_iO_2}}{\Omega_{Si}} \frac{R}{t}, \tag{5.33}$$

where $\Omega_{S_1O_2}$ and Ω_{S_1} are the volumes of an SiO_2 molecule and Si atom, respectively, R is the magnitude of the displacement vector associated with the precipitate, and t is the thickness. Bender (1984a) has estimated that for small precipitates forming at 650°C, w is 0.35. With n and ΔC_{OI} being -0.58 and 1×10^{18} cm^{-3}, C_I^g is estimated to be 2.0×10^{17} cm^{-3}. Mahajan et al. (1977) assume that this supersaturation plays a role in the formation of stacking faults. Their model is shown schematically in Figure 5.42.

The formation of silicon-oxygen clusters, whose habit plane is {111} and whose strain field is directed along the <111> direction, constitutes the first stage in the nucleation of stacking faults as shown in Figure 5.42*a*. Maher et al. (1976) have experimentally observed such clusters. The concominant emission of silicon interstitials occurs. When the supersaturation of the silicon interstitials reaches a critical value, they precipitate out into an embryonic Frank loop, Figure 5.42*b*. We emphasize

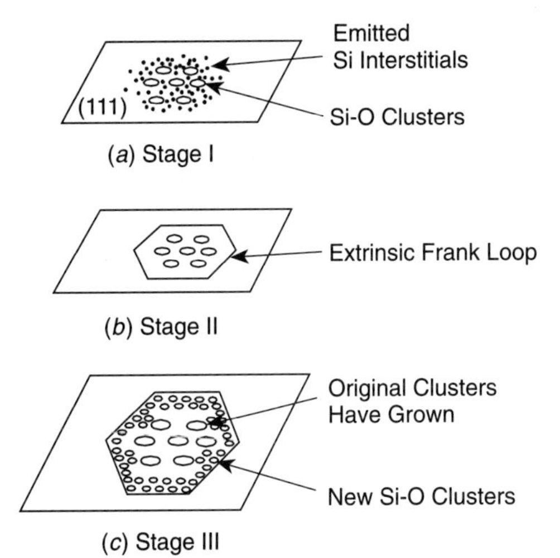

FIGURE 5.42
Schematic of the model proposed by Mahajan et al. (1977) to rationalize the role of SiO_x precipitates in the nucleation and growth of extrinsic stacking faults in annealed Si containing oxygen.

that this situation is analogous to the formation of microdefects covered in section 5.5. The presence of the strain field along the <111> direction facilitates this process. The fault expands either by the growth of the existing clusters during additional aging or by the formation of new silicon-oxygen clusters on the stacking fault as depicted in Figure 5.42*c*.

5.6.2.3 Mechanical properties

For certain devices we need to deposit layers on CZ substrates at high temperatures. If the substrates are mechanically weak, they may sag under their own weight during the deposition. This condition is very undesirable because it would be difficult to process them subsequently using photolithographic techniques covered in Chapter 11. With this objective in mind, we now discuss the influence of the loss of structural perfection discussed above on the mechanical properties of CZ silicon crystals.

Siethoff (1973) and Mahajan et al. (1979) have investigated the deformation behavior of as-grown and annealed CZ crystals. Figure 5.43 shows the representative stress-strain curves of crystals deformed at 800, 900, and 1000°C and also of the annealed crystals deformed at 800°C. Important observations are (1) the as-grown crystals exhibit a yield drop at 800°C; the magnitude of the drop decreases with the increasing deformation temperature, and the drop is absent at 1000°C; (2) the annealed crystals are weaker than the as-grown ones and do not show yield drops; and (3) the yield stress shows a very strong temperature dependence.

In the context of the present discussion, we need to explain the weakening of crystals on annealing. We know that the as-grown crystals are of high quality and have few mobile dislocations. When such crystals are deformed, the mobile dislocations move with a high velocity because an imposed strain-rate ε^* is related to the number of moving dislocations n and their velocity v by

$$\varepsilon^* = \text{nbv}, \tag{5.34}$$

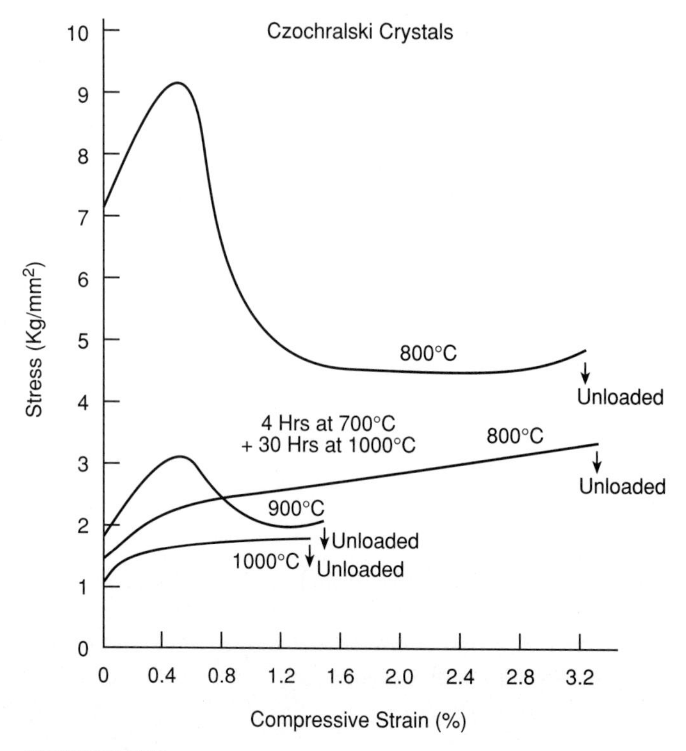

FIGURE 5.43
Representative stress-strain curves of CZ Si crystals, oriented for single slip, at different temperatures. The stress-strain curve of a heat-treated crystal, obtained at 800°C, is also included. Note the dramatic difference in the responses of as-grown and heat-treated crystals. (*After Mahajan et al. [1979].*)

where b is the Burgers vector. Furthermore, $v \alpha (\tau)^m$, where τ is the applied shear stress and m is the stress exponent. When v is high, τ must be high; that is, the strength of the crystal must be high. However, on annealing, additional mobile dislocations are introduced into the crystal; that is, n increases. Therefore, to produce the same strain-rate ε^*, v must decrease, resulting in a lower value of τ.

5.6.2.4 Internal gettering of impurities by silicon-oxygen precipitates

Since precipitates constitute strain centers in the silicon matrix, they can be used to remove fast-diffusing impurities from the silicon lattice. This is achieved by accumulating the impurities in regions contiguous to and at the precipitates and associated substructure. The concept of *internal gettering* was proposed by Tan and Tice (1976), and since then it has gained considerable acceptance in the silicon-integrated-circuit-fabrication technology.

The conditions conducive to internal gettering are created by a two-step anneal. A high-temperature oxygen out-diffusion anneal, typically above 1100°C, is performed first. During this step the concentration of oxygen in the surface regions where circuit elements are fabricated drops below its solubility limit in the silicon lattice. This high-temperature anneal is followed by a low-temperature anneal. During this anneal silicon-oxygen precipitates that is, the strain centers, form only in the central region of the wafer where the oxygen concentration is still in excess of the solubility limit. During subsequent device processing, impurities could migrate to the strain centers.

The efficacy of the precipitates and associated substructure as gettering sites has been demonstrated experimentally by Ourmazd (1986), Colas et al. (1986), and Jastrzebski et al. (1987). For additional details, the reader is referred to the review article by Ourmazd (1986) on internal gettering.

5.7 HIGHLIGHTS OF THE CHAPTER

The highlights of this chapter are as follows:

- Large-diameter, high-quality silicon crystals can be grown using the CZ process. A drawback is that these crystals contain oxygen up to $\sim 2 \times 10^{18}$ cm^{-3}, a result of the melts being contained in quartz crucibles. This problem does not exist for the FZ silicon crystals, but there is a limitation on their diameter. Crystals of various compound semiconductors are grown by the liquid-encapsulated Czochralski and Bridgman techniques.
- Several sources of dislocations in as-grown crystals have been identified. The addition of impurities to enhance the critical-resolved shear stress of the lattice is an attractive approach for lowering the dislocation densities in compound semiconductors.
- When the impurity distribution coefficient in the melt and the solid is different from one, the impurity atoms in the doped crystal are inhomogeneously distributed in a striated pattern that constitutes growth striations.
- The point defects, inherited by the CZ and FZ crystals during their growth from the melt, cluster together to form microdefects. These microdefects are also distributed in striated patterns that resemble those of the impurities.
- During the annealing of CZ crystals, the excess oxygen comes out of solution in the form of silicon-oxygen clusters and precipitates. These features introduce thermal donors, impair the perfection of crystals, and have deleterious effects on the mechanical strength.

REFERENCES

Abe, T. and S. Maruyama. *Derki Kagaku* 35, 149 (1967).
Ashby, M. F. and L. Johnson. *Phil. Mag.* 20, 1009 (1969).
Augustus, P. D. and D. J. Stirland. *J. Electrochem. Soc.* 129, 614 (1982).
Bauser, E. and G. Rozgonyi. *J. Electrochem. Soc.* 129, 1782 (1982).

Beam, E. A. III; S. Mahajan; and W. A. Bonner. *Mats. Sci. and Eng.* B7, 83 (1990).
Beam, E. A. III; H. Temkin; and S. Mahajan. *Semicond. Sci. Technol.* 7, A229 (1992).
Bean, A. R. and R. C. Newman. *J. Phys. Chem. Solids* 32, 1211 (1971).
Bender, H. *Phys. Stat. Sol.* (a) 86, 245 (1984a). Bender, H. Ph.D. dissertation, University of Antwerp (1984b).
Bender, H. and J. Vanhellemont. *Phys. Stat. Sol.* (a) 107, 455 (1988). Bender, H. and J. Vanhellemont. In *Handbook on Semiconductors:* "Oxygen in Silicon." ed. S. Mahajan, vol 3, 2 ed. Amsterdam, North Holland (1994) p. 163.
Bonner, W. A. *J. Cryst. Growth* 54, 21 (1981).
Bosomworth, D. W.; W. Hayes; A. R. L. Spray; and G. D. Watkins. *Proc. Roy. Soc.* A 317, 133 (1970).
Bourret, A. In *Defects in Semiconductors:* "Oxygen Aggregation in Silicon" ed. L. C. Kimerling and J. M. Parsey Jr., Warrendale, PA. The Metallurgical Society, 1985, p. 129. Bourret A., In *Oxygen, Carbon, Hydrogen and Nitrogen in Crystalline Silicon*: "Transmission Electron Microscope Studies of O, C, N Precipitation in Crystalline Silicon." edited J. C. Mikkelson, Jr., S. J. Pearton, J. W. Corbett, and S. J. Pennycook, Materials Research Society, Pittsburgh, 1986, p. 223.
Brasen, D. and W. A. Bonner. *Mats. Sci. and Eng.* 61, 167 (1983).
Brown, R. A.; T. A. Kinney; P. A. Sackinger; and D. E. Bornside. *J. Cryst. Growth* 97, 99 (1989).
Burton, J. A.; R. C. Prim; and W. P. Slichter. *J. Phys. Chem. Solids* 21, 1987 (1953).
Cazcarra, V. and P. Zunina. *J. Appl. Phys.* 51, 4206 (1980).
Cockayne, B.; G. T. Brown; and W. R. MacEwan. *J. Cryst. Growth* 64, 48 (1983).
Colas, E. G.; E. R. Weber; and S. Hahn. *Mats. Res. Symp. Proc.* 71, 13 (1986).
Cullis, A. G., P. D. Augustus; and D. J. Stirland, *J. Appl. Phys.* 51, 2556 (1980).
Dash, W. C. *J. Appl. Physics* 29, 705, 736 (1958).
Dash, W. C. *J. Appl. Phys.* 30, 459 (1959).
deKock, A. J. R. *Philips Research Report,* Supl. no. 1 (1973). In *Festkörper Probleme* vol. 16, ed. J. Truesch, 1976, p. 179. In *Semiconductor Silicon:* "Point Defect Condensation in Dislocation Free Silicon Crystals" 1977, ed. H. R. Huff and E. Sirtl. Pennington: Electrochemical Society, 1977 p. 508. In *Handbook on Semiconductors:* "Crystal Growth of Bulk Crystals: Purification, Doping and Defects," vol. 3, ed. S. P. Keller. Amsterdam: North-Holland, 1980 p. 247.
deKock, A. J. R.; W. T. Stacy; and W. M. van de Wijgert. *Appl. Phys. Lett.* 34, 611 (1979).
deKock, A. J. R. and W. M. van de Wijgert. *J. Cryst. Growth* 49, 718 (1980).
Ehrenreich, H. and J. P. Hirth. *Appl. Phys. Lett.* 46, 668 (1985).
Föll, H.; U. Gösele; and B. O. Kolbesen. *J. Cryst. Growth* 40, 90 (1977).
Föll, H. and B. O. Kolbesen. *Appl. Phys.* 8, 319 (1975).
Fuller, C. S. and F. H. Doleiden. *J. Appl. Phys.* 29, 1264 (1958).
Gass, J.; H. H. Muller; H. Stussi; and S. Schweitzer. *J. Appl. Phys.* 51, 2030 (1980).
Gault, W. A.; E. M. Monberg; and J. E. Clemens. *J. Cryst. Growth* 74, 491 (1986).
Gaworzewski, P. and K. Schmaiz. *Phys. Stat. Sol. (a)* 55, 699 (1979).
Harada, H.; T. Itoh; N. Ozawa; and T. Abe. In *VLSI Science and Technology 1985,* ed. W. M. Bullis and S. Broydo. Pennington, NJ: Electrochemical Society, 1985, p. 526.
Huff, H. In *The Concise Encyclopedia of Semiconducting Materials and Related Technologies:* "Silicon Properties and Materials Specifications," ed. S. Mahajan and L. C. Kimerling. Oxford: Pergamon Press, 1992, p. 488.

Jastrzebski, L.; R. Soydan; J. McGinn; R. Kleppinger; M. Blumenfeld; G. Gillespie; N. Armour; B. Goldsmith; W. Henry; and S. Vecrumba. *J. Electrochem. Soc.* 134, 1018 (1987).

Jordan, A. S.; R. Caruso; and A. R. von Neida. *Bell System Tech. J.* 59, 593 (1980).

Kafalas, J. In *Gallium Arsenide Technology,* ed. D. K. Ferry., In Howard W. Sams and Co., 1985.

Kaiser, W.; H. L. Frisch; and H. Reiss. *Phys. Rev.* 112, 1546 (1958).

Kaiser, W.; P. H. Kech; and C. F. Lange. *Phys. Rev.* 101, 1264 (1956).

Kimerling, L. C. and J. L. Benton. *Appl. Phys. Lett.* 39, 410 (1981).

Lagowski, J.; H. C. Gatos; J. M. Parsey; K. Wada; M. Kaminska; and W. Walukiewicz. *Appl. Phys. Lett.* 40, 342 (1982).

Lin, W. and D. W. Hill. *J. Appl. Phys.* 51, 5540 (1980).

Lin, W. and M. Stavola. *J. Electrochem. Soc.* 132, 1412 (1985).

Mahajan, S. *Progress in Mats. Sci.* 33, 1 (1989).

Mahajan, S.; D. Brasen; and P. Haasen. *Acta Met.* 27, 1165 (1979).

Mahajan, S.; G. Rozgonyi; and D. Brasen. *Appl. Phys. Lett.* 30, 73 (1977).

Maher, D. M.; A. Staudinger; and J. R. Patel. *J. Appl. Phys.* 47, 3813 (1976).

Mikkelsen, J. C. *Appl. Phys. Lett.* 40, 336 (1982a). Mikkelsen, J. C. *Appl. Phys. Lett.* 41, 871 (1982b).

Mikkelsen, J. C. *Mat. Res. Soc. Symp. Proc.* 59, 19 (1986).

Morizane, K.; A. F. Witt; and H. C. Fatos. *J. Electrochem. Soc.* 113, 51 (1966).

Morizane, K.; A. F. Witt; and H. C. Gatos. *J. Electrochem. Soc.* 114, 738 (1967).

Nozaki, T.; Y. Yatsurugi; and N. Akiyama. *J. Radioanalyt. Chem.* 4, 8 (1970).

Nygren, S. F. *J. Cryst. Growth* 19, 21 (1973).

Oehrlein, G. S. and J. W. Corbett. In *Defects in Semiconductors II:* "Early Stages of Oxygen Clustering and its Influence on the Electrical Behavior of Silicon," ed. S. Mahajan and J. W. Corbett. New York: North-Holland, 1983 p. 107.

Ourmazd, A. *Mats. Res. Symp. Proc.* 59, 331 (1986).

Ourmazd, A.; W. Schrötter; and A. Bourret. *J. Appl. Phys.* 56, 1670 (1984).

Pajot, B. and B. Cales. *MRS Symp. Proc.* 59, 39 (1986).

Parsey Jr., J. M.; J. Lagowski; and H. Gatos. In *III-V Opto-Electronics Epitaxy and Device Related Processes:* "Effects of Melt Stoichiometry and Impurities on the Formation of Dislocations in Bulk GaAs," ed. V. G. Keramidas and S. Mahajan, Pennington, NJ: The Electrochemical Society, 1983, p. 61.

Petroff, P. M. and A. J. R. deKock, *J. Cryst. Growth* 30, 117 (1975).

Petroff, P. M. and A. J. R. deKock. *J. Cryst. Growth* 35, 4 (1976).

Rai, R. S.; S. Guruswamy; K. T. Faber; and J. P. Hirth. *Phil. Mag A,* 60, 1339 (1989).

Rai, R. S.; S. Mahajan; D. J. Michel; H. H. Smith; S. McDevitt; and C. J. Johnson. *Mat. Sci. Eng.* B10, 219 (1991).

Ravi, K. V. *Imperfections and Impurities in Semiconductor Silicon.* New York: Wiley-Interscience, 1981, p. 237.

Ravi, K. V. and C. J. Varker. *J. Appl. Phys.* 45, 263 (1974).

Rea, S. N. *J. Cryst. Growth* 54, 267 (1981).

Roksnoer, P. J.; W. J. Bartels; and C. W. T. Bulle. *J. Cryst. Growth* 35, 245 (1976).

Seidensticker, R. G. In *Crystals: Growth, Properties and Applications:* "Dendritic Web Growth of Silicon," vol. 8, ed. H. C. Freyhardt. Berlin: Springer-Verlag, 1982, p. 145.

Seki, Y.; J. Matsui; and H. Watanabe. *J. Appl. Phys.* 47, 3374 (1976).

Seki, Y.; H. Watanabe; and J. Matsui, *J. Appl. Phys.* 49, 822 (1978).

Siethoff, H. *Acta Met.* 21, 1523 (1973).

Stavola, M. *Appl. Phys. Letts.* 44, 514 (1984).
Stavola, M. and K. M. Lee. In *Oxygen, Carbon, Hydrogen and Nitrogen in Crystalline Silicon:* "Electonic Structure and Atomic Symmetry of the Oxygen Donor in Silicon," ed. J. C. Mikkelsen Jr., S. J. Pearton, J. W. Corbett; and S. J. Pennycook. Pittsburgh: Materials Research Soc., 1986, p. 95.
Stavola, M.; J. R. Patel; L. C. Kimerling; and P. E. Freeland. *Appl. Phys. Lett.* 42, 73 (1983).
Stavola, M. and L. C. Snyder. In *Proceedings of the Symposium on Defects in Silicon,* ed. W. Murray Bullis and L.C. Kimerling. Pennington, NJ: The Electrochemical Soc., 1983, p. 61.
Stolte, C. A. In *Semiconductors and Semimetals:* "Ion Implantation and Materials for GaAs Integrated Circuits," vol. 20, ed. R. K. Willardson and A. C. Beer. New York: Academic Press, 1984, p. 89.
Swaminathan, V. and A. T. Macrander. *Materials Aspects of GaAs and InP Based Structures.* Englewood Cliffs, NJ: Prentice Hall, 1991, p. 53.
Takano, Y. and M. Maki. In *Semiconductor Silicon 1973,* ed. H. R. Huff and R. R. Burgers. Pennington NJ: The Electrochemical. Soc., 1973, p. 469.
Tan, T. Y. and W. K. Tice. *Phil. Mag.* 30, 615 (1976).
Tempelhoff, K.; F. Spiegelberg; R. Gleichmann; and D. Wruck. *Phys. Stat. Sol.*(a) 56, 213 (1979).
Terashima, K.; T. Katsumoto; F. Orito; and T. Fukuda. *Jap. J. Appl. Phys.* 23, L302 (1984).
Utech, H. P. and M. C. Flemings. *J. Appl. Phys.* 37, 2021 (1966).
Varker, C. J. and K. V. Ravi. *J. Appl. Phys.* 45, 272 (1974).
Wada, K.; N. Inoue; and K. Kohra. *J. Crystal Growth* 49, 749 (1980).
Wald, F. V. In *Crystals: Growth, Properties and Applications:* "Crystals Growth of Silicon Ribbons for Terrestrial Solar Cells by the EFG Method," vol. 5, ed. H. C. Freyhardt, Berlin: Springer-Verlag, 1981, p. 147.
Witt, A. F. and H. C. Gatos. *J. Electrochem. Soc.* 113, 808 (1967).
Yatsurugi, Y.; N. Akiyama; Y. Endo; and T. Nozaki. *J. Electrochem. Soc.* 120, 975 (1973).

PROBLEMS

5.1. The as-grown silicon web has the following crystallography: The growth axis is $[11\bar{2}]$, and the plane of the web is (111). Assuming that slip occurs in the web because of thermal-gradient-induced stresses, show schematically the distribution of etch pits that would be observed on the web surface and also indicate the directions along which the pits would be aligned.

5.2. *a.* Due to prevailing conditions during the growth of GaAs crystals by the horizontal Bridgman technique, $5 \times 10^{17}\ \text{cm}^{-3}$ As interstitials are introduced into the crystal. Assuming that these interstitials coalesce together to form circular Frank loops 10 nm in diameter, calculate the number of loops/$(\text{cm})^3$ in the crystal, given that the lattice parameter of GaAs is 0.565 nm.

b. Are gallium interstitials required in the formation of these loops? Also, comment on the probability of formation of dislocation loops versus As precipitates.

5.3. During CZ growth the first-to-freeze portion of the crystal is necked down to 3 mm in diameter to reduce the density of dislocations in the crystals. If the yield strength of silicon at high temperature is 5×10^6 g/cm^2, calculate the weight of an ingot that the neck can support. Also, calculate lengths of ingots that this weight can support when the diameters are 150 and 250 mm. The density of silicon is 2.328 g/cm^3.

5.4. After necking, a dislocation in the central portion of the neck propagates into a silicon crystal; see Figure 5.15*a*. Assuming that this dislocation is perfect edge in character and its Burger vector is a/2 <110>, calculate the number of point defects required to climb 1 centimeter length of this dislocation to the neck surface, given that the neck diameter is 3 mm. Lattice parameter of silicon is 0.543 nm.

5.5. During growth an InP crystal experiences thermal-gradient-induced stresses along the [111] direction. Assuming that resolved shear stresses on different slip systems are in excess of that required for plastic deformation of indium phosphide, show schematically the etch-pit patterns that are observed on slices having the following orientations: [001], [110], [111], and [112]. In each case indicate directions along which etch pits are aligned.

5.6. During crystal growth the addition of certain impurities to the host lattice can reduce the probability of dislocation multiplication. Using Table 3.1, suggest which of the following additions you would use or not use and why for the growth of p- and n-type cadmium telluride and p- and n-type aluminum nitride: beryllium, magnesium, zinc, aluminum, gallium, carbon, tin, phosphorus, antimony and sulphur.

5.7. CZ crystals, doped with boron to 5×10^{15} cm^{-3}, are required. What concentration of boron atoms in the melt would enable the crystals to obtain the desired carrier concentration? Assuming that the melt is 70 kg, how many grams of boron (atomic weight = 10.8) should be added?

5.8. Calculate the variation in concentration of iron along the length of an indium phosphide crystal, grown by LEC, when the initial concentration of iron in a small melt is 5×10^{18} cm^{-3}.

5.9. Assuming that nonplanar surfaces can be described in terms of steps bounded by planes normal to the surface, which of the following silicon growths are likely to exhibit type II striations and why: [001], [110], and [111].

5.10. *a.* Calculate the excess concentrations of vacancies and interstitials present in a silicon crystal after it is cooled from a temperature of 1410 to 900°C; the respective energies of formation are 2.3 and 1.1 eV.

b. If the vacancies and interstitials coalesce separately into circular loops of 10 and 50 nm in diameter on {111} planes, calculate the number of loops/cm^3 that would form.

5.11. *a.* Calculate the number of silicon atoms involved in the formation of a fully relaxed spherical silicon dioxide precipitate 50 nm in radius during aging of a CZ crystal, given that densities of silicon and silicon dioxide are 2.33 and 2.2 g/cm^3.

b. If the excess silicon generated during the relaxation process in (*a*) coalesces to form circular dislocation loops on {111} planes, calculate the number of loops generated during aging; assume that the loop diameter is 20 nm.

5.12. A prismatic dislocation 1 μm in length dissociates into a Shockley partial and a Frank partial. If the Frank partial is subjected to a flux of silicon interstitials generated during the formation of silicon dioxide precipitates and behaves as a Bardeen-Herring source, calculate the size of a fully relaxed silicon dioxide precipitate whose formation would yield enough silicon interstitials to produce a $\{111\}$ extrinsic fault 1 μm in diameter.

5.13. On aging, the strength of CZ crystals is reduced from 100 MPa to 10 MPa. Assuming that this decrease is due to the formation of a number of dislocations contiguous to SiO_2 precipitates, calculate the number of precipitates required to produce this effect, given that the number of dislocations per precipitate is five and the stress exponent m is two.

CHAPTER 6

Epitaxial Growth

This chapter covers the techniques that are used to deposit epitaxial layers of semiconductors. It also considers the issues associated with the growth on nonplanar surfaces and heteroepitaxy; that is, the composition of the layer or its crystal structure is different from that of the underlying substrate. The final topics are the origins of defects during epitaxial growth and microstructures of mixed III-V layers.

6.1 EPITAXIAL GROWTH TECHNIQUES

For optimal performance, devices such as detectors, light-emitting diodes, and lasers require multilayer structures. These structures are grown on substrates fabricated from bulk crystals that serve as templates. The overgrowth is referred to as an *epitaxial layer* and the process is termed *epitaxy.*

Epitaxy can be broadly classified into two categories: (1) homoepitaxy, isoepitaxy, or autoepitaxy and (2) heteroepitaxy. In *homoepitaxy* the composition of the layer is essentially the same as that of the underlying substrate. Typical examples of homoepitaxy are silicon epitaxial layers grown on silicon substrates, that is, Si/Si, and GaAs/GaAs. Even n-GaAs/p-GaAs growth is classified as homoepitaxy. When the composition of the layer differs from that of the substrate, epitaxy is termed *heteroepitaxy;* typical examples are Si/GaAs, InP/InGaAsP, and $(01\bar{1}2)$ sapphire/(001)Si.

Epitaxial growth technology has made great strides during the last three decades. Various techniques are currently available for depositing layers, and they can be broadly classified into three groups: (1) liquid phase epitaxy (LPE), (2) vapor phase epitaxy (VPE), and (3) molecular beam epitaxy (MBE). There are further derivatives of the VPE and MBE processes depending on the sources for atoms constituting the layer; the sources can be in the form of chlorides, hydrides, and organometallics (OM). The corresponding designations for VPE and MBE are chloride VPE, hydride VPE, OMVPE, gas source MBE, and OMMBE.

6.1.1 Liquid Phase Epitaxy

LPE is used to deposit layers of various semiconductors. The technique is simple, elegant, and has many advantages: (1) experimental setup is simple and inexpensive, (2) growth temperatures are low, 350 to 900°C, (3) relatively high growth rates, 0.1 mm/min, are possible, and (4) a low concentration of point defects is incorporated into the layers. The factors limiting the quality of the epitaxial layers are primarily related to morphological features and composition variations (Small and Potemski 1982). However, because of its simplicity and flexibility, LPE has been attractive for growing complex layered structures required for optoelectronic devices.

We can illustrate the principle of LPE by referring to Figure 6.1, which shows the Ga-As phase diagram. Imagine that we bring a Ga melt, saturated with As at 600°C, in contact with a GaAs substrate also maintained at 600°C. If we cool the melt and the substrate together within the two-phase field consisting of the Ga-rich liquid and the GaAs solid, the GaAs solid will form on cooling. This GaAs will deposit epitaxially on the underlying substrate.

Figure 5.12 showed that the Ga-As system has an existence region, which limits the highest temperature that can be used to deposit the GaAs layers. If we were to grow a GaAs layer from a Ga-melt at 1000°C, the layer would be Ga-rich. Likewise, the layer grown from an As-melt would be As-rich. In both cases the layers would be nonstoichiometric.

We now elaborate on the situation that prevails at the solid–liquid interface during LPE. We consider a model system A-B whose characteristics are identical to those of the Ga-As system. We deposit the solid AB from a liquid solution of B in

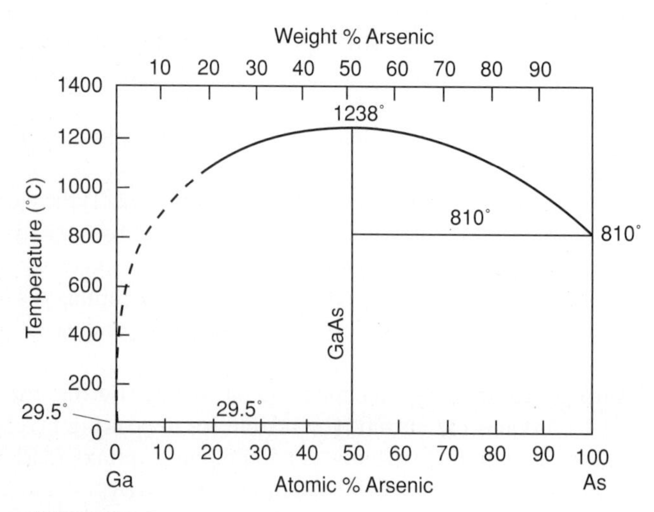

FIGURE 6.1
Phase diagram of the Ga-As system.

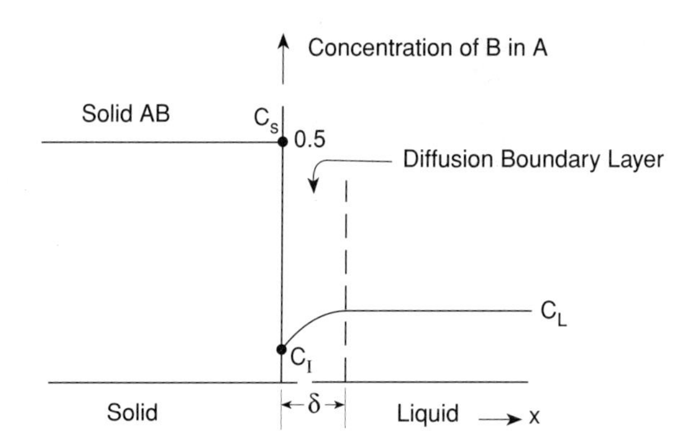

FIGURE 6.2
Schematic of a situation that may prevail at a melt-substrate interface during homoepitaxy of AB by LPE. The solid is deposited from the melt of A atoms, which contains a small amount of B atoms.

A. When the solution is cooled through the two-phase field in contact with the substrate AB, some of the B atoms combine with the A atoms to form AB. This process occurs at the substrate surface because of the ease of nucleation. As a result, the concentration of the B atoms in the melt is reduced in the vicinity of the substrate as illustrated in Figure 6.2. The concentrations of the solute in the liquid solvent A and at the melt–solid interface are, respectively, C_L and C_I. This concentration difference occurs over a diffusion boundary layer of thickness δ. To grow the layer AB, the B atoms must diffuse through the boundary layer and then combine with the A atoms to form AB on the substrate.

We can divide the growth of a layer into two stages: (1) the formation of stable nuclei of AB on the substrate surface, and (2) their subsequent growth into a continuous film. For the growth of a thick film, the two stages are repeated many times. The probability of formation p of a critical size nucleus is given by

$$p \propto \exp\left(\frac{-\Delta G_{crit}}{k_B T}\right), \tag{6.1}$$

where ΔG_{crit} is the free-energy change associated with the nucleation of a critical size nucleus and k_B and T have their usual meanings. In the case of homoepitaxy, ΔG_{crit} represents the energy required to create additional solid–liquid interfaces. However, the situation is much more complicated for heteroepitaxy involving misfits. In this case the energy is expended in creating the heterointerfaces and the structural defects.

When growth is initiated, nuclei of various sizes form. Only those nuclei whose sizes are equal to or greater than the critical size are stable and grow; the others are unstable and are dissolved into the liquid. The critical size increases with the increasing growth temperature because of the decrease in the driving force for the nucleation. On the other hand, the density of the nuclei increases with the decreasing growth temperature.

In the absence of steps on the growth surface (see section 3.3.3), the nuclei form randomly on the surface. Subsequently, they grow laterally by the attachment of atoms at their interfaces with the liquid. This lateral growth brings the nuclei together to form a continuous film. The processes of nucleation and lateral growth are repeated many times during the growth of a thick layer.

The layers are generally grown on vicinal substrates. As shown in Figure 3.24, the surfaces of such substrates consist of terraces, steps, and kinks. In the presence of steps, there is a competition between the attachment of atoms to the step edges and the formation of nuclei on the terraces. If half of the terrace length is smaller than the distance covered by the migrating atoms on the surface during the deposition of a monolayer or a bilayer, the step attachment dominates because of the availability of tighter bonding sites at the steps. The layer then grows by the lateral movement of steps, which is referred to as the *step-flow mechanism.*

The lateral growth of the nuclei occurs by the diffusion of solute atoms B through the boundary layer and their combining with A atoms at the surfaces of the nuclei; that is, the growth of the stable nuclei occurs by the diffusion of B atoms through the liquid. As an illustration, consider the case of GaAs growth that takes place by the diffusion of As in a Ga melt. In the temperature range of 600 to 800°C where the LPE of GaAs is generally carried out, the diffusion constant D is given by (Hsieh 1974):

$$\mathrm{D} = 8 \cdot 6 \cdot 10^{-4} \exp\left(\frac{-3240}{k_B T}\right) \mathrm{cm}^2/\mathrm{sec} \tag{6.2}$$

Assuming a linear composition gradient in the diffusion boundary layer shown in Figure 6.2, the flux of the solute atoms J can be written as

$$\mathrm{J} = \frac{\mathrm{D}(\mathrm{C_L} - \mathrm{C_I})}{\delta} \tag{6.3}$$

If ρ is the density of the grown layer, the growth rate R is given by the following expression (Bryskiewicz 1978):

$$\mathrm{R} = \frac{\mathrm{D}(\mathrm{C_L} - \mathrm{C_I})}{\rho\delta}, \tag{6.4}$$

because $\mathrm{J} = \rho\mathrm{R}$. The actual growth strategies determine the value of C_I because it depends on the solution temperature and its supersaturation. Two growth strategies are commonly used for LPE: step cooling and equilibrium cooling. Since supersaturations are different in the two strategies, the growth rates and the overall quality of the resulting layers are different.

6.1.1.1 Approximate mathematical treatment of step cooling and equilibrium cooling

We base the analysis of the step and equilibrium coolings on three assumptions: (1) The melt and the substrate are in equilibrium with each other so that the solute concentration in the melt is given by the liquidus curve; that is, the activity of solute in the melt is equal to its activity over the substrate; (2) the concentration of solute in the melt C_L and D do not change during the growth; and (3) the melt and the substrate are the same temperature. Since thin layers are generally grown from comparatively large melts and since the temperature range used for the growth is small, the preceding assumptions are reasonable.

Figure 6.2 shows that the solute concentrations in the melt and at the solid–liquid interface are C_L and C_I, respectively. As indicated earlier the layer of composition AB grows by the diffusion of the B atoms through the boundary layer whose thickness is δ. The magnitude of C_L in turn depends on the temperature at which the A melt is saturated with the B atoms. The following one-dimensional diffusion equation describes the situation shown in Figure 6.2:

$$\frac{\partial C}{\partial t} = D\frac{\partial^2 C}{\partial x^2} + v\frac{\partial C}{\partial x}, \tag{6.5}$$

where v is the velocity of the moving solid–liquid interface and x is the distance from the interface. Since only a thin layer is deposited from a relatively large volume of the melt, the interface can be assumed to be stationary; that is, $v\frac{\partial C}{\partial x}$ can be neglected in Equation (6.5). As a result, Equation (6.5) reduces to

$$\frac{\partial C}{\partial t} = D\frac{\partial^2 C}{\partial x^2}. \tag{6.6}$$

The amount of solute that is transported through the diffusion boundary layer and forms the deposit per unit area M is given by

$$M = \int_o^t D\left(\frac{\partial C}{\partial x}\right)_{x=o} dt. \tag{6.7}$$

Finally the layer thickness L_T is given by

$$L_T = \frac{M}{C_S}, \tag{6.8}$$

where C_S is the concentration of solute in the grown layer. For example, C_S is 0.5 for InP or GaAs.

Step cooling. The hypothetical phase diagram of the A-B system in Figure 6.3 shows the growth sequence imposed during step cooling. In the step-cooling

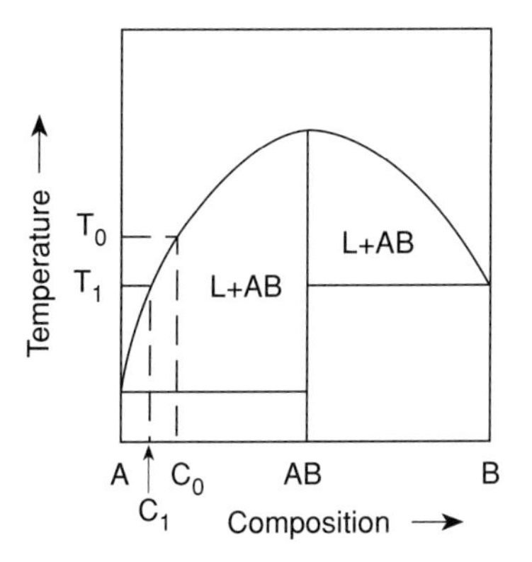

FIGURE 6.3
Schematic illustrating the step-cooling approach used in LPE.

approach, the melt A, which is saturated with the B atoms at temperature T_o, is rapidly cooled to T_1 and is then brought in contact with the substrate maintained at T_1. For the growth of thicker layers, it is necessary to cool the two together at a predetermined rate. The imposed boundary conditions during growth via step cooling are

$$C(x, 0) = C_o$$

and

$$C(0, t) = C_1, \tag{6.9}$$

where $C(x, t)$ is the concentration of the B atoms in the melt at a distance x from the substrate at time t. The imposed supercooling is $T_o - T_1$, and the driving force for the formation of AB is related to the slope of the liquidus curve $m = dT/dC$ around $T = T_o$. The lower the value of m, the higher the driving force for the formation of AB.

The solution to the diffusion Equation (6.6) is

$$\frac{C - C_1}{C_o - C_1} = \mathrm{erf}\left[\frac{x}{2\sqrt{Dt}}\right]. \tag{6.10}$$

Substituting for $\partial C/\partial x$ from Equation (6.10) in Equation (6.7) and integrating the resulting equation between 0 to t we have

$$M = 2(C_o - C_1)\left(\frac{Dt}{\pi}\right)^{\frac{1}{2}}. \tag{6.11}$$

The layer thickness from Equation (6.8) is $L_T = m/C_s$. Assuming that the liquidus curve between T_o and T_1 is essentially a straight line, m can be written as

$$m = \frac{T_o - T_1}{C_o - C_1}. \tag{6.12}$$

Substituting for M and $(C_o - C_1)$ from Equations (6.11) and (6.12) into Equation (6.8), the layer thickness as a function of time at temperature T_1 is given by

$$L_T = 2\,\frac{(T_o - T_1)}{m\,C_s}\left(\frac{D}{\pi}\right)^{1/2} t^{1/2}. \tag{6.13}$$

If we differentiate L_T with respect to t, we obtain the growth rate. We can thus deduce from Equation (6.13) that the growth rate is proportional to $t^{-1/2}$; that is, it decreases with time. The degree of supersaturation in the melt decreases with time, resulting in the reduced driving force for the formation of AB.

Equilibrium cooling. We now consider the equilibrium-cooling approach. Suppose we bring the melt of A atoms, saturated with the B atoms at temperature T_o, in contact with the substrate that is also maintained at T_o. We then cool the two together at a predetermined rate to deposit AB on the substrate. If we impose the cooling rate α, we can show that the layer thickness L_T is given by

$$L_T = \frac{4}{3}\left(\frac{\alpha}{C_s m}\right)\left(\frac{D}{\pi}\right)^{1/2} (t)^{3/2}. \tag{6.14}$$

where α is given by

$$C(o, t) = C_o - \left(\frac{\alpha}{m}\right)t \tag{6.15}$$

We see that in this case the growth rate dL_T/dt is proportional to $t^{1/2}$; that is, the growth rate increases with time because the driving force for the formation of AB increases with the decrease in temperature.

We now comment on the quality of the layers grown by the two approaches. Since the initial growth rate in the step-cooling technique is high, nonequilibrium concentrations of point defects and impurities are incorporated into the layers. The higher the degree of supercooling, the higher their concentrations. On the other hand, we maintain equilibrium between the layer and the growth solution during the equilibrium cooling. Its ramification is that the concentrations of point defects and impurities in the equilibrium cooled layers are lower than those in the step-cooled layers. Since the point defects are electrically active in the semiconductors (see section 3.4), the structural and electrical characteristics of layers grown by the equilibrium-cooling technique are superior to those of layers grown by the step-cooling technique.

EXAMPLE 6.1. Compare the perfection of GaAs epitaxial layers grown by LPE from Ga- and As-rich solutions.

Solution. Figures 5.12 and 6.1 help us in addressing this problem. The temperature range of 400 to 800°C appears suitable for the growth of GaAs layers from a Ga-rich solution. This temperature range is suitable because the tendency for the formation of GaAs per degree drop in temperature is lower than in other regions; that is, m is higher. Also, the atomic mobility at 400°C may not be adequate for depositing the equilibrium composition. Taking into consideration that thermal decomposition of the substrates will be severe above 650°C, the possible temperature range for the growth of the layer is 500 to 650°C. Referring to Figure 5.12, it is apparent that the existence region on the Ga-rich side does not extend to this temperature range. Therefore, in principle high-quality GaAs layers can be grown in this temperature range using LPE from Ga-rich solutions.

The lowest temperature for the growth of a layer from the As-rich side is 810°C, at which the decomposition of the substrate will be severe. Also, the tendency for the formation of GaAs per degree drop in temperature is high between 810 and 1238°C, implying high growth rates and therefore poor-quality layers. The presence of the existence region at high-growth temperatures that was discussed in section 5.2.2.3 would also produce As-rich layers. If we assume that these As atoms occupy the Ga sublattice, then the layers would contain a high density of antisite defects.

In conclusion, the perfection of GaAs layers grown from the Ga-rich solution will be superior to that of the layers deposited from the As-rich melt.

6.1.1.2 Choice of solvents

Several factors are important in the selection of a solvent for LPE: (1) The melting point of the solvent should be considerably lower than that of the semiconductor to be deposited, (2) the purity of the solvent should be very high, (3) the vapor pressure of the solvent at the growth temperature should be low, (4) the reactivity of the solvent with the boat material, such as high-purity graphite, should be minimal, and (5) the solid solubility of the solvent in the layer should be extremely low. Solvents, such as Ga and In, are nearly ideal and satisfy all the preceding requirements for the growth of Ga- and In-based compound semiconductors by LPE.

Solvents belonging to group IV, such as tin, cannot be used for the growth of III-V semiconductors because they act as dopants in the layer. However, they are ideal solvents for the LPE growth of group IV semiconductors. Baliga (1982) has grown high-quality silicon layers from the tin solution at 950°C. The minority-carrier lifetime in the resulting layers is ~100 msec, a fairly high value.

6.1.1.3 In situ etching

In the temperature range where we carry out the LPE growth of III-V semiconductors, the vapor pressures of the group V species are very high. As a result, the substrate undergoes thermal decomposition during the heating cycle, resulting in pits. We illustrate this situation schematically for a (001) GaAs substrate in Figure 6.4*a*. When we heat the substrate above 500°C, As is lost preferentially from the $(\bar{1}11)_{As}$ and $(1\bar{1}1)_{As}$ planes. We must add or remove {111} planes in pairs to maintain the stability of the structure (see section 3.2), which we accomplish by removing {111} planes containing the gallium atoms that are underneath the $(\bar{1}11)_{As}$ and $(1\bar{1}1)_{As}$ planes. Since the vapor pressure of gallium at the growth temperature is very low, the removal rate of the $(111)_{Ga}$ and $(11\bar{1})_{Ga}$ planes is considerably lower than that of the $(\bar{1}11)_{As}$ and $(1\bar{1}1)_{As}$ planes. This anisotropy in the thermally induced removal rates of the two types of {111} planes produces an asymmetric-shape decomposition pit shown in Figure 6.4*a*. These types of pits have been observed on the surfaces of GaAs and InP. Figure 6.4*b* shows the pits observed on the (001) GaAs surface after heating it to 850°C for 1 hr in dry hydrogen (Benz and Bauser 1980). We see a remarkable similarity between the schematic in Figure 6.4*a* and the actual pits shown in Figure 6.4*b*.

We need to remove the pits prior to the LPE growth because in the presence of melt droplets and nonplanar surfaces the layer is of poor quality. To remove the pits, we bring the decomposed substrate in contact with a Ga-melt containing As atoms. Its composition is such that the activity of the arsenic atoms in the melt is lower than that at the substrate surface; that is, the melt is slightly undersaturated with respect to the arsenic atoms. To equilibrate the activities of the As atoms in the melt and above the substrate, the underlying substrate becomes the source of As atoms and is dissolved in the melt, resulting in a smooth surface for the layer growth. We refer to the preceding step in the growth sequence as *meltback*, and it is essential for the growth of high-quality LPE layers.

The meltback step can also help us to remove impurities from the underlying substrates because liquid metals are known to getter impurities. Therefore, in principle we can use LPE to grow the layers having very low carrier concentrations. To date, indium phosphide layers having the lowest carrier concentration have been produced by LPE.

EXAMPLE 6.2. Compare and contrast the crystallographies of thermally induced decomposition pits that may form on the $(111)_A$ and $(\bar{1}\bar{1}\bar{1})_B$ surface of indium phosphide.

Solution. It is difficult to initiate thermally induced decomposition pits on a $(111)_A$ surface because it is bounded by In atoms that have a very low vapor pressure. However, if a pit is able to nucleate heterogeneously, its shape and crystallography may

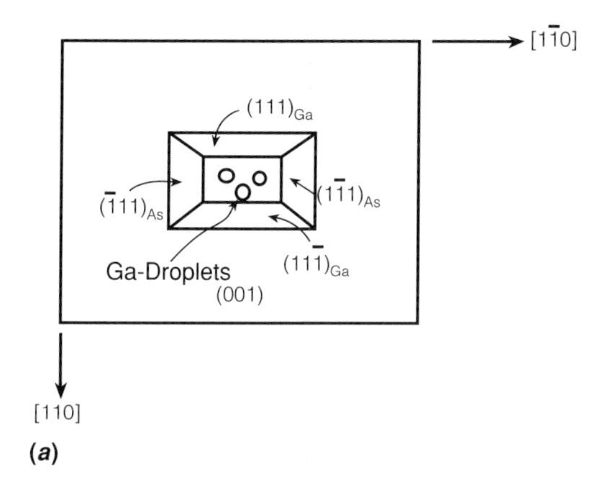

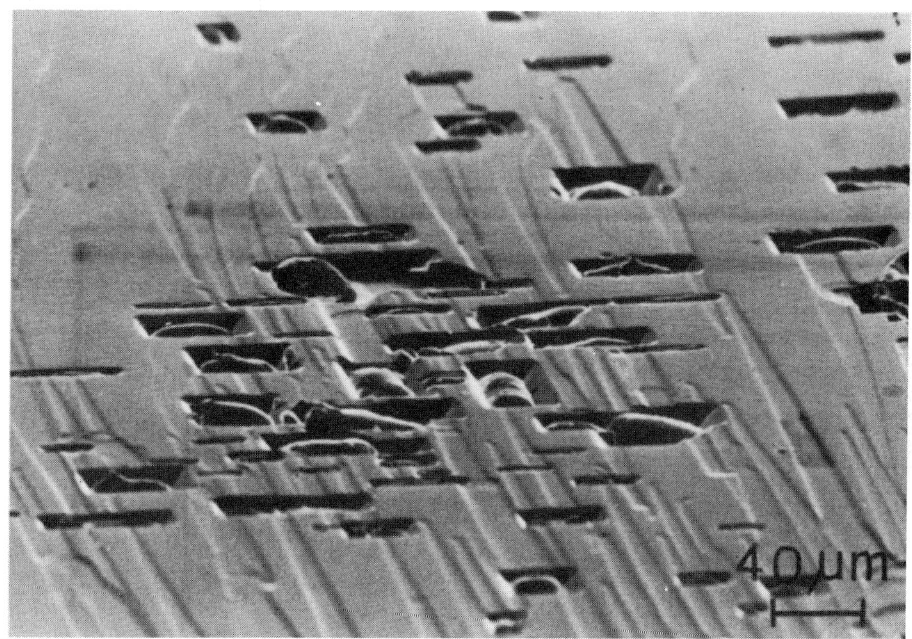

FIGURE 6.4
(*a*) Schematic illustrating the crystallography of a thermally induced decomposition pit on the (001) surface of GaAs. (*b*) Thermal decomposition pits observed on a (001) surface of GaAs after heating for 1h at 850°C in dry hydrogen. (*After Benz and Bauser [1980].*)

evolve as follows. The evaporation of In atoms bounding the (111) surface will expose the (111) layer of phosphorus atoms, which will readily evaporate. The repetition of this process will expose the $\{\bar{1}\bar{1}\bar{1}\}_B$ facets. At this stage the decomposition pit will expand by evaporation from all four {111} facets, resulting in either tetrahedron- or truncated tetrahedron-shape pits shown below:

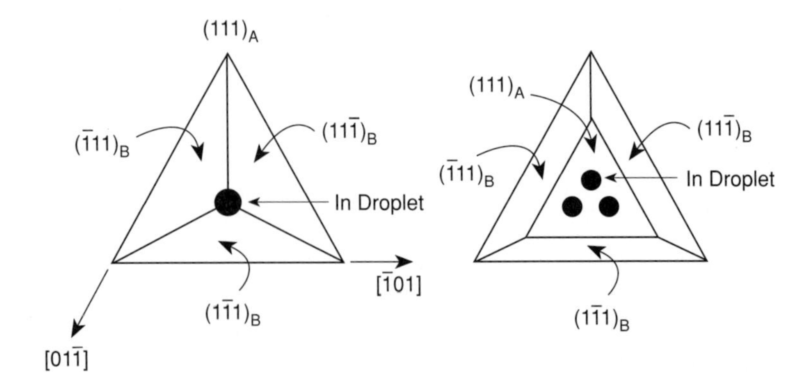

It will be easy to nucleate a thermally induced decomposition pit on a $(\bar{1}\bar{1}\bar{1})_B$ surface because it is bounded by phosphorus atoms whose vapor pressure is very high. Invoking arguments similar to those for the pits on the $(111)_A$ surface, the likely shape and crystallography of the tetrahedron-shape pits is shown below:

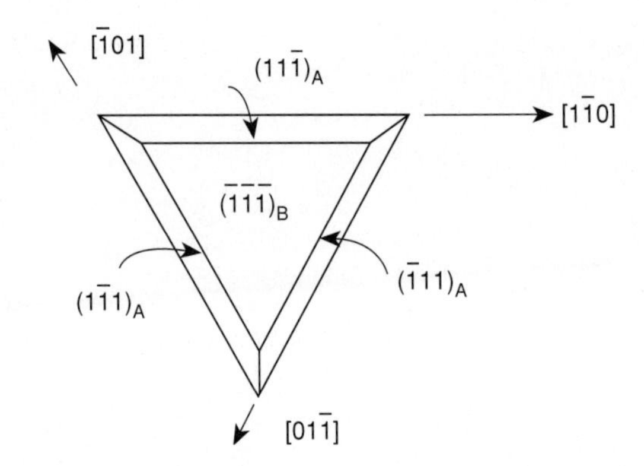

If we assume that the pits grow by the loss of atoms from surface regions, then the size of pits on the $(\bar{1}\bar{1}\bar{1})_B$ surface would be larger. On the other hand, the pits on the $(111)_A$ surface could be deeper because they are bounded by the $\{\bar{1}\bar{1}\bar{1}\}_B$ surfaces.

6.1.1.4 LPE systems and growth of multilayer structures

Since its invention by Nelson, LPE has been carried out in three different configurations: tipping (Nelson 1963), dipping (Rupprecht et al. 1967), and sliding (Panish et al. 1970), the latter being the most popular choice for the growth of multilayer structures. Two assemblies are commonly employed to slide a growth solution on and off a substrate: either a linear boat in a horizontal furnace or a rotating-drum arrangement in a vertical furnace. Figure 6.5 shows a sliding-boat apparatus that allows a meltback. We can use this setup to grow heterostructures consisting of up to four layers, the number of layers generally required for the fabrication of a variety of optoelectronic devices.

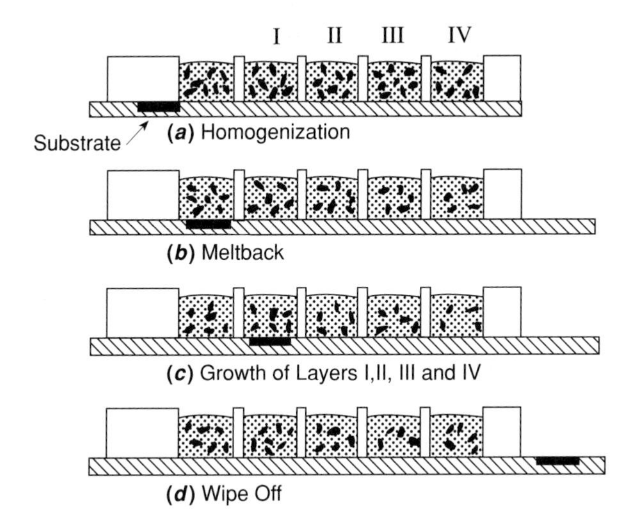

FIGURE 6.5
Schematics showing a sliding-boat arrangement used to grow multilayer structures by LPE.

The first step in the growth sequence is the synthesis of melts of different compositions. We carry out this step at high temperatures in high-purity hydrogen to prevent the oxidation of the melts. After the synthesis we cool the melts to ambient temperature and insert a substrate of appropriate orientation into the recessed portion of the slider shown in Figure 6.5*a*. We again homogenize the melts for a short duration at high temperatures. During the homogenization the substrate undergoes thermal decomposition as discussed in section 6.1.1.3. To eliminate the decomposition-induced damage, we subject the substrate to a meltback (Figure 6.5*b*). Subsequently, we deposit the layers of different compositions on the back-melted substrate by sliding it from one well to the next (Figures 6.5*c* and 6.5*d*). The critical step in the growth of the multilayers is a complete wipe off of the melt as the substrate is slid from one well to another (Mahajan et al. 1982). The ramifications of melt-carryover are discussed in detail in section 6.4.

For the growth of the multilayer structures involving ternary and quaternary compositions such as GaAlAs, InGaAs, and InGaAsP, we need to know the liquid composition that deposits a solid of the desired composition. Several authors (Stringfellow and Greene 1969; Panish and Ilegems 1972; Ilegems and Panish 1974; Jordan and Ilegems 1975) have computed the phase diagrams for ternary and quaternary systems using the simple solution models and the relevant thermodynamic data. Swaminalthan and Macrander (1991) have developed the relevant relations and the interested reader is referred to their work.

From the analysis of the extensive published data on the liquid-solid composition relationships, Kuphal (1984) has developed empirical expressions that relate the melt and solid compositions for the growth of $In_{1-x}\,Ga_x\,As_y\,P_{1-y}$ layers on (001) InP substrates in the temperature range of 570 to 660°C.

$$\begin{aligned} x^1_{Ga} &= \text{the mole fraction of Ga in the liquid} \\ &= \exp(-3584/T)\,[0.70694x + 3.4624x^2 - 8.7492x^3 \\ &\quad +36.554x^4 - 32.878x^5] \end{aligned} \tag{6.16}$$

$$x^{l}_{As} = \left(-\frac{7181}{T}\right)[3.8451 \times 10^{4} x^{l}_{Ga} - 5.6805 \times 10^{6}(x^{l}_{Ga})^{2} + 5.0985 \times 10^{8}(x^{l}_{Ga})^{3} - 2.6191 \times 10^{10}(x^{l}_{Ga})^{4} + 7.0231 \times 10^{11}(x^{l}_{Ga})^{5} - 7.6075 \times 10^{12}(x^{l}_{Ga})^{6}] \tag{6.17}$$

$$x^{l}_{p} = \exp\left(-\frac{11411}{T}\right) \times 10^{2}[13.305(1-y) - 4.7256(1-y)^{2} + 12.417(1-y)^{3} - 3.3953(1-y)^{4}], \tag{6.18}$$

where T is the liquidus temperature. Since the solubility of phosphorus exhibits the strongest temperature dependence, its amount in the solution essentially determines the liquidus temperature. Even though Equations (6.16) to (6.18) are empirical in nature, they provide a very good basis for the selection of liquid compositions to deposit $In_{1-x}Ga_xAs_yP_{1-y}$ layers of specific compositions on (001) InP substrates.

We observe that liquids of different compositions are required to deposit layers of the same composition on differently oriented substrates. Figure 6.6, reproduced from the study of Nakajima and Okazaki (1985), illustrates this effect for $In_{0.53}Ga_{0.47}As$ layers grown at different temperatures. It is clear that to deposit the same composition on $(\bar{1}\bar{1}\bar{1})_P$, $(111)_{In}$, and (100) substrates, we need the decreasing amount of Ga in the melt. The researchers do not understand this effect, but it could result from the orientation dependence of the atomic attachment kinetics.

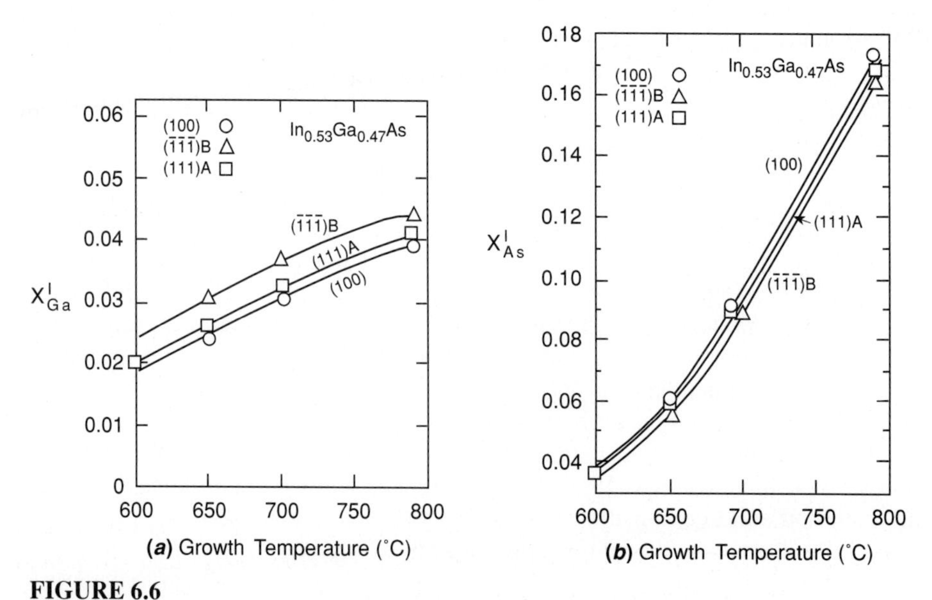

FIGURE 6.6
Mole fractions of (*a*) Ga (x^{l}_{Ga}) and (*b*) As (x^{l}_{As}) required for growth of lattice-matched $In_{0.53}Ga_{0.47}As$ layers on $(\bar{1}\bar{1}\bar{1})_B$, $(111)_A$, and (100) InP substrates as a function of the growth temperature. (*After Nakajima and Okazaki [1985].*)

6.1.1.5 Growth of doped layers

As we indicated in Chapter 2, many device applications require epitaxial layers of specific conductivity types and carrier concentrations. We can grow such layers by adding the appropriate amounts of dopants in the growth melts. These amounts are determined by the distribution coefficient k of the respective dopants (see section 5.4). If k is less than one, the layer is more heavily doped as the growth proceeds; the melts are small and the rejection of the dopant atoms into the melts enriches them.

Typical p- and n-type dopants in III-V epitaxial layers are cadmium and zinc and sulphur, selenium, tellurium and tin, respectively. Silicon behaves as an amphoteric dopant in GaAs. When the layer is grown in the temperature range of 600 to 700°C, silicon atoms are incorporated into the As sublattice. Therefore, it behaves as a p-type dopant. On the other hand, growth in the 900°C range produces n-type material, implying incorporation of silicon atoms into the Ga sublattice. We use the amphoteric behavior of silicon in GaAs to fabricate p-n junctions during LPE growth.

6.1.1.6 Surface morphology of layers

We generally grow LPE layers on vicinal surfaces, which we can describe in terms of the microscopic steps (see section 3.3.3). The separation between the steps depends on the angle of misorientation; the larger the angle of misorientation, the smaller the separation between the atomic steps. The LPE growth occurs by the attachment of atomic species to these steps and their subsequent propagation parallel to the surface (Benz and Bauser 1980), a suggestion consistent with the discussion in section 6.1.1. When step bunching occurs during growth (Bennema and Gilmer 1973), the average step height and the step spacing increase. The steps that become visible on the surfaces of thick LPE layers are referred to as *terraces* (Peters 1973; Saul and Roccasecca 1973; Bauser et al. 1974). The terraces observed on the surfaces of GaAs layers grown on differently misoriented (001) GaAs substrates by LPE are reproduced in Figure 6.7.

The formation of terraces is very sensitive to the initial orientation of the substrate (Peters 1973; Saul and Roccasecca 1973; Bauser et al. 1974). We can eliminate them on the surfaces of GaAs layers by depositing the layers on substrates whose misorientations are below 0.1° (Peters 1973; Bauser et al. 1974). The surfaces of such layers are smooth and exhibit only the growth steps of extremely small height. We show one such example in Figure 6.8 where the GaAs layer is grown on a nearly oriented (001) GaAs substrate (Benz and Bauser 1980). Since in this case the steps are very far apart, we believe that the layer grows by the repeated formation of the deposit nuclei on the surface, followed by their lateral growth (see section 6.1.1). The role of the steps in the growth is minimal.

We can also eliminate the terraces by misorienting the substrate by a critical angle. Figure 6.9 shows the surface of a 50 μm thick GaAs LPE layer grown on a GaAs substrate that is misoriented by 3 degrees from the (001) plane (Benz and Bauser 1980). The terraces are barely discernible. We argue that for step bunching to occur, different steps must move at different velocities. These variations in the velocities could result from the differences in the fluxes of the atomic species to the steps. Since the steps are close together in the case of the critically misoriented substrate,

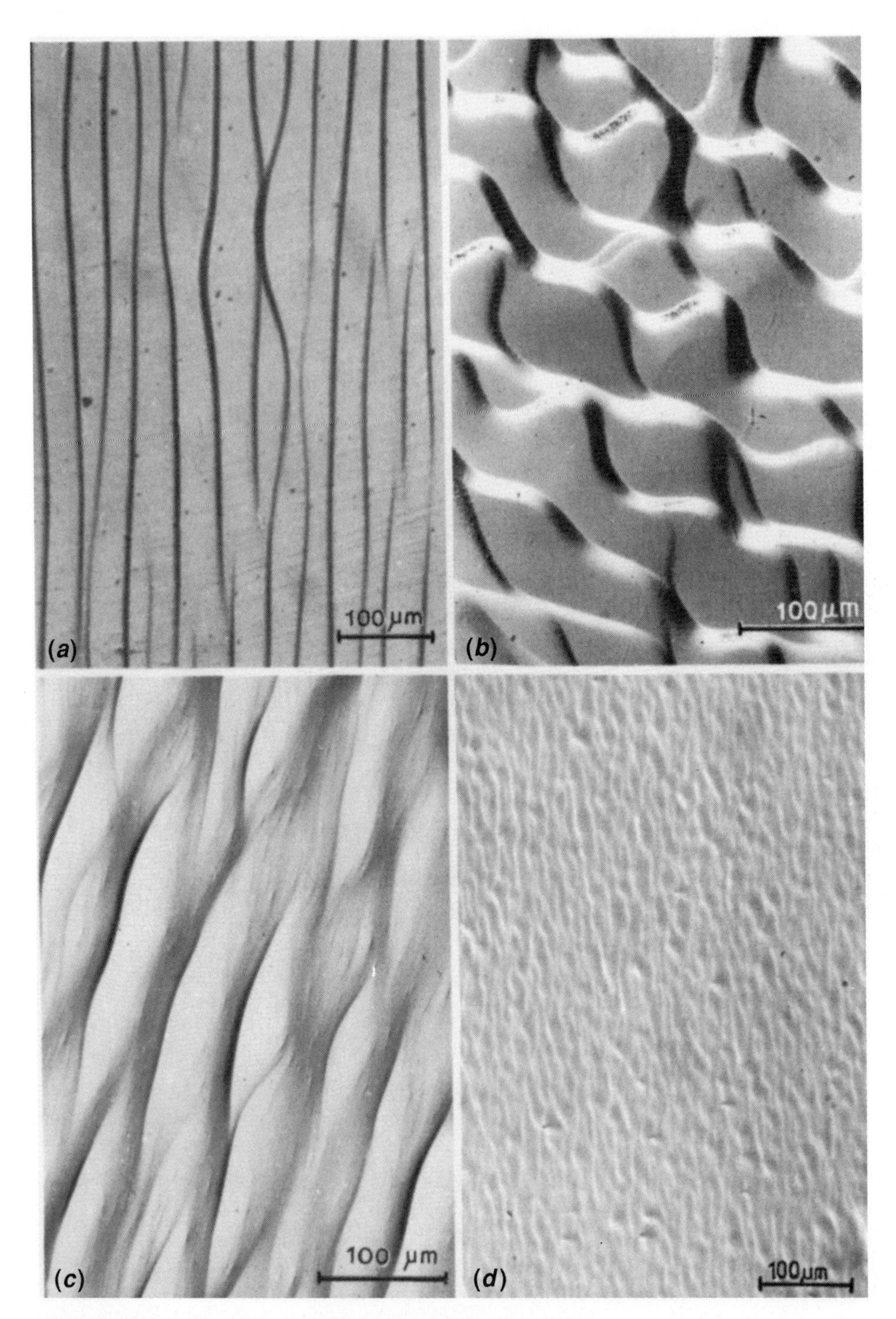

FIGURE 6.7
Surface terraces observed on nominally (001)-oriented, undoped GaAs layers grown by LPE: (*a*) grown on a smooth substrate, the substrate was misoriented by 28 minutes from the [001] direction; the growth temperature Tg is 745°C, and the layer thickness t is 24 μm; (*b*) grown on a rough substrate; Tg = 720°C and t = 16 μm; (*c*) grown on a smooth surface; Tg = 788°C and t = 110 μm; and (*d*) grown on a smooth substrate; Tg = 460°C and t = 1 μm. (*After Benz and Bauser [1980].*)

FIGURE 6.8
GaAs epitaxial layer grown by LPE on a carefully oriented (001) GaAs substrate. The misorientation from the (001) plane is less than 5 minutes. Note the absence of terraces. Terraces do develop near the edges, where the substrate surface is misoriented because of beveling by the polishing and etching processes. (*After Benz and Bauser [1980]*.)

the difference in the fluxes to the neighboring steps may not occur. Thus, step bunching is avoided, leading to the layers that are nearly free of terraces.

We can also affect the evolution of terraces by changing the dopant and the layer composition. Figure 6.10*a* shows the terraces on the surface of a Zn-doped (001) InP layer, whereas they are barely discernible on the surface of the n-type

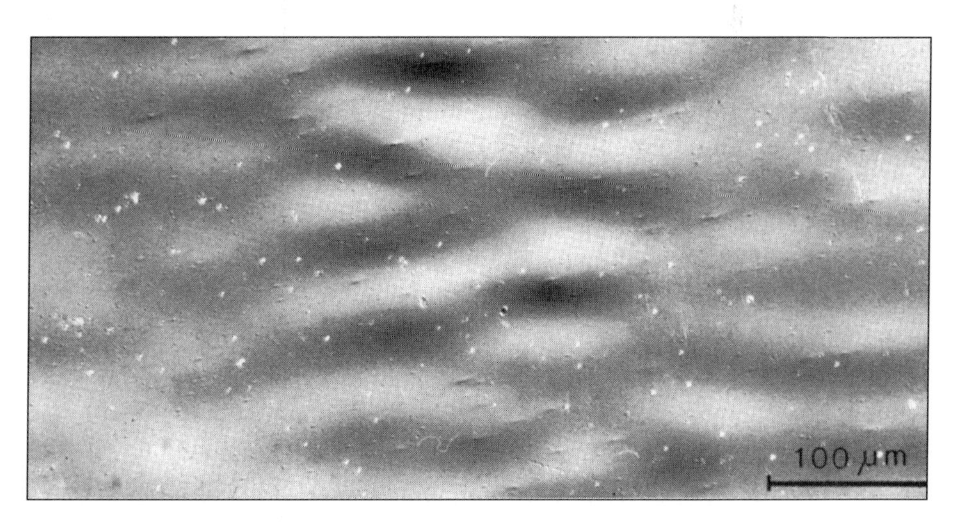

FIGURE 6.9
Surface of a 50 μm thick GaAs layer grown by LPE on a substrate that is misoriented by 3° from the (001) plane. (*After Benz and Bauser [1980]*.)

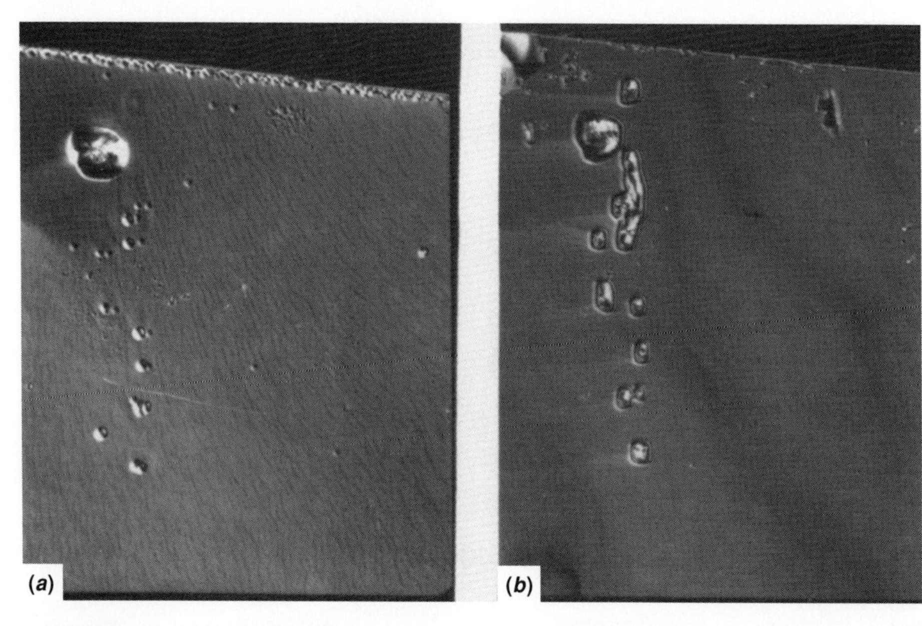

FIGURE 6.10
(*a*) Micrograph showing the surface of a Zn-doped (001) InP layer grown by LPE and (*b*) micrograph showing the surface of an n-type InGaAsP layer grown by LPE.

InGaAsP layer shown in Figure 6.10*b*. We attribute these effects to the influence of melt composition, that is, the atomic species, on the step flow.

Meniscus lines are another common feature observed on the surfaces of LPE layers, as shown in Figure 6.11. They are line-shape "defects" and develop when the melt is removed from the surface of an epitaxial layer (Small et al. 1975, 1977). Since the melts move with a "stick slip" motion during sliding, the layer may be dissolved in those places where the moving meniscus pauses momentarily, resulting in the meniscus lines.

6.1.2 Vapor Phase Epitaxy

As the name implies, the species required for the growth of an epitaxial layer are transported as vapors to the substrate using high-purity hydrogen. The species are in the form of either elemental or compound vapors. When the elemental species reach a substrate surface, they may get adsorbed. Since the substrate is kept at a high temperature, the atomic species undergo surface diffusion and combine with each other to form a layer. The situation regarding the compound sources is very similar except they must dissociate at the substrate surface to produce the atomic species. As discussed in section 6.1.1, the layer may grow either by the step-flow mechanism or by the formation and coalescence of the deposit nuclei.

Under certain growth conditions the atomic species may react with each other to form stable nuclei in the gas phase. These embryonic deposits are then transported to the substrate, where they coalesce to form a continuous film. It is unlikely

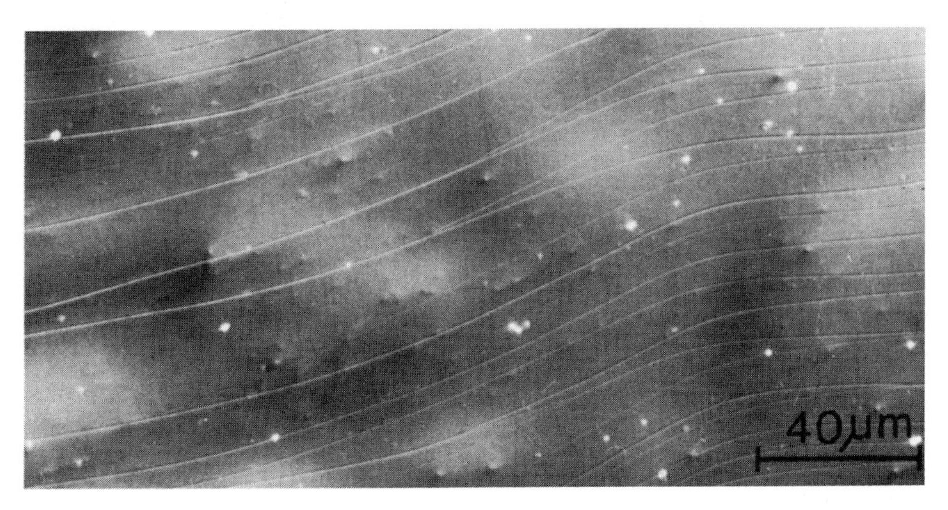

FIGURE 6.11
Meniscus lines observed on the surface of a GaSb layer grown by LPE. (*After Benz and Bauser [1980].*)

that the embryonic nuclei would be correctly oriented to coalesce to form a single crystalline film. This situation would therefore lead to polycrystalline growth, a highly undesirable feature because the presence of grain boundaries has deleterious effects on the electronic properties (see section 3.4).

Two types of reactors are used in VPE: horizontal and vertical. The respective configurations are shown in Figures 6.12*a* and 6.12*b*. In both configurations the reactants are delivered to the substrate by transport in the gas phase with a finite velocity. The reactant velocity must be close to zero in the vicinity of the substrate. As a result, a stagnant layer develops contiguous to the substrate, and the reactant species must diffuse through this layer to reach the substrate. We can show that for a horizontal reactor the thickness of the stagnant boundary layer δ is given by

$$\delta = \left(\frac{\mu x}{\rho v}\right)^{1/2}, \tag{6.19}$$

where μ is the absolute fluid viscosity, x, is the distance of the substrate from the susceptor edge as shown in Figure 6.12*a*, ρ is the fluid density, and v is the fluid velocity.

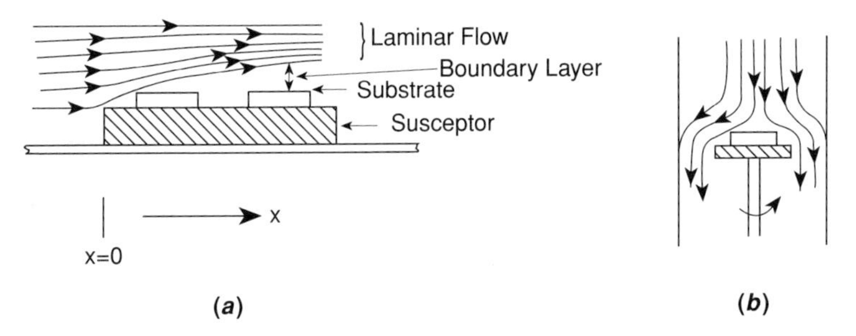

FIGURE 6.12
Schematics showing gas-flow patterns in (*a*) horizontal and (*b*) vertical reactors. Note the formation of a stagnant boundary layer in the vicinity of the substrate.

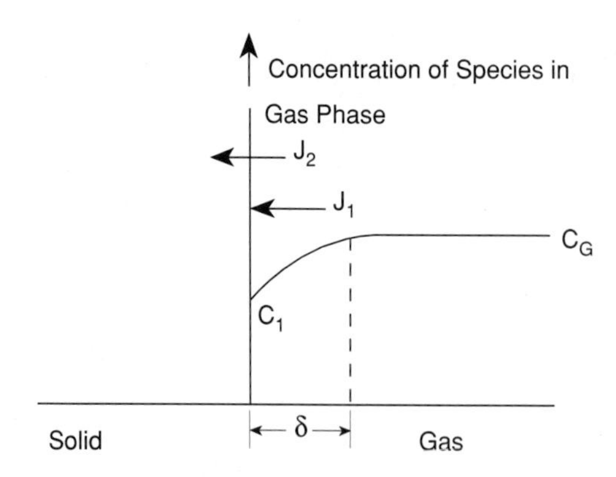

FIGURE 6.13
Schematic of a situation that may prevail at a gas-solid interface during VPE growth. C_G and C_l refer to the concentration distributions of the reactant gas within the gas phase and at the gas-solid interphase. J_1 is the flux of the reactant gas, that is, the number of atoms or molecules crossing a unit area in a unit time, whereas J_2 is the flux of the reactant gas consumed in forming the layer.

6.1.2.1 Basics of VPE growth

The following five steps are common to all VPE processes: (1) transport of the reactants to the substrate surface, (2) adsorption of the reactants on the substrate surface, (3) reaction between the adsorbed species to form an epitaxial layer, (4) desorption of some of the reaction products from the surface, and (5) transport of some of the by-products away from the substrate surface. It is very difficult to develop a mathematical description that correlates these five steps. Grove (1966) developed a model for VPE that incorporates the first three steps. In spite of its inherent simplicity, the model can explain many salient features of VPE.

Figure 6.13 shows the situation that prevails at the gas–solid interface during VPE. Figures 6.2 and 6.13 show that the two situations are conceptually identical. Assuming that the difference between the concentrations of the reactant in the gas phase and at the solid–gas interfaces varies linearly within the diffusion boundary layer, the flux J_1 into the boundary layer can be written as

$$J_1 = h_g(C_G - C_l), \tag{6.20}$$

where h_g, the proportionality constant, is termed the *gas-phase mass-transfer coefficient* and C_G and C_l are concentrations of species in the gas phase and at the substrate surface, respectively. If we assume that J_2 is linearly proportional to C_l, then flux J_2 at the gas-solid interface can be given by

$$J_2 = k_S\, C_l, \tag{6.21}$$

where k_S is the surface-reaction-rate constant. The linear dependence assumed in Equation (6.21) implies that the reaction follows the first-order kinetics. Under steady-state conditions the two fluxes must be equal. By equating Equations (6.20) and (6.21) and solving for C_l, we have

$$C_l = \frac{C_G}{1 + \dfrac{k_s}{h_g}}. \tag{6.22}$$

Two distinct situations can arise during VPE. If $h_g \gg k_S$, then C_l approaches C_G. Under this condition the growth of the layer is controlled by the surface reaction.

This condition is termed the *surface-reaction controlled case* and prevails at lower growth temperatures. When $h_g \ll k_S$, then C_1 approaches zero. This condition is known as the *mass-transfer controlled case* because the mass-transfer coefficient h_g controls the layer growth. This condition exists at high growth temperatures and implies that surface reactions occur quite readily, but the layer growth is limited by the arrival rate of species through the gas phase.

The growth rate dL_T/dt (cm/sec) is

$$\frac{dL_T}{dt} = \frac{J_2}{N}, \tag{6.23}$$

where N is the number of atoms incorporated into a unit volume of the film and L_T is the layer thickness. For silicon N is equal to 5×10^{22} atoms/cm^3. Substituting for C_1 from Equation (6.22) into Equation (6.21) and then substituting for J_2 into Equation (6.23), we obtain the following expression for the layer growth rate:

$$\frac{dL_T}{dt} = \frac{k_S h_g}{k_S + h_g} \cdot \frac{C_G}{N}. \tag{6.24}$$

The concentration of the reactant in the gas phase is given by

$$C_G = C_T M, \tag{6.25}$$

where C_T is the total number of molecules per (cm)^3 in the gas phase and M is the mole fraction of the reaction species. Substituting for C_G in Equation (6.24), we have

$$\frac{dL_T}{dt} = \frac{k_S h_g}{k_S + h_g} \left(\frac{C_T M}{N} \right). \tag{6.26}$$

There are two ramifications of Equation (6.26). First, it predicts that the growth rate should be proportional to the mole fraction of reacting species in the gas phase. We illustrate this situation using the growth of silicon layers from the $SiCl_4 + H_2$ mixtures ($SiCl_4 + 2H_2 \rightarrow Si + 4HCl$) as an example. Figure 6.14 shows the growth rate of silicon as a function of the mole fraction of $SiCl_4$. Up to $M = 0.1$, the growth rate at 1270°C varies linearly with the mole fraction, but it then begins to drop. We

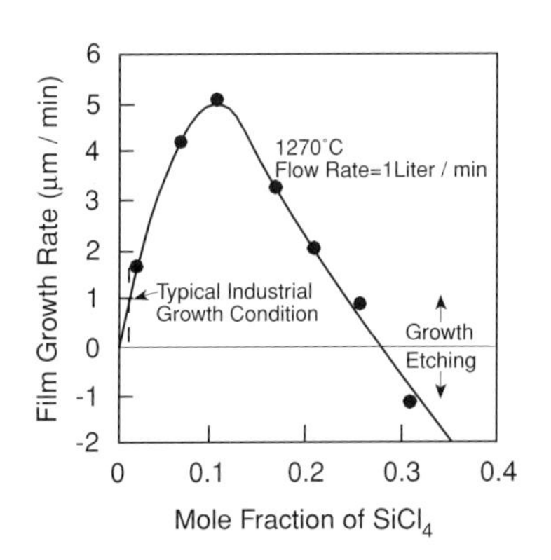

FIGURE 6.14
Growth rate of Si films as a mole fraction of $SiCl_4$ and H_2 mixtures. (*After Theuerer [1961].*)

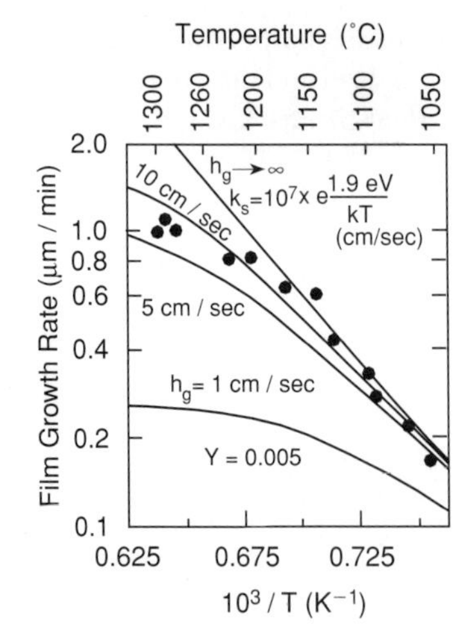

FIGURE 6.15
Growth of Si from $SiCl_4$ and H_2 mixture as a function of temperature. (*After Shepherd [1965].*)

attribute this change to the competition between growth and etching caused by HCl (see section 6.1.2.2). At M = 0.27 the negative growth, that is, etching, of the substrate starts. Second, Equation (6.26) indicates that at a constant value of M, the growth rate is determined by the smaller of k_S and h_g. In the two limiting cases, $h_g \gg k_S$ and $h_g \ll k_S$, the growth rates are given by

$$\frac{dL_T}{dt} = \frac{k_S C_T M}{N} \tag{6.27}$$

$$\frac{dL_T}{dt} = \frac{h_g C_T M}{N}. \tag{6.28}$$

The surface-reaction-rate constant k_S describes the kinetics of the chemical reaction at the substrate surface. Assuming that the chemical reaction is thermally activated, k_S can be written as

$$k_S = k_o \exp\left(\frac{-E_s}{k_B T}\right), \tag{6.29}$$

where k_o is a temperature-independent frequency factor and E_S is the activation energy associated with the step that controls the chemical reaction. On the other hand, h_g is essentially independent of temperature and depends principally on the gas-flow conditions in the reactor. Figure 6.15 shows the variation in growth rate as a function of reciprocal temperature for the deposition of Si from the $SiCl_4$ and H_2 mixture. At low temperatures the growth rate varies linearly with $1/T$, and E_S is estimated to be 1.9 eV. At higher temperatures the growth rate tends to become temperature insensitive because h_g controls the growth.

EXAMPLE 6.3. Calculate the growth rate of a silicon epitaxial layer deposited from a silicon tetrachloride source at 1150°C. Assume that the mass-transfer coefficient of the reactors is $h_g = 5$ cm/sec, the surface-reaction coefficient k_s is $10^7 \exp(-1.9\text{ eV}/k_B T)$ cm/sec, and $C_G = 5 \times 10^{16}$ cm^{-3}. By how much will the growth rate change if the growth temperature is reduced to 1100°C?

Solution

$$\text{The value of } k_S \text{ at } 1150^\circ\text{C} = 10^7 \exp\left(\frac{-1.9}{8.62 \times 10^{-5} \times 1423}\right) \text{ cm/sec}$$

$$= 1.96 \text{ cm/sec.}$$

Substituting the values of k_s, h_g, C_G, and $N(5 \times 10^{22}$ atoms/cm$^3)$ in Equation (6.24), the growth rate of the layer is

$$= \frac{1.96 \times 5 \times 5 \times 10^{16}}{6.96 \times 5 \times 10^{22}} \text{ cm/sec}$$

$$= 1.41 \times 10^{-6} \text{ cm/sec.}$$

$$\text{The value of } k_S \text{ at } 1100^\circ\text{C is } = 10^7 \exp\left(\frac{-1.9}{8.62 \times 10^{-5} \times 1373}\right)$$

$$= 1.02 \text{ cm/sec.}$$

$$\text{The growth rate of the layer at } 1100^\circ\text{C is } = \frac{1.02 \times 5 \times 5 \times 10^{16}}{6.02 \times 5 \times 10^{22}} \text{ cm/sec}$$

$$= 0.85 \times 10^{-6} \text{ cm/sec.}$$

The growth rate at 1100°C is about 60 percent of its value at 1150°C, implying a strong temperature dependence of the growth rate.

6.1.2.2 VPE of silicon and silicon-germanium layers

Chemistry of growth. The silicon epitaxial structures have numerous applications in state-of-the-art integrated circuits and bipolar transistors. The advantages of these epitaxial wafers are twofold. First, they give device designers more flexibility to control doping profiles because of the availability of differently doped regions on the same wafer. Second, epitaxial wafers do not suffer from the same problems as bulk wafers. For example, as discussed in section 5.6, oxygen can precipitate out in the Czochralski wafers during processing, whereas this cannot occur in epitaxial layers because their oxygen content does not exceed the solubility limit at the processing temperature. The availability of epitaxial materials allows optimization of certain circuit parameters.

Four sources of silicon are used for epitaxial deposition: silicon tetrachloride ($SiCl_4$), trichlorosilane ($SiHCl_3$), dichorosilane (SiH_2Cl_2), and silane (SiH_4). Figure 6.16 shows the vapor pressures of $SiCl_4$, $SiHCl_3$, and SiH_2Cl_2 as a function of temperature. At a given temperature the vapor pressure of SiH_2Cl_2 is greater than those of $SiHCl_3$ and $SiCl_4$. Alternatively, to obtain a given vapor pressure of silicon, each source has to be heated to a different temperature. We can conclude from these observations that $SiCl_4$ is the most stable silicon source, whereas SiH_2Cl_2 is the least stable. Therefore, for lower temperature depositions we prefer less stable sources, such as SiH_2Cl_2 and SiH_4, because they pyrolyze easily.

We extensively use the reduction of $SiCl_4$ by hydrogen, between 1150 and 1250°C, for depositing silicon layers. The overall chemical reaction controlling the formation of silicon can be written as follows:

$$SiCl_4 + 2H_2 \leftrightarrow Si + 4HCl. \tag{6.30}$$

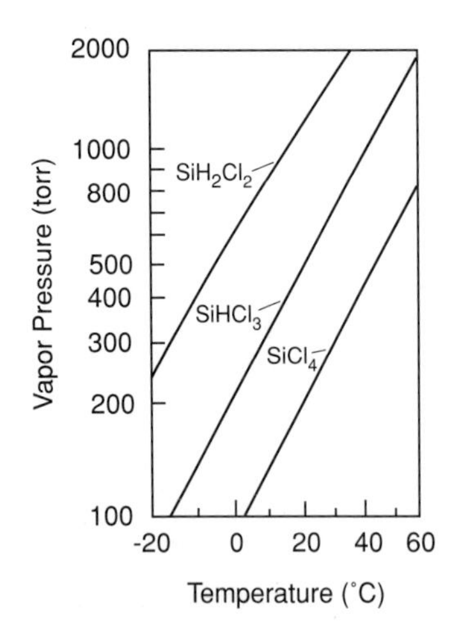

FIGURE 6.16
Vapor pressure versus temperature for different silicon sources used in epitaxy.

This reaction is reversible and can be driven in either direction by changing the mole fractions of different components in the gas phase. By removing HCl from the reaction zone, the formation of silicon is favored. On the other hand, by increasing the mole fraction of HCl, etching will be preferred over growth.

The overall reaction given by Equation (6.30) does not present a complete picture of the reaction sequence. Ban and co-workers (1975a, 1975b, 1975c, and 1978) have investigated the chemistry of the Si-H-Cl system using mass spectrometry. In addition to $SiCl_4$, HCl, H_2, and Si, they find other species, such as $SiCl_2$, $SiHCl_3$, and SiH_2Cl_2, in the growth system. These species may be either adsorbed on the substrate or are present in the gas phase. Based on the researchers' results, the reaction given by Equation (6.30) could proceed according to the following steps:

$$SiCl_4 + H_2 \leftrightarrow SiHCl_3 + HCl \tag{6.31}$$

$$SiHCl_3 + H_2 \leftrightarrow SiH_2Cl_2 + HCl \tag{6.32}$$

$$SiH_2Cl_2 \leftrightarrow SiCl_2 + H_2 \tag{6.33}$$

$$SiHCl_3 \leftrightarrow SiCl_2 + HCl \tag{6.34}$$

$$SiCl_2 + H_2 \leftrightarrow Si + 2HCl \tag{6.35}$$

If we define the efficiency of a deposition reaction as a ratio of the amount of silicon deposited to the amount of incoming silicon, the efficiency of conversion of $SiCl_4$ is lower than that of $SiHCl_3$ and SiH_2Cl_2. However, the advantage of $SiCl_4$ is its stability and its insensitivity to small amounts of impurities.

We can convert silane (SiH_4) into silicon by the following reaction:

$$SiH_4 \rightarrow Si + 2H_2. \tag{6.36}$$

The decomposition of SiH_4 could take place around 500°C, thereby permitting epitaxial growth at lower temperatures in the mass-transfer regime. However, there are three disadvantages with the silane process. First, the homogeneous gas-phase decomposition can occur, resulting in powdery deposits on the reactor walls, or even on wafers. Second, SiH_4 is more sensitive to oxidation than other chlorosilanes. Third, since HCl is not generated in the above reaction, the reverse-etching reaction cannot take place.

Meyerson (1986) and Greve (1993) have grown high-quality, undoped and doped silicon and silicon-germanium layers as well as silicon/silicon-germanium heterostructures having sharp interfaces between 500 and 600°C using ultralow-pressure VPE; the gaseous sources in their studies were silane and germane. The implication of their results is that if the growth environment is ultraclean, silicon and germanium atoms have sufficient surface mobility to occupy the correct sites during the growth, resulting in high-quality layers.

In situ etching prior to growth. We need to perform in situ etching on the substrates so that we have clean surfaces for epitaxial growth. This step is very critical in silicon epitaxy. In situ etching may also remove damage from the surfaces of polished wafers.

To carry out etching, we expose the substrates maintained at a temperature of ~1200°C to dry hydrogen gas. This step reduces the native oxide that may have formed during handling of the wafers and their transfer into the reactor. After the oxide removal, anhydrous HCl gas, diluted with hydrogen, is introduced into the growth system. Etching proceeds by the following reaction:

$$\mathrm{Si} + 4\mathrm{HCl} \leftrightarrow \mathrm{SiCl_4} + 2\mathrm{H_2}. \tag{6.37}$$

Reaction (6.37) is the reverse of the reaction given by reaction (6.30) and is driven to the right by increasing the mole fraction of HCl in the gas mixture; the substrate remains between 1100 and 1250°C. Etching produces high-quality, optically flat wafers. For etching we use a gas mixture consisting of 1 to 5 percent mole fraction of HCl in hydrogen, and etch rate is 0.5 to 2 μm/min. The higher concentrations of HCl cause pitting of the surfaces. The purity of etching gas is also very critical. Traces of water, nitrogen, and hydrocarbons can lead to the formation of oxides, nitrides, and carbides on the wafer surface. Since epitaxial growth cannot occur on these particles, the quality of the layer is impaired.

We are currently interested in the processing of silicon devices at low temperatures so that the interdiffusion between different materials in ICs is minimal. It is therefore desirable to carry out wafer etching at lower temperatures. Etchants, such as Cl_2, HBr, HI, SF_6, and H_2S, have been used for such applications. For example, device-quality surfaces can be produced by etching them with SF_6 at 1050°C. The governing etching reaction is

$$4\mathrm{Si} + 2\mathrm{SF_6(g)} \rightarrow \mathrm{SiS_2(g)} + 3\mathrm{SiF_4(g)}. \tag{6.38}$$

SF_6 is nontoxic and noncorrosive.

Growth of doped layers. Many device applications require layers of different carrier concentrations and conductivity type. We can meet these requirements by introducing the appropriate amounts of dopants in the gas phase during the layer growth. We can use both solid and liquid dopant sources, but the gaseous sources are preferable because they provide a finer control on the composition of the layer.

We use gaseous sources, such as diborane (B_2H_6), phosphine (PH_3), and arsine (AsH_3), for doping, and we use hydrogen gas as a diluent to permit an accurate control of the carrier concentration. The source molecules dissociate at the substrate surface to produce dopant atoms that are then incorporated into the growing layer. The dopant concentration varies linearly with the partial pressure of the source gas up to concentrations as high as 10^{18} cm^{-3}. However, at higher concentrations a slight fall off in the incorporation of P occurs, whereas the B incorporation exhibits superlinear characteristics.

The growth rate of the layer is slightly altered in the presence of dopant atoms. An enhanced growth rate is observed with diborane, whereas the rate is slightly lowered in the presence of phosphine and arsine. We attribute these effects to the differences between the lengths of the Si-Si, Si-B, Si-P, and Si-As bonds (see section 3.3). Since the tetrahedral radius of boron (0.088 nm) is very small, Si-B bond will be shorter and hence stronger than the Si-Si bond. The presence of B on the surface reduces the reevaporation rate of silicon atoms into the gas phase, resulting in a higher growth rate. On the other hand, the tetrahedral radii of phosphorous (0.110 nm) and arsenic (0.118 nm) are very close to that of silicon (0.117 nm). Therefore, the bond lengths in the doped layers and the undoped layers will be very similar, with the result that the reevaporation rate may be only slightly affected.

Redistribution of dopants during growth. Since we carry out epitaxy at high temperatures where the atomic mobility is substantial and the vapor pressures of dopants are high, significant redistribution of the dopants can occur during growth. Two distinct processes can bring about these changes: (1) diffusion in the gas phase and (2) solid state diffusion. Let us imagine a situation in which a lightly doped layer is being grown on a highly doped substrate. In the beginning of the growth, the dopant atoms evaporate from the substrate into the gas phase and are incorporated into the layer. This process is referred to as *autodoping,* and it has two consequences: (1) The carrier concentration in the layer next to the substrate is higher than the desired value, and (2) the carrier profile across the substrate–layer interface is not sharp. As the layer grows, the problem of autodoping is progressively reduced.

If there are differences in the concentrations and the types of dopants in the substrate and the epitaxial layer, atomic movement will occur from one to the other because of the concentration gradients (see Chapter 8). This movement takes place until the chemical potentials of different dopants in the substrate and the layer are equal. Its ramifications are that the doping profile close to the substrate-layer interface is not sharp and differs from the desired one. During growth the atomic movements also occur within the layer, leading to the changes in the dopant profile. Depositing the layers at lower temperatures can reduce the problems associated with autodoping and atomic movement.

6.1.2.3 VPE of compound semiconductors

The three major disadvantages of LPE are (1) the presence of terraces, (2) nonuniformity in the layer thickness, and (3) only small substrates can be processed. Dutt et al. (1984) have obviated the size problem by simultaneously depositing layers on a number of substrates, but the other two issues have not been resolved successfully. As a result, VPE of compound semiconductors has been explored by a number of researchers. Three VPE processes have emerged: halide VPE, hydride VPE, and OMVPE. These processes differ in the chemistries of sources for the groups III and V species. We highlight their salient features below.

Halide VPE. The growth process is called halide VPE because the group V source is a chloride. This process produces very high purity binary III-V compounds because all the starting materials are readily available in very high purity.

We can describe the process by referring to Figure 6.17 that shows a two-zone, resistance-heated, hot-wall reactor; the two zones are referred to as the *source* and *deposition* zones (Long et al. 1988). The group III sources are elemental melts and are contained in graphite boats located in the source zone, whereas the group V sources are $AsCl_3$ and PCl_3. Both are volatile liquids with vapor pressures similar to that of water. The substrates are placed on graphite or quartz carriers in the deposition zone.

In a typical InP growth run, a portion of high-purity hydrogen passes through a PCl_3 bubbler maintained at a low temperature. The transported PCl_3 and H_2 react with each other according to the following reaction:

$$PCl_3 + \frac{3}{2} H_2 \rightarrow \frac{1}{4} P_4 + 3HCl. \tag{6.39}$$

This reaction occurs upstream from the source zone, and the products flow over an indium source held at 750 to 850°C. Indium and phosphorous react to form InP.

$$In + \frac{1}{4} P_4 \rightarrow InP \tag{6.40}$$

The flow of phosphorous over the liquid In for an extended period produces a crust of polycrystalline InP. After the continuous layer of InP forms, the rate of the

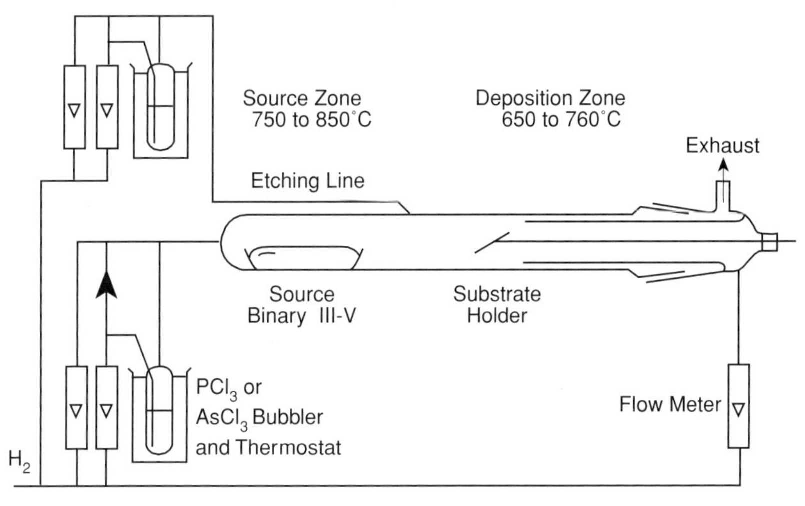

FIGURE 6.17
Schematic of a halide VPE reactor. (*After Long et al. [1988].*)

reaction between In and P slows down because the diffusion of P through InP is very slow. The HCl gas resulting from the reaction given by Equation (6.39) reacts with the InP crust to form InCl and P_4.

$$InP + HCl \leftrightarrow InCl + \frac{1}{2} H_2 + \frac{1}{4} P_4 \tag{6.41}$$

The reaction products on the right are transported to the substrate that is in the deposition zone. Since the temperature in the deposition zone is lower than the temperature in the source zone (see Figure 6.17), InCl, P, and hydrogen undergo the reverse reaction, leading to the deposition of InP.

The temperature in the deposition zone is fairly high. As a result, the substrates undergo thermal decomposition, resulting in pits on the surface (see section 6.1.1.3). To remove the pits, the substrates must therefore be etched in situ prior to the growth. We accomplish this step by raising the substrate temperature so that the reaction given by Equation (6.41) is driven to the right. Alternatively, we can inject HCl or PCl_3 in the reactor in the region between the source and deposition zones as shown in Figure 6.17. Both techniques produce satisfactory surfaces for the layer growth.

The main disadvantage of the halide VPE using liquid group III sources is that mixed III-V materials, such as InGaAs and InGaAsP, cannot be grown. Besides the differences between the growth temperatures of the binary constituents, the two-phase sources hamper compositional control. We can use the binary solid sources to deposit the mixed compositions (Vohl 1981; Cox et al. 1985; Johnston et al. 1989). Figure 6.18 shows a solid-source reactor for the growth of $In_{1-x} Ga_x As_y P_{1-y}$ layers having different values of x and y (Johnston et al. 1989). Conceptually, we can regard the growth of the quaternary layer as a simultaneous deposition of the appropriate amounts of InP, InAs, and GaAs. Polycrystalline, high-purity InP, InAs, and GaAs, contained in graphite boats, serve as the sources instead of the binary crusts of the two-phase sources in the conventional halide VPE. We tailor the layer composition by adjusting the flows of PCl_3 and $AsCl_3$ over the solid sources. Using this approach, we can grow InGaAsP layers of different compositions that are lattice matched to InP.

The purity of the binary layers, grown from the liquid sources by the halide VPE, is very high. We attribute this high purity to the in situ production of HCl that we use to transport the group III elements. The thickness uniformity of the layers is also fairly good. Figure 6.19 shows the thickness uniformity of GaAs epitaxial

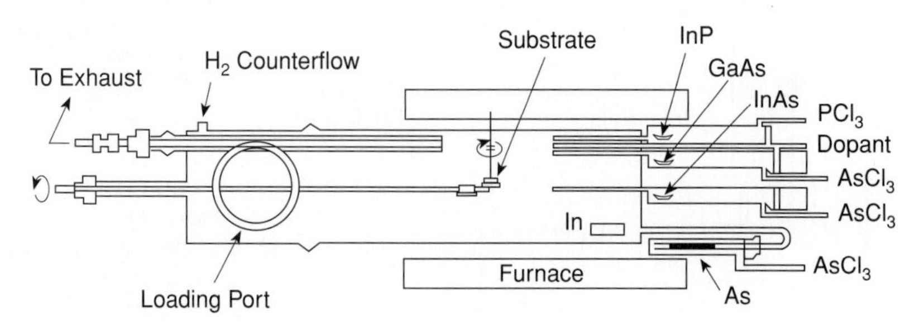

FIGURE 6.18
Schematic of a solid-source reactor for the growth of InGaAsP layers of various compositions. (*After Johnston et al. [1989].*)

layers as a function of the $AsCl_3$ mole fraction and the growth temperature (Komeno et al. 1981). For the $AsCl_3$ mole fraction between 10^{-3} and 10^{-2}, we achieve the thickness uniformity of less than 1 percent between 650 and 700°C.

Hydride VPE. The sources for the group V elements in this growth technique are in the form of hydrides, such as PH_3 and AsH_3. We independently control the fluxes of group III and V elements, permitting greater flexibility and control of growth. We can also deposit the layers on large as well as multiple wafers. We achieve smooth surfaces and can deposit mixed compositions.

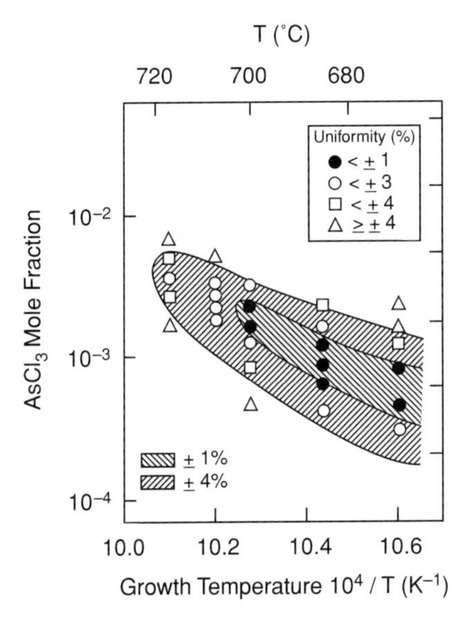

FIGURE 6.19
Thickness uniformity of GaAs epitaxial layers as a function of $AsCl_3$ mole fraction and temperature. (*After Komeno et al. [1981].*)

We show a basic hydride reactor in Figure 6.20 (Olsen 1982). The reactor has three zones: the source zone that contains the liquid metals; a mixing zone in which the hydrides and metal chlorides, produced by reactions between the source metals and HCl, are mixed; and a deposition zone where the epitaxial growth occurs. The respective temperatures for these zones for the growth of InP are 800 to 850°C, 825 to 900°C, and 650 to 725°C.

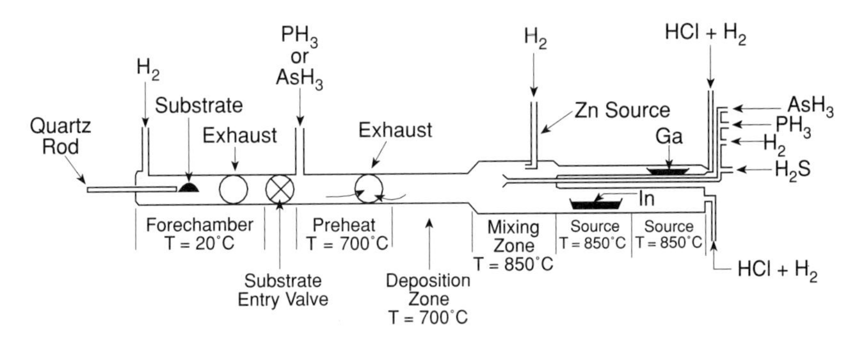

FIGURE 6.20
Schematic of a hydride VPE reactor for the growth of In- and Ga-based mixed III-V layers. (*After Olsen [1982].*)

To highlight differences between the hydride and halide VPEs, we consider the growth of InP by the hydride VPE. In this case the dissociation of PH_3 produces the flux of P.

$$PH_3 \rightarrow \frac{1}{4}P_4 + \frac{3}{2}H_2 \tag{6.42}$$

HCl, which is supplied externally, reacts with the In liquid to form InCl according to the following reaction:

$$\text{In} + \text{HCl} \leftrightarrow \text{InCl} + \frac{1}{2}H_2. \tag{6.43}$$

InCl and P_4 are thoroughly mixed in the mixing zone, and in the presence of hydrogen they combine to form a layer InP in the deposition zone. The governing reaction is

$$\text{InCl} + \frac{1}{4}P_4 + \frac{1}{2}H_2 \leftrightarrow \text{InP} + \text{HCl}. \tag{6.44}$$

To preserve the substrate surface during the warm-up, heating is carried out in a PH_3 environment. We can also etch the substrate in situ by driving reaction (6.44) to the left.

The background carrier concentrations of the layers grown by the hydride VPE are generally above 3×10^{15} cm^{-3}, whereas they are an order of magnitude lower for the halide VPE. This difference is due to the impurity pickup by HCl as it flows through the reactor tubes; HCl is produced in situ in the halide case.

EXAMPLE 6.4. Due to the nonoptimal growth conditions, In droplets are deposited on $(\bar{1}\bar{1}\bar{1})_B$ and $(111)_A$ surfaces of InP by the reverse of reaction (6.43). If substrates are being held at 600°C, show shapes and crystallographies of the resulting dissolution pits. Ignore the loss of phosphorous from the substrate at high temperature.

Solution. When In droplets come in contact with a $(\bar{1}\bar{1}\bar{1})_B$ surface, they dissolve phosphorous from the underlying regions. This process balances the activities of P in the droplets and at the InP surface. The removal of P from the substrate releases In, which becomes part of the droplet, and exposes low-energy side facets bounded by $\{111\}_A$ planes. The resulting dissolution pit may have the following crystallography:

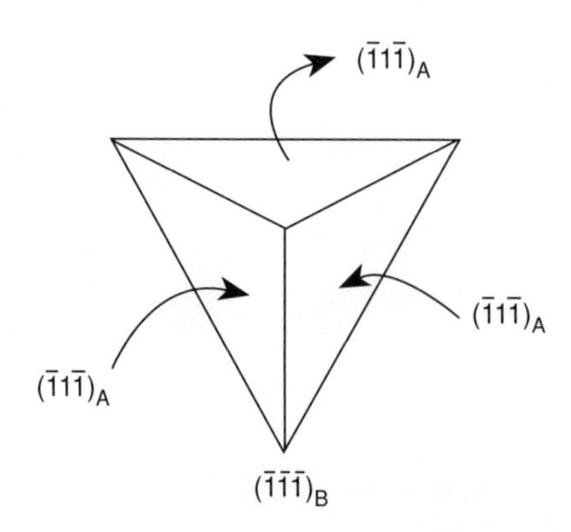

The $\{111\}_A$ facets in this case will be less developed than in the case of thermal-decomposition induced pits discussed in Example 6.2. The presence of In droplets in the pit mean that the evaporation rate of In atoms from the facets will be very low.
The In droplets will not react with a (111) surface that is bounded by In atoms because there is no driving for such a reaction. Hence no pits will form on this surface.
It is interesting to compare the origins of dissolution pits and thermal-decomposition-induced pits discussed in Chapter 3. Both pits originate from the removal of phosphorous. In the case of dissolution pits, P is removed by the In droplets, whereas in the case of decomposition pits, P is lost at high temperature because of its high vapor pressure.

OMVPE. OMVPE is a very promising epitaxial growth technique. It was pioneered by Manasevit in 1968 who demonstrated that in the presence of hydrogen the mixture of trimethylgallium or $(CH_3)_3Ga$, and arsine can be pyrolized between 600° and 700°C to produce layers of GaAs. Since then OMVPE has made great strides and is the technique of choice for most commercial applications.

OMVPE has two principal advantages over the growth techniques discussed earlier in this chapter. First, metal alkyls that serve as sources for group III elements are volatile liquids at room temperature. As a result, their vapor pressures are high and can be transported easily using a carrier gas. Second, the pyrolysis temperatures for the metal alkyls are lower than those for the metal halides. We can therefore grow layers at lower temperatures, resulting in heterostructures having sharper composition profiles at the interface.

Figure 6.21 shows an OMVPE reactor for the growth of GaAs. AsH_3 serves as the As source and is diluted with H_2. Hydrogen also passes through a bubbler containing $(CH_3)_3Ga$ and transports the vapors of the metal alkyl to heated substrates. Likewise, the dopants can be provided by incorporating appropriate sources in the

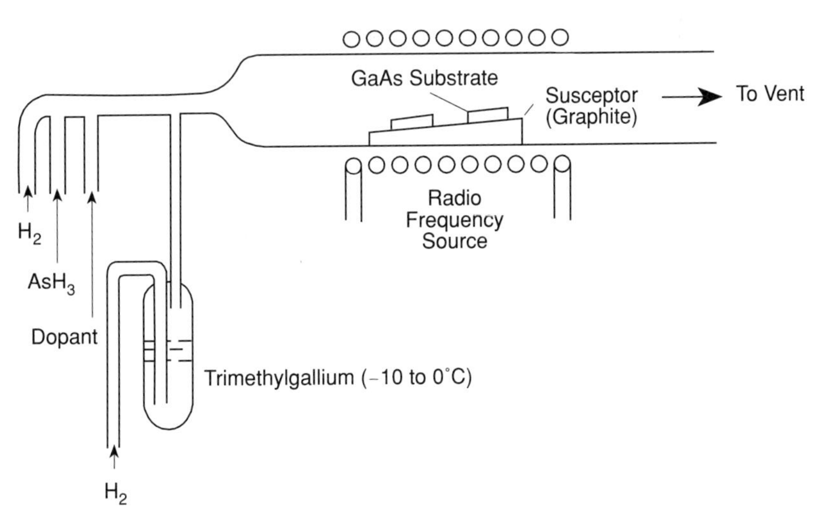

FIGURE 6.21
Schematic of a reactor used for OMVPE of GaAs.

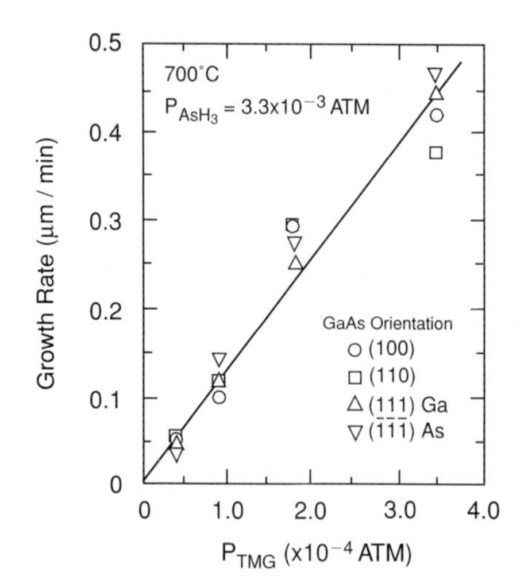

FIGURE 6.22
Growth rate of GaAs as a function of the partial pressure of trimethylgallium. Note that growth rate is independent of the orientation of the substrate. (*After Stringfellow [1984].*)

growth system. The species arrive at the substrate by diffusing through the boundary layer and reacting with each other to form GaAs.

$$(CH_3)_3Ga + AsH_3 \xrightarrow{600-700^\circ C} GaAs + 3CH_4 \tag{6.45}$$

Reaction (6.45) may have intermediate stages that we do not understand well. Recent study of Aspnes et al. (1988) indicates that the alkyls are first adsorbed at the surface of a substrate and are then pyrolyzed, releasing gallium atoms. The gallium atoms subsequently migrate on the surface and combine with the As atoms produced by the decomposition of arsine to form GaAs.

Figure 6.22 shows that the growth rate of GaAs varies linearly with the partial pressure of trimethylgallium and is independent of the orientation of the substrate (Stringfellow 1984). We expect this behavior if the diffusion of trimethylgallium across the boundary layer is rate limiting (see section 6.1.2.1). This behavior is also consistent with the observed independence of the growth rate of the substrate orientation because only the surface reaction kinetics should be orientation sensitive.

Carbon is a common contaminant in OMVPE layers due to the formation of hydrocarbons. Since reaction (6.45) is irreversible, we cannot use it to etch the substrates in situ. However, we can perform etching by the introduction of anhydrous HCl gas or $AsCl_3$ vapor into the reactor.

The OMVPE technique is very versatile. To date, a variety of materials have been grown using OMVPE. Mixed III-V materials like GaAlAs are easy to grow. InGaAs, InGaAsP, GaAsSb, InAsSb, GaInP, (GaAl)InP, AlN, CdTe, ZnSe, and so forth, layers have been grown successfully. The superlattices and quantum-well laser structures have also been deposited.

Surface morphology of VPE layers. The surfaces of VPE layers grown on vicinal substrates are fairly smooth. The features resembling the terraces of LPE layers are not seen (see section 6.1.1.6). As we indicated in section 6.1.1, a layer can grow in two ways: by the formation of the deposit nuclei on the surface followed by their coalescence, and by the step-flow mechanism. We believe that the step-flow mechanism dominates in the VPE growth for two reasons. Since the growth temperatures are relatively high, we can argue from the nucleation theory that the critical size of the nuclei may be large. The surface mobility of the atoms is also high. As a result, they can easily reach the step edges before clustering to form the nuclei on the terraces of the substrate. In VPE the fluxes of the species are also more uniform across the surface, minimizing the differences in the flow rates of the steps and preventing step bunching.

Doping of VPE layers. We dope the VPE layers in a controlled manner using gaseous sources. We use the hydrides of S, Se, and Te for n-type doping. It is generally difficult to grow n-type GaAs layers with a carrier concentration of less than 10^{16} cm^{-3}. We attribute this difficulty to the incorporation of the silicon atoms on the group III sites during the growth; the reactor walls are the chief source of this contamination. We also use a number of organometallic sources to dope various VPE layers. These sources include diethylzinc $[(C_2H_5)_2Zn]$, diethylcadmium $[(C_2H_5)_2Cd]$, tetramethlytin $[(CH_3)_4Sn]$, and diethylcadmium $[(C_2H_5)_2Te]$ for p-type and n-type doping, respectively.

The semi-insulating layers of InP and GaAs—resistivities in the range of 10^8 to 10^9 Ω-cm—have been grown by VPE. Ferrocene $Fe(CH_5H_5)_2$ has been used as a source for Fe doping of InP layers deposited by OMVPE (Long et al. 1984; Macrander et al. 1984). The semi-insulating GaAs layers have been grown using chromium doping by adding chromyl chloride (Cr_2Cl_2) to the gas stream.

Orientation effects. As discussed in section 6.1.2.1, we have two temperature regimes for the VPE growth. At low temperatures the kinetics of the surface reactions control layer growth, whereas the mass transfer through the diffusion boundary layer controls the growth at high temperatures. We therefore expect that the orientation of a substrate will affect the growth rate at low temperatures. The studies of Shaw (1968) on the halide VPE of GaAs at various temperatures on substrates of different orientations bear out these effects.

Figure 6.23 shows the temperature dependence of the growth rate of GaAs as a function of the growth temperature and the surface orientation. We see that in the low temperature, that is, kinetic controlled region, the growth rate is highly sensitive to the orientation. However, at high temperatures the growth rates are essentially independent of the orientations. We replot the data of Figure 6.23 in a different form in Figure 6.24 that shows the dependence of the growth rate at low temperatures on the substrate orientation. We see that the growth rates on the Ga-terminated surfaces are higher than those on the As-terminated surfaces. We can rationalize the observed behavior by amending Equation (6.41) as follows so that it applies to the halide VPE growth of GaAs.

$$\mathrm{GaCl} + \frac{1}{4}\mathrm{As_4} + \frac{1}{2}\mathrm{H_2} \leftrightarrow \mathrm{GaAs} + \mathrm{HCl} \tag{6.46}$$

If we assume that the crucial step in the synthesis of GaAs is the decomposition of the As_4 molecules into the elemental arsenic, the dissociation should occur readily

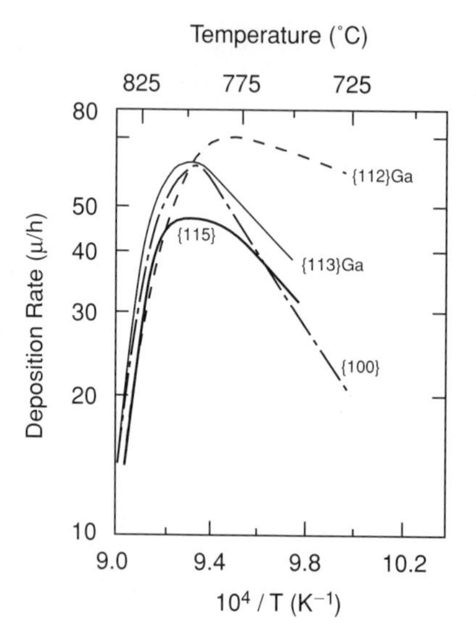

FIGURE 6.23
Temperature dependence of growth rate of halide VPE GaAs. (*After Shaw [1968].*)

on the Ga-terminated surfaces. We argue that the driving force for the dissociation of As_4 molecules on the As-terminated surfaces is weak.

The incorporation of dopants into GaAs layers grown by halide VPE also depends on the orientation of the substrate. This behavior is apparent from the results of Williams (1964) shown in Table 6.1. For example, the incorporation ratios of Zn and Te on $\{\bar{1}\bar{1}\bar{1}\}_B/\{111\}_A$ are 0.2 and 20, respectively. A plausible explanation is that since Zn atoms are incorporated into the Ga sublattice, the presence of $\{111\}_{Ga}$ surface should facilitate this process. To incorporate Zn atoms during the growth of GaAs on a $\{\bar{1}\bar{1}\bar{1}\}$ surface, the Zn atoms must replace some of the Ga atoms below the exposed $\{\bar{1}\bar{1}\bar{1}\}_{As}$ surface, a more difficult process. On the other hand, this situation is reversed for the Te atoms. These atoms are incorporated more easily into a

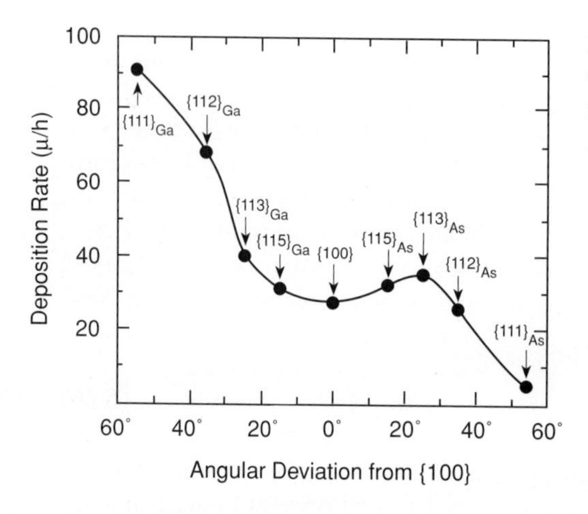

FIGURE 6.24
Orientation dependence of growth rate of halide VPE GaAs. (*After Shaw [1968].*)

TABLE 6.1
Orientation dependence of dopant incorporation during halide VPE of GaAs

Dopant	Dopant incorporation ratio $(\bar{1}\bar{1}\bar{1})$ As / (100)	$(\bar{1}\bar{1}\bar{1})$ As / (111) Ga	$(\bar{1}\bar{1}\bar{1})$ As / (110)
No intentional doping	11	15	28
Zn	0.43	0.2	0.49
Te	7.4	20	15
Se	6.3	—	4.4
Sn	2.3	—	—
S	1.4	—	—

Source: Williams 1964.

layer growing on a $\{\bar{1}\bar{1}\bar{1}\}_{As}$ surface than into a layer grown on a $\{111\}_{Ga}$ surface. We can similarly rationalize the other results in Table 6.1.

6.1.3 Molecular Beam Epitaxy

6.1.3.1 Compound semiconductors

According to Cho and Arthur (1975), the pioneers of this growth technique, molecular beam epitaxy (MBE) describes the epitaxial growth of semiconducting layers by the reaction between the molecular beams of its constituents and a crystalline surface held at a suitable temperature under ultrahigh vacuum (UHV) conditions. Conceptually, we can consider an MBE system as a very high vacuum evaporation apparatus.

MBE has several features that distinguish it from other epitaxial growth techniques: (1) a low growth rate (between 0.1 and 0.3 nm/sec for GaAs), (2) a low growth temperature (<630°C for GaAs), (3) the provision for the abrupt cessation or initiation of growth that enables us to grow heterostructures having chemically and structurally sharp interfaces, (4) the ability to smooth growth surfaces, and (5) the potential for in situ monitoring of the growth and chemical analysis. We can now produce unique multilayer structures suitable for "band gap engineering" devices using the growth technique.

Conventional molecular beam epitaxy. Figure 6.25 shows the essential elements for an MBE system used for the growth of doped GaAs and GaAlAs layers. The molecular beams of Ga, Al, As, and appropriate dopants are generated from the heated Knudsen effusion cells whose temperatures are accurately controlled to within ±1°C. Each source is provided with an externally controlled shutter. The operation of these shutters permits rapid changes in composition and doping of the layers. Beams emanating from the cells impinge on a rotating substrate mounted on a heated Mo block; a thin film of In is used for mounting purposes. The cells are

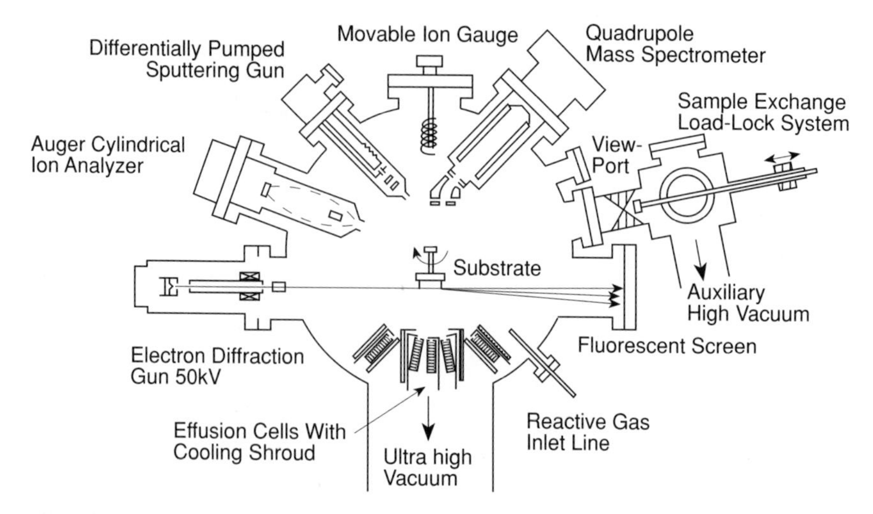

FIGURE 6.25
Schematic of essential elements for MBE used for the growth of doped GaAs and GaAlAs layers.

fabricated from pyrolytic boron nitride (PBN) or high-purity graphite because these materials are chemically inert and can withstand high temperatures.

Assuming that the cell aperture is smaller than the mean free path of vapor molecules within the cell, the flux density J (molecules cm^{-2} sec^{-1}) is given by

$$J = \frac{Ap\cos\theta}{\pi L^2 (2\pi m k_B T)^{1/2}}, \tag{6.47}$$

where A is the area of the aperture, p is the equilibrium vapor pressure in the cell, L is the distance between the cell and the substrate, m is the mass of the effusing species, k_B is the Boltzmann constant, and T is the cell temperature. Equation (6.47) serves only as a guideline, since in practice the aperture is enlarged to increase the growth rate. The fluxes emanating under such nonequilibrium conditions are measured experimentally.

After impinging on the substrate surface, different atomic species undergo adsorption and migration on the surface and combine to form the deposit. The parameter known as "sticking coefficient" S describes the overall process of adsorption, diffusion, and growth. For each species, S is defined as a fraction of the total number of impinging atoms or molecules that stick to the surface and are incorporated into the film. At temperatures where epitaxial layers are generally grown, only those group V atoms are incorporated into the layer that collide with the free group III atoms. As a result, S for the As atoms during the growth of GaAs can oscillate between zero and one depending on the surface population of free Ga atoms. Therefore, the growth rates of the GaAs and GaAlAs layers are essentially determined by the arrival rates of Ga and Al atoms.

We use the reflection high-energy electron diffraction (RHEED) to monitor the dynamics of growth during MBE. Figure 6.26*a* shows an example of the RHEED

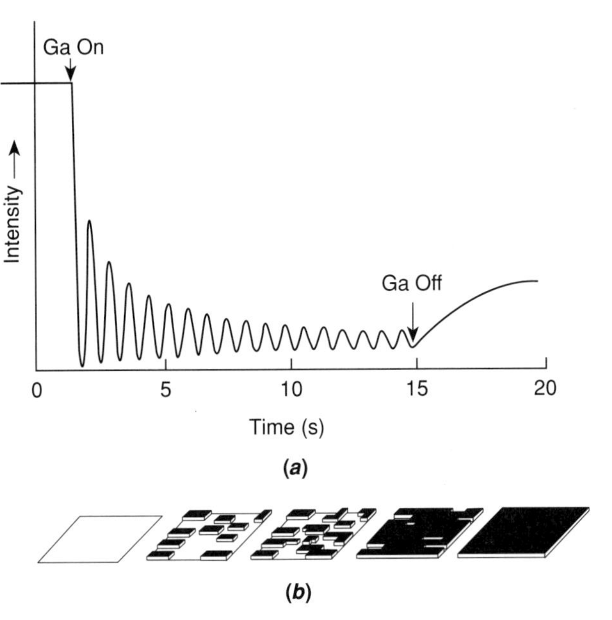

FIGURE 6.26
(*a*) Intensity oscillations of the RHEED specular beam from a GaAs (001) 2 × 4 surface viewed in the [110] azimuth during growth. The period exactly corresponds to the growth of a single molecular (GaAs) layer. (*b*) Schematic used to explain the origin of oscillations.

oscillations observed during the homoepitaxy of GaAs. Neave et al. (1983) and Van Hove (1983) explain these oscillations in terms of the layer-by-layer growth. They have suggested that each period of the oscillations represents a monolayer growth. Initially, the RHEED intensity is very high because the substrate is very smooth. Since the growth temperature is generally low in MBE, we expect the layer to grow by the formation of two-dimensional islands on the substrate surface, followed by their coalescence. We show this process in Figure 6.26*b*. The presence of the islands reduces the RHEED intensity because the layer is rough. This intensity is progressively reduced as more islands are nucleated. This reduction continues until the monolayer coverage exceeds 0.5, and then the intensity begins to recover. The highest intensity will be observed when the coverage is one; that is, the monolayer has been grown. We can thus determine the growth rate from the number of the RHEED oscillations observed during the growth.

Foxon and Joyce (1975, 1977) have investigated the kinetic models for the formation of GaAs from the elemental sources. We present their conclusions pictorially in Figures 6.27 and 6.28 for the two possible molecular species of arsenic, that is, As_2 and As_4. We see from Figure 6.27 that the basic process governing the As_2-Ga interaction on the GaAs surface is simple. The As_2 molecule undergoes dissociative chemisorption on the surface Ga atom to produce two As atoms. At lower temperatures (<330°C), the As_2 molecules may surface migrate and combine to form As_4 molecules that may be desorbed from the surface (see Figure 6.27).

The situation for the As_4 flux is certainly more complex (Foxon and Joyce 1975). Foxon and Joyce evaluated the correlation between the fluxes and sticking coefficients of As_4 and Ga and proposed a mechanism for the growth of GaAs from Ga and As_4 beams, as shown in Figure 6.28. Its critical feature is the pairwise dissociation of As_4 molecules adsorbed on the adjacent two gallium atoms. From the

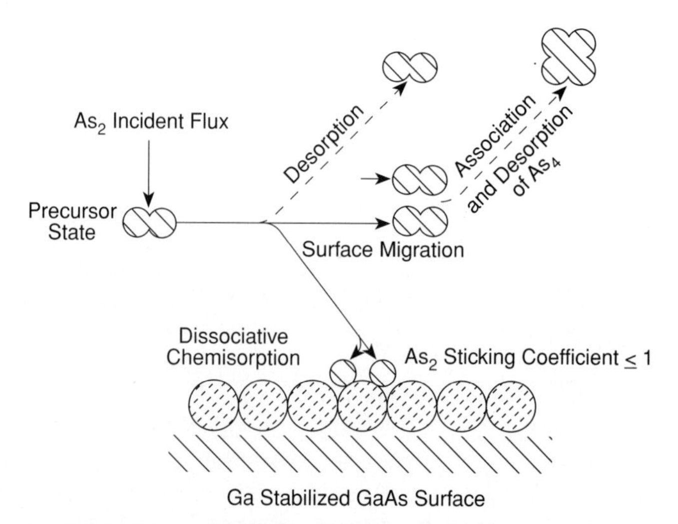

FIGURE 6.27
Model for the growth chemistry of GaAs from molecular beams of Ga and As_2. (*After Foxon and Joyce [1977].*)

two As_4 molecules, only four arsenic atoms are incorporated into the layer; the rest are desorbed as an As_4 molecule (see Figure 6.28).

Gas source and organometallic MBE. Using elemental sources in MBE, it is difficult to grow phosphorous containing III-V semiconductors because of the lack of control on the vapor pressures of P_4 and P_2 molecules. In 1980 Panish proposed an elegant solution that entailed the use of gaseous sources, such as PH_3 and AsH_3,

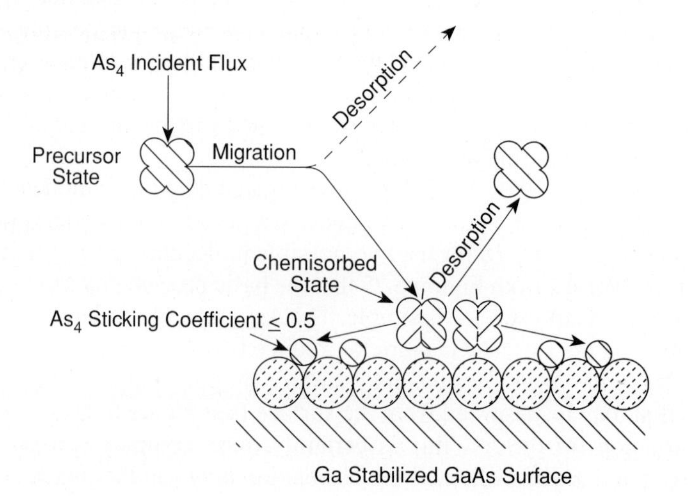

FIGURE 6.28
Model of the growth chemistry of GaAs from molecular beams of Ga and As_4. (*After Foxon and Joyce [1975].*)

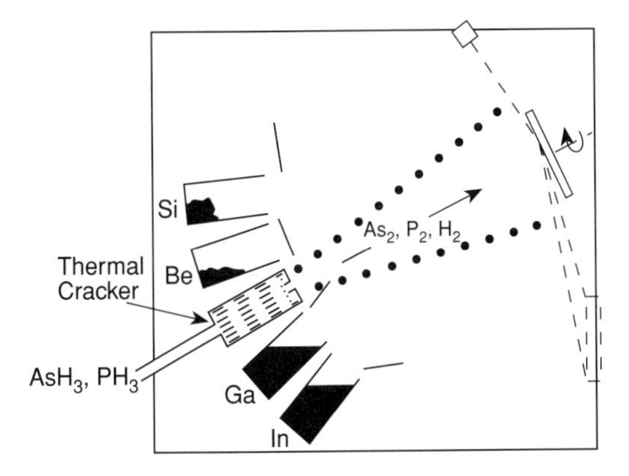

FIGURE 6.29
Schematic of source configuration for GSMBE. (*After Panish and Temkin [1989].*)

for the growth of InP and GaAs by MBE. The ultrahigh vacuum growth process where hydrides of group V elements are cracked to produce the beams of group V atoms is referred to as *gas-source molecular beam epitaxy* (GSMBE). With this technique, researchers have grown thin layers of InGaAsP using the elemental group III sources (Panish and Temkin 1989).

The schematic in Figure 6.29 shows a basic source configuration of GSMBE (Panish and Temkin 1989). The innovative feature of GSMBE is a thermal cracker for pyrolysing PH_3 and AsH_3, which is required because the hydrides do not decompose readily at the substrate surface. Virtually complete decomposition of these hydrides has been achieved using such a cracker. During GSMBE the hydride decomposition generates H_2, resulting in a higher pressure in the growth system than that which prevails during conventional MBE. This high pressure imposes an upper limit on the growth temperature.

Tsang (1989) has replaced the elemental sources for the group III atoms in GSMBE with the organometallic sources and has labeled the resulting growth technique as "chemical beam epitaxy." It may be more appropriate to use the term *organometallic molecular beam epitaxy* (OMMBE). In this technique a mixture of the organometallic compounds is injected into a single gas source that is maintained $\sim$50°C. The organometallics do not decompose at this temperature, but pyrolyse on encountering a heated substrate. A high degree of mixing of the gaseous species is achieved, leading to a good compositional uniformity of the mixed layers. Since the organometallics are not pyrolysed before impinging on the heated substrate, the chemical reactions occurring at the epi–gas interface in OMMBE are considerably more complex than those in GSMBE. The diffusion boundary layer may also be absent in OMMBE.

6.1.3.2 Silicon and silicon-germanium

The MBE of silicon has several advantages over VPE: (1) low growth temperatures, which reduce diffusion and autodoping effects, (2) precise control of layer thickness and doping profile on an atomic level, and (3) ability to grow novel structures, such as superlattices, quantum wells, and heterostructures.

We show a practical Si MBE system in Figure 6.30 (Bean 1981). The silicon beam is obtained by heating ultrapure silicon by an electron beam. Radiation damage

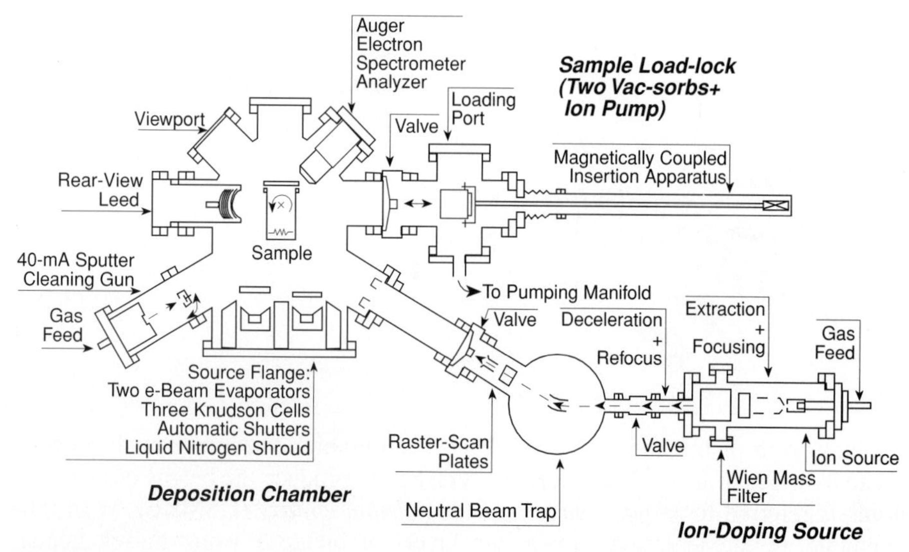

FIGURE 6.30
Schematic of a practical MBE system for silicon. (*After Bean [1981].*)

to substrates, caused by electron beam–induced X-rays, is minimized by shielding them from ions and backscattered electrons. Dopants are introduced into the films by producing dopant beams from effusion cells or by low-energy ion implantation. In the effusion-cell approach, gallium and antimony are used as p- and n-type dopants, respectively. Conventional dopants, such as P, As, and B, are not used because they have high vapor pressures. However, they can be incorporated into the film during MBE using low-current and low-energy ion implantation. As in the case of compound semiconductors, we can achieve very abrupt changes in doping profile by blocking the dopant fluxes with shutters positioned above the effusion cells. Since the sticking coefficients of the dopants at a typical growth temperature of 750°C are very low, higher dopants fluxes are required to incorporate the desired amount. We can also grow the silicon-germanium layers in a similar manner using MBE. We can use solid germanium source to grow silicon/silicon-germanium heterostructures.

6.2 EPITAXIAL GROWTH ON PATTERNED SUBSTRATES

We use the expitaxial growth on patterned substrates extensively for producing the multilayers that efficient III-V double-heterostructure lasers require. Figure 6.31 shows some of the structures used in the fabrication of state-of-the-art InP/InGaAsP lasers. For example, to grow wafers for a buried-head laser shown in the top-left panel in Figure 6.31, we first grow a four-layer structure consisting of n-InP/n-InGaAsP/p-InP/p^+-InP. We then etch the mesas by photolithography. We subsequently grow p- and n-type InP layers around the mesas. When the mesa is forward biased, the surrounding regions are reverse biased. As a result, the injected current is

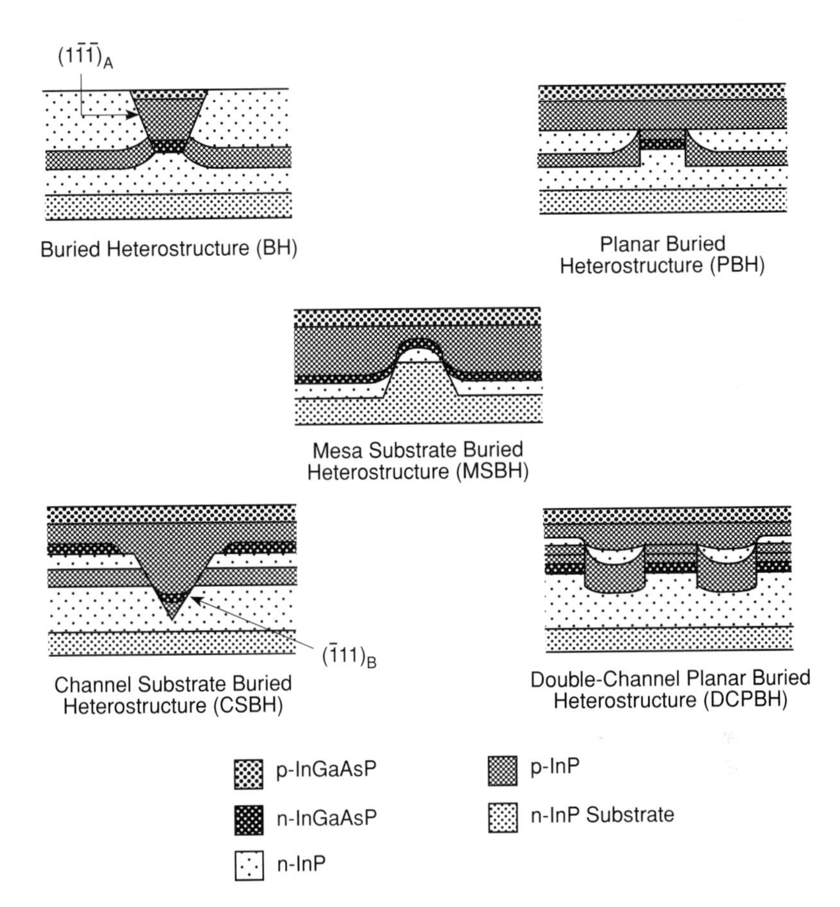

FIGURE 6.31
Schematics of heterostructures used in the fabrication of state-of-the art InP/InGaAsP lasers.

localized in the mesa area. Another advantage of the laser structures shown in Figure 6.31 is that the emitted light is confined on all sides because the refractive index of the active layer is lower than those of the surrounding layers, which facilitates the lasing operation.

We can pattern substrates in two ways. First, they are covered uniformly with a dielectric, and windows are opened selectively. This patterning is followed by the epitaxial growth. Second, the substrates are photolithographically patterned and etched to produce mesas and grooves and subsequently the layers are grown on such patterned substrates. The effects of the two types of patterning on the growth kinetics and the layer compositions are different and are discussed below.

To illustrate the first case of patterning, we have chosen closely spaced windows on the surface of a silicon separated by SiO_2. We can produce closely spaced silicon islands using VPE if the deposition of Si on SiO_2 can be avoided, that is, high selectivity. We observe that the higher the concentration of chlorine atoms in the silicon source, the higher the degree of selectivity (Wolf and Tauber 1987). We, therefore, achieve the highest degree of selectivity with $SiCl_4$ as the silicon source, whereas SiH_4 shows very little selectivity. The high selectivity implies that the desorption kinetics of silicon atoms from the oxide surface are fast in comparison to desorption from the silicon surface.

We see from Figure 6.31 that surfaces having different crystallographic orientations and curvatures are present after patterning. We can show from thermodynamic considerations that at a given temperature, the surfaces with different curvatures have different chemical potentials. To illustrate this difference, let us consider the case of LPE on patterned substrates. If a planar surface is in equilibrium with a solution of solute concentration C, then a curved surface is in equilibrium with a solution of concentration $C \pm \Delta C$; the choice of the sign depends on whether the surface is convex or concave. Cahn and Hoffman (1974) derived the following expression that relates C, ΔC, and the radius of a curved surface R:

$$\frac{\Delta C}{C} = \frac{2\gamma V_m}{k_B T R}, \tag{6.48}$$

where γ is the surface tension of the growth solution, V_m is molar volume of the crystal, k_B is Boltzmann constant, and T is growth temperature. This expression implies that a growth solution that is in equilibrium with a planar surface is undersaturated and supersaturated with respect to convex and concave surfaces. This condition would cause dissolution of convex regions and enhanced growth at concave regions. We illustrate the enhanced LPE growth of InP around a mesa on a (001) InP substrate in Figure 6.32.

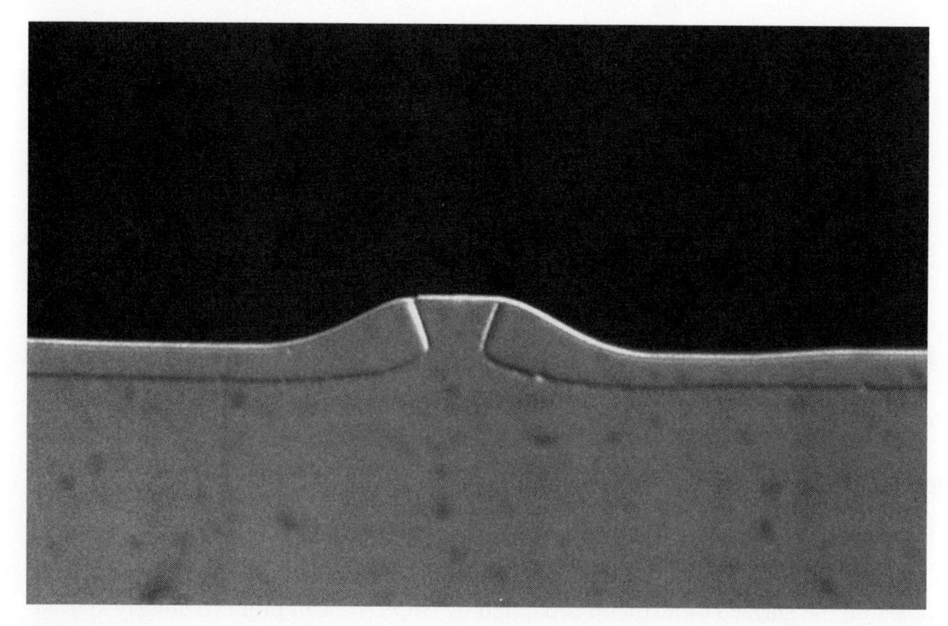

FIGURE 6.32
Morphology of an InP layer grown around an InP mesa by LPE.

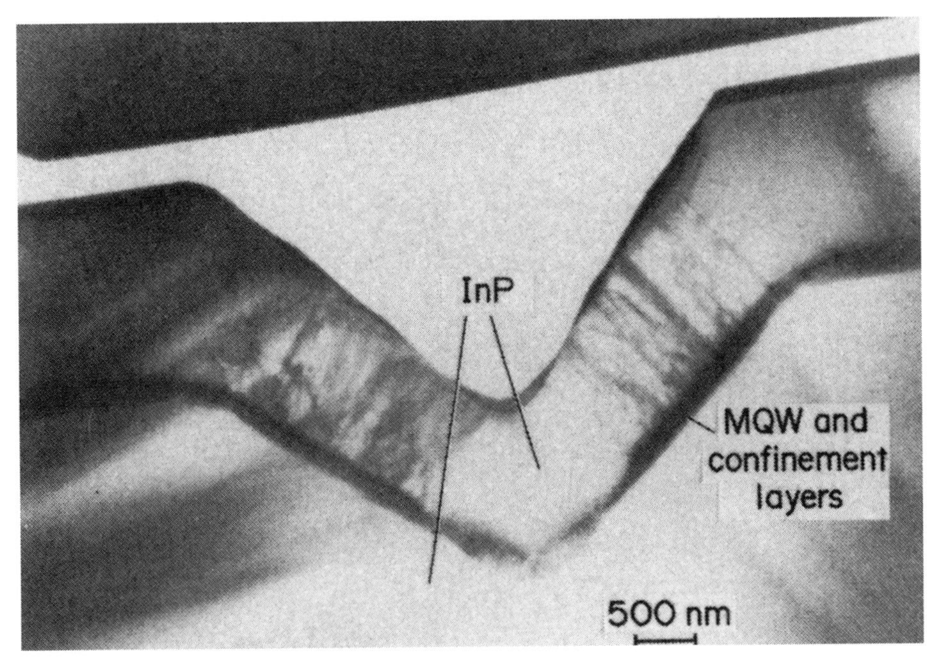

FIGURE 6.33
Micrograph showing defect structures of {111} and (001) InGaAsP layers grown simultaneously by OMVPE; MQW refers to multiple quantum wells. (*After Bhat and Mahajan [1992].*)

Since the orientation of the underlaying substrate affects the distribution coefficients of solutes in a liquid solution (see Figure 6.6 and section 6.1.1.4), the layers simultaneously deposited on planar regions of two different orientations will have different compositions. Identical effects must exist in layers grown by the vapor-phase techniques. We show in Figure 6.33 an electron micrograph obtained from an InGaAsP layer grown within a groove bounded by {111} planes and on a planar (001) InP surface (Bhat and Mahajan 1992). The layers deposited on the {111} planes are highly defective because they are not lattice matched with the underlying InP. We can rationalize only if we assume that the layers compositions on {111} and (001) surfaces are different. We visualize that the atomic species move from the {111} surfaces to the (001) surfaces. This migration affects the adsorption, desorption, and incorporation characteristics of different species, resulting in layers of different compositions on the surfaces having the two orientations.

6.3 HETEROEPITAXY

We can cite many examples where the successful heteroepitaxy of materials could permit the integration of different functions on a chip. For example, if we can grow III-V materials having low dislocation densities on silicon, we may be able to integrate the light-emitting devices with the integrated circuits. The integration of the light-emitting and magnetooptical materials would simplify the data storage technology. If we are able to grow CdTe layers on GaAs or silicon substrates, we can then deposit (Hg, Cd)Te layers on the composite substrates to fabricate large-area focal-plane arrays.

Several distinct situations arise during heteroepitaxy. First, the lattice parameters and crystal structures of the substrate and the overgrowth are the same. The growth of lattice-matched InGaAsP layers on InP substrates is a good example. Second, the crystal structures of the substrate and the layer are the same, but their lattice parameters are different. The growth of silicon-germanium layers on silicon substrates can be cited as an example. Third, the lattice parameters are almost the same, but their crystal structures are different. The growth of nickel silicides on silicon is a good example. Fourth, the lattice parameters as well as crystal structures are different. The growth of silicon on sapphire used for the fabrication of radiation hard circuits is a relevant example.

A number of generic issues arise during these four cases: (1) morphology of an overgrowth during the early and later stages of deposition, (2) effects of surface steps on the generation of domain boundaries during the growth of a polar semiconductor on a nonpolar semiconductor, and (3) generation of accommodation substructure, that is, how misfits are accommodated. We address the first two issues in this section and cover the third in section 6.4 where the origins of defects in epitaxial layers are discussed in detail.

When a lattice-mismatched layer is deposited, three distinct situations can develop, as shown in Figure 6.34. Under certain growth conditions, layer-by-layer growth, that is, the Frank-van der Merwe mode, may be achieved in reasonably thick layers. In some cases the layer may grow pseudomorphically initially, but nonplanar growth develops with additional growth. This is referred to as the Stranski-Krastanow mode. In the third case islands form right from the beginning of the growth (Volmer-Weber mode).

We can evaluate the stability of strained layers versus three-dimensional growth as follows. In cubic crystals, the strain energy (E_{ST}) for a film of area (A) and height (h) is

$$E_{ST} = 2G\left(\frac{1+\nu}{1-\nu}\right)Ah, \tag{6.49}$$

where G and ν are shear modulus and Poisson's ratio of the film. Assume that when the film undergoes a two-dimensional–to–three-dimensional transformation,

Frank-van der Merwe Mode
Layer-by-Layer Growth

Stranski-Krastanow Mode
Layer-Plus-Island Growth

Volmer-Weber Mode
Island Growth

FIGURE 6.34
Schematic showing three situations that can develop during the growth of a heteroepitaxial layer.

hemispherical islands of radius R_C form, where R_C is the critical radius of the islands at the growth temperature. If the density of the film is ρ, then the total number of atoms in the film is

$$\frac{6.23 \times 10^{23}\ \mathrm{Ah}\rho}{\mathrm{W}}, \tag{6.50}$$

where W is the atomic weight of the film. The number of atoms in each island or cluster is

$$\frac{(\mathrm{Rc})^3}{2(\mathrm{r})^3}, \tag{6.51}$$

where r is the tetrahedral radius of atom constituting the film. The number of clusters that would form is given by

$$\frac{2 \times 6.23 \times 10^{23} \times \mathrm{Ah}\rho(\mathrm{r})^3}{\mathrm{W}(\mathrm{R_C})^3}. \tag{6.52}$$

Total energy of a semispherical cluster is

$$\frac{1}{2} \times 4\pi(\mathrm{R_C})^2\gamma, \tag{6.53}$$

where γ is the surface energy/area of the clusters. Equation (6.53) neglects the interfacial energy between the clusters and the substrate. The total surface energy (E_{SUR}) of the cluster is

$$\mathrm{E_{SUR}} = \frac{24.8\pi 10^{23}\mathrm{AH}\rho(\mathrm{r})^3\gamma}{\mathrm{W}(\mathrm{R_C})} \tag{6.54}$$

$$\mathrm{R_C} = \frac{12.4\pi \times 10^{23}\rho\gamma(\mathrm{r})^3(1-\nu)}{\mathrm{G}(1+\nu)\mathrm{W}}. \tag{6.55}$$

When R_C is less than the value given by Equation (6.55), two-dimensional growth will be favored over the formation of three-dimensional clusters. In reality the deviation between the experimental results and values computed from Equation (6.55) is substantial because the interface energy between the dissimilar materials has been neglected in the above derivation.

Initially, lattice mismatched layers may grow pseudomorphically. However, with additional growth nonplanar morphology develops in InGaAs grown on GaAs (Berger et al. 1988), InAs on InP (Hollinger et al. 1992), InGaAs on InP (Gendry et al. 1992), and silicon-germanium layers on silicon (Cullis et al. 1992). We show in Figure 6.35 an example of the nonplanar Si-Ge growth on silicon under two different reflections (Cullis et al. 1992). In Figure 6.35*a*, the surface undulations are visible, but not the associated strain. However, both features are observed in Figure 6.35*b*. The observed undulations are caused by elastic distortions at the

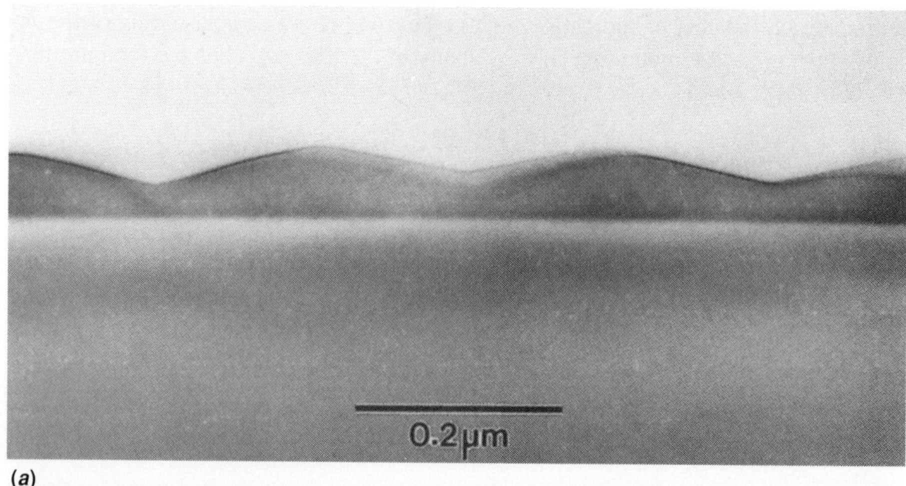

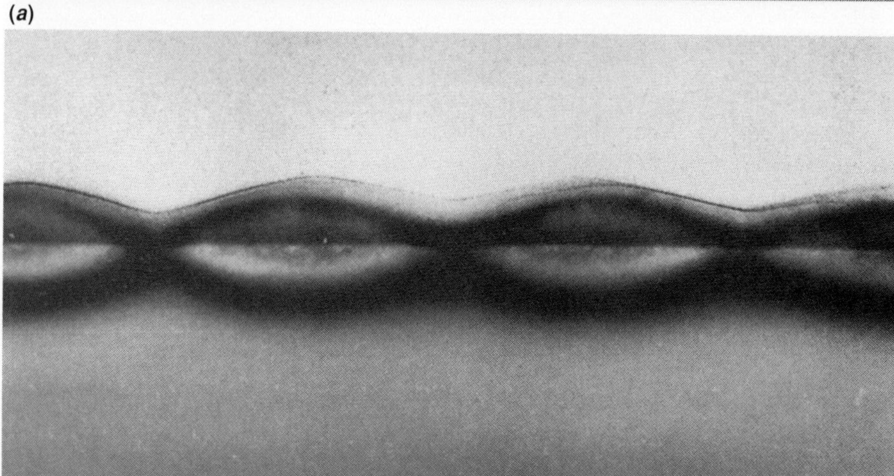

FIGURE 6.35
Undulations observed on the surface of a silicon-germanium layer grown on (001) silicon under different reflections: (*a*) only undulations are visible; and (*b*) both undulations and strain constrast are seen. (*After Cullis et al. [1992].*)

growing surface (Cullis et al. 1992), not by the nucleation of three-dimensional islands as suggested by Berger et al. (1988). The formation of undulations relaxes the elastic stresses within the heterostructure (Cullis et al. 1992).

Generally we carry out epitaxy on vicinal surfaces. The presence of steps on such surfaces affects the defect structure of a polar semiconductor grown on a nonpolar semiconductor. We illustrate this effect in Figure 6.36 where a polar zincblende crystal (AB) is deposited on a (001) nonpolar diamond-cubic crystal C. For the sake of discussion, we assume the lattice parameters of the two materials to be

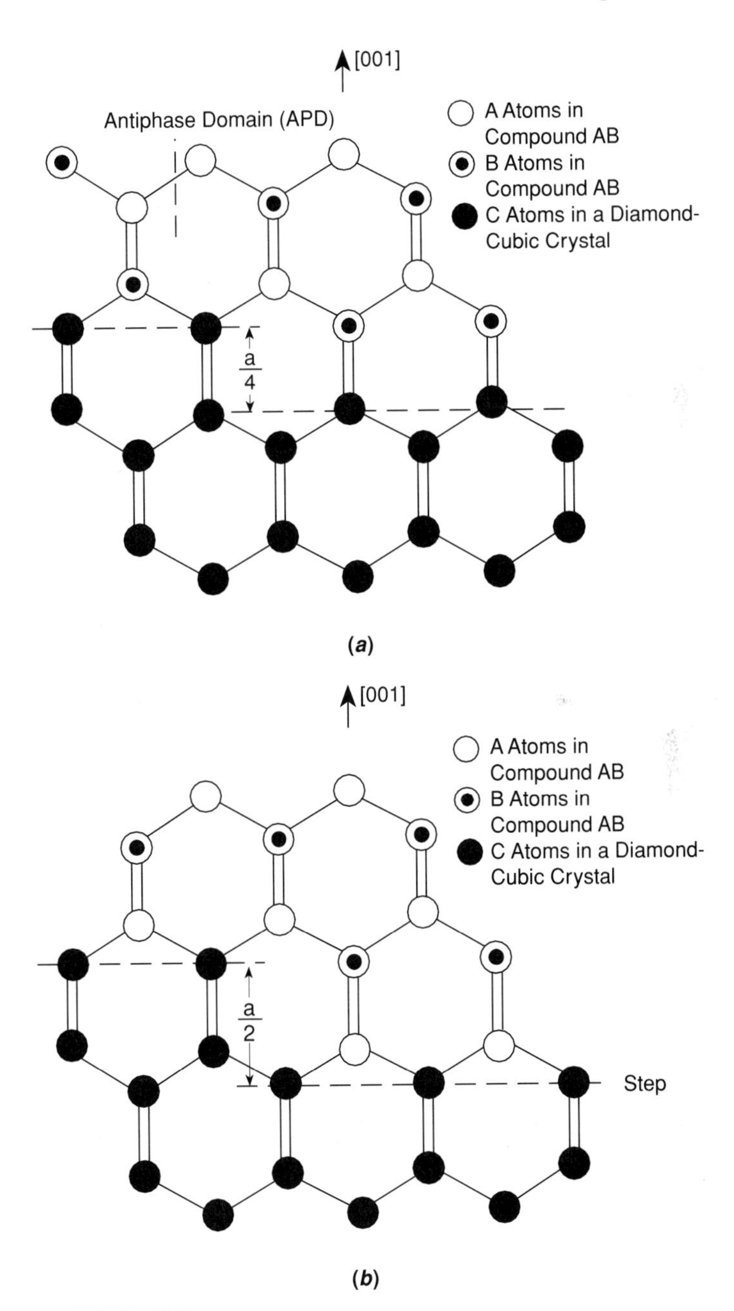

FIGURE 6.36
Schematics illustrating the role of steps in generating domain boundaries in a zinc-blende compound AB grown on a (001) diamond-cubic material C: (*a*) step height is (a/4) and a domain boundary is generated in AB, and (*b*) step is a/2 and a domain boundary is not generated in AB.

equal. Figure 6.36*a* shows a step of height (a/4) on the (001) surface of crystal C. If we assume that the B atoms attach preferentially to C atoms on the surface, an antiphase domain boundary (APD) would develop as indicated by the dashed line in Figure 6.36*a*. The APD would evolve even if the A atoms attach preferentially to the C atoms. We show in Figure 6.36*b* that the formation of APD can be avoided if the step of height (a/2) is present on the surface. The atomic arrangements within the overgrowth on either side of the step will then be in phase with each other, and the domain boundary does not form. The surfaces having double steps have been used successfully to grow a domain-free polar semiconductor like GaAs on silicon, a nonpolar semiconductor.

A variety of other defects, such as misfit dislocations, stacking faults and twins, are introduced in epitaxial layers during heteroepitaxy. Since under certain circumstances misfit dislocations and stacking faults may originate from threading dislocations, we address the formation of these features in section 6.4.

EXAMPLE 6.5. Calculate the critical radius of silicon clusters at which three-dimensional islands are favored over two-dimensional growth at 800°C on sapphire substrates.

Solution. The surface energy of clusters at 700°C may be estimated as follows:
The surface energy of silicon at the melting point = 730 ergs/cm^2.

The rate of change of surface energy with temperature = -0.10 ergs/cm^2/°C.

The melting point of silicon is 1410°C. Therefore, the surface energy of clusters of 700°C = 730 + (1410 − 700) × 0.1

$$= 730 + 71$$

$$= 801 \text{ergs/cm}^2.$$

R_C can be calculated using Equation (6.55)

$$= \frac{12.4 \times \pi \times 10^{23} \times 2.33 \times 801 \times (1.17)^3 \times 10^{-10} \times 0.67}{6.8 \times 10^{10} \times 1.33 \times 14}$$

$$= \frac{7802.8 \times 10^{-10}}{126.6} \text{ cm}$$

$$= 61.6 \times 10^{-10} \text{cm}.$$

6.4 DEFECTS IN EPITAXIAL LAYERS

We classify defects in epitaxial layers into two categories: growth-process-independent defects and growth-process-dependent defects. The respective examples of the two categories are threading dislocations, misfit dislocations, and contamination-induced stacking faults for the first category and macroscopic "defects" caused by melt-carryover during LPE (Mahajan et al. 1982) and oval-shape defects observed in GaAs layers grown by MBE (Wood et al. 1981) for the second. This section discusses the sources of various defects and mechanisms of their formation.

6.4.1 Growth-Process-Independent Defects

6.4.1.1 Threading dislocations

We generally use dislocated substrates for the epitaxial growth. This situation is especially true in the case of III-V epitaxy. Two distinct situations arise when a layer is deposited on a dislocated substrate. First the Burgers vector of the dislocation terminating at the substrate surface is inclined to the surface. This dislocation is replicated into the layer because it is associated with a spiral pattern at its emergence point at the surface. The spiral pattern is replicated into the overgrowth leading to the incorporation of the substrate dislocation into the overgrowth. Various experimental observations on InP layers (Keramidas et al. 1981; Mahajan et al. 1981) support this assessment.

In the second situation the Burgers vector of the substrate dislocation is parallel to the growth surface. To a first approximation when the Burgers vector is parallel to the surface, a step should not form at the emergence point of the dislocation at the surface because the component of Burgers vector normal to the surface is zero. However, Beam (1989) used computer simulations to show that protrusions and depressions develop at the emergence point of the dislocation. He attributes these features to the relaxation of different stress fields associated with the dislocation. These surface features are replicated during the layer growth, implying that these dislocations are also replicated. The studies of Bauser and Strunk (1981, 1984) and Beam et al. (1990) on silicon and InP homoepitaxial layers, respectively, are consistent with this assessment.

We conclude that insofar as threading dislocations are concerned, the dislocation density in a homoepitaxial layer can at best be equal to that in the substrate. Therefore, to achieve low dislocation densities in epitaxial layers, the densities of dislocations in substrates should be low; that is, the crystals from which the substrates are fabricated should be nearly perfect.

The issue of threading dislocations in the device structures is important because their presence affects the performance of those structures. We illustrate this situation using two examples. Figure 6.37 shows the dependence of the quantum efficiency of Si-doped GaAlAs light-emitting diodes on the threading-dislocation densities. The efficiency (η) is high at low dislocation densities and decreases substantially when the density increases to 10^6cm^2. We attribute this change to the loss of carriers due to recombination at dislocations (see section 3.4).

Threading dislocations also play a role in the rapid degradation of GaAs/GaAlAs lasers. Figure 6.38 shows a scanning electron microscope image of nonluminescent regions observed in a degraded GaAs/GaAlAs laser (Ishida and Kamejima 1979). These regions are labeled as DLDs (for dark line defects) in Figure 6.38 and are oriented along the <100> and <110> directions in the active region. The dark spot defects (DSDs) in the figure have been attributed to the migration of the contact metal into the device (Ishida and Kamejima 1979). When we examine DLDs by transmission electron microscopy, we can see the dislocation structure shown in Figure 6.39. Dislocation clusters D_1, D_2, and D_3 appear to form from the existing threading dislocations. We envisage that during the laser operation,

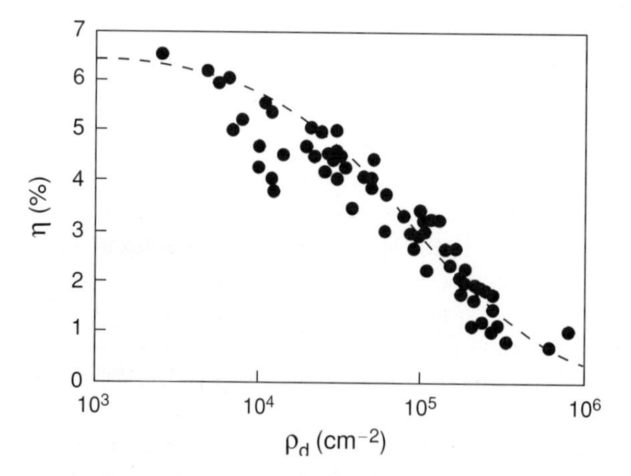

FIGURE 6.37
Variation in efficiency with dislocation density for a large number of Si-doped GaAlAs LEDs. (*After Roedel et al. [1979].*)

dislocations in the active layer undergo nonradiative, recombination-enhanced glide and climb (Petroff and Hartman 1974; Ishida and Kamejima 1979; Mahajan et al. 1979), resulting in the observed structure.

We need to understand whether or not the correspondence between dislocations in the substrate and the layer discussed above is observed at higher dislocation densities ($>10^6 cm^{-2}$). We find that under such conditions the density of dislocations in the layer is generally lower than that in the substrate (Saul 1971; Mahajan et al. 1982). We can rationalize this observation by referring to Figure 6.40 that shows the formation of a closed loop by the coalescence of two edge-type threading dislocations in an epitaxial layer; Beam (1989) has observed such closed loops in InP homoepitaxial layers deposited on highly dislocated InP substrates. Since the coalescence of the dislocations is effected by their movement

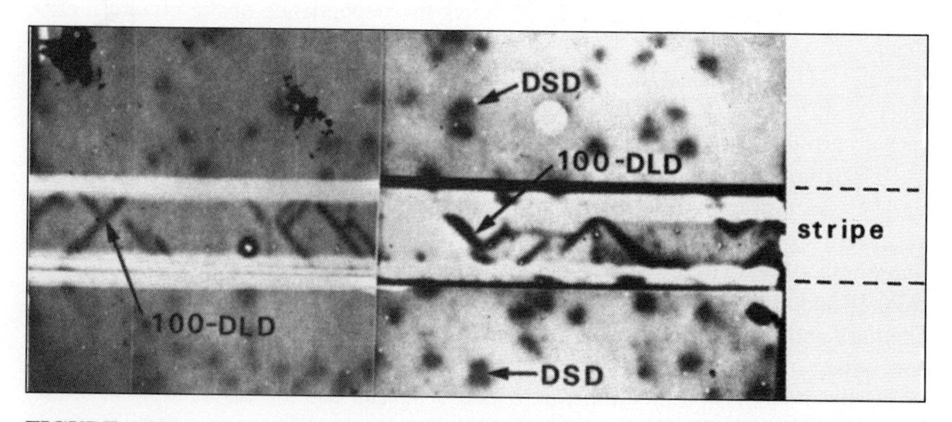

FIGURE 6.38
Scanning electron microscopy images of a degraded GaAs/GaAlAs laser. <100>- and <110>-oriented dark line defects (DLDs) appear within the active laser stripe. Many dark spot defects (DSDs) appear outside this region. (*After Ishida and Kamejima [1979].*)

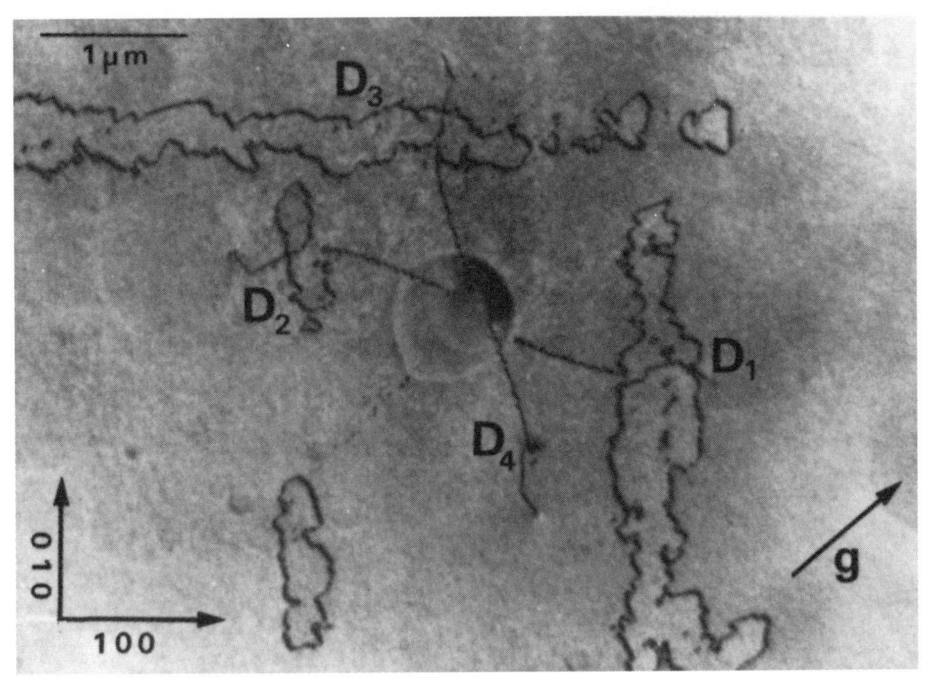

FIGURE 6.39
High-voltage electron micrographs obtained from a degraded GaAs/GaAlAs laser. Dislocation clusters D_1, D_2, D_3, and D_4 develop from existing dislocations. (*After Ishida and Kamejima [1979].*)

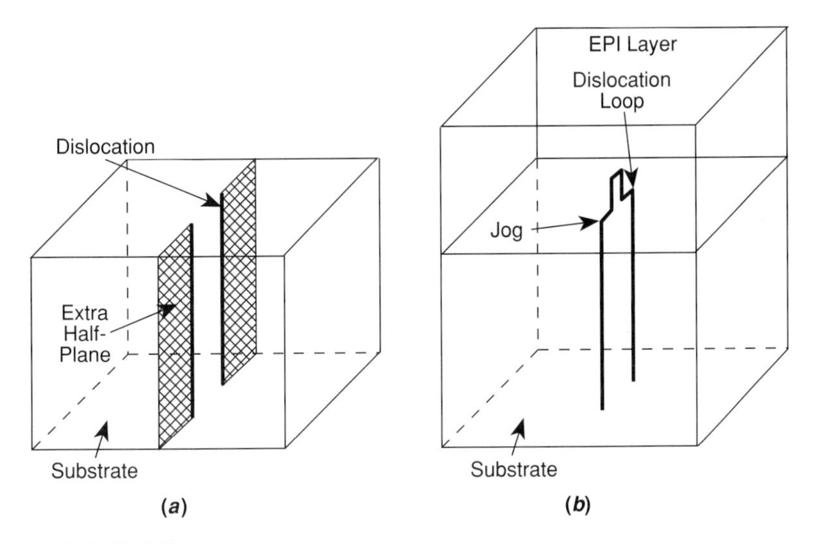

FIGURE 6.40
Schematic illustrating the formation of a closed loop from the coalescence of two edge dislocations of opposite Burgers vectors: (*a*) two edge dislocations within the substrate and (*b*) after they have coalesced within the layer to form a closed loop. Self-glide and climb facilitate the coalescence process.

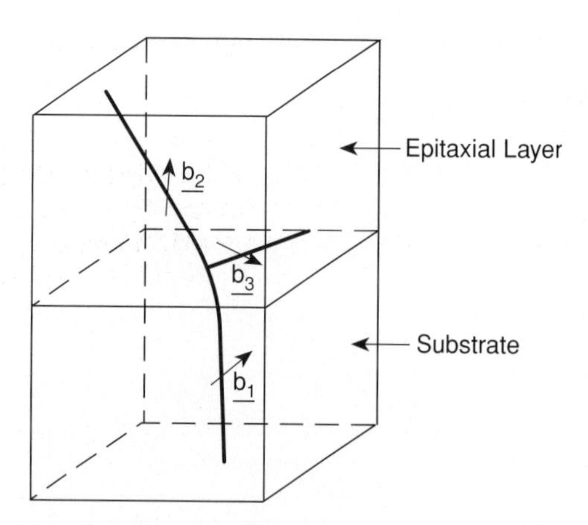

FIGURE 6.41
Schematic illustrating the dissociation of a substrate dislocation with Burgers vector $\vec{b}_1$ into two dislocations with Burgers vector $\vec{b}_2$ and $\vec{b}_3$ in the epitaxial layer. The Burgers vectors are referred to as *layer-base vectors*. The lattice parameters of the substrate and the layer are equal but have different crystal structures.

under the influence of their mutual stress fields, that is, self-glide and climb, dislocations should be closely spaced so that these processes can occur. A dislocation density of $\sim 5 \times 10^6$ cm^{-2} may constitute the lower limit for the self-glide and climb to occur; at lower densities the elastic interactions between the widely separated dislocations are too weak for self-glide because they vary inversely with the dislocation separation.

Let us now consider a case where an epitaxial layer is lattice matched to a dislocated substrate whose crystal structure is different from that of the layer. As discussed earlier spiral steps, protrusions, and depressions are produced at the emergence points of dislocations at the substrate surface. During epitaxy these features are replicated into the overgrowth, resulting in the incorporation of dislocations into the layer. Since the crystal structures of the layer and the substrate are different, Burgers vectors of dislocations in the layer may not have lower energy. We can lower the energy by the following dislocation reaction:

$$\vec{b}_1 \rightarrow \vec{b}_2 + \vec{b}_3, \tag{6.56}$$

where $\vec{b}_1$, and $\vec{b}_2$, and $\vec{b}_3$ are Burgers of dislocations in the substrate and the layer, respectively. All these Burgers vectors are referred to the *layer-base vectors*. We show the resulting situation in Figure 6.41.

EXAMPLE 6.6. Two dislocations oriented along the [001] and [011] directions are contained in a (001) substrate. Their emergence points at the substrate surface are labeled A and B. If distance AB is 2 μm, calculate the separation between the emergence points of the [011]- oriented dislocation from that of the [001] dislocation. Assume that the thickness of the layer is 5 μm.

Solution. Since the [001] dislocation is oriented normal to the surface, its emergence points at the substrate and epitaxial layer surfaces will lie on top of each other. However, the emergence points of the inclined dislocation will undergo lateral shift that can be calculated using the following schematic:

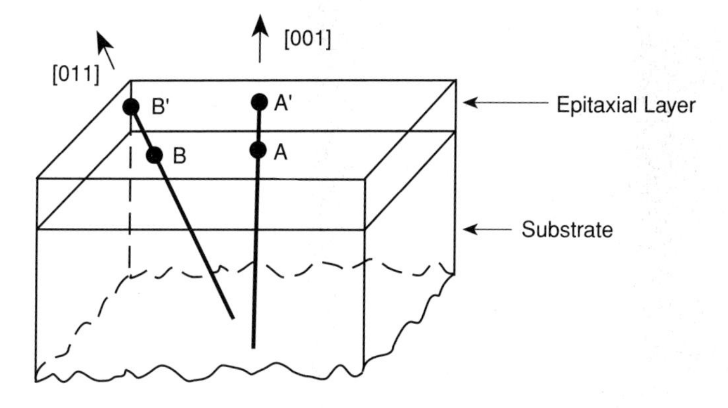

[011]-oriented dislocation

The cosine of the angle between dislocations oriented along the directions [001] and [011] is $\frac{1}{\sqrt{2}}$.

The lateral shift of the dislocation during growth of a 5 μm epitaxial layer = 5 × tangent of the above angle = 5 μm.

Therefore, the emergence point of the [011] dislocation B′ will be at a distance of 7 μm from the emergence point of the [001] dislocation A′ in the preceding figure.

EXAMPLE 6.7. Imagine a situation in which two perfect-edge dislocations of opposite Burgers vectors are contained in a $(11\bar{2})$ InP substrate. If these dislocations are replicated into the layer during epitaxy, calculate the number of point defects required per unit length of the dislocation to bring the two perfect edge location together to annihilate each other. Assume that the Burgers vector of the dislocations is $a/2\ [1\bar{1}0]$, that the glide plane is (111), and that the dislocations lie on the same $(1\bar{1}0)$ plane and are separated from each other by 50 nm.

The following schematic shows this situation:

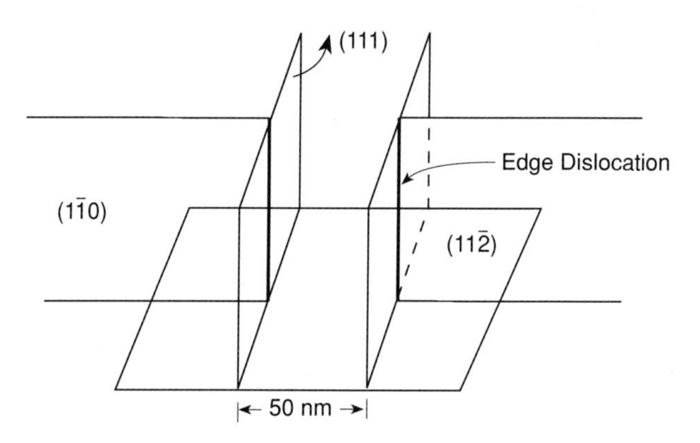

Solution. It can be shown that the separation between atoms along the edge dislocation in Figure 3.20 $= \frac{\sqrt{6}a}{4}$, where a is the lattice parameter.

Therefore, the number of atoms present along a dislocation length of 1 cm $= \frac{4}{\sqrt{6}a}$

The number of interstitials required for the dislocations to climb a distance of $\frac{a}{\sqrt{3}} = \frac{4}{\sqrt{6}a} \times 2$.

Therefore, the number of interstitials required to coalesce the two dislocations $= \frac{8\sqrt{3} \times 250 \times 10^{-8}}{\sqrt{6}a^2}$.

The lattice parameter of InP = 0.585 nm.

Substituting for a in the above relation, the number of interstitials required $= \frac{8 \times 250 \times 10^{-8}}{\sqrt{2} \times (5.85)^2 \times 10^{-16}}$

$= 4.13 \times 10^9$.

The computed value is fairly high, implying that it will be difficult for the dislocations to annihilate each other exclusively by climb.

6.4.1.2 Misfit dislocations

In the preceding subsection on threading dislocations, we assumed that the lattice parameters of the substrate and the overgrowth are equal. There are many interesting situations where this condition is not satisfied, for example during the growth of silicon on sapphire and GaAs on silicon. In these cases misfits exist between the layer and the substrate. Since the presence of misfits increases the strain energy of the system as the layer grows (see Equation [6.49]), at a certain thickness the layer relaxes to its unconstrained lattice parameter. This thickness is referred to as a *critical thickness*.

Phenomenonlogically, the layer can acquire its unconstrained lattice parameter by either the insertion or the removal of extra-half planes that terminate at the layer-substrate interface. The lines along which these planes terminate are termed *misfit dislocations*. If the unconstrained lattice parameters of the substrate and the layer are, respectively, a_1 and a_2, the separation S between the misfit dislocations can be computed as shown below in Equation 6.59.

Consider the situation where (n − 1) planes of the material with the larger lattice parameter a_2 match with n planes of the material with the smaller lattice parameter a_1. Under this condition the following relationship should hold:

$$(n - 1)a_2 = na_1. \tag{6.57}$$

Solving for n we have,

$$n = \frac{a_2}{|a_2 - a_1|}. \tag{6.58}$$

One misfit dislocation is introduced for a distance of na_1. Therefore, the separation between the dislocations S is given by the following expression:

$$S = \frac{a_2 a_1}{|a_2 - a_1|}. \tag{6.59}$$

The linear density of the misfit dislocations D, that is, the number per unit length, is given by

$$D = \frac{|a_2 - a_1|}{a_2 a_1}. \tag{6.60}$$

The preceding discussion assumes a one-dimensional misfit. However, the result is generic and can be extended to a two-dimensional situation.

This phenomenological discussion does not shed any light on the mechanisms of formation of misfit dislocations in different situations. To develop a mechanistic explanation, let us first consider the case where the overgrowth and the substrate have a misfit, but their crystal structure is the same. Initially the layer grows pseudomorphically, and the substrate dislocations are replicated into the layer. We illustrate this behavior in Figure 6.42*a*. When the layer reaches a critical thickness, a threading dislocation BC in the layer undergoes glide. As a result of the glide, the misfit

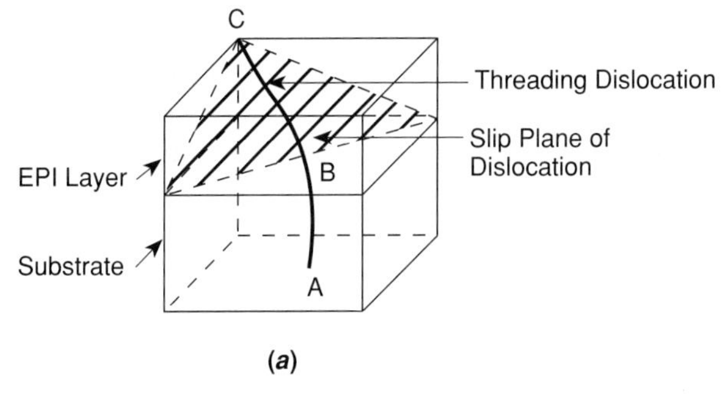

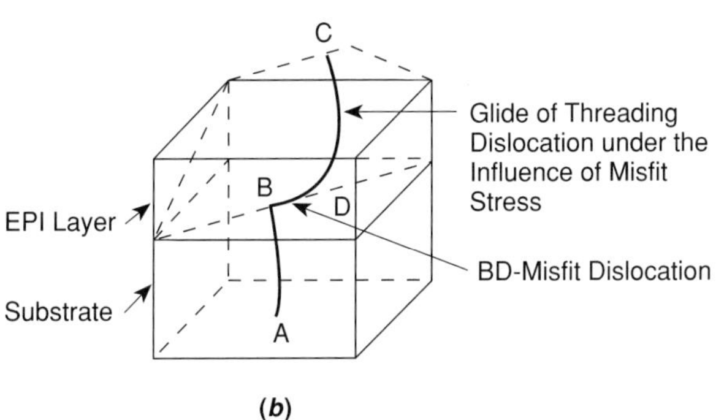

FIGURE 6.42
Schematic illustrating how the glide of (*a*) a threading dislocation ABC under the influence of misfit stress leads to the formation of (*b*) a misfit dislocation BD that lies along the line of intersection of the glide plane within the layer with the substrate surface.

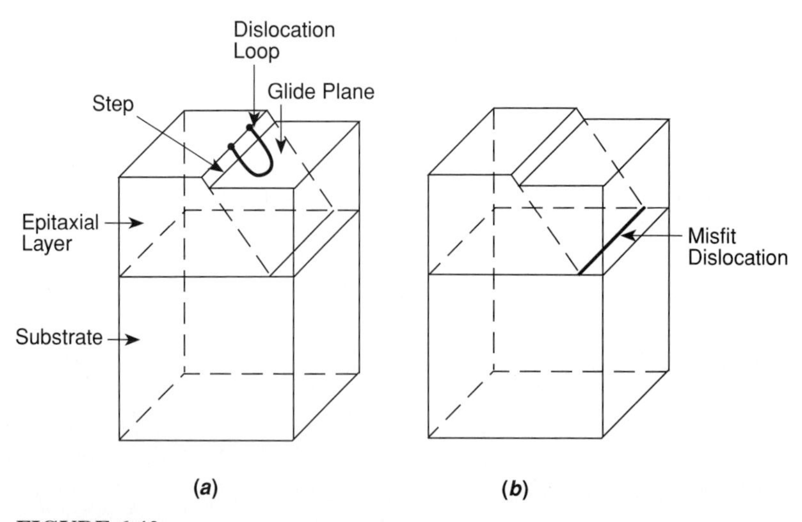

FIGURE 6.43
Schematic illustrating the nucleation of a dislocation half-loop from a step; (*b*) its subsequent propagation on its glide plane leads to the formation of a misfit dislocation.

dislocation BD forms along the line of intersection of the glide plane in the overgrowth with the substrate surface; see Figure 6.42*b*. This model is due to Matthews (1966) and operates in a number of material systems.

For dislocation BC to glide, the resolved component of the misfit stress must either be equal to or exceed the critical resolved shear stress of the layer. In other words, the critical resolved shear stress is reached at the critical thickness in lattice-mismatched epitaxial systems. Since the critical resolved shear stress depends on temperature, stoichiometry, and composition of the layer, these materials parameters affect the critical thickness; that is, the critical thickness is not a unique parameter of the material. Recent results on mismatched InGaAs layers deposited on InAlAs by MBE are consistent with the preceding assessment.

These ideas on the formation of misfit dislocations are applicable only if the underlying substrate is dislocated. In some situations, for example, during the growth of GaAs and silicon-germanium layers on macroscopically dislocation-free silicon, substrate dislocations are not available for the generation of misfit dislocations. For misfits to be accommodated in this case, dislocation loops could nucleate at steps present on the layer surface and subsequently propagate on their glide planes toward the layer-substrate interface (Narayan et al. 1988). This situation is shown in Figures 6.43*a* and 6.43*b*. Comparing Figures 6.42*b* and 6.43*b*, it is evident that the final arrangements of misfit dislocations in the two cases are nearly similar.

Therefore, it would be difficult to assess experimentally the operating mechanism. To make such an assessment, we must examine the samples at a stage where the glide loops have barely expanded into the epitaxial layer so that threading segments are still visible; see Figure 6.43*a*.

One of the challenges is to grow lattice-mismatched heterostructures having low densities of threading dislocations. Since the density of misfit dislocations is dictated by the materials involved, the density of misfit dislocations is intrinsic in nature and cannot be altered by the growth conditions. However, we may be able to independently alter the density of threading dislocations. Two approaches can be used to achieve this objective. One of the schemes entails the growth of a graded buffer layer. For example, to grow GaAs on silicon, a silicon-germanium buffer layer graded between silicon and germanium was grown first. Subsequently, GaAs was deposited on a lattice-matched Ge layer (Xie et al. 1992). As the misfit between the successive layers is small, only a few loops may nucleate from step edges and propagate toward the layer-substrate interface. During this time threading dislocations can glide sideways over substantial distances. This process is repeated during various stages of the growth of the buffer layer. The net effect is that the density of misfit dislocations is the same in the graded heterostructure and the nongraded structure, but the density of threading dislocations is considerably lower in the graded heterostructure.

In the second approach the overgrowth is annealed immediately after its thickness exceeds the critical thickness. The prevailing situation when the critical thickness is exceeded is shown in Figure 6.44*a*. During subsequent annealing, positive- and negative-threading dislocations belonging to different half-loops annihilate each other, resulting in an arrangement shown in Figure 6.44*b*. As a result, the density of threading dislocations is considerably reduced. If an additional thickness is grown after the anneal, the density of dislocations in the overgrowth is considerably less. The validity of this idea has not been demonstrated experimentally.

EXAMPLE 6.8. Assuming that the Matthews model is responsible for the generation of misfit dislocations in III-V heteroepitaxy, determine the directions along which the misfit dislocations will be aligned for the following substrate orientations: (001) and (112).

Solution. Since III-V materials deform by slide on the a/2 <110> $\{\bar{1}11\}$ systems (see Figure 6.42), misfit dislocations generated by the Matthews mechanism will lie along the lines of the intersections of the glide planes within the layer with the substrate surface.

(001) Substrate

The (111), $(11\bar{1})$, and $(\bar{1}11)$, $(1\bar{1}1)$ planes intersect the (001) surface along the $[1\bar{1}0]$ and [110] directions. Therefore, misfit dislocations will be aligned along these directions and will form a cross-hatch pattern.

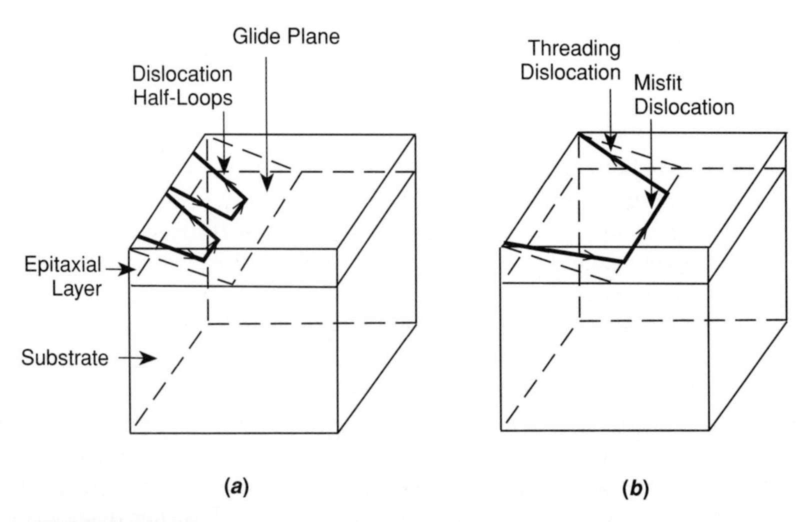

FIGURE 6.44
Schematic illustrating the reduction of threading dislocations in mismatched heteroepitaxy: (*a*) when the layer thickness is slightly in excess of the critical thickness and (*b*) after a short anneal of the situation shown in (*a*).

(112) Substrate

The (111), $(11\bar{1})$, and $(\bar{1}11)$, $(1\bar{1}1)$ planes intersect the (112) surface along the $[1\bar{1}0]$, $[1\bar{1}0]$, $[13\bar{2}]$, and $[\bar{3}\bar{1}2]$ directions, respectively. Therefore, misfit dislocations will be aligned along these directions.

EXAMPLE 6.9. Consider the heteroepitaxial growth of InP layers on (001) GaAs substrates having a dislocation density of 10^4 cm^{-2}. Can the operation of the Matthews model fully relax the misfit between the two materials? The lattice parameters of InP and GaAs are 0.586 and 0.565 μm.

Solution. The separation between the misfit dislocations in a fully relaxed heterostructure can be calculated from Equation (6.59)

$$= \frac{0.586 \times 0.565}{0.021} \text{ nm}$$

$$= 15.8 \text{ nm}.$$

The separation between the threading dislocations in the GaAs substrate

$$= \frac{1}{\sqrt{10^4}} \text{ cm}$$

$$= 10^{-2} \text{ cm}$$

$$= 100 \; \mu\text{m}.$$

The separation between the misfit dislocations is considerably smaller than the separation between the threading dislocations. Therefore, without invoking extensive dislocation multiplication during epitaxy, misfit cannot be fully relaxed by the Matthews model.

6.4.1.3 Stacking faults and twins

Stacking faults and twins in epitaxial layers can form in two ways: (1) by growth accidents on facets and (2) by misfit stress-induced deformation. In both cases faults and twins lie on {111} planes and are bounded by a/6 $<11\bar{2}>$ partials. (See sections 3.2 and 3.3 for additional details on the crystallography of faults and twins.) The presence of contaminants either on surfaces of substrates or in the growth system generally enhances the occurrence of growth mistakes. Figure 4.17 showed the stacking fault clusters observed on the surface of a lattice-matched GaAlAs layer grown on a (001) Syton-polished GaAs substrate by LPE (Dutt et al. 1981). Dutt et al. (1981) indicated that the majority of the faults nucleate at the substrate–layer interface, and attributed these faults to the presence of calcium on the substrate surface.

The mechanism governing the formation of the above growth faults is not well understood. Matthews (1966) suggested that in the presence of contaminants, the layer may nucleate in the form of islands. During the additional growth these islands coalesce. If the islands are not in perfect registry with each other because of poor communication between the overgrowth and the substrate, the formation of stacking faults may accommodate the lack of registry.

The formation of misfit-induced stacking faults and twins in heteroepitaxial layers is easier to understand. In the case of the diamond-cubic and zinc-blende structures, a/2 $<110>$ misfit dislocations at the layer-substrate interface could dissociate into two Shockley partials (see section 3.3.2). Under the influence of the misfit stress, one partial can glide toward the surface and generate a stacking fault while the second partial stays at the interface and accommodates a portion of the misfit. Overlapping faults and twins can be generated in a similar fashion. Since the magnitude of the Burgers vectors of the Shockley partials is considerably smaller than that of perfect dislocations, the partials are not very effective in accommodating the misfit.

6.4.2 Growth-Process-Dependent Defects

Certain types of macroscopic "defects" in epitaxial layers are specific to a particular growth technique. This situation is not surprising because the underlying approach and the conditions prevailing at the growth interface in each growth technique are different. It is impossible for us to discuss in detail every defect that has been observed. Instead, we have chosen three examples to emphasize the relationship between the origins of defects and prevailing growth conditions: (1) melt-carryover-induced defects in heterostructures grown by LPE, (2) formation of hillocks during VPE, and (3) evolution of oval-shape defects during MBE.

6.4.2.1 Melt-carryover-induced defects

We indicated in section 6.1.1.4 that multiple melts are required to grow multilayer heterostructures by LPE. Since the melt composition determines the layer composition, complete wipe off of the melt is essential when we slide the wafer from one well to the next. Otherwise, the composition of the next melt is altered locally, resulting in local changes in the composition of the layer. Kopf and Sumski (1975) and Ladany et al. (1981) have observed such changes in the composition of GaAs layers. In some cases holes and dissolution pits can form in epitaxial layers by the melt-carryover (Mahajan 1989).

For the sake of illustration, let us consider a case where the wipe off of the melt-back liquid A from the surface of a (001) substrate of AB, crystallizing in the zinc-blende structure, is poor. We show this condition schematically in Figure 6.45*a*. Since the droplets do not contain any B atoms, they start dissolving the underlying substrate to balance the activities of the B atoms in the A droplet and solid AB, resulting in a dissolution pit as shown in Figure 6.45*b*. If the layer AB is grown on the pitted surface in Figure 6.45*b*, the growth would not occur on the dissolution pits containing the liquid A. This lack of growth occurs because in and around the dissolution pit the melt is undersaturated with respect to B atoms at the growth temperature. As a result, holes would develop in the layer. During the wipe off after the layer growth, the melt is trapped in these holes; see Figure 6.45*c*. If additional layers are grown on the surface in Figure 6.45*c*, the hole may be replicated into various layers. We have observed such situations in the InP/InGaAsP heterostructures grown by LPE (Mahajan 1989).

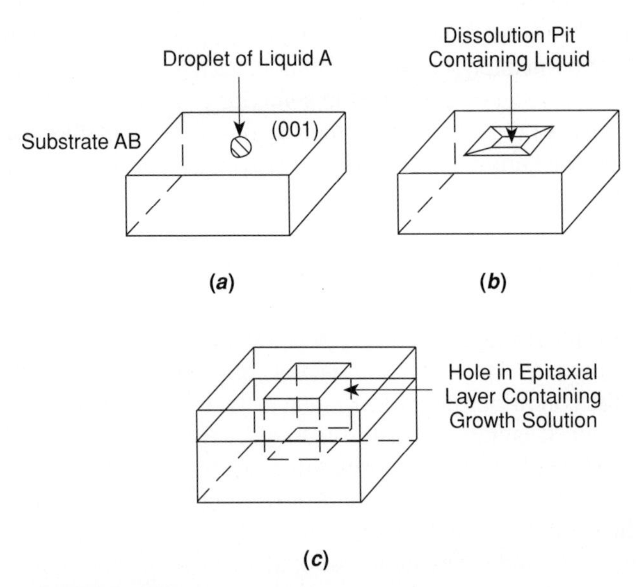

FIGURE 6.45
Schematics illustrating the formation of a hole in an AB epitaxial layer due to poor wipe off of liquid A after the melt back: (*a*) droplet of liquid A after the melt back, (*b*) formation of a dissolution pit, and (*c*) formation of a hole in the layer.

6.4.2.2 Formation of hillocks in VPE layers

If the growth conditions, that is, the growth temperature, V/III or VI/II ratio, substrate misorientation, and so forth, are not optimal, hillocks have been observed on the surfaces of epitaxial layers (Dilorenzo 1972; van de Ven et al. 1987; Gleichmann et al. 1990). Figure 6.46 reproduced from the work of Dilorenzo (1972) on halide VPE of GaAs shows hillocks on the surfaces of homoepitaxial layers deposited on GaAs substrates of different orientations. The formation of hillocks is much more pronounced at lower growth temperatures and at higher $AsCl_3$ mole fraction in the gas stream. The hillocks have also been seen in InP (Gleichmann et al. 1990), CdTe (Snyder et al. 1991), and HgCdTe layers (Bhat and Ghandhi 1986; Capper et al. 1989; Bevan et al. 1990).

The study of hillocks by transmission electron microscopy indicates that they contain stacking faults and dislocations (van de Ven et al. 1987; Gleichmann et al. 1990). We do not understand the origins of these microstructural features. A plausible source could be the clusters of group V or VI atoms formed during the growth. If we assume that these clusters form under nonoptimal growth conditions, the layer grown over these clusters could be highly defective. During additional deposition, the vertical growth rate at the defective regions is higher than the rate at nondefective areas because the dislocations terminating at the growth surface serve as growth spiral sources.

6.4.2.3 Oval-shape defects in MBE layers

Macroscopic defects in the shape of an oval, referred to as *oval defects,* have been observed in GaAs layers grown by MBE (Chai and Chow 1981; Kirchner et al. 1981; Wood et al. 1981; Bafleur et al. 1982; Petit et al. 1984). Figure 6.47, taken from the work of Bafleur et al. (1982), shows a typical oval defect. A line, oriented along the <110> direction lying in the (001) plane, separates black and white lobes associated with the defect. Bafleur et al. (1982) have also investigated the oval defects in detail by selective chemical etching and transmission electron microscopy. Their study shows two distinct situations contiguous to the oval defect: (1) crystal defects are not noticeable, and (2) ovals consist of dislocations, stacking faults, twins, and polycrystalline material.

If we assume that the emission of Ga droplets from the cells is the main cause of ovals (Wood et al. 1981), we can rationalize their formation. As discussed in section 6.4.2.1, the presence of Ga droplets on GaAs epitaxial layers or substrates produces dissolution pits. Concomitantly, these droplets are exposed to the As flux in an MBE system. The concentration of As in the Ga droplets increases until the two-phase field on the Ga-rich side of the Ga-As phase diagram is reached. At this stage GaAs forms and deposits epitaxially on the exposed facets of the dissolution pits. This process continues until the Ga droplets are totally consumed. However, if some of the droplets are not fully converted into GaAs prior to the cessation of the growth, they could solidify into defective polycrystalline aggregates.

The three types of macroscopic defects discussed under section 6.4.2 illustrate some of the complexities that are encountered in various epitaxial techniques. To

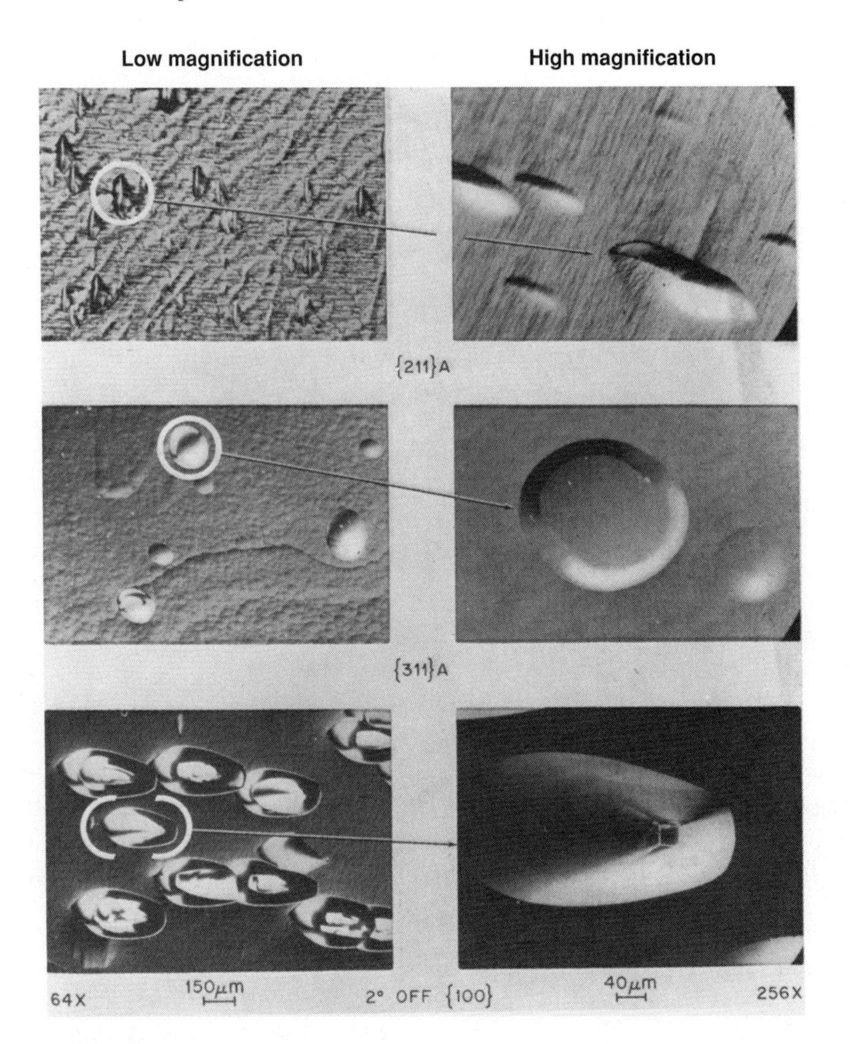

FIGURE 6.46
Hillocks observed on surfaces of GaAs epitaxial layers grown on ~(100)-, (211)- and (311)-oriented GaAs substrates; the layers were grown by halide VPE. (*After Dilorenzo [1972].*)

grow nearly perfect layers, we must not only optimize the growth conditions but also minimize the operational problems. For example, in LPE growth the edges of graphite pieces used to wipe off the melts must be smooth and sharp. For MBE of GaAs, the Ga cell must be kept clean to avoid Ga spitting.

6.5 MICROSTRUCTURES OF MIXED III-V EPITAXIAL LAYERS

The binary III-V and II-VI semiconductors have interesting applications in solar cells, gamma-ray detectors, MESFETS, and light emitters. The range of accessible band gaps for the binary materials is rather limited. However, we can broaden this

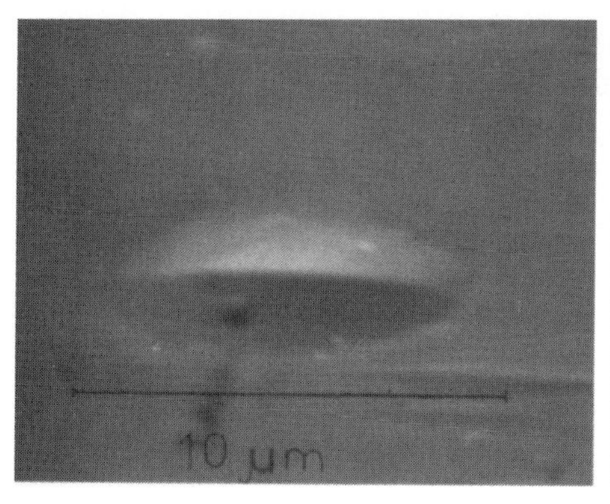

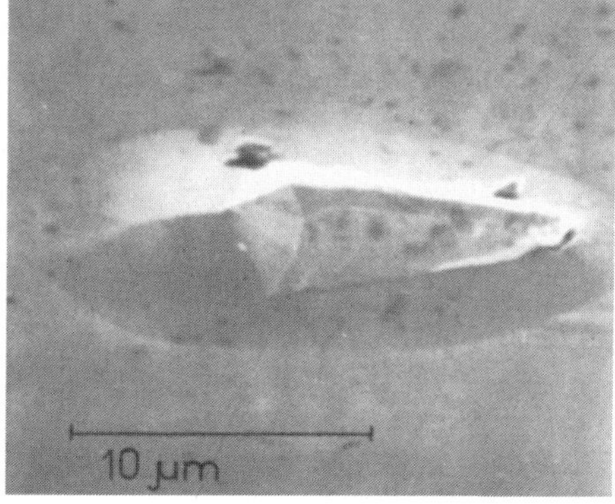

FIGURE 6.47
Scanning electron micrographs showing an oval-shape defect observed on the surface of GaAs homoepitaxial layer grown by MBE. (*After Bafleur et al. [1982].*)

range by introducing another degree of freedom, that is, mixing two semiconductors belonging to the same family. We illustrate this situation in Figure 6.48 that shows band gaps, lattice constants, and emission wavelengths of various semiconductors. The vertical line from InP indicates that the lattice-matched layers of different compositions, and thus different band gaps, can be deposited on InP substrates. This approach forms the basis for light emitters and detectors based on the InP/InGaAsP system. By mixing two semiconductors, we can also obtain the range of band gaps spanning those of the end members. Conceptually, this process entails mixing different types of atoms on the sublattices of groups II, III and VI, V atoms. We want to know whether or not this mixing occurs in a random fashion. Observations to date indicate that the atomic species in the mixed layers are not distributed at random on the respective sublattices if the tetrahedral radii of the atomic species occupying a particular sublattice are different. We observe two types of deviations from randomness: phase separation and atomic ordering.

Figures 6.49*a-d*, reproduced from the study of McDevitt et al. (1992), show the microstructure of an InGaAsP layer grown on a (001) InP substrate by LPE under different operating reflections. Figure 6.50 shows the electron-diffraction pattern observed from the layer in Figure 6.49. The well-developed satellite reflections are present contiguous to the main Bragg spots in Figure 6.50; see the enlargement in the lower-right corner. These observations are consistent with the occurrence of phase separation, and its wavelength is of the order of 8 nm. A plausible explanation is that the layer separates into In- and Ga-rich regions. The compositions of such regions in InGaAs layers grown by OMVPE have been analyzed by field-ion microscopy (Liddle et al. 1989). The compositional differences between the two types of regions are ± 5 atomic percent. We also know that the phase separation is two-dimensional and occurs at the surface while the layer is growing.

Phase separation is a common feature of the mixed III-V epitaxial layers. We see it in the layers grown by LPE (Henoc et al. 1982; Norman and Booker 1985; Mahajan 1989), VPE (Chu et al. 1985), OMVPE (Mahajan et al. 1989), and MBE (McDevitt et al. 1992). The wavelengths of the phase-separated regions are affected

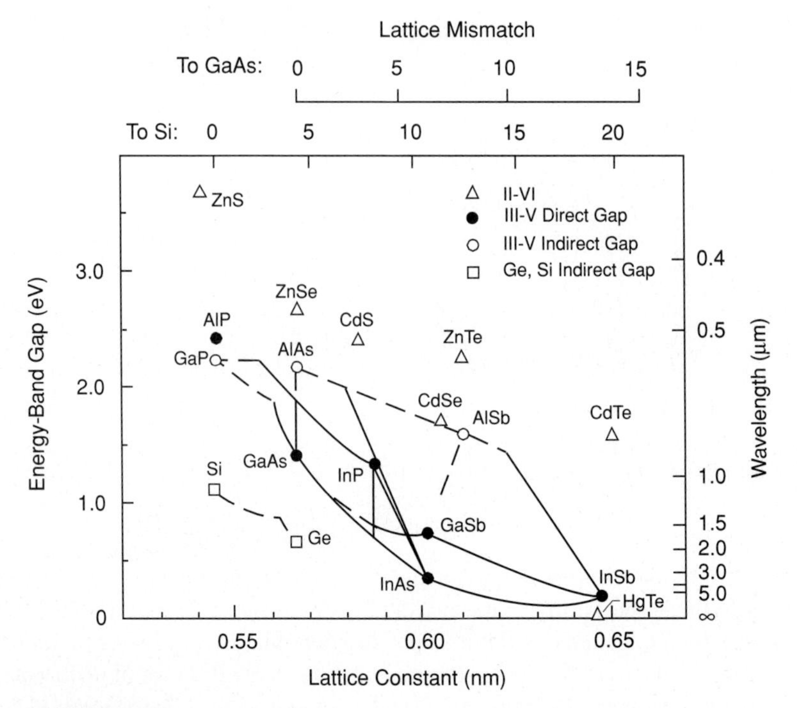

FIGURE 6.48
Band gaps versus lattice constants for III-V, II-VI, and IV semiconductors. The lines joining the III-V compounds give the band gap and lattice constant of ternary layers. Open and closed symbols denote indirect- and direct-gap materials.

by the growth technique because the growth temperature as well as the mechanism governing the transport of species along the growth interface are different.

In a mixed layer XYZ_2, the two bond lengths, that is, X-Z and Y-Z, are unequal. Therefore, the stacking of tetrahedra to build the solid must involve bond stretching and bending, a high-strain energy situation. This energy can be lowered if the layer undergoes phase separation at the surface because most of the bonds within phase-separated regions have the same length.

In addition to phase separation, the mixed layers grown on ~(001) substrates by vapor-phase techniques show atomic ordering. Figure 6.51 shows a (110) cross-sectional electron micrograph obtained from a (001) InP/InGaAsP/InP heterostructure grown by VPE. The features characteristic of phase separation are seen. In addition, we see two sets of extra spots halfway between the <111> spots in the diffraction pattern. The extra spots are due to atomic ordering. The absence of extra spots in the $(1\bar{1}0)$ pattern implies that ordering occurs only on two of the possible four {111} planes, that is, $(\bar{1}11)$ and $(1\bar{1}1)$. The metallurgical literature refers to this type of ordered structure as *CuPt type*.

For the sake of discussion, let us consider the occurrence of CuPt-type ordering in a mixed layer $GaInP_2$. If the atomic species were distributed at random, their

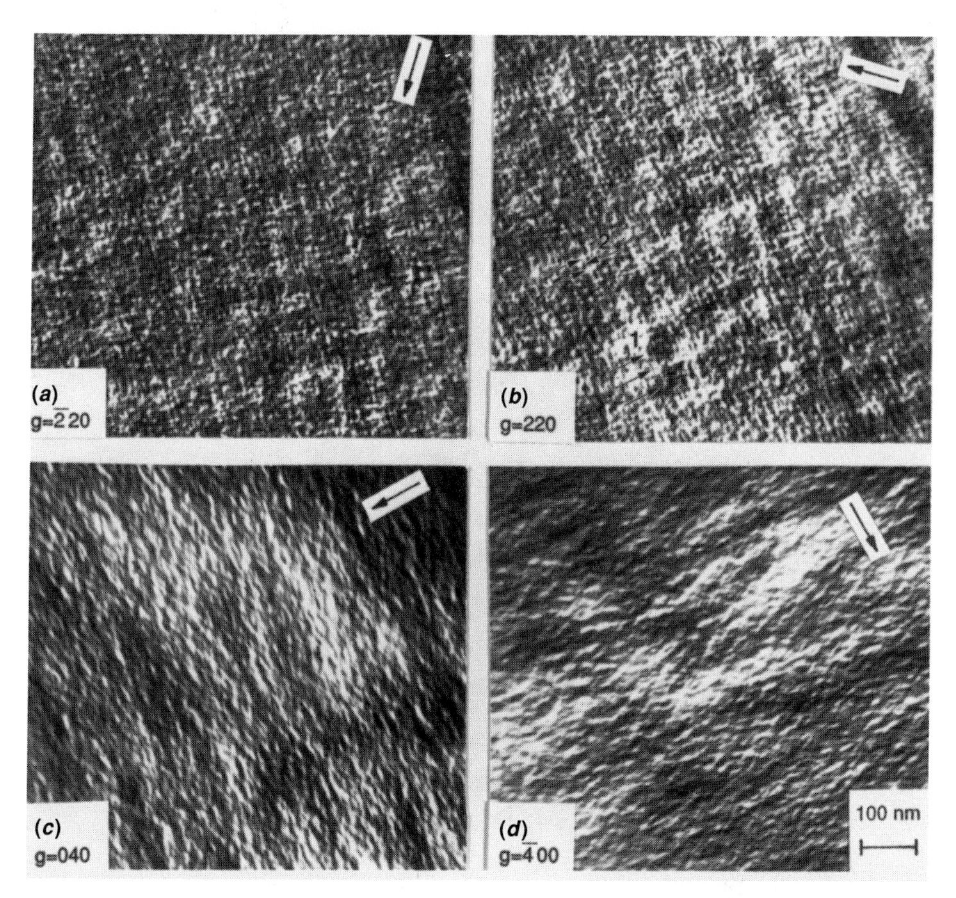

FIGURE 6.49
Electron micrographs (a)-(d) showing the contrast behavior of a phase-separated microstructure observed in a (001) InGaAsP epitaxial layer grown by LPE under different reflections. (*After McDevitt et al. [1992].*)

arrangement within the {111} planes would be A(III) a(P) B(III) b(P) C(III) c(P) A(III) a(P) . . . ; Aa Bb Cc . . . refer to the stacking arrangement of the {111} planes in the zinc-blende lattice. When the layer undergoes ordering, this arrangement is changed to A(Ga-rich) a(P) B(In-rich) b(P) C(Ga-rich) c(P) A(In-rich) a(P) B(Ga-rich) b(P) C(In-rich) c(P) A(Ga-rich) a(P) . . . The schematic in Figure 6.52 shows the resulting structure. The doubling of the direct-space periodicity along the <111> direction is evident. Table 6.2 lists the layer compositions, growth techniques, and growth temperatures for which atomic ordering has been observed. From Table 3.1 we can infer that the compositions in Table 6.2 that undergo ordering, contain atomic species differing in their tetrahedral radii.

We showed in section 3.3.3 that the occurrence of (2×4) reconstruction on a (001) group V terminated surface involved the formation of dimers. We can show that this reconstruction produces modulating subsurface stresses. The stress is

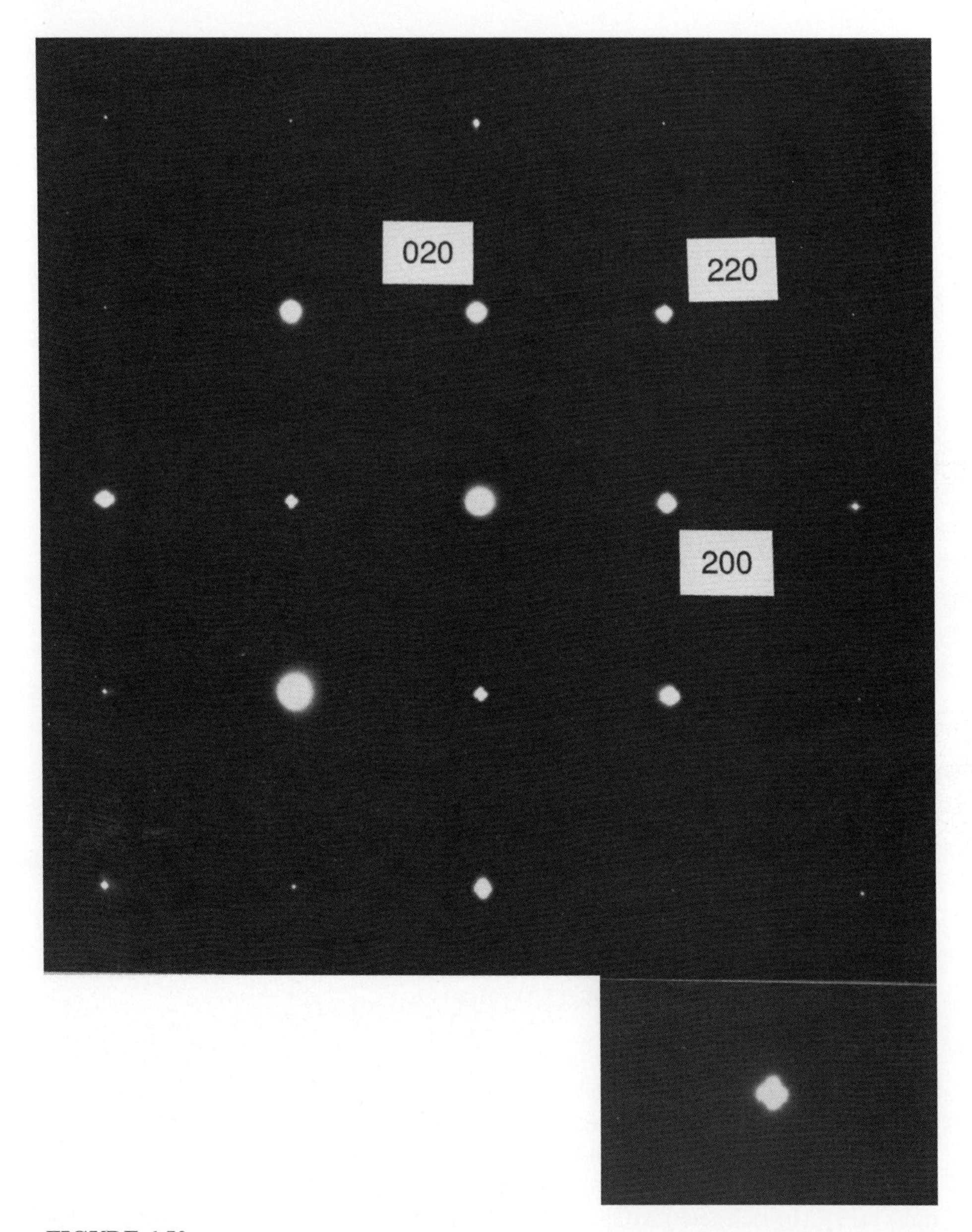

FIGURE 6.50
Electron-diffraction pattern obtained from the layer shown in Figure 6.49. Enlargement in the lower-right corner shows the presence of a couple of satellite spots. (*After McDevitt et al. [1992].*)

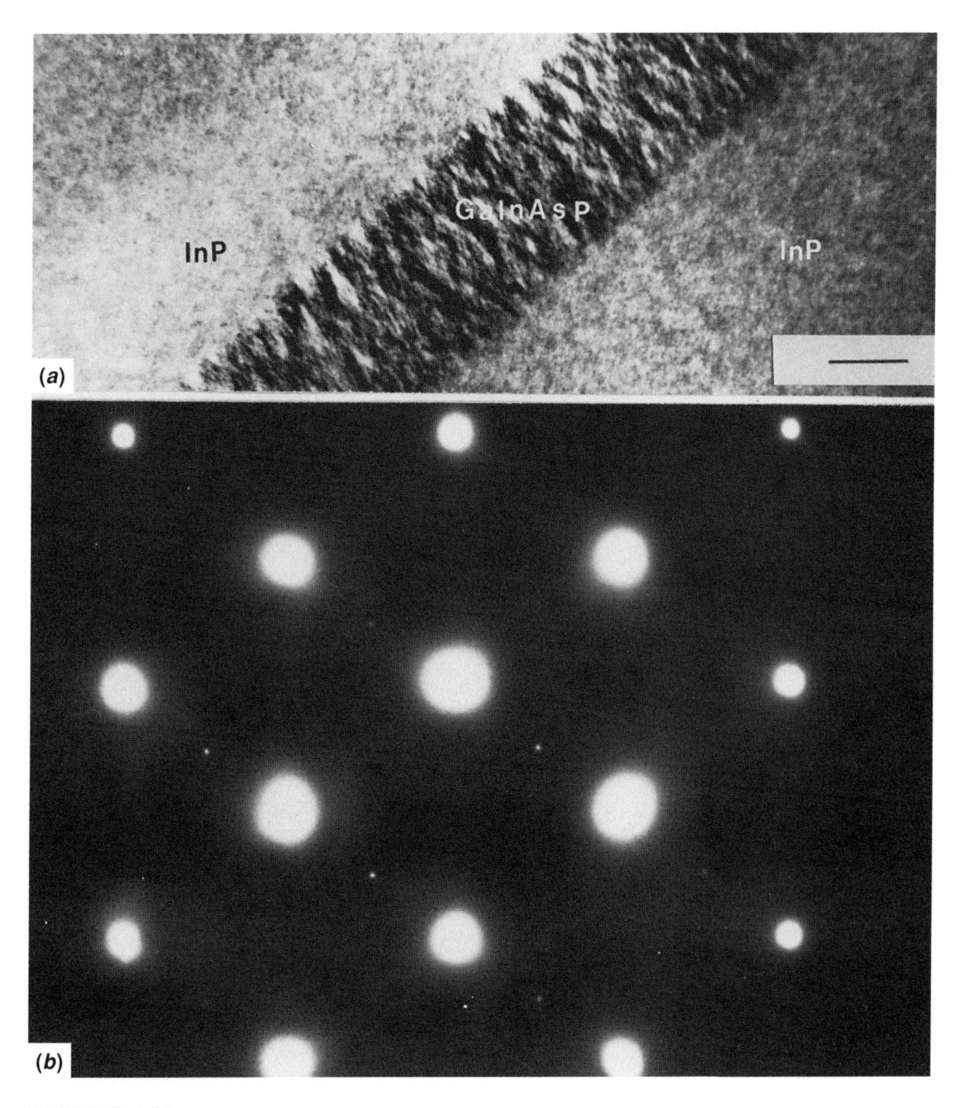

FIGURE 6.51
(*a*) Electron micrograph showing the (110) section of a heterostructure grown by the vapor-levitation technique and (*b*) the diffraction pattern. (*After Shahid et al. [1987].*)

compressive in nature below the dimers, whereas it is tensile between the dimers. As a result, the atomic species in the phase-separated microstructure undergo rearrangement. The larger atoms tend to occupy the tensile regions, while smaller atoms segregate in the compressive regions, leading to CuPt-type ordering.

Phase separation and atomic ordering affect the electronic properties of the layers. Figure 6.53 shows the changes in Hall mobility of a phase-separated InGaAsP layer grown by LPE as a function of the annealing temperature. The

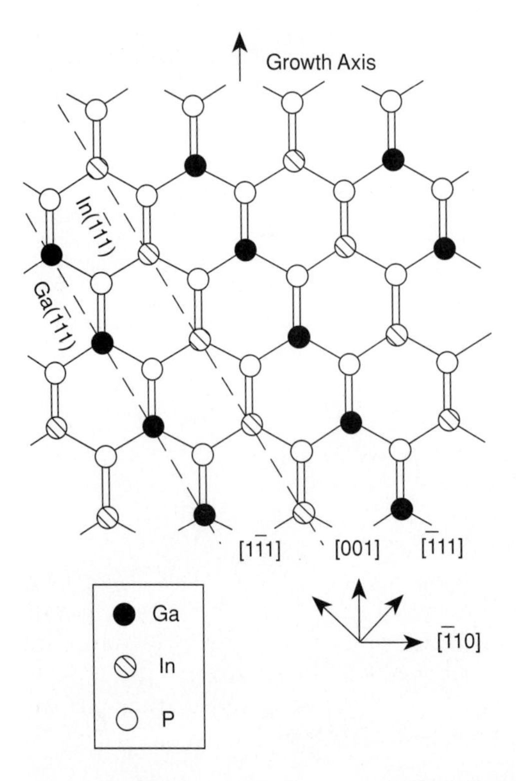

FIGURE 6.52
Schematic showing the atomic arrangement in a $Ga_{0.5}In_{0.5}P$ layer that has undergone CuPt-type ordering.

carrier mobility increases with the increasing annealing temperature. We visualize that the phase-separated layers contain many mini-heterobarriers because the regions have different compositions and therefore different band gaps. Annealing reduces the differences in compositions between the regions because of diffusion-induced mixing reduces or eliminates the barriers and leads to higher mobility than

TABLE 6.2
Some of the layer compositions exhibiting CuPt-type ordering

Layer composition	Substrate and its orientation	Growth technique	Growth temperature °C
$GaAs_{0.5}Sb_{0.5}$	(001) GaAs and InP	MBE	480–580
$In_{0.53}Ga_{0.47}As$	~(001) InP	VPE	600–700
InGaAsP	~(001) InP	VPE	640–660
$Ga_{0.5}In_{0.5}P$	~(001) GaAs	OMVPE	600–700
AlGaInP	~(001) GaAs	OMVPE	550–700
$GaAs_{0.5}P_{0.5}$	~(001) GaAs	OMVPE	670
InAlAs	~(001) InP	MBE	560
InP Sb	~(001) InAs	OMVPE	375–480
AlInP	~(001) GaAs	OMVPE	610–740

Source: Zunger and Mahajan 1994.

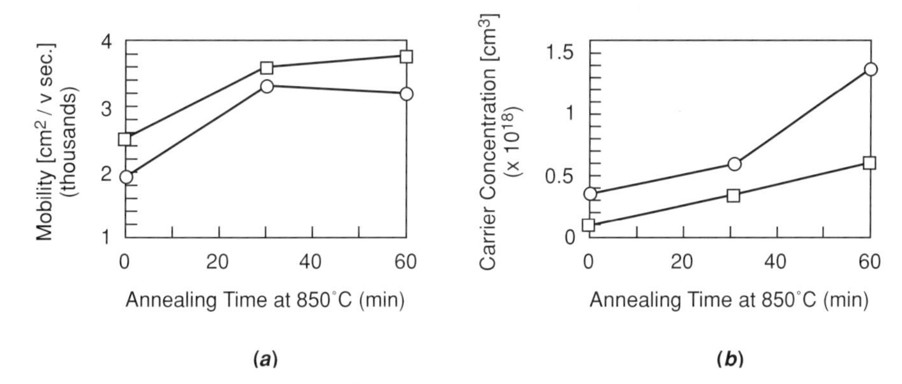

FIGURE 6.53
Changes in Hall mobility of a phase-separated InGaAsP layer grown by LPE as a function of annealing temperature. (*After McDevitt [1990].*)

that on the phase separated layer. In addition, the ordered layers have lower band gaps (Gomyo et al. 1987). Figure 6.54 shows the dependence of band gap on growth temperatures in GaInP layers grown by OMVPE. The band gap is lowest at the growth temperature where ordering is most pronounced (Gomyo et al. 1987). Zunger and Mahajan (1994) critically reviewed the status of phase separation and atomic ordering in mixed III-V layers. The interested reader can refer to this review for additional details on this topic.

We envisage that the presence of phase separation and atomic ordering in mixed III-V layers enhances the degradation resistance of light-emitting devices containing such active layers. For a given dislocation density, the light-emitting devices containing InGaAsP active layers appear to be more degradation resistant than the GaAlAs devices. Two explanations have been proposed for this difference. Since the band gap of InGaAsP is lower than that of GaAlAs, the energy available for nonradiative, recombination-enhanced defect reactions will be lower in the InGaAsP material. Consequently, defect generation and migration and dislocation glide and climb may not occur readily in InGaAsP materials. This argument assumes that the energies required for similar reactions in the two materials are nearly the same, a condition which is highly unlikely.

An alternative suggestion is that the multiplication of dislocations by nonradiative, recombination-enhanced glide and climb is difficult in phase-separated and atomically ordered InGaAsP layers (Mahajan 1989). Since the phase-separated regions may have different shear moduli, dislocations may tend to lie in the regions having lower shear modulus. During glide, the dislocations must traverse the high-shear modulus regions, which requires additional energy and makes glide difficult. Likewise, the motion of dislocations through ordered regions produces antiphase domain boundaries, and this too and the formation of boundaries is a high-energy situation. The climb of dislocations in ordered structures is also difficult because its occurrence creates antiphase domain boundaries. In addition, the activation energy for climb is increased in the presence of atomic ordering (Vardya and Mahajan 1995).

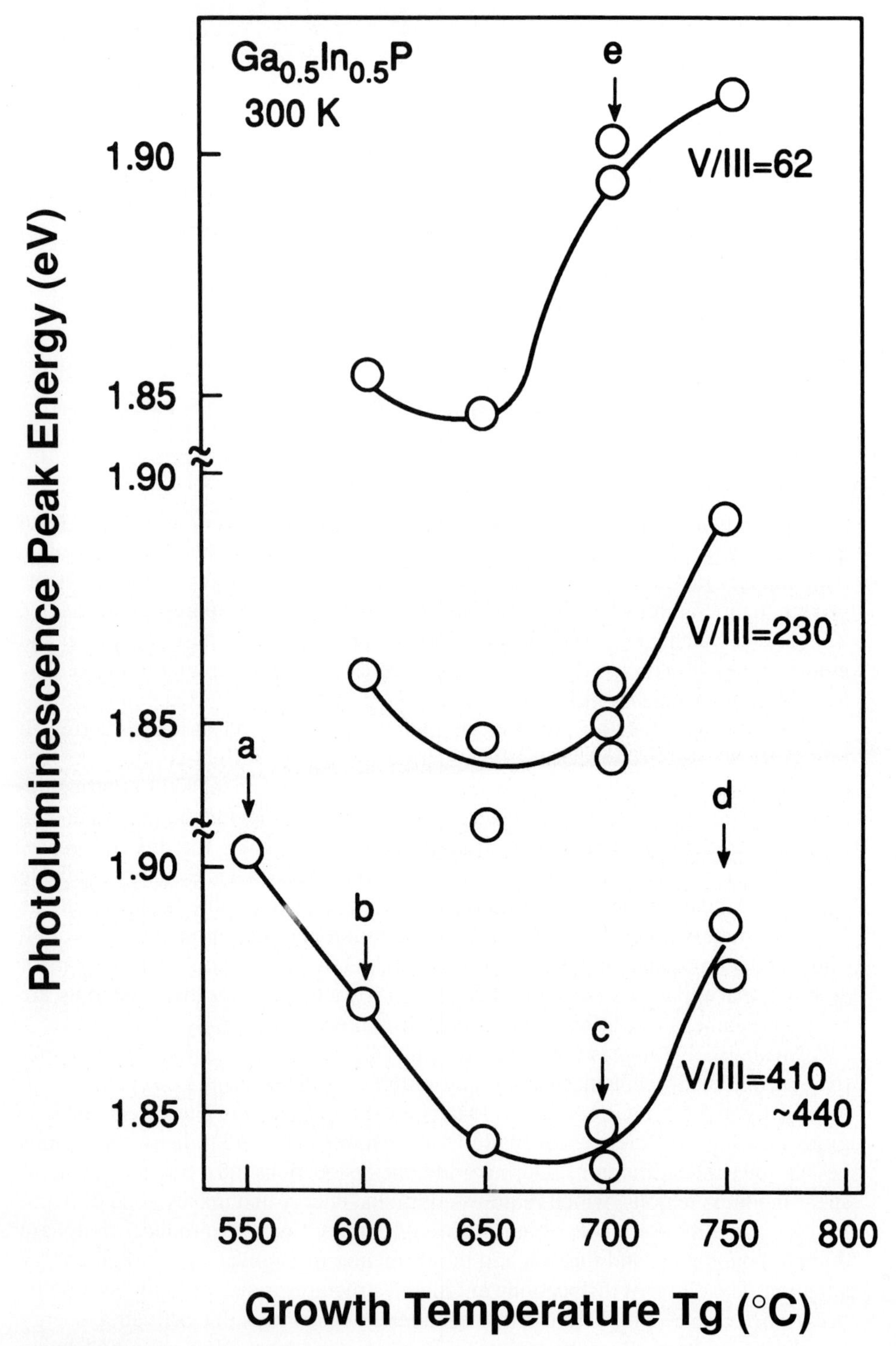

FIGURE 6.54
Variations in band gaps with growth temperatures observed in GaInP layers grown by OMVPE. (*After Gomyo et al. [1987].*)

Let us now consider a situation where a conducting channel in a MESFET consists of a mixed III-V layer. For the sake of discussion, assume it to be an InGaAs layer. As discussed above, the mixed layer undergoes phase separation and atomic ordering. Phase separation reduces the carrier mobility, whereas ordering is known to enhance it. Therefore, to fabricate high-speed FETs, it will be desirable to eliminate phase separation while preserving order. On the other hand, we have argued above that in the presence of phase separation and atomic ordering the light-emitting devices exhibit a higher degradation resistance. *This thereby implies that the microstructure of an active layer in a device must be tailored for optimal performance.*

6.6 HIGHLIGHTS OF THE CHAPTER

The highlights of this chapter are as follows:

- Several epitaxial growth techniques, such as LPE, VPE, MBE, and their modifications, are available for depositing layers. These techniques differ in how they deliver the layer constituents to the substrate surface. The VPE techniques have a high throughput, whereas MBE can produce heterostructures having structurally and compositionally sharp interfaces.
- The layer grows either by the attachment of atoms to the surface steps, that is, by the step-flow mechanism, or by the formation and coalescence of the two-dimensional nuclei at the surface. At higher temperatures, the step-flow mechanism is operative.
- The growth on patterned substrates could lead to local variations in the growth rate and the composition of the layer.
- The surfaces of heteroepitaxial layers acquire undulations to accommodate the elastic stresses that exist at the substrate-layer interface.
- Two types of defects can form in epitaxial layers: growth-process-independent defects, which are generic to all the growth techniques, and growth-process-dependent defects, which are characteristic of the growth technique.
- The atomic species within the mixed III-V layers are not distributed at random within their respective sublattices. Two types of deviations from randomness occur: phase separation and atomic ordering. The driving force for phase separation is the difference in bond lengths in the tetrahedral units in the random alloy. In addition, the atomic ordering is induced by the subsurface stresses produced by surface reconstruction. Both microstructural features affect the electronic properties of the layer.

REFERENCES

Aspnes, D. E.; E. Colas; A. A. Studna; R. Bhat; M. A. Koza; and V. G. Keramidas. *Phys. Rev. Lett.* 61, 2782 (1988).

Bafleur, M.; A Munoz-Yagne; and A. Rocher. *J. Cryst. Growth* 59, 531 (1982).

Baliga, B. J. *J. Electrochem. Soc.* 129, 665 (1982).
Ban, V. S. *J. Electrochem. Soc.* 122, 1389 (1975c).
Ban, V. S. *J. Electrochem. Soc.* 125, 317 (1978).
Ban, V. S. and S. L. Gilbert. *J. Cryst. Growth* 32, 284 (1975a).
Ban, V. S. and S. L. Gilbert. *J. Electrochem. Soc.* 122, 1382 (1975b).
Bauser, E.; M. Frik; K. S. Lochner; L. Schmidt; and R. Ulrich. *J. Cryst. Growth* 27, 148 (1974).
Bauser, E. and H. Strunk. *J. Cryst. Growth* 51, 362 (1981).
Bauser, E. and H. Strunk. *J. Cryst. Growth* 69, 561 (1984).
Beam, III, E. A., Ph.D. dissertation, Carnegie Mellon University, Pittsburgh (1989).
Beam, III, E. A.; S. Mahajan; and W. A. Bonner. *Mats. Sci. & Eng.* B7, 83 (1990).
Bean, J. C. *IEEE Proc. Int. Electron Device Met.* 6 (1981).
Bennema, P. and G. H. Gilmer. In *Crystal Growth: An Introduction,* ed. P. Hartman. Amsterdam: North-Holland, 1973, p. 263.
Benz, K. W. and E. Bauser. *Crystals: Growth, Properties and Applications* vol. 3. New York: Springer Verlag, 1980, p. 1.
Berger, P. R.; K. Chang; P. Bhattacharya; J. Singh; and K. K. Bajaj. *Appl. Phys. Lett.* 53, 684 (1988).
Bevan, M. J.; N. J. Doyle; J. Greggi; and D. W. Snyder. *J. Vac. Sci. & Tech.* A8, 1049 (1990).
Bhat, I. B. and S. K. Ghandhi. *J. Cryst. Growth* 75, 241 (1986).
Bhat, R. and S. Mahajan. In *The Concise Encyclopedia of Semiconducting Materials and Related Technologies:* Organometallic Vapor, Phase Epitaxy, ed. S. Mahajan and L. C. Kimerling. Oxford: Pergamon Press, 1992, p. 349.
Bryskiewicz, T. *J. Cryst. Growth* 43, 101 (1978).
Cahn, J. W. and D. W. Hoffman. *Acta Met.* 22, 1205 (1974).
Capper, P.; C. D. Maxey; P. A. C. Whiffen; and B. C. Easton. *J. Cryst. Growth* 96, 519 (1989).
Chai, Y. G. and R. Chow. *Appl. Phys. Lett.* 38, 796 (1981).
Cho, A. Y. and J. R., Arthur. *Prog. Solid State Chem.* 10, 157 (1975).
Chu, S. N. G.; S. Nakahara; K. E. Strege; and W. D. Johnston Jr. *J. Appl. Physics* 57, 4610 (1985).
Cox, H. M.; M. A. Koza; V. G. Keramidas; and M. S. Young. *J. Cryst. Growth* 73, 523 (1985).
Cullis, A. G.; D. J. Robbins; A. J. Pidduck; and P. W. Smith. *J. Cryst. Growth* 123, 333 (1992).
Dilorenzo, J. V. *J. Cryst. Growth* 17, 189 (1972).
Dutt, B. V.; S. Mahajan; R. J. Roedel; G. P. Schwartz; D. C. Miller; and L. Derick. *J. Electrochem. Soc.* 128, 1573 (1981).
Dutt, B. V.; D. D. Roccasecca; H. Temkin; and W. A. Bonner. *J. Cryst. Growth* 66, 525 (1984).
Foxon, C. T. and B. A. Joyce. *Surf. Sci.* 60, 434 (1975).
Foxon, C. T. and B. A. Joyce. *Surf. Sci.* 64, 293 (1977).
Gendry, M.; V. Drout; C. Santinelli; and G. Hollinger. *J. Vac. Sci. Technol.* B10, 1829 (1992).
Gleichmann, R.; C. Frigeri; and C. Pelosi. *Phil. Mag.* A 62, 103 (1990).
Gomyo, A.; T. Suzuki; T. Kobayashi; S. Kawata; I. Hino; and T. Yuasa. *Appl. Phys. Lett.* 50, 673 (1987).
Greve, D. W. *Mats. Sci. & Eng.* B 18, 22 (1993).
Grove, A. S. *Ind. & Eng. Chem.* 58, 48 (1966).
Henoc, P.; A. Izrael; M. Quillec; and H. Launois. *Appl. Phys. Lett.* 40, 963 (1982).

Hollinger, G.; M. Gendry; J. L. Duvault; C. Santinelli; P. Ferret; C. Miossi; and M. Pitaval. *Appl. Surf. Sci.* 56–58, 665 (1992).
Hsieh, J. J. *J. Cryst. Growth* 27, 49 (1974).
Ilegems, M. and M. B. Panish. *J. Phys. Chem. Solids* 35, 409 (1974).
Ishida, K. and T. Kamejima. *J. Electron. Mat.* 8, 57 (1979).
Johnston Jr., W. D.; M. A. DiGiuseppe; and D. A. Wilt. *AT&T Technical Journal* 60, 53 (1989).
Jordon, A. S. and M. Ilegems. *J. Phys. Chem. Solids* 36, 329 (1975).
Keramidas, V. G.; S. Mahajan; H. Temkin; and W. A. Bonner. *Inst. Phys. Conf.* ser no. 56, 95 (1981).
Kirchner, P. D.; J. M. Woodall; J. F. Freeous; and G. D. Petit. *Appl. Phys. Lett.* 38, 427 (1981).
Komeno, J.; M. Nogami; A. Shibatomi; and S. Ohkawa. *GaAs and Related Compounds, Inst. Phys. Conf. Soc.* 56, (1981), p. 9.
Kopf, L. and S. Sumski. *J. Cryst. Growth* 28, 365 (1975).
Kuphal, E. *J. Cryst. Growth* 67, 441 (1984).
Ladany, I.; R. T. Smith; and C. W. Magee. *J. Appl. Phys.* 52, 6064 (1981).
Liddle, J. A.; R. A. D. Mackenzie; C. R. M. Grovenor; and A. Cerezo. *Inst. Phys. Conf.* ser. no. 100, 81 (1989).
Long, J. A.; R. A. Logan; and R. F. Karlisek, Jr. In *Optical Fiber Telecommunications II,* ed. S. E. Miller and J. P. Kaminow. New York: Academic Press, 1988, chapter 16.
Long, J. A.; V. G. Riggs; and W. D. Johnston Jr. *J. Cryst. Growth* 69, 10 (1984).
Macrander, A. T.; J. A. Long; V. G. Riggs; A. F. Bloemeke; and W. D. Johnston Jr. *Appl. Phys. Lett.* 45, 1297 (1984).
Mahajan, S. *Prog. in Mats. Sci.* 33, 1 (1989).
Mahajan, S.; D. Brasen; M. A. DiGiuseppe; V. G. Keramidas; H. Temkin; C. L. Ziffel; W. A. Bonner; and G. P. Schwartz. *Appl. Phys. Lett.* 41, 266 (1982).
Mahajan, S.; W. D. Johnson Jr.; M. A. Pollock; and R. E. Nahory. *Appl. Phys. Lett.* 34, 717 (1979).
Mahajan, S.; V. G. Keramidas; and W. A. Bonner. *J. Electrochem. Soc.* 129, 1556 (1982).
Mahajan, S.; V. G. Keramidas; A. K. Chin; W. A. Bonner; and A. A. Ballman. *Appl. Phys. Lett.* 38, 255 (1981).
Mahajan, S.; M. A. Shahid; and D. E. Laughlin. *Inst. Phys. Conf.* ser. no. 100, 143 (1989).
Matthews, J. W. *Phil Mag.* 13, 1207 (1966).
McDevitt, T. L.; Ph.D. Dissrtation. Carnegie Mellon University (1990).
McDevitt, T. L.; S. Mahajan; D. E. Laughlin; W. A. Bonner; and V. G. Keramidas. *Phys. Rev.* B 45, 6614 (1992).
Meyerson, B. *Appl. Phys. Lett.* 48, 797 (1986).
Nakajima, K. and J. Okazaki. *J. Electrochem. Soc.* 132, 1424 (1985).
Narayan, J.; S. Sharan; A. R. Srivatsa; and A. S. Nandedkar. *Mats. Sci. & Eng.* B 1, 105 (1988).
Neave, J. H.; B. A. Joyce; P. J. Dobson; and N. Norton. *Appl. Phy.* A 31, 1 (1983).
Nelson, H. *RCA Review* 24, 603 (1963).
Norman, A. G. and G. R. Booker. *J. Appl. Phys.* 57, 4715 (1985).
Olsen, G. H. In *GaInAsP Alloy Semiconductors,* ed. T. P. Pearsall. New York: John Wiley, 1982, p. 1.
Panish, M. B.; I. Hayashi; and S. Sumski. *Appl. Phys. Lett.* 16, 326 (1970).
Panish, M. B. and M. Ilegems. In *Progress in Solid State Chemistry,* ed. H. Reiss and J. O. McCaldin. New York: Pergamon Press, p. 39 (1972).
Panish, M. B. and H. Temkin. *Ann. Rev. Mat. Sci.* (1989).
Peters, R. C. *Inst. Phys. Conf.* ser. no. 17, 55 (1973).

Petit, G. D.; J. M. Woodall; S. L. Wright; P. D. Kirchner; and J. L. Freeous. *J. Vac. Sci. & Technol.* B2, 241 (1984).
Petroff, P. M. and R. L. Hartman. *J. Appl. Phys.* 45, 3899 (1974).
Roedel, R. J.; A. R. Von Neida; R. Caruso; and L. R. Dawson. *J. Electrochem. Soc.* 126, 637 (1979).
Rupprecht, H.; J. M. Woodall; and G. D. Pettit. *Appl. Phys. Lett.* 11, 81 (1967).
Saul, R. H. *J. Electrochem. Soc.* 118, 793 (1971).
Saul, R. H. and J. Roccasecca. *J. Appl. Phys.* 44, 1983 (1973).
Shaw, D. W. *J. Electrochem. Soc.* 115, 405 (1968).
Small, M. B.; K. H. Bachem; and R. M. Potemski. *J. Cryst. Growth* 39, 216 (1977).
Small, M. B.; A. E. Blakeslee; K. K. Shih; and R. M. Potemski. *J. Cryst. Growth* 30, 257 (1975).
Small, M. B. and R. M. Potemski. *SPIE* vol. 323. Washington, DC: International Society for Optical Engineering, 1982, p. 423.
Snyder, D. W.; S. Mahajan; E. I. Ko; and P. J. Sides. *Appl. Phys. Letts.* 58, 848 (1991).
Stringfellow, G. B. *J. Cryst. Growth* 58, 192 (1982).
Stringfellow, G. B. *J. Cryst. Growth* 68, 111 (1984).
Stringfellow, G. B. and P. E. Greene. *J. Phys. Chem. Solids* 30, 1779 (1969).
Swaminathan, V. and A. T. Macrander. *Materials Aspects of GaAs and InP Based Structures.* Englewood Cliffs, NJ: Prentice Hall, 1991, p. 83.
Tsang, W. T. In *Beam Processing Technologies* ed. N. G. Einspruch, S. S. Cohen, and R. N. Singh. New York: Academic Press, 1989.
van de Ven, J.; J. L. Weyher; H. Ikink; and J. L. Giling. *J. Electrochem. Soc.* 134, 989 (1987).
Van Hove, J. M.; C. S. Lent; P. R. Pekite; and P. I. Cohen. *J. Voc. Sci. & Technol.* B1, 741 (1983).
Vardya, R. and S. Mahajan. *Phil. Mag.* A71, 465 (1995).
Vohl, P. *J. Cryst. Growth* 54, 101 (1981).
Williams, F. W. *J. Electrochem. Soc. III,* 886 (1964).
Wolf, S. and R. N. Tauber. *Silicon Processing.* Sunset Beach, CA: Lattice Press, 1987, p. 155.
Wood, C. E. C.; L. Rathburn; H. Ohno; and D. DeSimone. *J. Cryst. Growth* 51, 299 (1981).
Xie, J. H.; E. A. Fitzgerald; P. J. Silverman; A. R. Kortan; and B. E. Wei. *Mats. Sci. Eng.* B 14, 322 (1992).
Zunger, A. and S. Mahajan. In *Handbook on Semiconductors,* Atomic Ordering and Phase Separation in Epitaxial III-V Alloys, vol. 3, ed. S. Mahajan. Amsterdam: North-Holland, 1994, p. 1399.

PROBLEMS

6.1. Use Figure 6.1 to calculate the thickness of a GaAs layer that would be deposited when a Ga-As melt saturated at 650°C is step cooled by 10°C and then a layer is grown for 30 mins. Assume that the diffusion coefficient of As in Ga melts is given by $8.6 \times 10^{-4} \exp\left(\frac{-3240}{k_B T}\right) \text{cm}^2/\text{sec}$.

6.2. Imagine a situation where InP homoepitaxial layers are grown by LPE on (110) and (112) substrates. If you were to examine these substrates prior to an In–melt back, what would be shape and crystallography of thermal-decomposition-induced pits for the two orientations? Assume that pits in both cases are bounded by {111} facets.

6.3. *a.* Imagine that a gallium droplet placed on a (001) GaAs surface is subjected to a temperature gradient along the x-direction only as shown below. Ignoring the loss of As from the surface due to heating, discuss the positional stability of the droplet.

b. Repeat problem 6.3(a) for a (110) substrate when the temperature gradient is along the $[1\bar{1}1]$ direction.

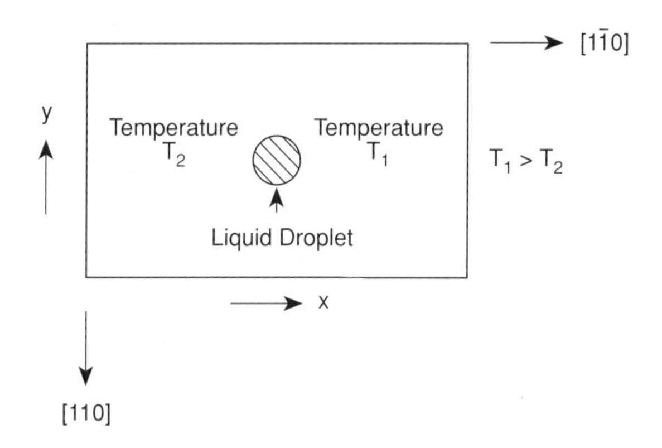

6.4. *a.* Calculate the possible changes in tetrahedral volume when InAs layers are doped p- and n-type with Be, Cd, Zn and Te, S, and Se, respectively.

b. Which of the above dopants would be most and least potent in strengthening the layer? Explain your answer.

6.5. Discuss the influence of dopants and layer composition on the formation of terraces shown in Figure 6.10.

6.6. Epitaxial stacking faults lying on {111} planes and forming squares on the (001) surface of silicon layers are observed. If the length of the fault square is one μm on the (001) surface, calculate the distance from the layer surface at which these pyramidal stacking faults nucleated.

6.7. Assuming that the mass-transfer coefficient for a silicon VPE reactor is 2.5 cm/sec, calculate the growth temperature below which the mass-transfer coefficient will begin to dominate the growth, given that the surface-reaction coefficient is $10^7 \exp(-2\text{eV}/k_B T)$ cm/sec.

6.8. If in situ etching of silicon by anhydrous HCl is nonoptimal, etch pits may form. Assuming that the pits are bounded by the low-energy {111} facets, show schematically the shapes of pits that may be observed on (001), (110), and (111) surfaces. Label all directions and facets.

6.9. The formation of hillocks in VPE GaAs can be avoided by tilting the (001) GaAs substrates around the [110] direction towards $(1\bar{1}1)_B$ or $(\bar{1}11)_B$. Assuming that As atoms can migrate a distance of 5 nm during the deposition of a bilayer, growth begins from step edges, and the growth surface is As-terminated, determine the misorientation angle above which hillocks may not form.

6.10. The steps present on vicinal surfaces can be used to control lateral growth of layers. Under appropriate growth conditions it is possible to grow half monolayers that nucleate from step edges and propagate along the terraces. Assuming that InGaAs and InP layers can be grown by OMMBE under such conditions on vicinal (001) InP substrates, show schematically the resultant structure after the growth of five layers and comment on its electronic properties.

6.11. Suppose you are employed as a process engineer in a semiconductor company and are asked for recommendations for the following technological issues: (i) the low-cost epitaxial growth technique; (ii) the growth technique that has high throughput, produces layers of uniform thickness, and can handle large-size wafers; and (iii) the growth technique that is suitable for depositing multilayer structures having very sharp interfaces. List your recommendations, giving reasons for your choice. Also, describe in your own words the principle of the each growth technique that you selected.

6.12. *a.* Imagine a situation in which 19 percent of the area of one $(\text{cm})^2$ GaAs wafer is covered with a dielectric and the remainder has 100 μm wide square openings. Assuming that during epitaxy Ga and As atoms do not adhere to the dielectric, are not desorbed into the gas phase, and can surface migrate a distance of 100 μm during the deposition of bilayers, sketch the thickness profile of the resulting deposit.

b. The (001) InP substrate has been patterned as shown below.
If the patterned substrate is subjected to a flux of In, Ga, and As to deposit InGaAs layers, sketch the profile of the deposit assuming that there is only facet-to-facet migration of the atomic species, but they do not desorb.

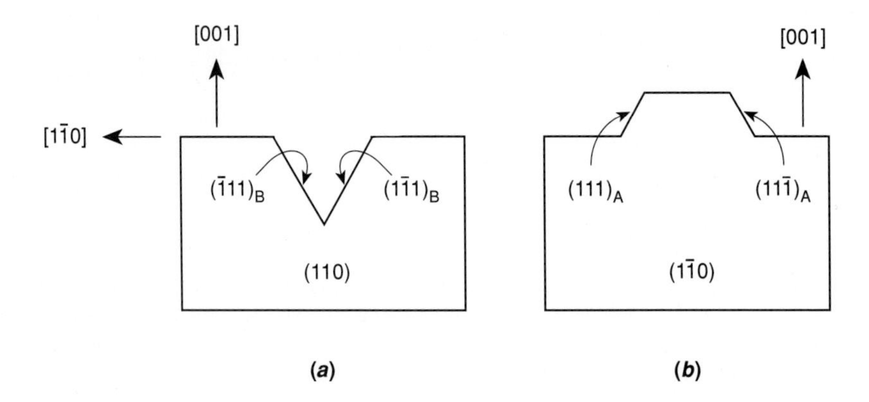

6.13. Show schematically the deposition of GaAs on vicinal (111) silicon substrates and ascertain which step structures would lead to domain-free and domain-containing growths. Assume that the lattice parameters of the substrate and the overgrowth are equal.

6.14. The nickel aluminide (NiAl) layers are being investigated as potential metal-base regions in the GaAs-based devices. Show how dislocations in GaAs are replicated into the NiAl overgrowths. NiAl has the B_2 structure and the potential slip systems are <010> {100} and <1$\bar{1}$0> {110}. The lattice parameter of GaAs is twice that of NiAl.

6.15. *a*. GaAs homoepitaxial layers, isoelectronically doped with In to $10^{19}cm^{-3}$, and undoped layers have been deposited on (001) substrates. Show schematically the distributions of etch pits on the surfaces of the layers.

b. Comment on the possibility of reducing dislocation densities in epitaxial layers over those in the substrates by isoelectronic doping.

6.16. *a*. Calculate the volumes of possible tetrahedral units in an InGaAs layer.

b. Discuss the influences of phase separation on glide and climb of perfect dislocations in the zinc-blende structure.

CHAPTER 7

Oxidation

This chapter covers the thermodynamics and kinetics of thermal oxidation of silicon and the factors affecting oxidation rate. The structure of the resulting oxide, the volume change accompanying oxidation, and the electrical charges within the oxide and at the Si/SiO_2 interface are also considered. Other topics in this chapter are the generation of stacking faults during oxidation, their effects on the electrical characteristics of diodes, and the nonthermal methods of oxidation.

7.1 THERMODYNAMICS OF OXIDATION

During the last three decades, integrated circuit technology based on silicon has matured from the production of discrete devices in the early 1960s to state-of-the-art ultra-large-scale integrated (ULSI) circuits. Our ability to produce high-quality SiO_2 films of various thicknesses has played an important role in the evolution of this technology. Thermal oxide has varied applications in the ULSI technology. It (1) serves as a mask during dopant diffusion and ion implantation, (2) provides electrical isolation between different devices on a chip, (3) is used as a gate oxide and capacitor dielectric in metal-oxide semiconductor devices, and (4) provides passivation of silicon surfaces. In addition, SiO_2 layers provide electrical isolation in multilevel metallization schemes. The various thermal SiO_2 thicknesses used in ULSI are listed in Table 7.1. Thermal oxidation can produce controlled thicknesses of SiO_2 films and Si/SiO_2 interfaces having desirable electronic properties.

During the fabrication of integrated circuits, we form SiO_2 films by the oxidation of Si at elevated temperatures. Either dry oxygen or steam is used for this purpose. Subsequent to the formation of the SiO_2, device wafers may undergo high-temperature processing steps. We want to know whether or not the oxide would be

TABLE 7.1
Range of thermal SiO_2 thicknesses used in ULSI processing

SiO_2 thickness (nm)	Application
6–10	Tunneling oxide.
15–50	Gate oxide; capacitor dielectric.
200–500	Masking oxide; surface passivation oxide.
300–1000	Field oxide.

stable at these processing temperatures. We can gain insight into this issue by assessing the free-energy change accompanying the oxidation of Si.

The reactions governing the oxidation of Si by dry oxygen and steam can be written as

$$Si(s) + O_2(g) \rightarrow SiO_2(s) \tag{7.1}$$

or

$$Si(s) + 2H_2O(g) \rightarrow SiO_2(s) + 2H_2(g). \tag{7.2}$$

For the oxidation of Si with pure oxygen at a constant temperature and pressure, the thermodynamic criterion for the formation of SiO_2 is that the change in the Gibbs free energy ΔG be negative. We can relate ΔG with various thermodynamic quantities as follows:

$$\Delta G = \Delta G^0 + RT \ln Q, \tag{7.3}$$

where

$$\Delta G^0 = G^0_{SiO_2} - G^0_{o_2} - G^0_{Si}, \tag{7.4}$$

and

$$Q \equiv \frac{a_{SiO_2}}{a_{O_2} \cdot a_{Si}}, \tag{7.5}$$

where a_{SiO_2}, a_{O_2}, and a_{Si} denote activities of SiO_2, O_2, and Si, respectively. If we assume that SiO_2 and Si are pure and that oxygen behaves like an ideal gas, we can write

$$Q = \frac{1}{P_{o_2}}, \tag{7.6}$$

where the activities of Si and SiO_2 are taken as unity and P_{o_2} is the partial pressure of oxygen in the environment. The thermodynamic equilibrium condition is given by

$$\Delta G = O. \tag{7.7}$$

Hence at equilibrium

$$\Delta G^0 + RT \ln Q^{eq} = 0 \tag{7.8}$$

or

$$\Delta G^0 = -RT \ln K, \tag{7.9}$$

where we have used the special symbol K to denote Q^{eq} which is equal to

$$K = \frac{1}{P^{eq}_{O_2}}, \tag{7.10}$$

where the $P^{eq}_{O_2}$ denotes the partial pressure of oxygen at the equilibrium. Hence

$$\begin{aligned} \Delta G &= -RT \ln\left(\frac{1}{P^{eq}_{O_2}}\right) + RT \ln\left(\frac{1}{P_{O_2}}\right) \\ &= RT \ln\left(\frac{P^{eq}_{O_2}}{P_{O_2}}\right). \end{aligned} \tag{7.11}$$

If $P_{O_2} > P^{eq}_{O_2}$, then ΔG is negative and the oxidation reaction is expected to occur spontaneously. The value of $P^{eq}_{O_2}$ may be obtained at any temperature from the following expression:

$$\Delta G^0 = -907{,}091 + 175.728\ T\ (\text{J/g} \cdot \text{mole}). \tag{7.12}$$

$P^{eq}_{O_2}$ is found to be of the order of 10^{-46} and 10^{-22} atm at 600° and 1250°C, respectively, implying that for all practical purposes silicon oxidizes readily. We know that a film of SiO_2 forms instantaneously even at room temperature where the mobilities of atoms are expected to be low; Si left in the air will develop an oxide of thickness around 4 nm.

A similar analysis can be carried out for the oxidation of Si by water vapor. We have in this case

$$\begin{aligned} \Delta G^0 &= G^0{}_{SiO_2} + 2G^0{}_{H_2} - G^0{}_{Si} - 2G^0{}_{H_2O} \\ &= -412{,}307 + 64.029\ T. \end{aligned} \tag{7.13}$$

The equilibrium constant is given by

$$K = \left(P^2_{H_2} \Big/ P^2_{H_2O}\right)^{eq}. \tag{7.14}$$

The ratio of the equilibrium partial pressures $(P_{H_2}/P_{H_2O})^{eq}$ at 600 and 1250°C are of the order of 10^{10} and 25,000. Thus at exceedingly low values of P_{H_2O}, Si can be oxidized.

In the presence of O_2 and H_2O, the reactions (7.1) and (7.2) can both proceed independently. We can therefore consider the system consisting of Si, SiO_2, H_2, O_2, and H_2O species as a three-component system consisting of Si, H_2 and O_2. According to the phase rule, there are two degrees of freedom. We can thus determine the extent to which each reaction occurs at any given temperature and pressure.

To reduce the effects of interdiffusion between different materials in the integrated circuits, we would like to produce SiO_2 films at lower temperatures. We could achieve this by oxidizing Si at high pressures. We know from thermodynamics that the change in free energy with the change in pressure at a constant temperature is related to the change in volume by the following expression:

$$\left(\frac{\partial \Delta G}{\partial P}\right)_T = \Delta V, \tag{7.15}$$

where ΔV is the excess of molar volumes of the products over the reactants. Since ΔV for oxidation reactions with pure oxygen is negative, the increase in pressure favors the formation of SiO_2, implying that lower temperatures can be utilized for oxidation. However, if we require extremely thin oxides, very low pressure oxidation in pure oxygen may be desirable because the thickness control under high-pressure conditions is difficult.

The solubility of oxygen in SiO_2 can also be treated thermodynamically. We have

$$O_2(\text{gas}) = \underline{O_2}(\text{in solution in } SiO_2). \tag{7.16}$$

The equilibrium constant for the above reaction is

$$K = \frac{a_{\underline{O_2}}}{a_{O_2}} = \frac{\gamma C^*}{P_G}, \tag{7.17}$$

where P_G is the partial pressure of O_2 in the gas phase, C^* is the solubility of O_2 in SiO_2, and γ is the activity coefficient. For low values of solubility, γ is constant, so that the solubility of O_2 is proportional to its partial pressure in the environment. The preceding relationship is known as Henry's law and implies that the oxygen molecule does not dissociate at the gas-oxide interface. The relationship expressed by Equation (7.17) has been experimentally verified. A similar relationship is also found for H_2O. Furthermore, the solubility of water is influenced by impurities in SiO_2. For example, solubility increases with the increase in sodium concentration in SiO_2.

As indicated in Chapter 2, we use doped silicon for the fabrication of integrated circuits. When this material is oxidized, two situations may develop. The dopant atoms may be either incorporated into the oxide or rejected into the semiconductor, resulting in a pileup at the Si/SiO_2 interface. We consider these situations next.

The prevailing equilibrium between the oxide and silicon imposes an important constraint on the redistribution of impurities during oxidation. Impurities originally present in silicon must be divided after oxidation so that the chemical potentials of component (i) in the oxide and silicon are the same. For example, we can write

$$\mu_i^{ox} = \mu_i^{Si}. \tag{7.18}$$

μ_i^{Si} can be expressed as

$$\mu_i = \mu_i^* + RT\ln(\gamma_i x_i), \tag{7.19}$$

where μ_i^* is the chemical potential of (i) in the standard state, γ_i is its activity coefficient, and x_i is the mole fraction of component (i). We can therefore write

$$\mu_i^{ox} = \mu_i^{*,ox} + RT(\gamma_i x_i)^{ox} = \mu_i^{*,Si} + RT\ln(\gamma_i x_i)^{Si}. \tag{7.20}$$

Hence the segregation coefficient m is

$$m = \frac{X_i^{Si}}{X_i^{ox}} = \frac{\gamma_i^{ox}}{\gamma_i^{Si}}\exp(\mu_i^{*,ox} - \mu_i^{*,Si}). \tag{7.21}$$

If m is greater than one, the impurity has a higher concentration in silicon than in the oxide. If m is less than one, the impurity has a higher concentration in silica than in silicon. The chemical potential in the standard state depends only on temperature and pressure. It is known experimentally that m is less than one for B, whereas m is greater than one for P, As, and Sb (Grove et al. 1964).

The equilibrium concentrations of impurities in the Si/SiO_2 system are modified at the interface according to the Gibbs adsorption equation (Lupis 1983). Qualitatively, if the impurity lowers the interface energy, it is then adsorbed at the interface. On the other hand, the impurity that tends to raise the interfacial energy is depleted at the interface under equilibrium conditions.

EXAMPLE 7.1. State the equilibrium conditions according to thermodynamics that prevail during oxidation of Si.

Solution. The Gibbs free energy of SiO_2 (s) may be expressed as

$$G = \overline{G}_{Si} + 2\,\overline{G}_o.$$

The Gibbs free energy of formation of $SiO_2(s)$ from Si(s) and $O_2(g)$ is given as

$$\Delta G = \Delta\overline{G}_{Si} + 2\,\Delta\overline{G}_o,$$

where

$$\Delta\overline{G}_{Si} = \overline{G}_{Si} - G^0_{Si}$$

and

$$\Delta\overline{G}_o = \overline{G}_o - {G^0}_{o_2}.$$

For any component (i)

$$d\overline{G}_i = RT\,d\ln f_i,$$

where f_i is the fugacity of component (i). The activity a_i of component (i) is defined by

$$a_i \equiv \frac{f_i}{f_i^0}$$

where f_i^0 is the fugacity of component (i) in the standard state. At the temperature T of interest in oxidation, we shall take pure Si at 1 atm pressure to be the standard state. Similarly pure O_2 at 1 atmosphere and temperature of interest is taken to be the standard state. The partial pressure of the atom of oxygen may be obtained from the thermodynamics of the dissociation reaction

$$\frac{1}{2}\,O_2(g) = O(g),$$

where we assume P_{o_2} is one atmosphere. We have therefore

$$\Delta G = RT \sum_i n_i \ln \frac{f_i}{f_i^0},$$

where n_i is the stoichiometric coefficient in the reaction that forms the oxide from its elements. Assuming that the fugacities can be written in terms of partial pressures, we write

$$\Delta G = RT \ln\left(\frac{P_{Si(in\ SiO_2)}\,P^2_{o(in\ SiO_2)}}{P^0_{Si}\,(P^0_o)^2}\right),$$

where $P_{o(in\ SiO_2)}$ is the pressure of oxygen atom in the gas phase over some particular composition of SiO_2. A similar interpretation is given for $P_{Si(in\ SiO_2)}$.

Under equilibrium conditions

$$\Delta G = 0,$$

and no transfer of material takes place between the gas and the solid phase. At the boundary of Si and the gas phase, the fugacity of each component is the same in both phases. At the boundary of Si and SiO_2, the equilibrium is different. Maintaining local equilibrium at these two interfaces contributes to variations in oxidation rates especially at small oxide thicknesses. Even though overall ΔG is little affected, the activities of the components at the interfaces are sensitively related to one another. For example, at the oxide-gas interface, the equilibrium may be governed by the E' center (represents local oxygen deficiency or silicon excess), whereas the equilibrium at the Si/SiO_2 interface may be controlled by P_b center (trivalent Si atom bonded to three Si atoms and a single dangling orbital perpendicular to the interface). The equilibrium constants for these point defects will be needed before we can make any quantitative predictions.

7.2 KINETICS OF OXIDATION

We indicated in Table 7.1 that various thicknesses of the oxide are used in the processing of circuits. The variation in the thickness is achieved by changing the duration that Si is exposed to O_2 at a given temperature. To generate the necessary data on the thickness variations and to also understand the influence of wafer orientation on the oxidation rate, we now consider the kinetics of oxidation.

A linear-parabolic model developed by Deal and Grove (1965) to explain the growth of oxide as a function of time and temperature adequately explains most of the experimental data on the growth of SiO_2. In this model the oxidant is assumed to participate in three serial steps, which are illustrated in Figure 7.1. The first stage involves the transport of the oxidant from the gas phase to the immediate vicinity of the gas-oxide interface, and its flux is given by J_1. The model assumes that an initial layer of oxide exists on the silicon surface, so the oxidant moves by diffusion towards silicon with a flux J_2. At the Si/SiO_2 interface the oxidant reacts with the Si to form the oxide, and its flux is J_3.

In the gas phase flux J_1 can be written as

$$J_1 = h_G (C_G - C_S), \tag{7.22}$$

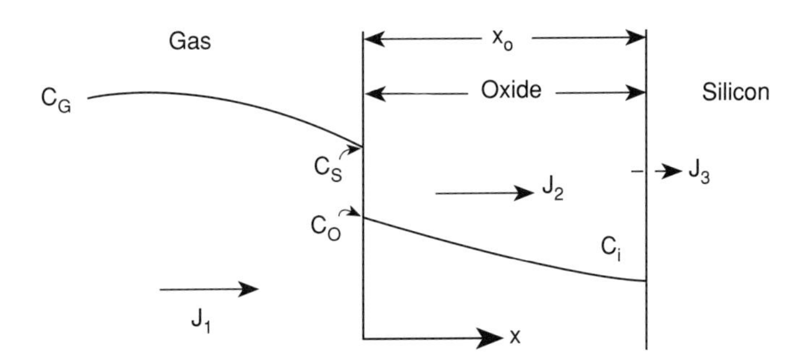

FIGURE 7.1
Schematic showing various fluxes for the linear-parabolic model.

where h_G is the gas-phase mass-transfer coefficient, C_G is the concentration of the oxidant in the gas phase, and C_S is the concentration of the oxidant at the gas-oxide interface. Assuming that Henry's law is valid, we can express the concentration of the oxidant at the outer surface of the oxide C_0 as

$$C_0 = H P_s, \tag{7.23}$$

where H is the Henry's law constant and P_s is the partial pressure of the oxidant right next to the oxide surface. If C* is the concentration of the oxidant at the oxide surface that is associated with partial pressure P_G, then C* is given by

$$C^* = H P_G. \tag{7.24}$$

Assuming that the oxidant obeys ideal gas law, we can write

$$PV = N_m k_B T, \tag{7.25}$$

where N_m is the number of molecules of the oxidant in the gas phase so that the concentration of the oxidant in the gas phase may then be expressed as

$$C = \frac{N_m}{V} = \frac{P}{k_B T}. \tag{7.26}$$

We can therefore write

$$C_G = \frac{P_G}{k_B T} \tag{7.27}$$

and

$$C_S = \frac{P_S}{k_B T}. \tag{7.28}$$

We can now rewrite the flux J_1 only in terms of the concentration of the oxidant in the oxide as

$$\begin{aligned} J_1 &= h_G(C_G - C_S) = h_G\left(\frac{P_G}{k_B T} - \frac{P_S}{k_B T}\right) \\ &= \frac{h_G}{k_B T}(P_G - P_S) = \frac{h_G}{H k_B T}(C^* - C) \\ &= h(C^* - C_0), \end{aligned} \tag{7.29}$$

where

$$h \equiv \frac{h_G}{H k_B T}. \tag{7.30}$$

To reach the silicon, the oxidant must diffuse through the oxide (see Chapter 8 on diffusion). Assuming steady-state conditions, the flux of the oxidizing species is assumed to be given by Fick's first law. Therefore, we can write

$$J_2 = -D\frac{dC}{dx_0}, \tag{7.31}$$

where x_0 is the distance of any point in the oxide layer from the gas-oxide interface. Assuming C_i to be the concentration of the oxidant adjacent to the Si/SiO_2

interface, we can express J_2 as

$$J_2 = -D\frac{(C_0 - C_i)}{x_0}, \tag{7.32}$$

where D is the diffusion coefficient of the oxidant in the oxide.

The incorporation of the oxidant into the Si leads to the formation of the oxide, and the flux corresponding to this surface reaction is

$$J_3 = k_s C_i, \tag{7.33}$$

where k_s is a rate constant for the chemical surface reaction for silicon oxidation.

Under steady-state conditions,

$$J_1 = J_2 = J_3 \equiv J. \tag{7.34}$$

We can solve for C_0 and C_i and obtain

$$C_i = \frac{C^*}{1 + \frac{k_s}{h} + \frac{k_s\, x_0}{D}} \tag{7.35}$$

and

$$C_0 = \frac{\left\{1 + \frac{k_s x_0}{D}\right\} C^*}{1 + \frac{k_s}{h} + \frac{k_s\, x_0}{D}}. \tag{7.36}$$

In thermal oxidation the flow of oxidant to the oxide surface is rarely a rate-limiting step so that $h \gg k_s$. When diffusivity is very small, that is, $D \ll k_s x_0$, we find

$$C_i \to 0,\ C_0 \to C^*. \tag{7.37}$$

The preceding situation is known as the *diffusion-controlled case*, since the supply of oxidant to the interface rather than the reaction at the interface controls the oxidation process. When $D \gg k_s x_0$, we find

$$C_i = C_0 = \frac{C^*}{1 + \frac{k_s}{h}}, \tag{7.38}$$

and there is an abundant supply of oxidant at the silicon interface, but oxidation is controlled by the rate at which the oxidation reaction occurs at the Si/SiO_2 interface.

The flux J is given by

$$J = \frac{k_s\, C^*}{1 + \frac{k_s}{h} + \frac{k_s\, x_0}{D}}. \tag{7.39}$$

The rate of growth of the oxide layer is given by

$$\frac{dx_0}{dt} = \frac{J}{N} = \frac{k_s\, C^*}{N\left(1 + \frac{k_s}{h} + \frac{k_s\, x_0}{D}\right)}, \tag{7.40}$$

where N is the number of oxidant molecules incorporated into a unit volume of the oxide layer. For pure oxygen N is 2.3×10^{22} molecules/cm^3, and for water N is 4.6×10^{22} molecules/cm^3. We note that two H_2O molecules are required for one molecule of SiO_2. We can obtain the general solution of Equation (7.40) by using the following boundary condition:

$$x_0(t = 0) = x_i, \tag{7.41}$$

This solution is

$$x_0^2 + Ax_0 = B(t + \tau), \tag{7.42}$$

where

$$A \equiv 2D\left\{\frac{1}{k_s} + \frac{1}{h}\right\}, \tag{7.43}$$

$$B \equiv \frac{2DC^*}{N}, \tag{7.44}$$

and

$$\tau \equiv \frac{x_i^2 + Ax_i}{B}. \tag{7.45}$$

The introduction of x_i allows us to take into account not only the initial layer of oxide thickness but also multiple oxidation steps.

The thickness of oxide as a function of time may be expressed in the following form:

$$x_0 = \frac{A}{2}\left\{\left(1 + \frac{(t + \tau)}{A^2/4B}\right)^{1/2} - 1\right\}. \tag{7.46}$$

For short oxidation times $(t + \tau) \ll A^2/4B$, we obtain

$$x_0 = \frac{B}{A}(t + \tau), \tag{7.47}$$

which represents a linear-oxidation law. For long oxidation times $t \gg A^2/4B$, we obtain

$$x_0^2 = Bt, \tag{7.48}$$

which represents a parabolic-oxidation law. The measurement of oxide thickness versus oxidation time enables us to determine the rate constants B and B/A. A typical set of experimental data showing oxide thickness as a function of time at various temperatures for (111)- and (001)-oriented silicon appears in Figure 7.2. The general form of relationships for Si oxidation and its two limiting forms can be plotted in a normalized form as shown in Figure 7.3.

The experimental data of Deal and Grove (1965) support their linear-parabolic oxidation model and shed light on a number of factors that control the oxidation process. The parabolic rate constant B is proportional to both the diffusion coefficient D and the solubility of the oxidant in the oxide C*. The latter is proportional to the oxidant pressure so that B is linearly proportional to the ambient pressure.

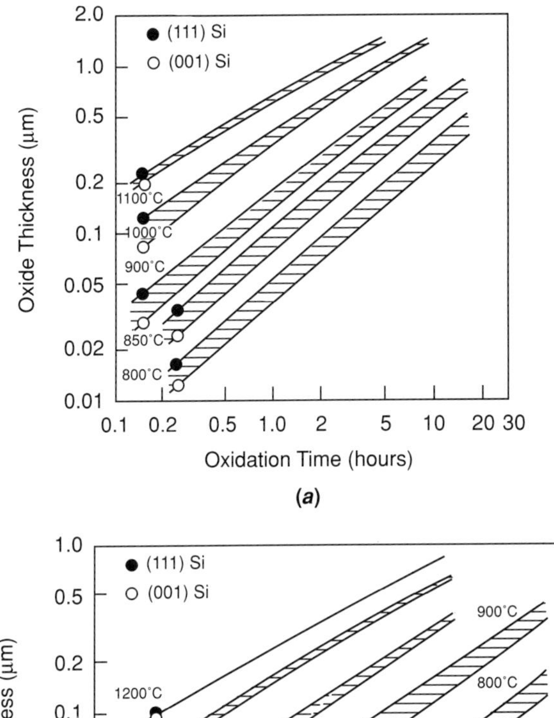

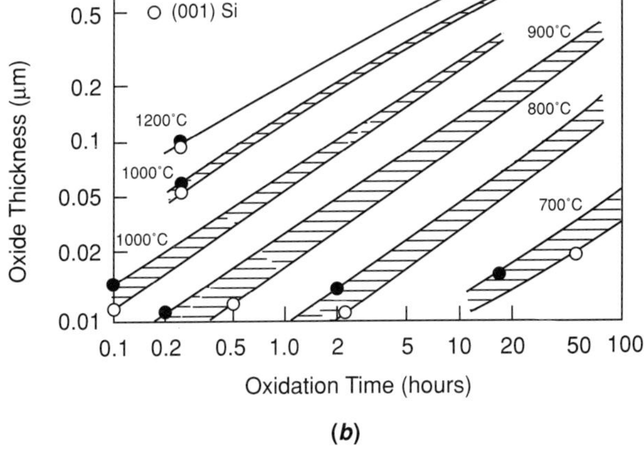

FIGURE 7.2
Oxide thickness versus time for silicon in (*a*) pyrogenic H_2O (640 torr) and (*b*) dry oxygen. (*After Deal and Grove [1965].*)

The diffusion coefficient of the oxidant is temperature dependent, so we can write

$$B = C_1 \exp\left(-\frac{E_1}{k_B T}\right). \tag{7.49}$$

The values of B determined at different temperatures can yield activation energy E_1. The activation energies for oxidation of silicon in dry oxygen and steam are, respectively, 1.23 and 0.78 eV. These values correspond to the activation energies for diffusion of oxygen and water in fused silica. Note that the parabolic rate constant is a product of D and C*. Even though for H_2O D is smaller than it is for O_2, C* for H_2O is three orders of magnitude larger than C* for O_2. Hence the oxidation in steam is much faster than the oxidation in oxygen.

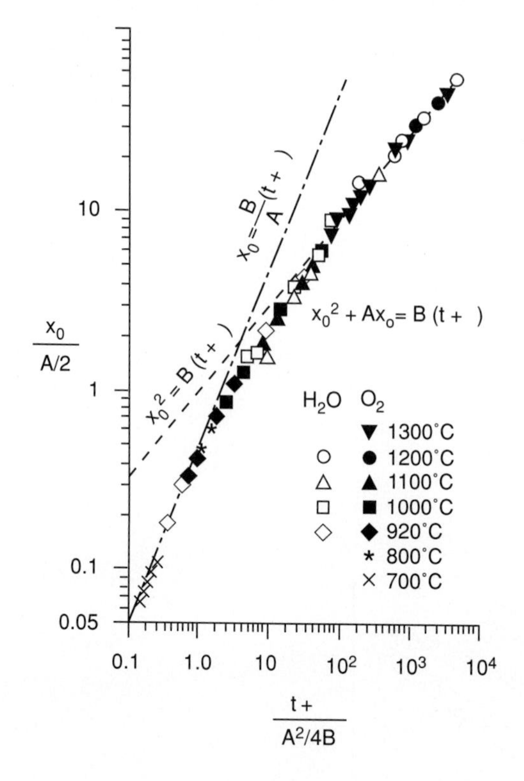

FIGURE 7.3
Normalized plot for linear-parabolic model of oxidation of silicon. (*After Deal and Grove [1965].*)

The linear rate constant B/A is given by

$$\frac{B}{A} = \frac{C^*}{N\left(\frac{1}{k_s} + \frac{1}{h}\right)}. \tag{7.50}$$

The constant A is not expected to depend on pressure, so the pressure dependence of B/A is due to that of B alone. The linear rate constant reveals a temperature dependence given by

$$\frac{B}{A} = C_2 \exp\left(-\frac{E_2}{k_B T}\right). \tag{7.51}$$

The activation energy E_2 determined from experiments for both oxygen and steam oxidation is 2.05 eV, which corresponds closely to the energy required to break a Si-Si bond. The gas-flow rate has no effect on B/A, since this is determined only by the reaction occurring at the Si/SiO_2 interface.

The empirical fit of experimental data to obtain B and B/A has some pitfalls (Irene and van der Meulen 1976). In particular, the value of τ has to be treated as an adjustable constant. Furthermore, the model cannot satisfactorily account for the initial rapid rate of oxidation of silicon. It is not at all clear that a single activation energy is an appropriate way to represent the diffusion of oxidant in an amorphous structure (Irene 1988). The model proposed by Deal and Grove (1965) treats the oxidation of silicon in one dimension. In the fabrication of devices, numerical modeling is necessary to deal with the growth of oxide in other directions. In practice, the topology of device structures has to be taken into consideration during the growth of oxides because it affects the local oxidation rates (Bassous et al. 1976; Marcus and Sheng 1982).

7.3 THE STRUCTURE OF SILICON DIOXIDE

To understand the behavior of impurities during oxidation, charges within the oxide, electronic characteristics of the Si/SiO_2 interface, and the diffusion of impurities in the oxide, we need to understand the structure of SiO_2. The following section summarizes the highlights.

Silicon dioxide obtained by the thermal oxidation of Si is amorphous in nature. It consists of silicon-oxygen tetrahedra; one of the tetrahedra is shown in Figure 7.4. Two tetrahedra can be linked by sharing the corner oxygen atoms, which is referred to as a *bridging oxygen*. Substantial rotation of a tetrahedron around this bridging oxygen can occur without drastic changes in bond lengths or bond angles. During this rotation it is easy to lose long-range order so that many of the corner oxygen atoms belong to only one tetrahedron. These oxygen atoms are termed *nonbridging*

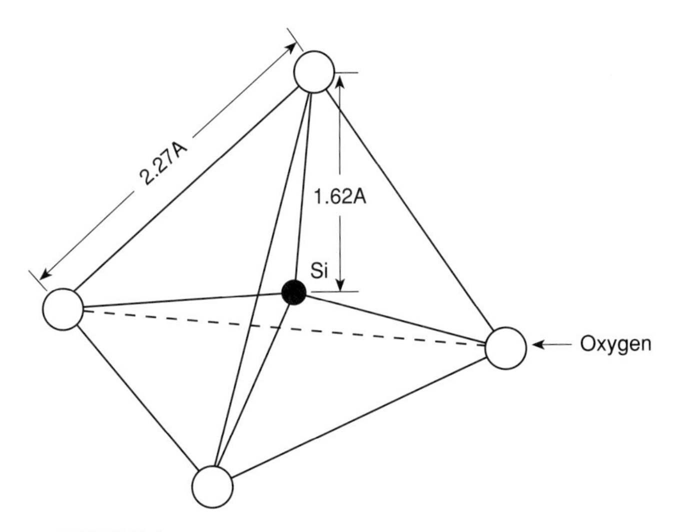

FIGURE 7.4
Schematic showing a SiO_4 tetrahedron. The mean value of Si-O-Si angle is 153° in vitreous silica. The Si atom has a hybrid sp^3 configuration, and the oxygen atom has a p^4 configuration. There is thus one unpaired electron on each oxygen atom. The structural formula is often represented as SiO_4^{4-}.

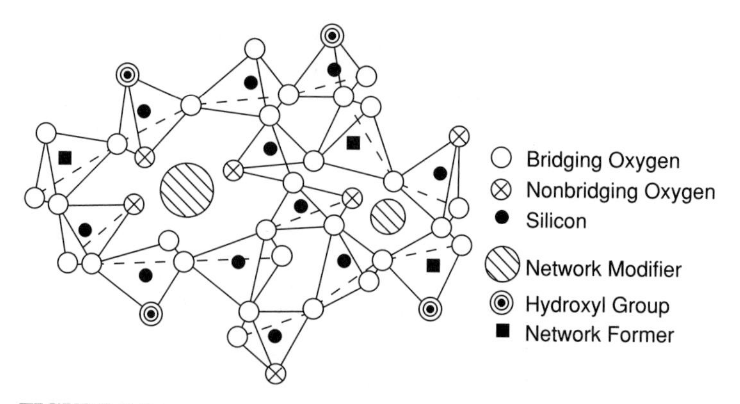

FIGURE 7.5
Structure of amorphous silica containing network modifier and network former. After Revesz (1965).

oxygen atoms. The higher the concentration of the nonbridging oxygen atoms, the weaker the cohesion of the oxide and the more open its structure.

The structure of amorphous silica is shown in Figure 7.5 (Revesz 1965). This figure helps us to understand why it is easier for oxygen to diffuse in silica than in silicon. The movement of silicon atoms requires the rupture of four Si-O bonds. However, the oxygen atom can move if only two Si-O bonds are broken. If the bridging oxygen were to move from its site, it would create an oxygen vacancy. Vacancies at nonbridging oxygen atoms are also possible but are unlikely based on energetic considerations. The preceding assessment is consistent with the movement of oxygen in silica investigated in a number of studies, for example Ligenza and Spitzer (1960).

The oxygen is assumed to migrate in silica as molecular oxygen. The solubility and the permeation of oxygen through silica depend linearly on O_2 pressure (Lie et al. 1982), confirming the preceding assumption. The O_2 molecule is thus not expected to interact with the amorphous structure and migrates via the interconnected open spaces in the structure.

Amorphous silica contains point defects, such as vacancies, interstitials, impurities, and three-dimensional defects like pores, channels, and cracks. Dielectric-breakdown histograms, rapid initial oxidation kinetics, and transmission electron microscopic observations of chemically etched SiO_2 films support the existence of micropores in oxide films (Gibson and Dong 1980). Furthermore, when the reaction between the silicon and the oxidant is incomplete, the silicon atoms are incorporated into the oxide.

Impurities that replace silicon in the Si-O_2 polyhedron are termed *network formers*. Boron and phosphorus are network formers and result in charged defects in the oxide. The column V elements give rise to an excess of nonbridging ions, whereas column III elements reduce the nonbridging ion concentration.

Large metal ions, located in interstitial sites, are called *network modifiers*. The introduction of sodium oxide is equivalent to producing two nonbridging oxygens,

Na_2O + Si-O-Si → Si-O + O-Si + $2Na^+$. The amorphous structure is thus weakened and becomes more porous to other diffusing species. Na^+, K^+, Pb^{2+}, and Ba^{2+} are among the common network modifiers whereas Al^{3+} can be either a modifier or a network former.

Water vapor combines with a bridging oxygen to form a pair of stable, nonbridging hydroxyl groups OH^- : H_2O + Si ·· O · Si → Si ·· OH + ·· Si. As a result, diffusivity of various atoms in SiO_2 is increased when water weakens the amorphous structure.

EXAMPLE 7.2. The basic bonding unit in silica is the SiO_4 tetrahedron, where each silicon atom is surrounded by four oxygen atoms, the Si-O distance is 0.164 nm, and the tetrahedral O-Si-O bond angle is 109.18°. Draw two of these bonding units and show how they are attached to each other in amorphous silica. If the next-nearest-neighbor interaction is important, what type of bond angle for Si-O-Si is favored?

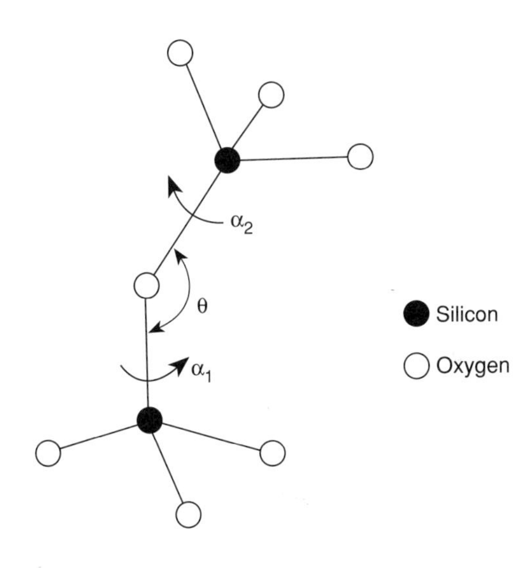

Solution. In amorphous silica the O-Si-O bond angle is presumed to remain accurately at 109.18°, and the Si-O distance and O-O distance are fixed at 0.162 nm and 0.26 nm, respectively. The degree of freedom necessary to form an amorphous structure is provided by the flexibility of Si-O-Si bond angle θ and the dihedral angles α_1 and α_2. The dihedral angles α_1 and α_2 may vary continuously. If θ is 90°, the oxygen from one tetrahedra gets very close to the oxygen from the neighboring tetrahedra, resulting in repulsion between tetrahedra. If θ is 180°, the silicon of one tetrahedra is the next-nearest neighbor to the silicon of the adjoining tetrahedra. Clearly there is a bond angle at which the second-nearest-neighbor O-O distance equals that of the nearest-neighbor O-O distance. The Si-O-Si angle in amorphous glass can vary with a most probable value around 150°, whereas the dihedral angles α_1 and α_2 can vary continuously. These features give the silica tetrahedra the necessary flexibility to form the amorphous structure.

7.4 VOLUME CHANGE ON OXIDATION AND STRESSES AT SILICON-SILICON DIOXIDE INTERFACES

We discussed in section 5.6.2.2 that the volume of SiO_2 precipitates formed on annealing of Czochralski silicon is greater than the volume of Si consumed in their formation. An identical situation develops during the thermal oxidation. Since oxidation is carried out at high temperatures and since the coefficients of thermal expansion of silicon and the oxide are different, stresses must develop at the Si/SiO_2 interface. We now elaborate on these issues.

Consider a situation in which a thickness t_1 of silicon is consumed in producing a thickness t_2 of the oxide as shown in Figure 7.6. If the density of silicon is $\rho(Si)$ and that of silicon dioxide is $\rho(SiO_2)$, then one mole of the oxide is formed for every mole of silicon consumed. Hence

$$t_1 = \frac{\rho(SiO_2)}{\rho(Si)} \frac{(MW)_{Si}}{(MW)_{SiO_2}} t_2, \tag{7.52}$$

where MW denotes the appropriate molecular weight. Substituting for $\rho(SiO_2)$ and $\rho(Si)$ 2.22 and 2.33 g/cm^3, respectively, and inserting the appropriate molecular weights, we find that $0.445\ t_2$ of silicon is consumed for the growth of an oxide of thickness t_2. This finding implies that the oxidation produces an increase in volume of about 124 percent. Since most of the expansion occurs perpendicular to the silicon surface, the volume change is easily accommodated. However, since there is very little expansion in the Si/SiO_2 interface, compressive stresses develop at the interface. One way to relieve these stresses is by viscous flow of the oxide. The viscosity of the oxide in turn depends on the oxidation temperature and composition of the oxide. Therefore, high oxidation temperatures (>950°C) and silica with dissolved moisture have low viscosities, resulting in low values of compressive stress in the oxide.

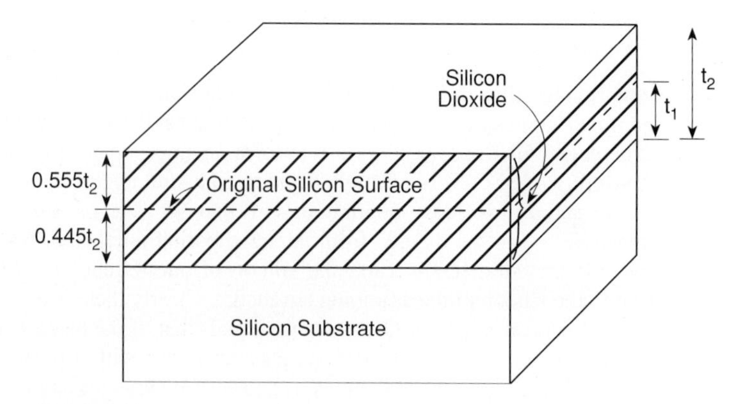

FIGURE 7.6
Planar growth of silicon dioxide on Si. For a growth of oxide of thickness t_2, a $0.445t_2$ thick layer of Si is consumed.

Dobson (1972) and Hu (1974) have suggested that silicon interstitials are injected into silicon during oxidation. This suggestion is consistent with the fact that extrinsic stacking faults are observed in silicon after oxidation. We discuss this situation further in section 7.6. The injection of the interstitials may also partially accommodate the volume change discussed above.

During cooldown from the oxidation temperature, stresses can develop at the Si/SiO_2 interface because Si and SiO_2 have different thermal expansion coefficients. Si contracts more than silicon dioxide contracts. For the two materials to remain attached, Si near the interface experiences a tensile stress, whereas the oxide is under compression. The associated differential strain ε can be estimated from the following expression:

$$\varepsilon = -\Delta\alpha\,(T_2 - T_1), \tag{7.53}$$

where $\Delta\alpha$ is the difference in the thermal expansion coefficients of Si and SiO_2, T_1 is the temperature used for the oxide growth, and T_2 is the final temperature after cooldown. Assuming that no viscous flow occurs in the oxide, stresses can be estimated from the observed strains using appropriate expressions.

7.5 FACTORS AFFECTING OXIDATION RATES

The thermal oxidation rates of Si depend on many parameters, and the following factors are most significant: (1) orientation of the Si surface, (2) dopants, (3) presence of halogen impurities in the gas phase, and (4) pressure.

7.5.1 Effects of Surface Orientation

The surface orientation influences only the linear rate constant B/A because the parabolic rate constant B is determined by the diffusion of oxidant through the oxide. These effects are shown in Table 7.2 where B/A and B are compared for (001) and (111) silicon at three temperatures. At a given temperature, B is essentially independent

TABLE 7.2
Variations in parabolic and linear rate constants with temperature and orientation during oxidation of silicon

Oxidation temperature °c	Orientation	Parabolic rate constant (μm^2/hr)	Linear rate constant (μm/hr)
900	(001)	0.143	0.150
	(111)	0.151	0.252
1000	(001)	0.314	0.664
	(111)	0.314	1.163
1100	(001)	0.521	2.977
	(111)	0.517	4.926

of the orientation, whereas B/A exhibits strong orientation dependence. In addition, the differential between the values of B/A for (111) and (001) orientation increases with the oxidation temperature. Both dry and wet oxidations exhibit similar orientation dependence of the linear rate constant.

The observed orientation dependence of B/A is attributed to two factors. First, the areal atomic densities of different crystallographic planes are different. For silicon the areal density increases as the orientation changes from <001> toward <110> and reaches a maximum for a <111> orientation. This observation is consistent with the observed values of B/A shown in Table 7.2. Second, the activation energy for thermal oxidation exhibits orientation dependence, which arises from a geometric effect called *steric hindrance*. Since the bond between silicon and oxygen is directional in nature, the ability of oxygen to bond to silicon depends on the angle that the Si-Si bond, that is, a <111> direction, makes with the surface, the separation between adjacent silicon atoms, and the size of the reacting molecule.

7.5.2 Influence of Dopants

Two distinct situations may arise during the oxidation of doped silicon crystals. First, the dopant atoms may be incorporated into the oxide during oxidation. Second, the dopant atoms may segregate at the Si/SiO_2 interface. Boron and phosphorus in silicon constitute the respective examples.

The segregation of boron atoms into the oxide during oxidation weakens the bond structure of the glassy network. As a result, both oxygen and water molecules can diffuse more readily through the oxide, leading to higher oxidation rates and thus to higher growth rates of the oxide. The effect is particularly significant when the diffusion of the oxidant through the oxide controls the oxide growth.

The phosphorus atoms tend to segregate at the Si/SiO_2 interface during oxidation. Therefore, there is no significant increase in the growth rate for high P concentrations when oxidation is primarily diffusion controlled. On the other hand, the effect is quite significant when oxidation is reaction-rate limited. Thus the effect of P on oxidation rate is significant at lower oxidation temperatures and for shorter oxidation times.

Generally the oxidation rate is increased in the presence of high concentrations of commonly used group III and V dopants (Ho et al. 1978). A plausible explanation could be that the vacancy concentration is higher at high doping levels. These vacancies could annihilate silicon interstitials that are injected into the crystal during oxidation (Dobson 1972; Hu 1974). As a result, the supersaturation of silicon interstitials may not occur and thus prevent a slowdown in oxidation.

7.5.3 Effects of Halogens

The introduction of halogens in the oxidation ambient enhances the oxidation rate because halogen species produce hydrogen, which in turn reacts with oxygen to form water vapor. Therefore, oxidation essentially occurs in steam. Furthermore, if chlorine is located at or near the Si/SiO_2 interface, it can passivate the interface. However, when there is too much chlorine, the oxide may blister and peel off.

7.5.4 Effects of Pressure

The oxidation rate can be increased with the increase in pressure at any given temperature (Ligenza 1961). The pressure influences both parabolic and linear rate constants. Since the solubility of the oxidant is directly proportional to the pressure, the time required to produce a given oxide thickness should be inversely proportional to pressure. Shorter oxidation times can have other beneficial effects by reducing dopant redistribution, minimizing junction movement, and reducing the density of structural defects introduced during processing. The production of thin oxide films requires reduced pressure for processing. When the pressures are in the range of 0.1 atm or when the thickness of the oxide grown is less than a few hundred angstroms, the pressure dependence of oxidation rate is not clearly understood.

EXAMPLE 7.3. Obtain a relationship between the linear rate constant for oxidation for intrinsic silicon as a function of dopant concentration. Assume that dopant-induced enhancement of the oxidation rate is due to silicon-vacancy concentration.

Solution. The vacancies in silicon can exist in a neutral state V^0_{Si}, in a positively charged state V^+_{Si}, and in negatively charged states V^-_{Si} and V^{--}_{Si}. The association of charge with vacancies makes the concentration of these levels sensitive to Fermi energy and hence to dopant concentration and temperature.

We can apply the law of mass action to the relation

$$V^0_{Si} + e^- = V^-_{Si}$$

to obtain

$$K_{Si^-} = \frac{[V^-_{Si}]}{[V^0_{Si}] \cdot n},$$

where the concentrations are symbolized in brackets. For an intrinsic semiconductor,

$$K_i = \frac{[V^-_{Si}]_i}{[V^0_{Si}]_i\ n_i}.$$

For the same defect reaction, the equilibrium constant is a function of temperature only so that

$$V^-_{Si} = [V^-_{Si}]_i \frac{n}{n_i}.$$

In an analogous fashion, we can obtain

$$[V^+_{Si}] = [V^+_{Si}]_i \frac{n_i}{n}$$

and
$$[V_{Si}^{--}] = [V_{Si}^{--}]_i \left(\frac{n}{n_i}\right)^2 .$$

The total concentration of vacancies $[V_{Si}]_T$ is $[V_{Si}]_T = [V_{Si}^0] + \frac{n_i}{n}[V_{Si}^+]_i + \frac{n}{n_i}[V_{Si}^-]_i + \left(\frac{n}{n_i}\right)^2 [V_{Si}^{--}]_i$.

For heavily doped n^+ or p^+ material, the total number of vacancies is considerably higher than the intrinsic concentration of vacancies. The linear oxidation rate may be written

$$\frac{B}{A} = \left(\frac{B}{A}\right)_i + k_v \{[V_{Si}]_T - [V_{Si}]_i\},$$

where $\left(\frac{B}{A}\right)_i$ is the linear oxidation rate for intrinsic semiconductor, k_v is a constant, and $[V_{Si}]_T$ is the total vacancy concentration.

7.6 OXIDATION-INDUCED DEFECTS

The thermal oxidation of silicon can introduce wafer warpage, dislocations, and stacking faults. If the oxide is produced only on one side of the wafer, the existence of oxidation-induced compressive stress in the oxide can cause wafer warpage. In addition, if the oxide is patterned and is not continuous, stresses develop at the oxide edge–wafer interface. These stresses can be substantial and can introduce dislocations in the underlying silicon (Ravi 1981).

The thermal oxidation of silicon frequently results in the formation of stacking faults (SFs) in the surface regions of the crystal. Queisser and van Loon (1964) studied these defects by etching and optical microscopy in specimens oxidized at 1100 to 1300°C in steam, wet oxygen, and dry oxygen. Subsequently, Booker and Stickler (1965), Sanders and Dobson (1969), Ravi and Varker (1974), Ravi (1975), and Shevlin and Demer (1979) investigated SFs in oxidized silicon. Their observations can be summarized as follows: (1) SFs are extrinsic in nature, lie on {111} planes, and are bounded by $a/3$ <111> Frank partials; (2) the formation of SFs is accentuated in the presence of surface damage; (3) in the absence of surface damage, faults are observed only in wafers that exhibit the banded or "swirl" type distribution of microdefects discussed in section 5.5.

When the surface of the silicon wafer is abraded or if damage resulting from the crystal-slicing operation is not entirely removed by chemical polishing, SFs are generated during oxidation (Ravi and Varker 1974; Shevlin and Demer 1979). Figure 7.7 shows an optical micrograph of an etched surface of an oxidized wafer, illustrating the formation of SFs along two scratches (Ravi and Varker 1974). The corresponding situation observed by transmission electron microscopy is produced as Figure 7.8. The close association of SFs with surface damage is apparent.

Shevlin and Demer (1979) systematically investigated the role of surface damage in the formation of SFs during the oxidation of silicon. They examined by TEM unidirectionally abraded specimens before and after short oxidation treatments.

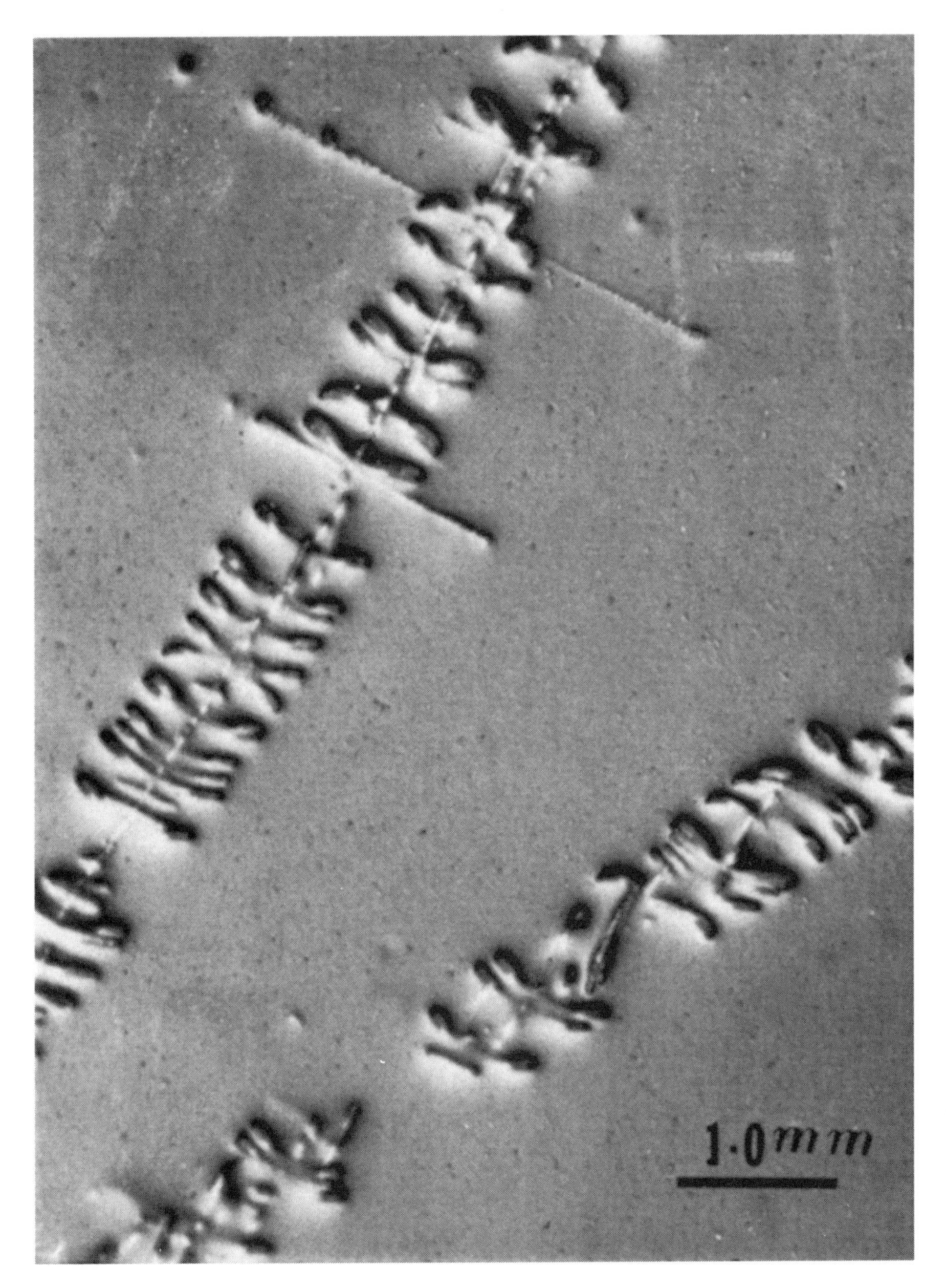

FIGURE 7.7
Optical micrograph of an etched Si surface showing the formation of oxidation-induced stacking faults along regions of mechanical damage. (*After Ravi and Varker [1974].*)

FIGURE 7.8
Electron micrograph showing the association of stacking faults with regions of mechanical damage in Si. (*After Ravi and Varker [1974].*)

They observed a/2 <110> Lomer dislocations at the damage sites. Figure 7.9 shows Lomer dislocations produced by [110] abrasions followed by steam oxidation of 1 min at 1050°C. Figure 7.10 shows SFs of various lengths observed along the [110] abrasion direction after steam oxidation for 4 min at 1050°C.

In the absence of surface damage, microdefects distributed in the swirl pattern in FZ silicon (see section 5.5.1) appear to be the sites where SFs form (Ravi and

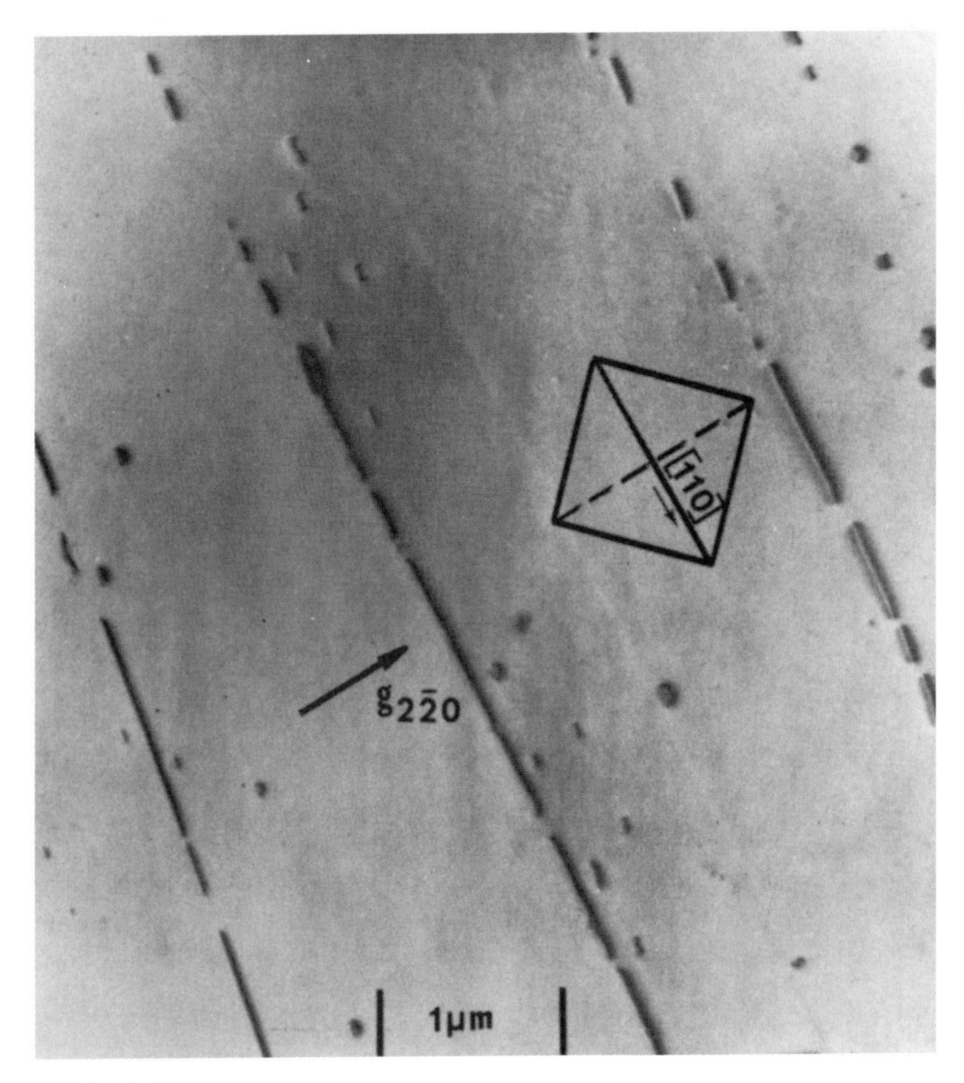

FIGURE 7.9
Lomer dislocations produced by [110] abrasion followed by steam oxidation of Si for 1 min at 1050°C. (*After Shevlin and Demer [1979].*)

Varker 1974). Figure 7.11, reproduced from the work of Ravi and Varker (1974), shows the distribution of microdefects observed in FZ silicon before and after oxidation. Figures 7.11*a* and 7.11*b* depict the swirl patterns before and after oxidation, whereas Figures 7.11*c* and 7.11*d* show higher magnification micrographs obtained from the same regions. It is apparent from Figure 7.11 that the distribution of SFs is closely correlated with that of the microdefects.

The preceding observations regarding the formation of oxidation-induced SFs can be rationalized as follows. When the (001) silicon is abraded along the [110] direction, the imposed shape change can be accommodated by the activation of a/2 $[10\bar{1}]$ $(\bar{1}1\bar{1})$, a/2 $[0\bar{1}\bar{1}]$ $(\bar{1}1\bar{1})$, a/2 $[0\bar{1}1]$ $(1\bar{1}\bar{1})$, and a/2 [101] $(1\bar{1}\bar{1})$ slip systems.

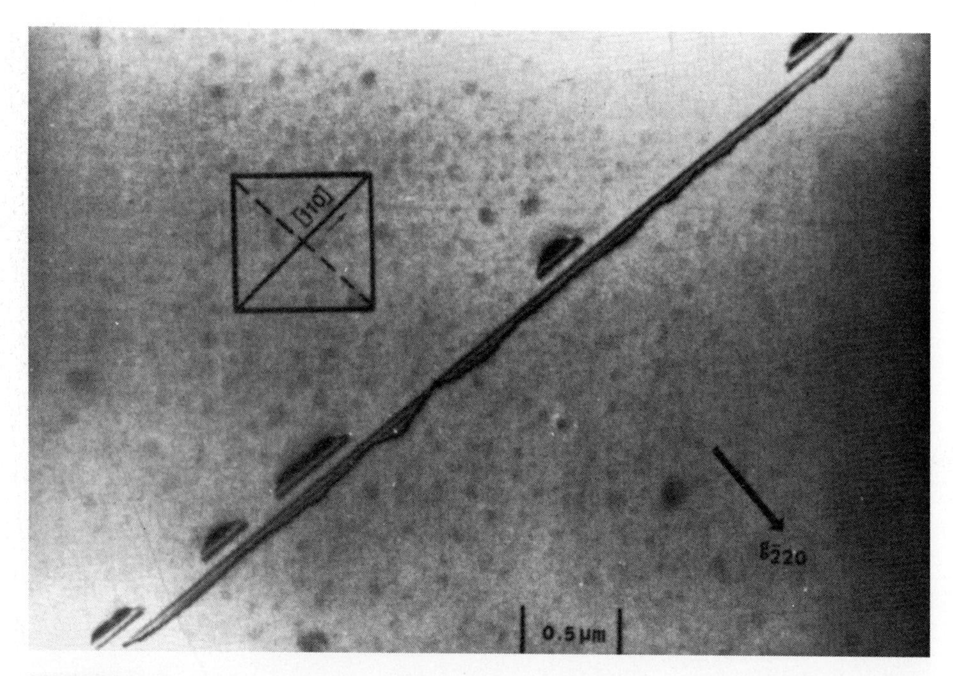

FIGURE 7.10
Long-stacking faults of various sizes produced by [110] abrasion followed by steam oxidation of Si for 4 min at 1050°C. (*After Shevlin and Demer [1979].*)

The dislocations involved in these systems can interact with each other, and the resulting dislocations would lie along the line of intersection of the $(\bar{1}1\bar{1})$ and $(1\bar{1}\bar{1})$ planes, that is, the [110] direction. The reaction governing the formation of a Lomer dislocation is

$$a/2\left[0\bar{1}1\right]_{(1\bar{1}\bar{1})} + a/2\left[10\bar{1}\right]_{(\bar{1}1\bar{1})} \rightarrow a/2\left[1\bar{1}0\right]. \tag{7.54}$$

The reaction is energetically favorable. Furthermore, as discussed in section 5.5.1, a number of investigators have shown that A-type swirl defects are perfect dislocation loops interstitial in character, their Burgers vector is a/2 <110>, and they lie on the {111} and {110} planes.

We visualize that a/2 <110> dislocations are involved in the nucleation of SFs. The role of these dislocations can be rationalized in terms of the following dislocation reaction proposed by Hirsch (1962):

$$\frac{a}{2}\langle 110\rangle \rightarrow \frac{a}{3}\langle 111\rangle + \frac{a}{6}\langle 11\bar{2}\rangle \tag{7.55}$$

As indicated earlier Dobson (1972) and Hu (1974) have suggested that silicon interstitials are injected into silicon during the oxidation. This occurs because the volume of SiO_2 formed is considerably larger than the volume of Si consumed. The faults resulting from reaction (7.55) can grow by the absorption of the injected interstitials at the core of the a/3 <111> Frank partial bounding the fault. This situation is analogous to the Bardeen-Herring source discussed in section 3.3.2.

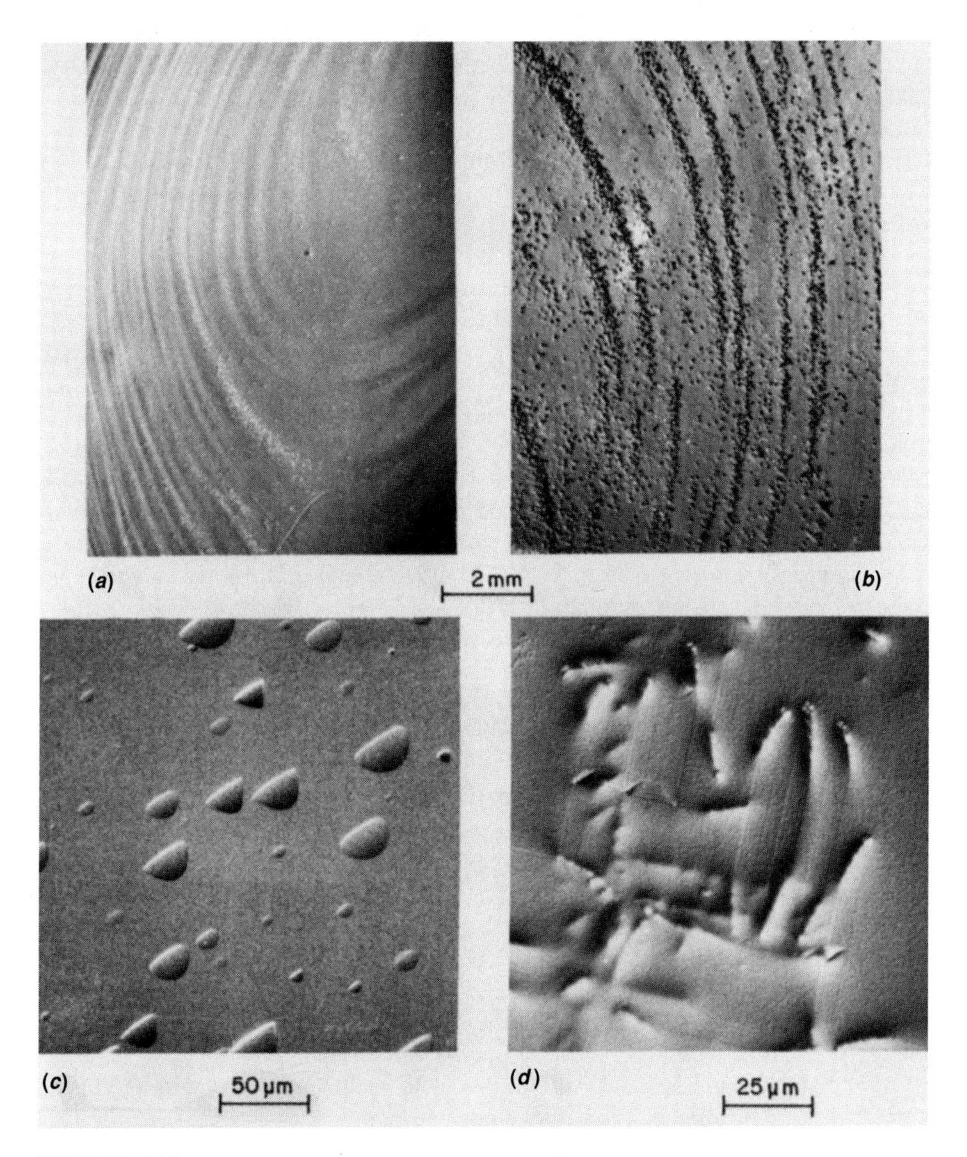

FIGURE 7.11
Microdefects observed in a swirl pattern in float zone Si: (*a*) before and (*b*) after oxidation; (*c*) and (*d*) are higher magnification micrographs from regions shown in (*a*) and (*b*). (*After Ravi and Varker [1974]*.)

It is pertinent to comment on the similarities between the oxidation-induced SFs and SFs resulting from the precipitation of oxygen in CZ silicon discussed in section 5.6.2.2. The two situations differ in the nucleation stage. In the case of oxidation-induced faults, an embryonic fault results from the dissociation of an a/2 <110> dislocation. The precipitation-induced fault involves the condensation of interstitials emitted during the formation of Si-O precipitates. In both cases SFs are extrinsic in nature, lie on the {111} planes, and are bounded by a/3 <111> Frank partials. In the case of oxidation-induced SFs, oxidation is responsible for generating silicon interstitials, whereas the formation of Si-O clusters in the second case produces interstitials required for the climb of the Frank partial bounding the SF. The common underlying reason in both cases is the volume expansion associated with the formation of either SiO_2 or Si-O clusters.

The following factors affect the formation of SFs during oxidation: the type of dopant in the crystal (Murarka and Quintana 1977), crystal orientation (Hu 1974; Murarka 1980), and the presence of HCl in the oxidizing ambient (Murarka 1980). These material parameters affect the generation rate of silicon interstitials (Murarka 1980), which in turn control the growth of SFs.

An interesting question is whether or not SFs can be eliminated after they form? Two approaches can be used to eliminate them. When the silicon wafers containing SFs are annealed in a vacuum, the faults shrink (Sanders and Dobson 1969). The consequence of annealing under reduced pressure could be that the equilibrium concentration of vacancies at the crystal surface is higher than it is contiguous to the fault. As a result, the vacancies migrate to the faulted region and get absorbed at the bounding partial, causing the fault to shrink.

The faults could also be converted into loops by unfaulting. Ravi (1974) has shown that when faults impinge on each other during growth, one of the faults undergoes an unfaulting reaction discussed in detail by Ruedl et al. (1962). Ruedl et al. emphasize that the unfaulting changes the nature of the defect, but does not eliminate it.

As discussed in Chapter 3, we expect the stacking faults to be electrically active along the bounding partials because of the dangling bonds. Figure 7.12 shows electron beam induced current (EBIC) displays of SFs in a p^+-n silicon diode (Varker and Ravi 1974). The recombination and generation events occurring at the faults are clearly evident. Figures 7.12*a*, through 7.12*c* show EBIC micrographs when the diode is at 0, 1, and 10 V reverse bias, respectively. Four faults are visible in the micrograph on the left. Two of them have been enlarged and their behaviors under different biases are shown on the right in Figure 7.12. The SFs behave as recombination centers at zero and 1 V reverse bias, whereas they act as generation centers at the reverse bias of 10 V.

We discussed in section 4.3.4 that the EBIC technique involves rastering the surface of a p-n junction or a Schottky diode with a high-energy electron beam. This process produces excess electron-hole pairs (EHPs) in the diode, and the carriers are separated from each other by the internal electric field, resulting in a current in the external circuit. If we assume that the fault partials, like the perfect dislocations, are associated with a half-filled band (see section 3.4), then the electrons can be captured locally by the partials to fill the band. As a result, the faults behave as recombination centers (see Figures 7.12*a* and 7.12*b*). When the reverse bias is increased to 10 V, the depletion region is substantially widened. It is conceivable that the SFs now lie within the depletion region. This location would require the half-filled band to be free of electrons; that is, the partials would emit electrons. This behavior produces a local increase in the collection current.

That SFs have a significant effect on the electrical characteristics of p-n junctions is further substantiated by Figure 7.13 that shows the reverse I-V characteristics of silicon diodes containing a different number of electrically active SFs (Ravi et al.

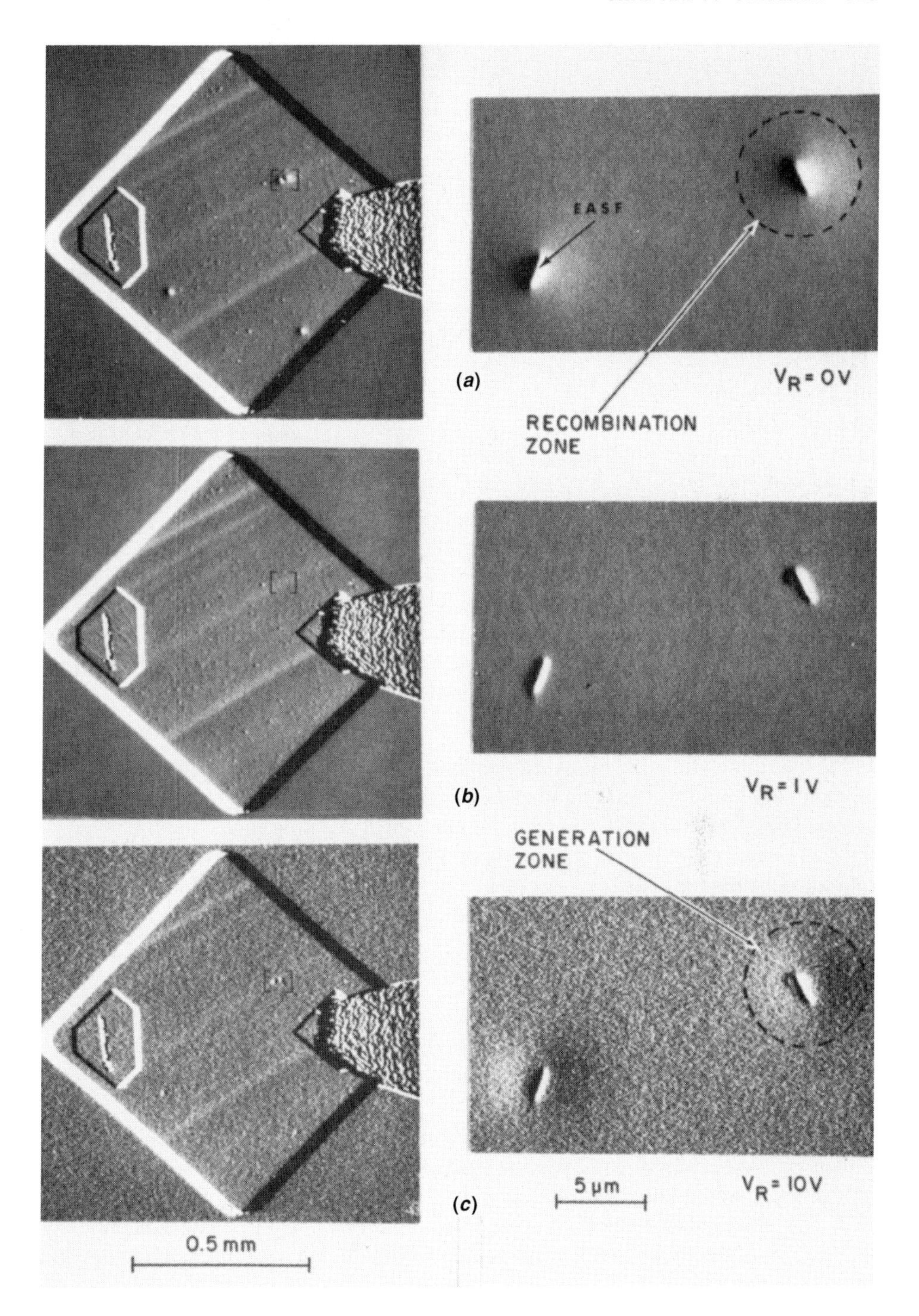

FIGURE 7.12
EBIC images obtained from a p$^+$-n diode showing recombination and generation zones around the electrically active stacking faults: (*a*) no reverse bias, (*b*) reverse bias of 1V, and (*c*) reverse bias of 10 V. (*After Varker and Ravi [1974].*)

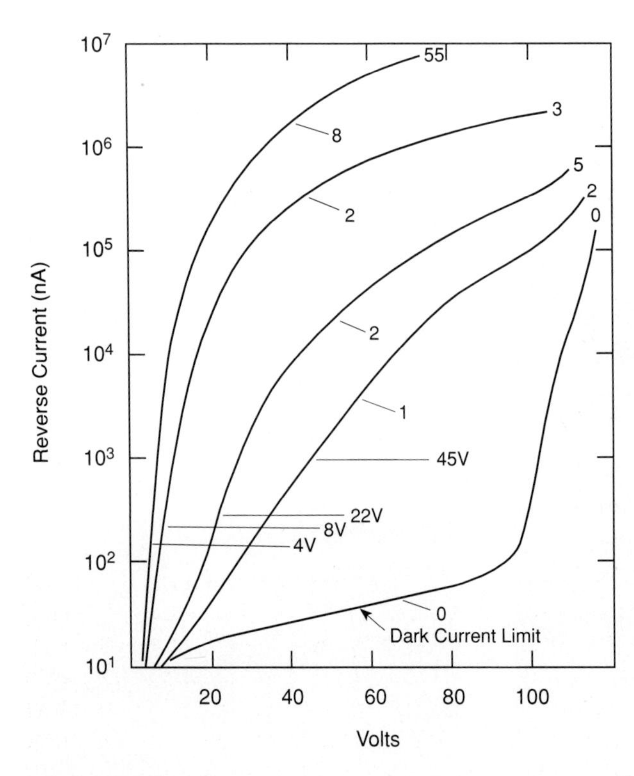

FIGURE 7.13
I-V characteristics of five diodes illustrating the effects of electrically active stacking faults on reverse currents. The total number of faults within the perimeter of the diode is shown at the ends of the curves, and the total number of the electrically active faults is indicated at the center. Voltages indicate the minimum threshold voltages for the group of active faults in a given diode. (*After Ravi et al. [1973].*)

1973). The total number of faults within the perimeter of the diode is shown at the end of each curve, and the total number of electrically active faults is indicated at the center. The correlation between the total number of stacking faults and the reverse-leakage current at a given reverse bias is very poor. However, we observe a better correlation between the density of electrically active faults and the I-V characteristics, a very interesting result. The need for SF-free thermal oxidation is apparent from the two examples.

7.7 CHARGES ASSOCIATED WITH THE SILICON-SILICON DIOXIDE SYSTEM

The unwanted charges within the oxide and at the Si/SiO_2 interface have deleterious effects on the performance of semiconducting devices. One defect in 1000 surface atoms can affect device characteristics. It is therefore extremely important to produce stable Si/SiO_2 interfaces. As discussed in section 7.1, the oxide is chemically stable, but its electrical characteristics depend on a number of intrinsic factors and process variables.

According to Deal (1980) four types of charges are associated with the Si/SiO_2 system, and they are schematically shown in Figure 7.14. The interface-trapped charges Q_{it} are located at the Si/SiO_2 interface. These charges have energy states within the band gap of silicon, can affect the electronic properties of nearby silicon, and can be positive or negative. These states are attributed to the disruption of the

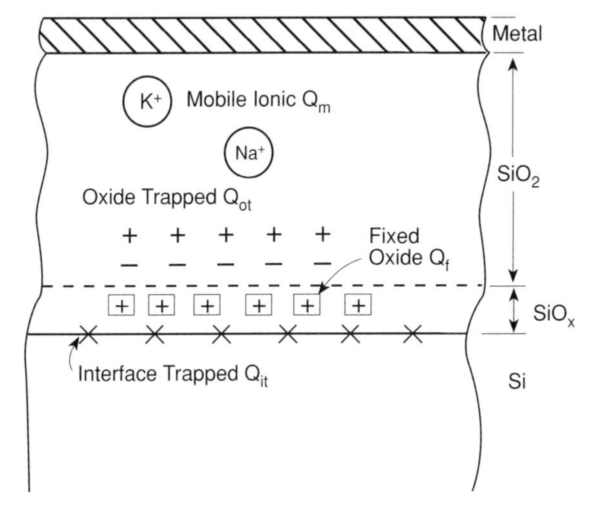

FIGURE 7.14
Charges associated with the Si/SiO_2 system. (*After Deal [1980].*)

periodic silicon lattice at the silicon-silica interface, resulting in dangling bonds. A hydrogen anneal at 450°C can effectively neutralize these trapped charges. Typical values of Q_{it} range from 10^{10} to $10^{13}/cm^2$. The energy levels are distributed throughout the band gap and do not interact with one another to form a band. However, if the levels are randomly distributed in the energy gap, the surface potential varies from one location to another across the interface. Furthermore, the density of these charges for the (111) silicon is three times that of the (001) surface because of the difference in areal densities of the two surfaces.

The fixed-oxide charges Q_f are usually positive and are located within approximately 3 nm of the Si/SiO_2 interface. Q_f cannot be charged or discharged and is not greatly affected by oxide thickness or type and doping level in silicon. It varies with the orientation in the same way as Q_{it}. It is related to the oxidation process and is thought to be associated with the incomplete oxidation of silicon. Its density decreases with the increasing temperature of oxidation. Annealing in an inert ambient also reduces Q_f. Furthermore, the presence of Q_f implies that some of the applied field in MOSFETs may be neutralized by these charges. Therefore, higher voltages are needed to achieve the same value of total capacitance.

Mobile-ionic charges Q_m are attributed to the presence of alkali ions such as sodium, potassium, and lithium, as well as negative ions and heavy metals. The presence of these mobile ions at room temperature leads to instabilities in threshold voltages of MOSFETs and deterioration of the gate-oxide reliability. The use of chlorine in the oxidizing ambient, gettering with phosphosilicate glass, and using a silicon nitride mask are some of the methods employed to control Q_m.

The oxide-trapped charges Q_{ot} may be positive or negative and are due to electrons or holes trapped in the bulk oxide. These defects are caused by ionizing radiation and high current in the oxide or by avalanche injection. The density of these charges may be reduced by a low-temperature anneal. Typical values of Q_{ot} range from 10^{10} to 10^{13} cm^{-2}.

The low-surface states produced in a thermally grown oxide must be related to the nature of the interface between silicon and amorphous silica. High-resolution electron micrographs reveal an abrupt transition from Si to amorphous SiO_2.

However, the exact nature of bonding between the Si atoms at the interface and the SiO_2 is still not clear.

7.8
OTHER TOPICS ON OXIDATION

Silicon is invariably covered with a native oxide ~2-4 nm thick. The question is how to grow the very high quality, thin SiO_2 films that are essential for higher device densities. MOS devices that are designed to operate at certain applied electric fields demonstrate the need for thin oxides. The capacitance C of a MOS structure is

$$C = K\frac{A}{L} \tag{7.56}$$

where K is the dielectric constant, A is the gate area, and L is the thickness of the dielectric. To achieve higher densities of MOSFETs, A has to be reduced. Consequently, L must be reduced to maintain C. We are considering ~20 nm thick oxide in the current designs. These small thicknesses are in the range where the linear-parabolic model for oxidation, discussed in section 7.2, does not accurately describe the oxidation behavior. Irene et al. (1986) have suggested that the initial oxidation rate in this case may be expressed as

$$\frac{dx_{ox}}{dt} = \frac{B}{(2x_{ox} + A)} + C_2 \exp\left(-\frac{x_{ox}}{L_2}\right), \tag{7.57}$$

where C_2 is a preexponential constant and L_2 is a characteristic decay length. Expression (7.56) was obtained from the results shown in Figure 7.15. The term C_2 has an Arrhenius temperature dependence with an activation energy of about 2.37 eV for (001) Si in dry oxygen. The rapid oxidation rate is attributed to several factors, none of which has been experimentally substantiated. The model invokes the space-charge effects where oxidation is increased by field-assisted diffusion, existence of micropores in the oxide providing additional channels for diffusion, or increase in diffusivity due to stress in the film or increased solubility of the oxidant in the thin oxide.

Polycrystalline silicon has a variety of applications in the VLSI technology, such as dynamic RAMs, EPROMs and CCDs. In these technologies the polycrystalline silicon has to be isolated by an insulator, which is accomplished by a thermally grown silicon dioxide layer. The polycrystalline silicon consists of grains of several orientations and is usually heavily doped. Since the oxidation rate depends on orientation (see Table 7.2), the oxidation of polycrystalline silicon produces rough surfaces. The degree of roughness depends on the distribution of orientations in the polycrystalline silicon and the doping level.

Essentially, two methods are employed to isolate devices from one another within an integrated circuit. One technique uses a p-n junction, whereas the second approach uses side-wall dielectrics. This process is referred to as *local oxidation of silicon* (LOCOS). It exploits the fact that Si_3N_4 is impervious to oxygen and can be used as a mask against oxidation. This masking is accomplished by first covering silicon with a thin layer of the nitride, then opening windows in it by etching, and

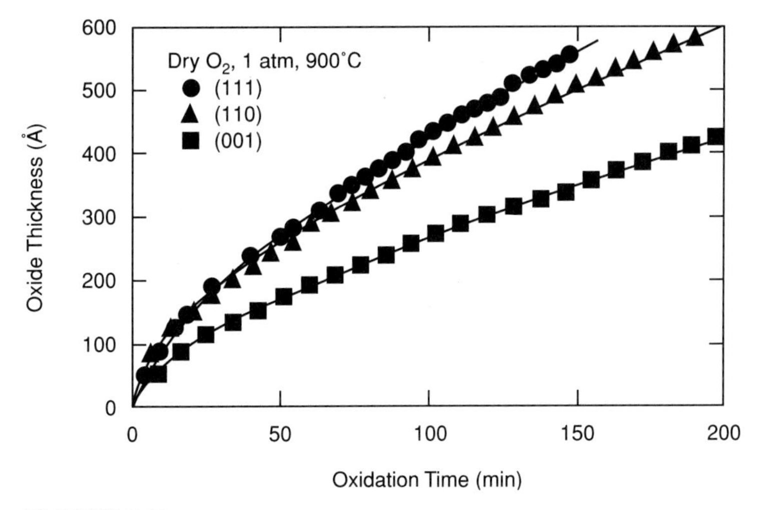

FIGURE 7.15
Oxide thickness versus time during initial stages of oxidation of Si. (*After Irene et al. [1986].*)

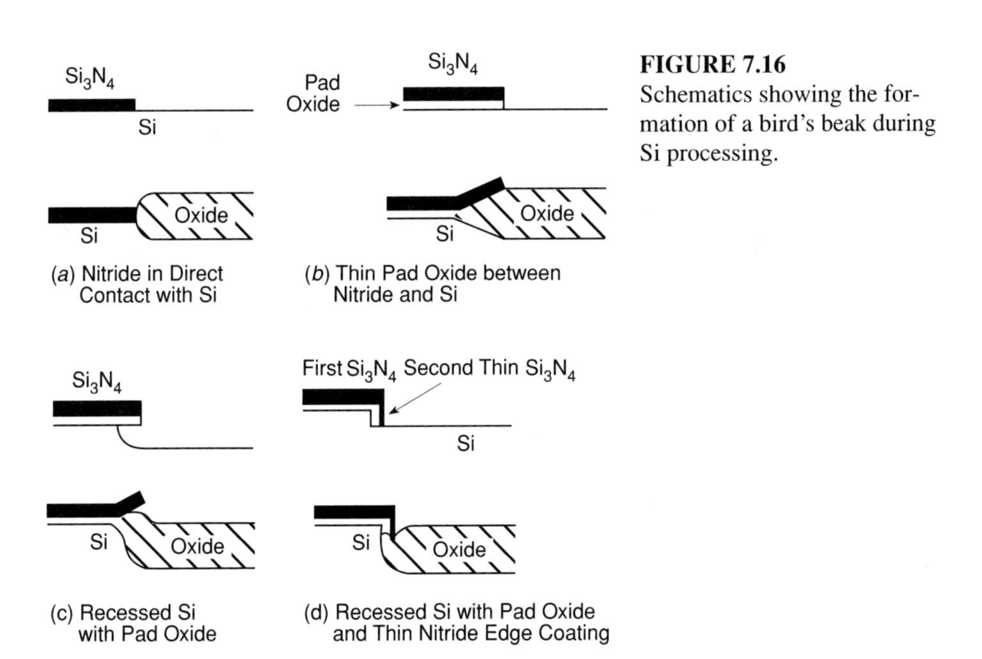

FIGURE 7.16
Schematics showing the formation of a bird's beak during Si processing.

subsequently growing an oxide in the windows. The process theoretically provides self-alignment between the active regions and the field regions of the device. However, oxide serves as a path for the lateral diffusion of oxygen. This process creates oxide underneath the nitride, producing a feature known as the bird's beak (Tamaki et al. 1983); its formation is illustrated in Figure 7.16.

7.9 NONTHERMAL OXIDATION METHODS

The need for an insulator at different stages in the processing of electronic circuits requires methods for producing an oxide film that are compatible with the manufacturing process. To maintain junction boundaries and to avoid redistribution of dopants during processing, the oxidation process must have a low thermal energy input. In addition, devices based on compound semiconductors require the deposition of silicon dioxide, since many of these semiconductors do not possess native oxides.

One process that has found some acceptance, especially in preparing insulating layers on compound semiconductors, is anodization. In this approach a semiconductor forms the anode in an electrolytic cell, and the cathode is usually platinum. The electrolyte is generally water, which produces the hydrogen and hydronium ions. Since the concentration of ions in pure water is very small and as a result has very low conductivity, attempts are made to change the resistance by adding acidic, basic, or neutral modifiers. Other nonaqueous media may also be used. The reaction that leads to oxidation is thought to be

$$Si + 2h^+ + 2H_2O \rightarrow SiO_2 + 2H^+ + H_2. \tag{7.58}$$

In a similar fashion the oxidation of GaAs probably occurs according to the following reaction:

$$GaAs + 12h^+ + 10H_2O \rightarrow Ga_2O_3 + As_2O_3 + 4H_2O + 12H^+. \tag{7.59}$$

A battery that injects electrons into the cathode provides the energy required for producing holes. The electrons combine with hydrogen ions and liberate hydrogen at the cathode. The delivery of holes to a p-type semiconductor surface by a battery is straightforward. However, for an n-type semiconductor, a depletion layer is created, and it acts as a barrier to the flow of holes. By illuminating the electrode or by operating above the breakdown voltage of the Schottky diode, it is possible to complete the oxidation process.

The uniformity of current across the electrode surface, the nucleation of oxide and its subsequent growth, the conditions in the electrolyte with respect to uniformity in composition, and the development of double layers adjacent to the electrodes are among the factors that control the formation of oxide. Since anodization is carried out at room temperature, low atomic mobility results in poor-quality oxide. Many of the anodically grown oxide films are generally porous and have much poorer quality than the thermally grown oxides.

Attempts to grow oxides at lower temperatures have prompted the use of photons, electron beams, microwaves, and plasmas. These techniques have thus far not resulted in acceptable-quality oxides. Similarly, rapid thermal-annealing techniques, wherein a high dose of thermal energy is provided for a very short time, have similarly failed to produce high-quality oxides. The demands on the silicon-silicon dioxide system are so severe that alternative methods to thermal oxidation have not yet found commercial acceptance.

7.10 HIGHLIGHTS OF THE CHAPTER

The highlights of this chapter are as follows:

- Silicon dioxide films, produced by thermal oxidation of silicon in dry oxygen and steam, are thermodynamically very stable.
- The oxidation kinetics exhibit two regimes: linear and parabolic. In the earlier stages of oxidation, that is, the linear regime, the reaction of oxygen with silicon at the Si/SiO_2 interface controls the growth of the oxide. In the parabolic regime, the diffusion of oxidant through the oxide controls oxidation.
- The volume of the oxide is substantially larger than the volume of silicon consumed. Depending on the oxidation temperature, this change may produce stresses at the Si/SiO_2 interface.
- During oxidation, silicon interstitials are injected into silicon. These interstitials interact with the abrasion-induced damage and swirls to form stacking faults.
- The oxide is amorphous in nature. Several types of charges are associated with it and the Si/SiO_2 interface.

REFERENCES

Bassous, E.; H. N. Yu; and V. Maniscalco. *J. Electrochem. Soc.* 123 1726 (1976).
Booker, G. R. and R. Stickler. *Phil. Mag.* 11, 1303 (1965).
Deal, B. E. *J. Electrochem. Soc.* 127, 979 (1980).
Deal, B. E. and A. S. Grove. *J. Appl. Phys.* 36, 3770 (1965).
Dobson, P. S. *Phil. Mag.* 26, 1301 (1972).
Gibson, J. M. and D. W. Dong. *J. Electrochem. Soc.* 127, 2722 (1980).
Grove, A. S.; O. Leistiko; and C. T. Sah. *J. Appl. Phys.* 35, 2695 (1964).
Hirsch, P. B. *NPL Conference on the Relation Between Structure and Strength in Metals and Alloys.* London: Her Majesty's Stationary Office, 1962, p. 440.
Ho, C. P.; J. D. Plummer; J. D. Meindl; and B. E. Deal. *J. Electrochem. Soc.* 125, 1146 (1978).
Hu, S. M. *J. Appl. Phys.* 45, 1567 (1974).
Irene, E. A. *CRC Critical Reviews in Solid State and Materials Sciences* 14, 175 (1988).
Irene, E. A. H. Z. Massoud; and E. Tierney. *J. Electrochem Soc.* 133, 1253 (1986).
Irene E. A. and Y. J. van der Meulen. *J. Electrochem. Soc.* 123, 1380 (1976).
Lie, L. N.; R. R. Razouk; and B. E. Deal. *J. Electrochem. Soc.* 129, 2828 (1982).
Ligenza, J. R. *J. Phys. Chem.* 65, 2011 (1961).
Ligenza, J. R. and W. G. Spitzer. *J. Phys. Chem. Solids* 14, 131 (1960).
Lupis, C. H. P. *Chemical Thermodynamics of Materials.* New York: North-Holland, 1983.
Marcus, R. B. and T. T. Sheng. *J. Electrochem. Soc.* 129, 1278 (1982).
Murarka, S. P. *Phys. Rev.* B21, 692 (1980).
Murarka, S. P. and G. Quintana. *J. Appl. Phys.* 48, 46 (1977).
Queisser, H. J. and P. G. G. van Loon. *J. Appl. Phys.* 35, 3066 (1964).
Ravi, K. V. *Phil. Mag.* 30, 1081 (1974).
Ravi, K. V. *Imperfections and Impurities in Semiconductor Silicon,* New York: Wiley-Interscience (1981), p. 237.

Ravi, K. V. *Phil. Mag.* 31, 405 (1975).
Ravi, K. V. and C. J. Varker. *J. Appl. Phys.* 45, 263 (1974).
Ravi, K. V.; C. J. Varker; and C. E. Volk. *J. Electrochem. Soc.* 120, 533 (1973).
Revesz, A. G. *IEEE Trans. Electron Dev.* ED-12, 97 (1965).
Ruedl, E.; P. Delavignette; and S. Amelinckx. *IAEA Symposium, Venice,* (1962) part 1, p. 363.
Sanders, I. R. and P. S. Dobson. *Phil. Mag.* 20, 881 (1969).
Shevlin, C. M. and L. J. Demer. *Phil. Mag.* A40, 685 (1979).
Tamaki, Y. et al. *J. Electrochem. Soc.* 130, 2266 (1983).
Varker, C. J. and K. V. Ravi. *J. Appl. Phys.* 45, 272 (1974).

PROBLEMS

7.1. A silicon wafer is oxidized for one hour at an unknown temperature where oxidation follows a parabolic relationship and 200 nm of oxide is grown. How long would it take to grow another 100 nm of the oxide under the same conditions?

7.2. A device having the following structure is fabricated in a silicon epi layer:

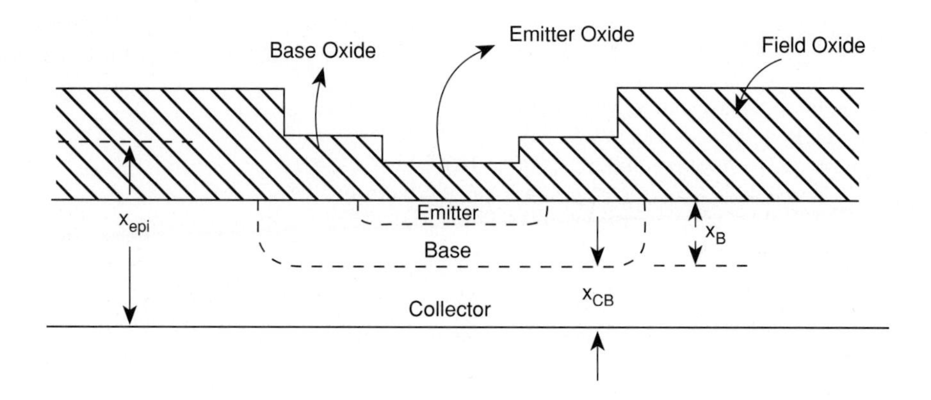

a. After epitaxial growth an initial oxide 600 nm thick is grown in steam at 1050°C. How long does this take?

b. After opening a window in the oxide, a boron-base predep is performed followed by a dip in 10:1 HF and a drive-in oxidation at 1100°C for one hour in steam. Assume that no oxidation of silicon occurs in predep but 200 nm of field oxide is removed by the HF dip. What are the oxide thicknesses in the base and field region after this step?

c. An emitter window is opened and a phosphorus predep ($C_S = 10^{21}$) is performed followed by an HF dip to remove excess phosphorus glass. Again assume no oxidation of silicon occurs in the predep but 100 nm of oxide is lost from the base and the field regions in the dip. A final oxidation is now performed at 920°C in steam for 30 minutes. What are the final oxide thickness over emitter, base and field regions?

d. If $x_B = 1.5\ \mu m$ and x_{CB} must be larger than 1 μm to maintain the required collector-to-base breakdown, what is the minimum epitaxial thickness to be specified for the product?

7.3. Calculate the maximum temperature variation that can be permitted in a furnace during oxidation of silicon at 1000°C in dry oxygen if the silicon dioxide thickness must be obtained within 100 ± 0.3 nm.

7.4. Silicon is doped locally with phosphorus to 8×10^{20} atoms/cm^3, and is surrounded by undoped silicon. If the surface oxide was initially 20 nm thick, calculate the height of the oxide step that develops at the undoped and doped interface when the wafer is oxidized in dry oxygen for five hours at 900°C.

7.5. What would be the thickness of silicon dioxide that grows on a silicon wafer when it is exposed to air at 300 K for one year.

7.6. A voltage V_g is applied to the metal (Al) electrode and is negative with respect to silicon. A thin layer of SiO_2 (30 nm) separates the metal and semiconductor. Sodium ions in the SiO_2 have mobility given by

$$\mu = \mu_o \exp(-E_d/k_B T),$$

where μ is 1.1×10^{-9} cm^2/v.s at 100°C and 7×10^{-8} cm^2/v.s at 200°C. Calculate the voltage V_g required to transport the sodium ions through the oxide layer at 300 K.

7.7. Consider a system consisting of Si(s), O_2(g), H_2(g), H_2O(g), and SiO_2(s). Determine the number of components. What are the degrees of freedom for this system? Explain.

7.8. Using the free-energy data for the reaction

$$4HCl + O_2 = 2H_2O + 2Cl_2,$$

explain how the oxidation of silicon is influenced by incorporating chlorine into the oxidizing ambient.

7.9. Silicon is dissolved in aluminum. The solid solution of aluminum and silicon are exposed to air at one atmosphere and 400°C. Which oxide forms first and why? How much, if any, silicon can be dissolved in aluminum without the danger of silicon being oxidized?

7.10. A <111> silicon wafer is oxidized at 1200°C in steam for 30 minutes. What is the thickness of oxide? How long does it take to produce the same oxide in steam at 920°C? How long does it take to produce the same oxide in dry O_2 at 1200°C?

7.11. Write the general expression to calculate the thickness of silicon dioxide formed during oxidation of silicon. What are the limiting forms of this relationship when the oxidation is (*a*) diffusion controlled and (*b*) surface reaction controlled? How are rate constants influenced by (1) orientation of silicon, (2) temperature at which oxidation is carried out, (3) thickness of oxide, (4) partial pressure of the oxidant, and (5) gas-flow rate?

7.12. Calculate the (*a*) change in volume and (*b*) change in enthalpy when silicon of area A is oxidized to a depth d by dry and wet oxidations.

CHAPTER 8

Diffusion

This chapter covers the diffusion of dopants and impurities into semiconductors. It emphasizes both the atomistic and phenomenological aspects of diffusion. It also considers the effects of concentration and external fields on diffusion, self-diffusion in semiconductors, influence of concomitant oxidation of silicon on diffusion, and evolution of diffusion-induced dislocation networks in silicon. The ideas developed in this chapter form a basis for selective doping that is required for the fabrication of many devices, such as LEDs, lasers, and FETs.

8.1 DESCRIPTION OF DIFFUSION

Diffusion refers to the movement of atoms by random jumps. In a crystalline material each of these jumps involves some type of point defect. An energy barrier is associated with these jumps because they produce the displacement of adjacent atoms and broken bonds. At a given temperature atoms in a solid vibrate about their mean position with a certain frequency, and they attempt to overcome the barrier during one of their excursions. Some atoms overcome the barrier during one of their many attempts and move into new locations. By proper control of this diffusion process, the dopant atoms deposited at or near the surface of a semiconductor can be moved inward to produce a desired impurity profile.

Two approaches have been developed to describe the diffusion phenomena: atomistic and phenomenological. The atomistic description considers the atomic nature of the diffusing species and host lattice explicitly. Since various dopant-dopant and dopant–point defect interactions complicate diffusion in semiconductors, the development of atomistic understanding is quite challenging. In the phenomenological approach a continuum replaces the solid, and coupled diffusion describes the flux of atoms.

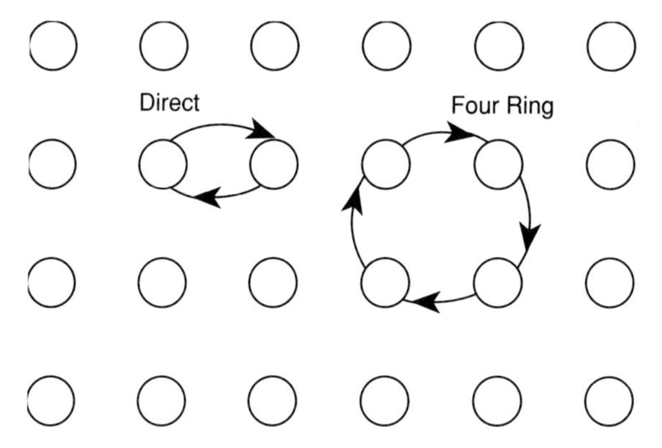

FIGURE 8.1
Diffusion mechanisms without involving defects: direct and four-ring exchange mechanisms.

8.1.1 Atomic Diffusion Mechanisms

The simplest mechanism for diffusion requiring no point defects involves the exchange of two contiguous atoms. This exchange is unlikely in tightly packed structures, because the neighboring atoms have to be compressed before they can squeeze past one another and exchange positions, which would require the rupture of six bonds. Alternatively, we can have a ring of atoms (three or more) where all the atoms in the ring rotate as a whole, enabling the atoms to move from one lattice site to another. The schematic in Figure 8.1 illustrates these two mechanisms. In the diamond-cubic structure, the motion of the ring consisting of six atoms requires the rupture of 12 bonds. We, therefore, suggest that diffusion cannot occur via the mechanisms shown in Figure 8.1.

Various processes involving defects in a solid describe atomic motion. The defects may be interstitials, vacancies, dislocations, and two-dimensional defects. As discussed in section 3.4, the point defects in semiconductors are generally charged. This implies that internal and external electric fields can influence diffusion (see section 8.6). The nature and number of charge defects are determined by the Fermi level, which in turn depends on the dopant concentration.

We have a number of diffusion mechanisms that involve interstitial point defects. The schematic in Figure 8.2 shows the motion of an interstitial atom. A smaller interstitial atom can move in the structure more rapidly than a self-interstitial, for example, Cu and Li atoms in Si. One advantage of the interstitial mechanism is the existence of a vast number of interstitial sites in the diamond-cubic and zinc-blende structures (see Figure 3.9). If a relatively large atom occupies an interstitial site, the atom can move if it is able to push its neighbor into an interstitial site and itself occupy the substitutional position in the crystal. This process is known as the *interstitialcy mechanism,* and we show it schematically in Figure 8.3. This mechanism requires two atomic jumps to move an atom from one site to another in the crystal.

Above zero Kelvin, we have an equilibrium number of vacancies whose concentration can be computed from Equation (3.11). An atom can move into the adjacent site by exchanging its place with a vacancy as we show in Figure 8.4. As indicated in section 3.3.1, the formation energy for a vacancy in silicon is fairly large. As a result, the vacancy concentration is low. For diffusion to occur, the diffusing

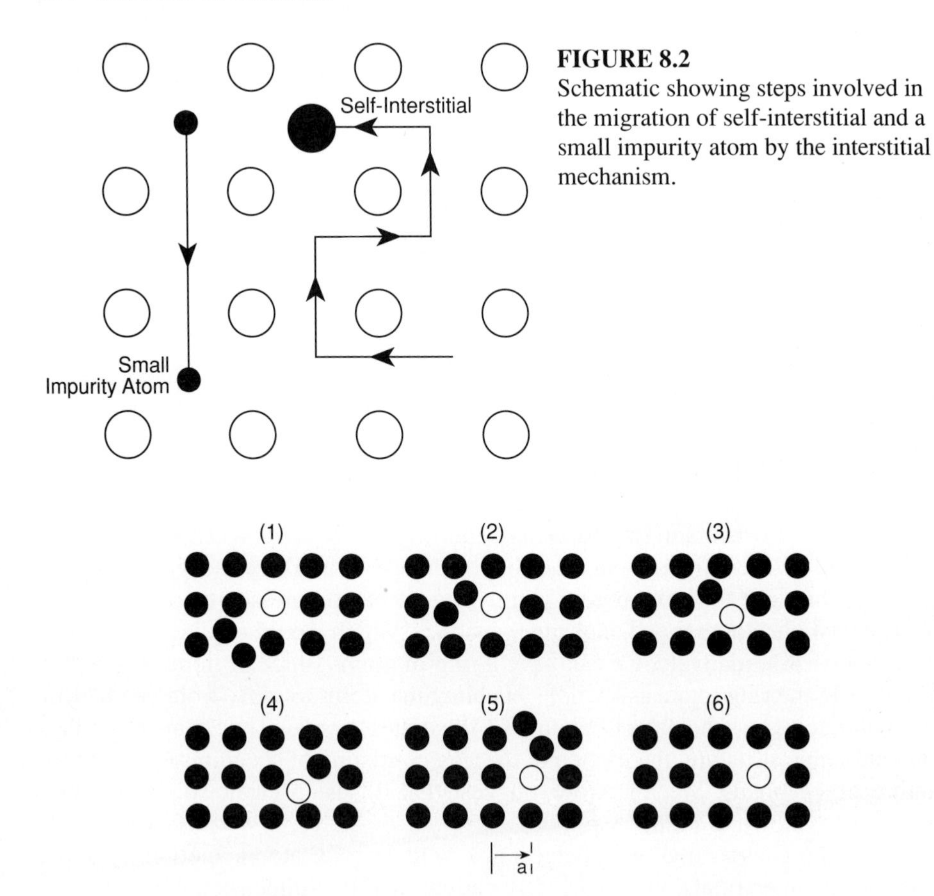

FIGURE 8.2
Schematic showing steps involved in the migration of self-interstitial and a small impurity atom by the interstitial mechanism.

FIGURE 8.3
Schematic depicting the interstitialcy diffusion mechanism.

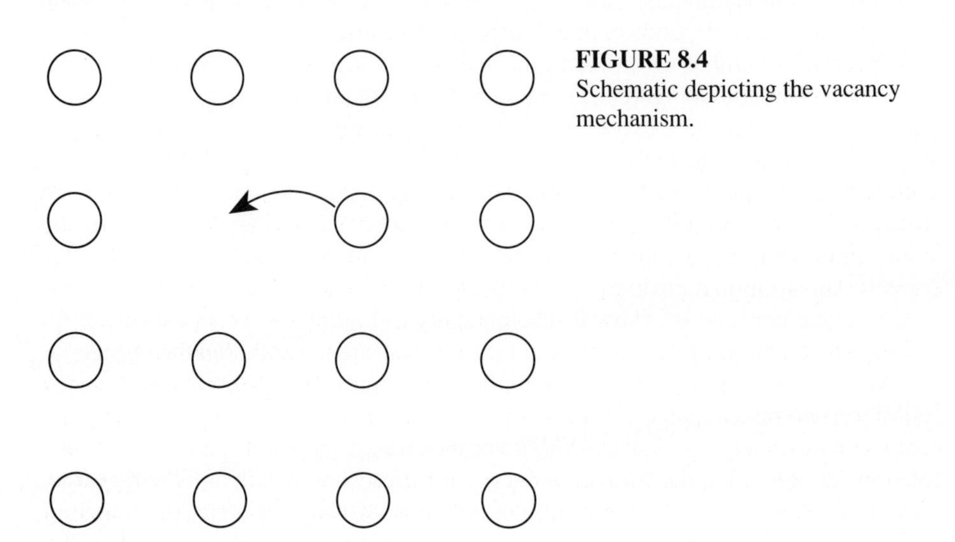

FIGURE 8.4
Schematic depicting the vacancy mechanism.

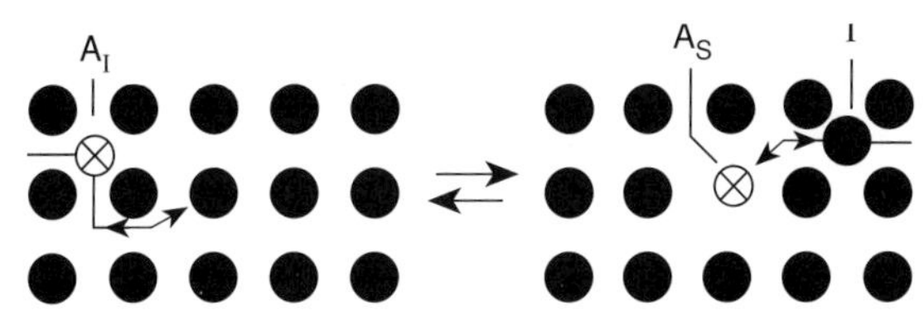

(a) Kick-out Mechanism

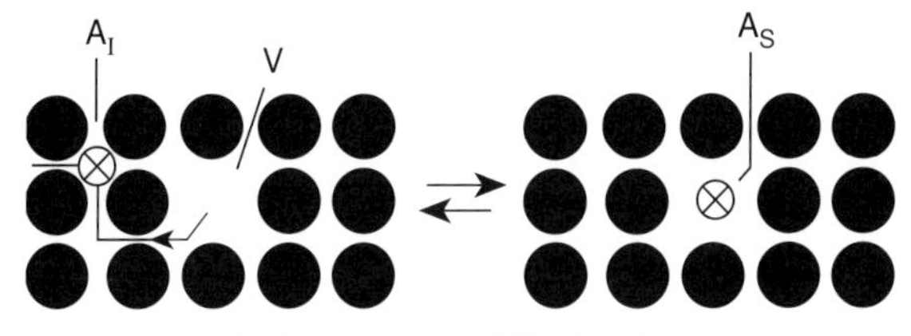

(b) Frank-Turnbull Mechanism

FIGURE 8.5
Schematics of (*a*) kick-out and (*b*) Frank-Turnbull mechanisms.

atom must wait until a vacancy becomes available. Because this period can be long, diffusion via the vacancy mechanism can be slow.

The elastic and/or coulombic forces between intrinsic point defects and impurities can lead to higher or lower diffusion coefficients, depending on whether the interaction is repulsive or attractive in nature. Impurities may occupy either substitutional or interstitial sites in a semiconductor. The schematic in Figure 8.5 illustrates two mechanisms that have been proposed for the diffusion of such impurities. In the *kick-out mechanism,* the exchange of atoms is described by the following reaction:

$$A_I \longrightarrow A_{Si} + Si_I, \tag{8.1}$$

where A_I, A_{Si}, and Si_I represent, respectively, atom A occupying an interstitial site, atom A occupying a substitutional site, and Si_I occupying the interstitial silicon atom. An alternative process known as the *dissociative mechanism* or *Frank-Turnbull mechanism* involves a vacancy. The atomic exchange during diffusion via this mechanism may be written as

$$A_I + V_{Si} \longrightarrow A_{Si}, \tag{8.2}$$

where V_{Si} represents a vacancy in the silicon lattice. The dissociative mechanism thus involves the annihilation of a vacancy by an interstitial impurity atom to place it at the substitutional site.

Diffusion of atoms can occur rapidly along the dislocation cores, (see section 3.3.2 for information on dislocation cores). The core acts as a pipe having high diffusivity. If the dislocation line extends along the length of the crystal, high-diffusivity paths are provided for impurity atoms. Even if dislocations are in the form of loops, solute atoms tend to segregate along the dislocations because of the attractive interaction between the impurity and dislocation strain fields. In addition to acting as pathways for diffusion, dislocations can act as sinks and sources for point defects. As a result, the point defect concentrations are affected. We discussed in section 3.3.3 that grain boundaries can be described in terms of dislocations. Therefore, grain boundaries also act as fast diffusion paths. The diffusion of atoms along the grain boundaries dominates at low temperatures where bulk diffusion is slow.

8.1.2 Phenomenological Description of Diffusion

According to Fick (1855) the flux of atoms is directly proportional to their concentration gradient. At any given time t, we can write

$$J = -D\,\frac{\partial C(x,t)}{\partial x}, \tag{8.3}$$

where J denotes the flux of atoms expressed in number of atoms passing perpendicular to a reference surface of unit area in unit time (atoms per unit area per sec). Equation 8.3 is known as *Fick's first law of diffusion.* We assume the flow of atoms to be along the x-direction. The concentration of atoms C(x,t) is a function of position and time and is expressed in terms of the number of atoms per unit volume. The concentration, in general, depends on the position, as we show in Figure 8.6*a*. The flux is assumed to be proportional to the concentration gradient, which is expressed as a derivative, rather than as a difference between the concentrations divided by the

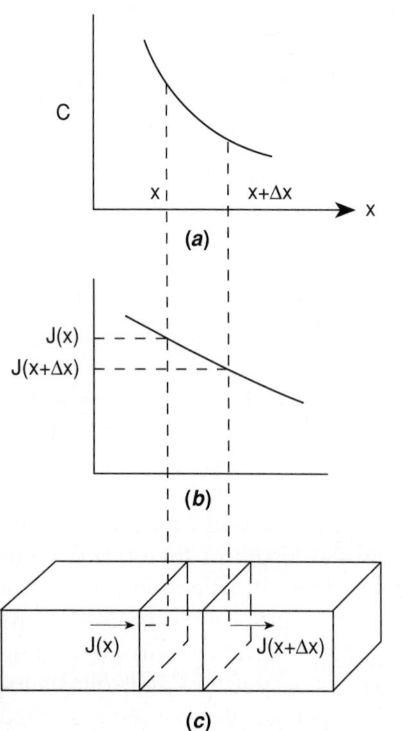

FIGURE 8.6
Schematics describing Fick's laws of diffusion: (*a*) concentration as a function of depth at any given time, (*b*) the flux associated with the variation in concentration, and (*c*) volume element in which the flux is shown entering and leaving.

distance, and is shown in Figure 8.6*b*. The implication is that the atoms move by a random process, not in a directed, straight-line motion. The negative sign in Equation 8.3 indicates that the net diffusion is occurring from a region of high concentration to a region of low concentration. The coefficient of proportionality D is termed the *diffusion coefficient* and is positive.

Assuming that the mass is conserved, consider the slab of material in the region between x and $x + \Delta x$ as shown in Figure 8.6*c*. For the sake of simplicity, we assume that the dimensions of the slab perpendicular to the x-axis are unity. At time t the concentration of atoms in the slab is C(t), so the total number of atoms in the volume element $\Delta x \cdot 1.1$ is $C(t)\,\Delta x$. If the concentration of atoms changes with time, then at time $t + \Delta t$ the concentration is $C(t + \Delta t)$ and the number of atoms in the slab is $C(t + \Delta t) \cdot 1.1\Delta x$. Therefore, the change in the number of atoms during an interval Δt in the volume element chosen is $\Delta x[C(t + \Delta t) - C(t)]$. This change must eventually be due to the difference in the flux of atoms J(x) entering the volume element at x and the flux of atoms leaving the volume element at $x + \Delta x$. The increase in the number of atoms in the volume element considered during the time Δt is $\Delta t\,[J(x + \Delta x) - J(x)]$. Since the mass is conserved, any increase in the concentration of atoms must be due to the decrease in the flux of atoms going out of the system. We must therefore have

$$\Delta x[C(t + \Delta t) - C(t)] = -\Delta t[J(x + \Delta x) - J(x)],$$

$$\frac{C(t + \Delta t) - C(t)}{\Delta t} = -\frac{J(x + \Delta x) - J(x)}{\Delta x},$$

or

$$\frac{\partial C}{\partial t} = -\frac{\partial J}{\partial x}, \tag{8.4}$$

which is known as the *continuity equation*. Substituting for J from Equation (8.3) into Equation (8.4) yields

$$\frac{\partial C(x,t)}{\partial t} = \frac{\partial}{\partial x}\left(D\,\frac{\partial C(x,t)}{\partial x}\right). \tag{8.5}$$

If the diffusion coefficient is a constant and independent of concentration, we may write

$$\frac{\partial C(x,t)}{\partial t} = D\,\frac{\partial^2 C(x,t)}{\partial x^2}. \tag{8.6}$$

Equation (8.6) is often referred to as *Fick's second law of diffusion*.

8.2 SELECTIVE DOPING BY DIFFUSION

In semiconductor-device processing, we carry out dopant diffusion under two different conditions. Suitably masked regions of a wafer are exposed to a diffusant whose concentration at the surface is constant. This step is referred to as *predeposition*.

Subsequently, the predeposited wafer is annealed at high temperatures to diffuse dopant into the semiconductor. This anneal is termed *drive-in anneal.*

We achieve a constant dopant concentration at the surface during predeposition by bringing a dopant source into equilibrium with a semiconductor surface. For selective doping of silicon, the source can be in the form of a solid such as phosphosilicate glass (PSG), liquid such as BBr_3 or $POCl_3$, or gas such as AsH_3 and PH_3. The dopant concentration at the surface is determined by its solubility limit in the semiconductor at the diffusion temperature. On the other hand, for a gaseous source the partial pressure of the dopant determines its concentration at the surface: the higher the pressure, the higher the concentration. Of course, when the solubility limit is reached, further increases in pressure have no effect on the surface concentration.

At high diffusion temperatures many dopants react with clean silicon surfaces, which causes pitting. For this reason we carry out predepositions in a nitrogen atmosphere containing a small amount of oxygen. As a result, a thin layer of oxide is grown before admitting dopants into the diffusion chamber. Selective doping of group III-V semiconductors via diffusion at high temperatures is cumbersome because the surfaces deteriorate by preferential evaporation of group V species (see section 6.1.1.3) because of their high vapor pressures at the diffusion temperature. Capping layers or counterpressures are used to prevent the out-diffusion of group V atoms.

The boundary conditions for the predeposition conditions are

$$\begin{aligned} C(0,t) &= C_S \\ C(\infty,t) &= 0, \end{aligned} \tag{8.7}$$

where C_S is the surface concentration. The solution of Equation (8.6) under the above boundary conditions is

$$C(x,t) = C_S \operatorname{erfc}\left(\frac{x}{2\sqrt{Dt}}\right), \tag{8.8}$$

where erfc is the complementary error function. We recall the definition of error function as

$$\operatorname{erf}(x) \equiv \frac{2}{\sqrt{\pi}} \int_0^x e^{-\eta^2}\, d\eta, \tag{8.9}$$

and

$$\operatorname{erfc}(x) \equiv 1 - \operatorname{erf}(x). \tag{8.10}$$

By definition $\operatorname{erf}(0) = 0$ and $\operatorname{erf}(\infty) = 1$. Normalized concentration versus distance curves for different times on linear and semilogarithmic scales are shown in Figures 8.7*a* and 8.7*b*, respectively. As expected, for a given dopant concentration at the surface, impurity atoms penetrate deeper at longer diffusion times.

The total number of dopant atoms per unit area of a semiconductor is given by

$$Q(t) = \int_0^\infty C(x,t)\, dx. \tag{8.11}$$

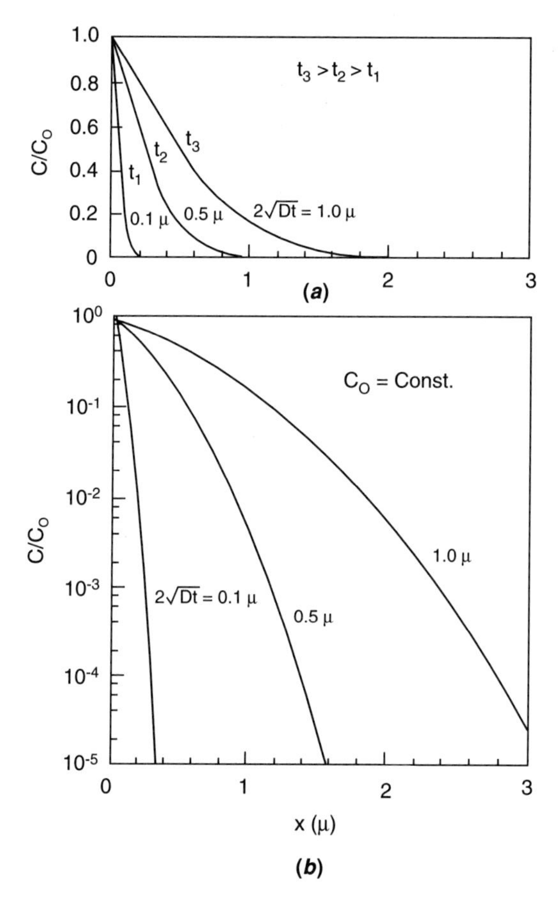

FIGURE 8.7
The complementary error function erfc: normalized concentration versus distance for successive times on (*a*) linear and (*b*) semilogarithmic scale.

For the predeposition diffusion, Q(t) is given by

$$Q(t) = \int_0^\infty \frac{C_S}{2} \operatorname{erfc}\left(\frac{x}{2\sqrt{Dt}}\right) dx = \frac{2}{\sqrt{\pi}}\sqrt{Dt}\, C_S. \tag{8.12}$$

In most predeposition processes, the goal is to achieve a certain specified concentration of dopant atoms in the silicon substrate. Furthermore, if the semiconductor contains background impurity C_b and has a conductivity opposite to that introduced by the diffusant, then the net dopant concentration at a distance x is

$$C_{net} = C_S \operatorname{erfc}\left(\frac{x}{2\sqrt{Dt}}\right) - C_b. \tag{8.13}$$

A p-n junction will form at a distance where C_{net} is zero. Using this condition, the junction depth x_j can be determined for Equation (8.13) and is given by

$$x_j = 2\sqrt{Dt}\, \operatorname{erfc}^{-1} \frac{C_b}{C_S}. \tag{8.14}$$

Typical junction depths after predeposition are in the range of a few thousand angstroms. Both temperature and time affect the depth of penetration through the parameter $\sqrt{Dt}$. Predeposition is normally carried out at high temperatures so that a large number of dopant atoms can be introduced into the semiconductor because of the high solubility. Since predeposition is a high-temperature process, it is not suitable for fabricating very shallow junctions.

The surface concentration of dopants obtained during predeposition cannot be maintained at room temperature, where devices operate, because the solubility limit is exceeded. It is therefore essential to reduce the surface concentration of a dopant below its solubility limit at room temperature. This procedure is accomplished by a second diffusion anneal in which the dopant atoms incorporated during predeposition are moved inwards. This step is referred to as *drive-in diffusion*. In the case of silicon, drive-in diffusion occurs in an oxidizing atmosphere where the surface oxide tends to prevent out-diffusion of dopant atoms.

The boundary conditions for drive-in diffusion are

$$\int_0^\infty C(x,t)\,dx = Q$$

and

$$C(\infty,t) = 0. \tag{8.15}$$

The solution of Equation (8.6) under the above boundary conditions is given by

$$C(x,t) = \frac{Q}{\sqrt{\pi Dt}} \exp\left(-\frac{x^2}{4Dt}\right). \tag{8.16}$$

The solution in (8.16) is a Gaussian function and is shown as normalized concentration versus distance curves for successive times on linear and semilogarithmic scales in Figures 8.8*a* and 8.8*b*, respectively. It is evident that the surface concentration of the dopant drops with time. This concentration is given by

$$C_S = C(0,t) = \frac{Q}{\sqrt{\pi Dt}}. \tag{8.17}$$

As before, the junction depth x_j is given by

$$x_j = \left\{4Dt \ln \frac{Q}{C_b\sqrt{\pi Dt}}\right\}^{1/2}. \tag{8.18}$$

The drive-in diffusion causes both the dopants and the junction depths to move further into the semiconductor. The importance of $\sqrt{Dt}$ as a length-scaling factor is evident. Furthermore, the Gaussian solution given by Equation (8.16) assumes that dopant atoms are confined to an extremely thin region after predeposition, so we can treat their distribution essentially as a delta function.

So far we have treated diffusion as if it proceeds only in one dimension. In reality, Fick's laws need to be generalized to three dimensions. For this purpose, we can write the flux of atoms $\vec{J}$ as a vector and relate it to the concentration gradient $\vec{\nabla}c$, also written as a vector. If we assume that the directions of these two vectors are opposite to one another, we can write

$$\vec{J} = -D\vec{\nabla}C \tag{8.19}$$

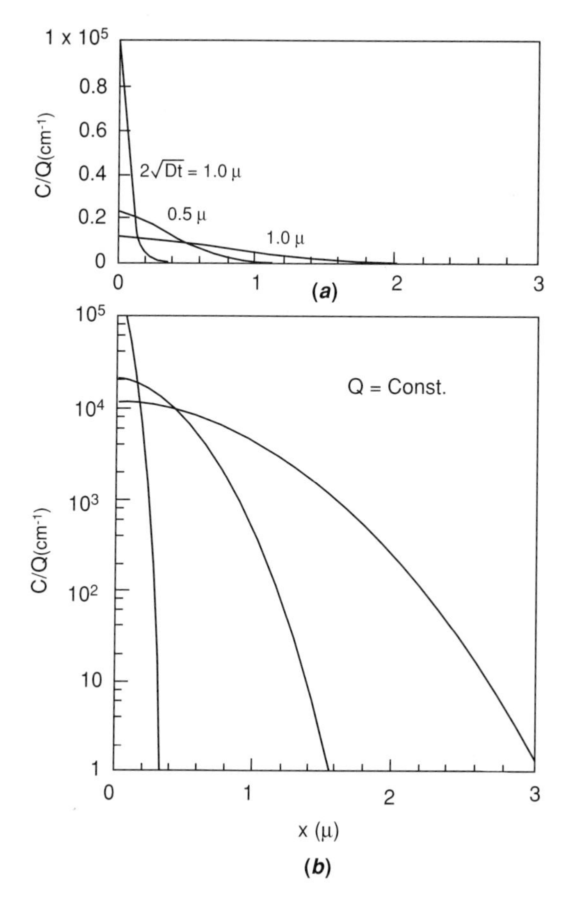

FIGURE 8.8
The Gaussian function: normalized concentration versus distance for successive times on (*a*) linear and (*b*) semilogarithmic scale.

as one form of generalized Fick's law, where

$$\vec{\nabla}C = \vec{i}\frac{\partial C}{\partial x} + \vec{j}\frac{\partial C}{\partial y} + \vec{k}\frac{\partial C}{\partial z} \tag{8.20}$$

and $\vec{i}$, $\vec{j}$, and $\vec{k}$ are orthonormal unit vectors. In a more generalized form of Fick's law we assume that each component of the flux is linearly related to all of the components of the concentration gradient. We can then write

$$J_i = \sum_{j=1}^{3} - D_{ij}\frac{\partial C}{\partial x_j}, \tag{8.21}$$

where the subscripts i and j assume each of the three directions. There is thus the possibility of more than one diffusion coefficient, which is the case for nonisotropic crystalline solids having symmetry lower than cubic.

The three-dimensional nature of diffusion becomes apparent when we are concerned with diffusion starting from a localized source. We encounter this situation when selective doping is carried out by diffusion through windows in a mask that is in intimate contact with a semiconductor wafer. Equations (8.8) and (8.16) give correct values of C(x,t) at all locations except near the edge of the mask window. The dopant atoms in the window region act as a dopant source, resulting in their lateral migration into the region contiguous to the window edge, which is underneath the mask. This effect is illustrated in Figure 8.9*a*, which shows the contours of constant doping concentration obtained during predeposition (Kennedy and O'Brien 1965). These contours have been computed assuming that the diffusion constant is independent of the concentration. Figure 8.9*b* shows the contours of constant doping concentration for drive-in diffusion. It is apparent from Figures 8.9*a* and 8.9*b* that the lateral penetration by the dopant atoms is ~75 to 85 percent of the vertical penetration.

When we fabricate bipolar transistors using diffusion, two distinct situations can arise if dislocations penetrate the emitter-base and base-collector junctions, as shown in Figure 8.10. Figure 8.10*a* shows that a space-charge cylinder may develop around the dislocation. Its evolution is due to interactions between the dopant ions and the dislocation core (see section 3.4). The space-charge region may serve as a

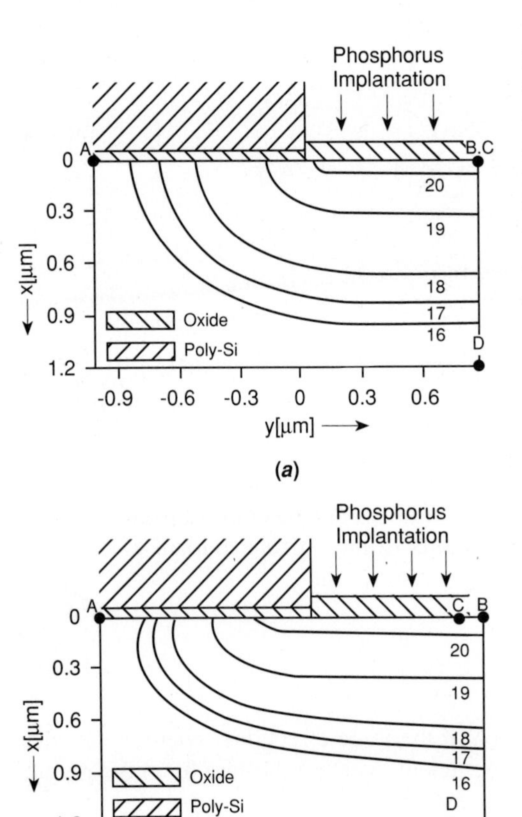

FIGURE 8.9
Concentration contours obtained from solutions of Fick's equation in two dimensions: (*a*) isoconcentration lines for P diffusion in two dimensions and (*b*) same as in (*a*) with the assumption that gate oxide acts as a perfect sink for supersaturated self-interstitials. (*After Kennedy and O'Brien [1965].*)

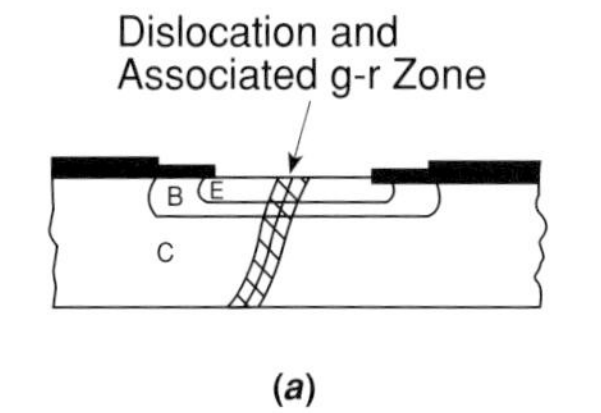

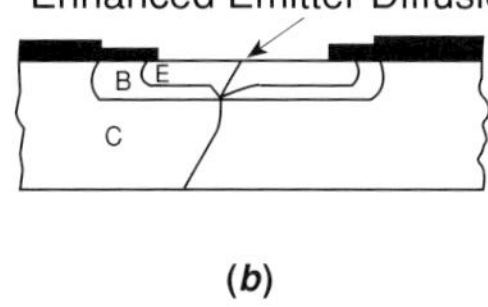

FIGURE 8.10
Two possible mechanisms whereby dislocations can cause emitter-collector pipes in bipolar junction transistors. (*After Ravi [1981].*)

generation-recombination (g-r) zone for carriers. In addition, accelerated diffusion along the dislocation during the fabrication of the emitter region may eliminate the emitter-base region, as shown in Figure 8.10*b*. As a result, the emitter layer is directly in contact with the collector region. We refer to this situation as an *emitter-collector pipe*.

EXAMPLE 8.1. Assuming that the diffusion coefficient in a one-dimensional diffusion depends on position, assess its influence on the dopant concentration.

Solution. If the diffusion coefficient D is dependent on location, we can write

$$\mathrm{D} = \mathrm{D(x)}.$$

We can write the generalized form of Fick's first law as follows:

$$\mathrm{J} = -\mathrm{D(x)}\,\frac{\partial \mathrm{C(x)}}{\partial \mathrm{x}} - \mathrm{C(x)}\,\frac{\mathrm{dD(x)}}{\mathrm{dx}}.$$

The second term represents the flux of atoms to regions where the diffusion coefficient is relatively low. Therefore, the dopant concentration is higher in the region with a smaller diffusion coefficient. Spatial dependence of a diffusion coefficient can arise if, as a result of processing, the surface region of a semiconductor and the bulk have different point defect concentrations. By employing the equation of continuity, we can write the generalized second law of Fick as follows:

$$\begin{aligned}\frac{\partial \mathrm{C(x,t)}}{\partial \mathrm{t}} &= \frac{\partial^2}{\partial \mathrm{x}^2}\,[\mathrm{D(x)C(x,t)}],\\ &= \mathrm{D(x)}\,\frac{\partial^2 \mathrm{C(x,t)}}{\partial \mathrm{x}^2} + 2\,\frac{\partial \mathrm{D(x)}}{\partial \mathrm{x}}\,\frac{\partial \mathrm{C(x,t)}}{\partial \mathrm{x}} + \mathrm{C(x,t)}\,\frac{\partial^2 \mathrm{D(x)}}{\partial \mathrm{x}^2}.\end{aligned}$$

If we ignore the last term, which is usually negligible, and consider the steady-state situation for which $\dfrac{\partial \mathrm{C(x,t)}}{\partial \mathrm{t}} = 0$, then

$$2\,\frac{\partial \mathrm{D(x)}}{\partial \mathrm{x}}\,\mathrm{C(x)} = -\mathrm{D(x)}\,\frac{\partial \mathrm{C(x)}}{\partial \mathrm{x}}.$$

Therefore, the diffusion constant that decreases along some spatial direction is accompanied by an increase in dopant concentration along that direction.

8.3 DEPENDENCE OF DIFFUSION COEFFICIENT ON TEMPERATURE

The diffusion coefficient D varies exponentially with temperature and can be expressed as

$$D = D_0 \exp\left(-\frac{E_A}{k_B T}\right), \tag{8.22}$$

where D_0 is a preexponential constant, E_A is the activation energy, k_B is the Boltzmann constant, and T is the temperature. If we plot lnD as a function of $1/T$, we obtain a straight line whose slope is given by $-E_A/k_B$, as shown in Figure 8.11 for the diffusion of Fe in γ-Fe, Fe in α-Fe, Cu in Cu, Zn in Cu, C in γ-Fe, Al in Al, and C in α-Fe. The intercept of this straight line at $1/T = 0$ gives the value of ln D_0. The importance of temperature control on diffusion can be ascertained from

$$\frac{dD}{D} = \left(\frac{E_A}{k_B T^2}\right) dT. \tag{8.23}$$

If we assume that the activation energy is 4 eV, an uncertainty in the temperature of about 5 K around 1000°C produces an error of about 14 percent in D.

We can understand the physical significance of D_0 and E_A by referring to Figure 8.12*a*. This figure shows the variation in the potential energy of atoms along a row. The minimum energy positions represent the stable equilibrium positions of

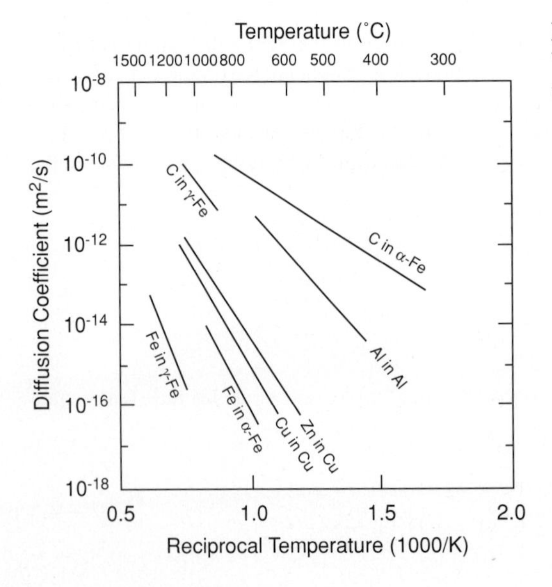

FIGURE 8.11
Variations in diffusion coefficient with temperature for several systems.

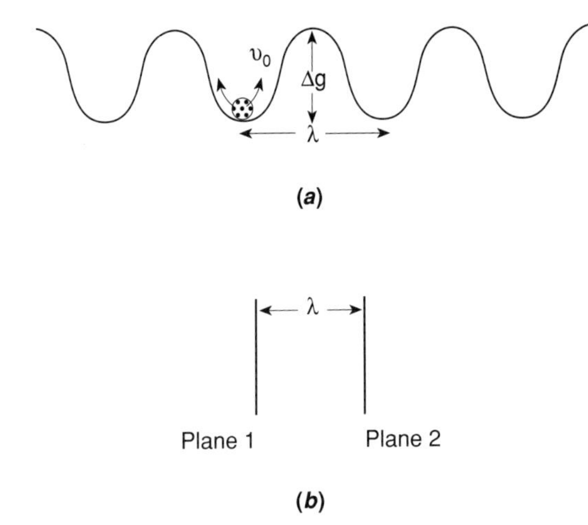

FIGURE 8.12
Schematic illustrating the diffusion of atoms along a periodic potential: (*a*) potential energy along a row of atoms in a crystalline solid and (*b*) the flux of atoms from one plane to an adjacent plane.

the atoms. As we indicated in section 8.1, atoms vibrate about their equilibrium position at any finite temperature because of thermal energy. Occasionally some of the atoms may have enough vibrational energy to overcome the potential barrier. If ν_0 is the mean frequency of oscillation of the atom and p is the probability that the jump occurs from site 1 to site 2, then the total jump rate Γ is given by

$$\Gamma = z\nu_0 p, \tag{8.24}$$

where z is the number of equivalent sites into which the atom can move. The probability of success for an atom to jump from site 1 to site 2 is given by

$$p = \exp\left(-\frac{E_A}{k_B T}\right). \tag{8.25}$$

We shall now consider two adjacent planes, designated 1 and 2, which are a distance λ apart, as shown in Figure 8.12*b*. Let n_1 and n_2 be the numbers of diffusing atoms per unit area in planes 1 and 2, respectively. If we assume the average frequency of jumping of each atom is Γ per second, the number of atoms jumping out of plane 1 in a time interval dt is $n_1\Gamma dt$. Since these atoms can jump in either direction, only half will be jumping from plane 1 to plane 2; the other half will be moving in the opposite direction. The net flux of atoms from plane 2 to plane 1 is therefore given by

$$J = \frac{1}{2}(n_1 - n_2)\Gamma = \frac{1}{2}(C_1 - C_2)\lambda\Gamma, \tag{8.26}$$

where C_1 and C_2 are concentrations of the diffusing species in planes 1 and 2.

$$n_1 = \lambda C_1 \quad \text{and} \quad n_2 = \lambda C_2, \tag{8.27}$$

thus

$$J = \frac{1}{2}(C_1 - C_2)\lambda\Gamma. \tag{8.28}$$

Assuming the concentration changes slowly with distance and time, we can write

$$C_1 - C_2 = -\lambda \frac{\partial C}{\partial x}. \tag{8.29}$$

Substituting for $C_1 - C_2$ from Equation (8.29) into Equation (8.28), we have

$$J = -\frac{1}{2}\lambda^2 \Gamma \frac{\partial C}{\partial x}. \tag{8.30}$$

Comparing Equation (8.30) with Equation (8.3), that is, Fick's first law of diffusion, we have

$$D = \frac{1}{2}\lambda^2\Gamma = \frac{1}{2}Z\nu_0\lambda^2 \exp\left(-\frac{E_A}{k_B T}\right). \tag{8.31}$$

We have assumed that diffusing species are permitted to hop between sites without considering the occupancy of those sites. This assumption implies that the diffusion coefficient has to be modified by the concentrations of point defects that may be responsible for diffusion. In addition, if we have a field acting on the diffusing species in a certain direction, the impurity may move preferentially in that direction. We expect the atoms to have an average velocity v due to a force F acting on them; v is usually expressed in terms of mobility B, which is v/F. Einstein has shown that mobility B is related to the diffusion coefficient by $D/k_B T$. Thus the flux of atoms in Fick's first law needs to be supplemented by the term CBF in the presence of an external field.

EXAMPLE 8.2. Obtain the Einstein relationship for the diffusion coefficient of atoms under a driving force F.

Solution. The flux J_i of atoms of species i is given by

$$J_i = n_i v_i,$$

where n_i is the number of atom per m^3 of species i and v_i is their average velocity. Let us suppose that the atoms are moving in response to a driving force F. The driving force on the atom i is the negative derivative of potential energy V with respect to distance so that

$$F = -\frac{\partial V}{\partial x}.$$

If we define mobility B_i of species i as the average velocity per unit force, we have

$$B_i = \frac{v_i}{F}$$

so that

$$B_i = -\frac{v_i}{\left(\frac{\partial V}{\partial x}\right)}.$$

Hence

$$J_i = -n_i v_i = -n_i B_i \frac{\partial V}{\partial x}.$$

Therefore

$$\frac{\partial V}{\partial x} B_i n_i = D_i \frac{\partial n_i}{\partial x}$$

so that on integration we obtain

$$n_i = (\text{constant}) \exp\left[-\frac{\partial V}{\partial x}\frac{B_i x}{D_i}\right].$$

If we assume further that the concentration of species i is also given by the Boltzmann formula, then

$$n_i = (\text{constant}) \exp\left[\frac{-\frac{\partial V}{\partial x} x}{k_B T}\right].$$

By equating the exponents, we obtain the Einstein relation

$$D_i = B_i k_B T.$$

8.4 DEPENDENCE OF DIFFUSION COEFFICIENT ON CONCENTRATION

If the concentration of a diffusant in a semiconductor is low, we can assume that the interaction between the diffusant atoms is negligible and the diffusion coefficient is not expected to depend on the concentration of the diffusing species. However, when the concentration is high, the diffusion coefficient depends on concentration. Therefore, we must write Equation (8.6) as

$$\frac{\partial C}{\partial t} = \frac{\partial}{\partial x}\left(D\frac{\partial C}{\partial x}\right) = \frac{\partial D}{\partial x}\frac{\partial C}{\partial x} + D\frac{\partial^2 C}{\partial x^2}. \tag{8.32}$$

Equation (8.32) can be solved by the method of Boltzmann and Matano. We can thus obtain the diffusion coefficient as a function of concentration.

To solve Equation (8.32), we combine the variables x and t into a variable η, where

$$\eta = \frac{x}{t^{1/2}}. \tag{8.33}$$

We can now express $\frac{\partial C}{\partial t}$, $\frac{\partial C}{\partial x}$, $\frac{\partial D}{\partial x}$, and $\frac{\partial^2 C}{\partial x^2}$ in terms of C and η.

$$\frac{\partial C}{\partial t} = \frac{dC}{d\eta}\frac{\partial \eta}{\partial t} = -\frac{1}{2}\frac{x}{t^{3/2}}\frac{dC}{d\eta} = -\frac{\eta}{2t}\frac{dC}{d\eta}. \tag{8.34}$$

$$\frac{\partial C}{\partial x} = \frac{dC}{d\eta}\cdot\frac{\partial \eta}{\partial x} = \frac{1}{t^{1/2}}\frac{dC}{d\eta}. \tag{8.35}$$

$$\frac{\partial^2 C}{\partial x^2} = \frac{\partial}{\partial x}\left(\frac{\partial C}{\partial x}\right) = \frac{\partial}{\partial \eta}\frac{\partial \eta}{\partial x}\left(\frac{\partial C}{\partial x}\right) = \frac{1}{t}\frac{d^2 C}{d\eta^2}. \tag{8.36}$$

$$\frac{\partial D}{\partial x} = \frac{dD}{d\eta}\cdot\frac{\partial \eta}{\partial x} = \frac{1}{t^{1/2}}\frac{dD}{d\eta}. \tag{8.37}$$

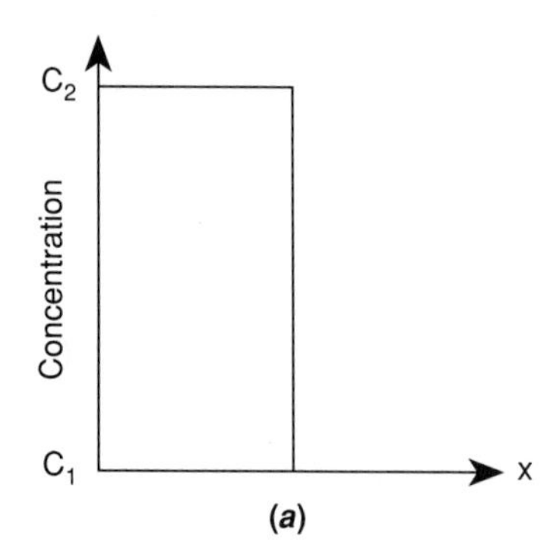

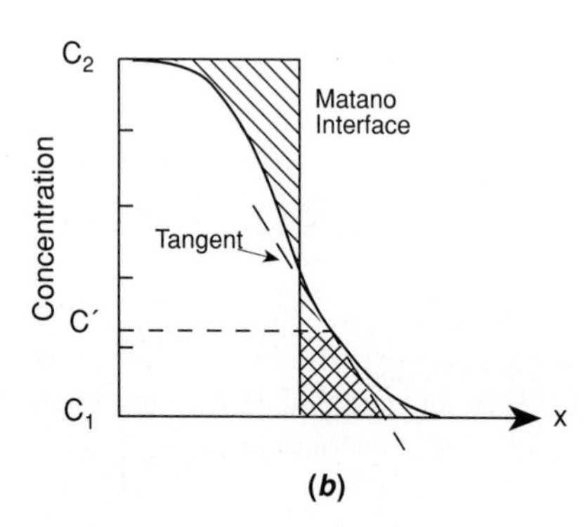

FIGURE 8.13
Determination of diffusion coefficient at concentration C′ according to Boltzmann–Matano analysis: (*a*) concentration profile at t = 0 and (*b*) concentration profile after diffusion has occurred for a certain time.

Substituting various terms from above into Equation (8.32), we have

$$-\frac{\eta}{2t}\frac{dC}{d\eta} = \frac{1}{t^{1/2}}\frac{dD}{d\eta}\cdot\frac{1}{t^{1/2}}\frac{dC}{d\eta} + \frac{D}{t}\frac{d^2C}{d\eta^2}. \tag{8.38}$$

Multiplying both sides, we have a differential equation in one variable η given by

$$-\frac{\eta}{2}\frac{dC}{d\eta} = \frac{dD}{d\eta}\frac{dC}{d\eta} + D\frac{d^2C}{d\eta^2} = \frac{d}{d\eta}\left(D\frac{dC}{d\eta}\right). \tag{8.39}$$

Let us now consider a situation where the concentration of the diffusing species at time t = 0 is shown in Figure 8.13*a*. We can now integrate Equation (8.39) between C_1 and C'

$$-\frac{1}{2}\int_{C_1}^{C'} \eta\, dC = \int_{C_1}^{C'} d\left(D\frac{dC}{d\eta}\right) = \left[D\frac{dC}{d\eta}\right]_{C_1}^{C'}, \tag{8.40}$$

where C' is the arbitrary concentration of interest. After diffusion has occurred for a certain period t, the concentration profile of the diffusing atom is shown in Figure 8.13*b* if we assume that the ends of the diffusion couple are still at their original concentration levels. We can utilize the following boundary condition at any given time t:

$$\frac{dC}{d\eta} = 0 \quad \text{at} \quad C = C_1 \quad \text{and } C = C_2. \tag{8.41}$$

Rewriting Equation (8.40) in terms of concentrations C_1 and C_2, we have

$$-\frac{1}{2}\int_{C_1}^{C_2} \eta \, dC = D\left.\frac{dC}{d\eta}\right|_{C=C_2} - D\left.\frac{dC}{d\eta}\right|_{C=C_1} = 0, \tag{8.42}$$

which implies that

$$\int_{C_1}^{C_2} \eta \, dC = 0. \tag{8.43}$$

Since we are considering diffusion for a fixed time, Equation (8.43) is equivalent to

$$\int_{C_1}^{C_2} x \, dC = 0. \tag{8.44}$$

Equation (8.44) defines the so-called Matano interface, where the number of diffusing atoms removed from the left of the interface is equal to the number of atoms added to the right of the interface. The preceding condition is equivalent to a boundary that separates the concentration versus distance curve so that the shaded areas in Figure 8.13*b* are equal.

The Matano interface does not coincide with the original interface of the couple and serves to define the origin $x = 0$, so the following integration may be performed graphically:

$$-\frac{1}{2}\int_{C_1}^{C'} \eta \, dC = D\left.\frac{dC}{d\eta}\right|_{C'} \tag{8.45}$$

Hence

$$D(C) = -\frac{1}{2t}\left(\frac{dx}{dC}\right)_{C=C'} \int_{0}^{C'} x \, dC. \tag{8.46}$$

We measure the concentration of the diffusing species as a function of distance. Using the slope and the shaded area shown in Figure 8.13*b*, we can obtain the diffusion coefficient as a function of concentration (Shewmon 1963).

We cannot use the preceding approach directly to analyze concentration profiles obtained from predeposition and drive-in diffusions, since the profiles do not satisfy the boundary condition requiring the concentration at the far left of $\eta = 0$ to be invariant with time. However, we can invoke the constancy of the total amount of dopant with time:

$$Q = \int_{0}^{\infty} C(x,t) \, dx. \tag{8.47}$$

The expression relating the diffusion coefficient with the concentration profile is given by

$$D\left[\frac{C(x_0,t)}{C_s}\right] = \frac{-C(x,t)x_0}{2t\left.\left(\frac{dC}{dx}\right)\right|_{x=x_0}}, \tag{8.48}$$

where C_S is the surface concentration and x_0 is the location at which D is determined. We should note that diffusion under an oxidizing environment can incorporate

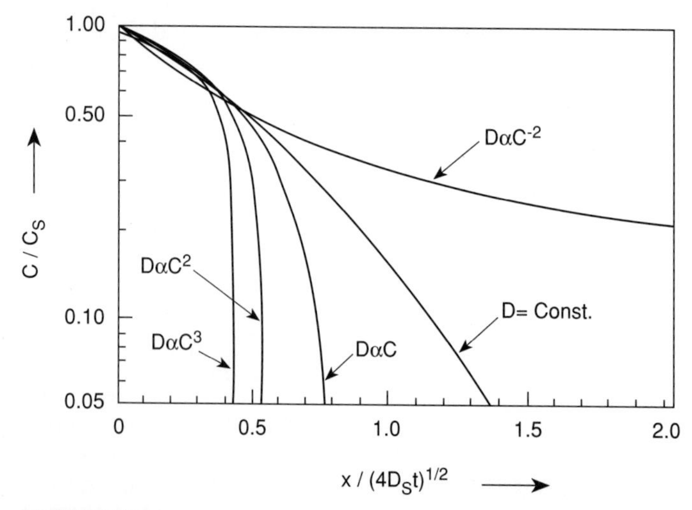

FIGURE 8.14
Normalized diffusion profiles for different concentration dependencies of diffusion coefficient. (*After Gösele [1988].*)

impurities in the oxide layer, so much more elaborate procedures are needed to determine the dependence of the diffusion coefficient on the composition.

The diffusion coefficient of an impurity in a semiconductor (Fair 1981) is generally considered to be independent of the concentration when its concentration is less than the intrinsic carrier concentration at the temperature of interest (Fair 1981). When the total impurity concentration exceeds the intrinsic carrier concentration, the diffusion coefficient depends on the ratio of extrinsic to intrinsic carrier concentrations. The dependence of diffusion coefficient on concentration is often written in the form

$$D = D_S \left(\frac{C}{C_S} \right)^{\gamma}, \tag{8.49}$$

where D_S is the diffusivity at the surface, C_S the surface concentration of the dopant, and γ is a parameter that describes the concentration dependence. Normalized diffusion profiles for different concentration dependencies are easily generated on a computer: Boxlike concentration profiles result for $\gamma > 0$, whereas for $\gamma < 0$ the diffusivity increases with decreasing concentration, as shown in Figure 8.14.

8.5 DEPENDENCE OF DIFFUSION COEFFICIENT ON EXTERNAL FIELDS

According to Fick's laws, equilibrium in the distribution of a given species (i) is achieved when its concentration gradient becomes zero. However, according to the general thermodynamic principles, at constant temperature and pressure the chemical potential μ_i of each component throughout the system should be uniform for diffusion to stop. Since two-phase alloy systems exist in the solid state, $\vec{\nabla} C_i$ being equal to zero does not automatically imply $\vec{\nabla}\mu_i$ to be zero.

Fick's laws of diffusion can be appropriately modified as

$$J_i = -L_i \text{ grad } \mu_i, \tag{8.50}$$

where the gradient of chemical potential rather than the concentration gradient is used. L_i is defined in Equation (8.55). The chemical potential μ_i is given by

$$\mu_i = \left(\frac{\partial G}{\partial n_i}\right)_{T,P}, \tag{8.51}$$

where G is the Gibbs free energy. When μ_i varies gradually with position in a system, a force F_i acting on the species (i) is given by

$$F_i = -\text{grad } \mu_i. \tag{8.52}$$

In response to this force, the atoms of the species (i) move with a certain average velocity v_i, where

$$v_i = -B_i \text{ grad } \mu_i \tag{8.53}$$

and the proportionality constant B_i is called the mobility. We can thus write

$$J_i = N x_i v_i = -N x_i B_i \text{ grad } \mu_i, \tag{8.54}$$

where x_i is the mole fraction of (i) and N is the total number of atoms per unit volume. Comparing Equations (8.50) and (8.54) we have

$$L_i \equiv N x_i B_i. \tag{8.55}$$

The chemical potential of the species (i) is given by

$$\begin{aligned} \mu_i &= \mu_i^0(T,P) + RT \ln a_i, \\ &= \mu_i^0(T,P) + RT \ln \gamma_i x, \end{aligned} \tag{8.56}$$

where a_i is the activity and γ_i is the activity coefficient of (i). Therefore,

$$\text{grad } \mu_i = RT\left(\frac{1}{x_i} \cdot \frac{dx_i}{dx} + \frac{d \ln \gamma_i}{dx}\right). \tag{8.57}$$

Since the concentration of the species (i) is defined as

$$C_i = N x_i, \tag{8.58}$$

we write

$$J_i = -RTB_i\left(1 + \frac{\partial \ln \gamma_i}{\partial x}\right)\frac{\partial C_i}{\partial x}. \tag{8.59}$$

Comparing Equations (8.59) and (8.3), we have

$$D_i = RTB_i\left(1 + \frac{\partial \ln \gamma_i}{\partial x}\right). \tag{8.60}$$

We should note that the term in the parentheses in Equation (8.60), known as the *thermodynamic factor,* depends on the knowledge of the solution behavior. The activities of any component in solution in a semiconductor are, however, largely unknown. The term RT B_i, often termed the *tracer diffusion coefficient,* denotes the intrinsic atomic mobility that is a function of the concentration and mobility of point defects. The thermodynamic factor can either increase or decrease the diffusion and depends on the behavior of component (i) in solution. If $\gamma_i > 1$ and $\gamma_i < 1$, the component (i) repels and attracts the host lattice atoms, respectively, resulting in the increase and decrease in the diffusion rate.

This generalization of Fick's laws permits us to include any type of externally superimposed fields. Examples of such fields are electric fields, temperature differences, differences in stress, and so on. The phenomenological description of many of these external fields and their influence on diffusion are available in the literature, see for example Ghandhi (1983).

8.6 SELF-DIFFUSION IN SEMICONDUCTORS

Self-diffusion refers to mass transport by the movement of a species constituting the host lattice. For example, the diffusion of silicon atoms in the silicon lattice or Ga atoms in the GaAs lattice are examples of self-diffusion. The self-diffusion measurements employ radioactive tracers. The self-diffusion coefficient is obtained by comparing the observed depth of distribution of tracer atoms after diffusion with a calculated profile obtained by solving Fick's laws under appropriate boundary conditions. For silicon, the isotope ^{31}Si is used for self-diffusion measurements and has a half-life of 2 1/2 hours. As a result, measurements are restricted to short times. The high-temperature data, typically above 1050°C, are thus experimentally accessible because diffusion is fast. The available data for silicon are plotted in Figure 8.15. The calculated activation energies are 5 and 4.2 eV at high and low temperatures, respectively. We attribute the continual decrease in activation energies and the dependence of diffusion coefficient on dopant concentrations to the variations in concentration of vacancies and their charge states (see section 3.4). The presence of interstitials adds a new complication. We cannot discern the role of interstitials in silicon, since direct measurements cannot differentiate between the contributions of vacancies and interstitials because of their small concentrations. The measured doping dependence of self-diffusion coefficient is consistent with the idea that neutral as well as positively and negatively charged intrinsic point defects are involved in self-diffusion.

Under equilibrium conditions we expect to have both self-interstitials and vacancies in concentrations [I] and [V], respectively. If we assume that self-diffusions D^{SD} by the vacancy and interstitials mechanisms occur independently, then D^{SD} can be written as

$$D^{SD} = D_I[I] + D_V[V], \tag{8.61}$$

where D_I and D_V are diffusion coefficients of interstitials and vacancies, respectively, and take into account all possible defect-charge states. To illustrate the complexities involved, we consider the relatively simple situation in which $[V] \gg [I]$. Four known silicon vacancy states correspond to V_{Si}^{x}, V_{Si}^{-}, $V_{Si}^{=}$, and V_{Si}^{+}, where V_{Si}^{x}

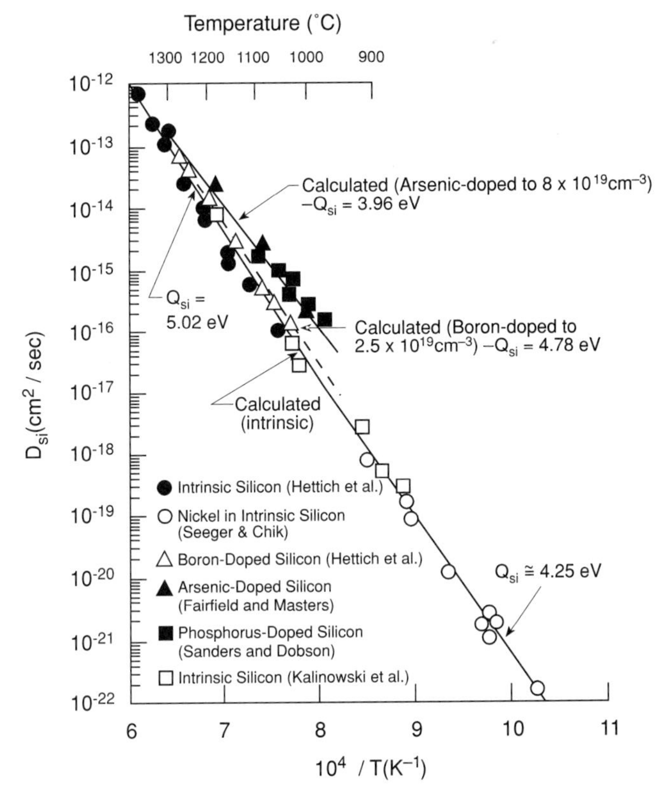

FIGURE 8.15
Silicon self-diffusion data in intrinsic and extrinsic silicon.

is a neutral vacancy and each charge state is indicated by the superscript. This defect identification follows the notation advanced by Kroger (1964) in which the main symbol for the point defect denotes the specific atom type or, in the case of a vacancy, the symbol V is used. The subscript for the point defect denotes the specific site in the crystal occupied by the defect, and the symbol I denotes the interstitial site. The dependence of concentrations of each type of vacancy as a function of the Fermi level calculated by Van Vechten (1975) for silicon at 300 K and 1000 K is shown in Figure 8.16. We see that at low temperatures V_{Si}^{x} is the dominant species in intrinsic silicon, but at high temperatures V_{Si}^{-} and $V_{Si}^{=}$ become numerous, and V_{Si}^{x} does not dominate at any value of E_f. Each time an ionized vacancy forms, the neutral vacancy population of the crystal is returned to the equilibrium value by generating an additional vacancy. Thus with increased doping regardless of the type, the total vacancy concentration increases with increasing population of charged vacancies. If diffusion occurs by the vacancy mechanism, the diffusion coefficient must therefore increase with doping. In particular, the effects of dopants begin to dominate when the doping level exceeds the intrinsic carrier concentration n_i.

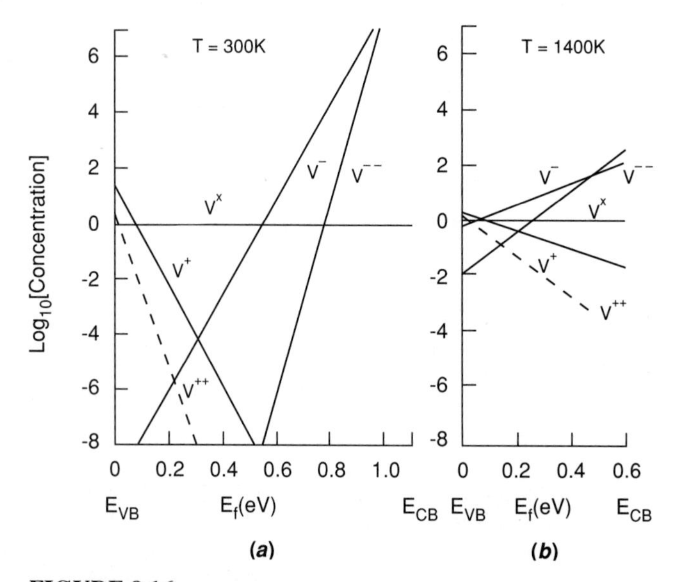

FIGURE 8.16
Calculated changes in the ratios of ionized-to-neutral vacancies as a function of Fermi level at two temperatures: (a) 300K and (b) 1400K (*After Van Vechten [1975]*).

We can assess the role of charged vacancies in self-diffusion in a semiconductor by comparing the dopant-induced changes in the vacancy concentration to those under intrinsic conditions. Consider, for example, a reaction between a neutral vacancy and a positively charged hole h^+ in a semiconductor, resulting in a positively charged vacancy:

$$V^x + h^+ = V^+. \tag{8.62}$$

The equilibrium constant K_1 for the preceding reaction is

$$K_1 = \frac{[V^+]}{p[V^x]}, \tag{8.63}$$

with the square brackets denoting concentrations of the two types of vacancies and p denoting a concentration of holes. As discussed in section 1.5.3.2, we have $p = n = n_i$ for an intrinsic semiconductor. Therefore,

$$K_1 = \frac{[V^+]_{int}}{n_i[V^x]_{int}} = \frac{[V^+]}{p[V^x]}, \tag{8.64}$$

where the subscript int on the concentrations emphasizes that the semiconductor is intrinsic. By noting that the concentration of V^x is the same in both the extrinsic and intrinsic semiconductors, we obtain

$$\frac{[V^+]}{[V^+]_{int}} = \frac{p}{n_i}. \tag{8.65}$$

For a vacancy that has −r charges, we can write

$$V^x + re^- = V^{-r}. \tag{8.66}$$

Consequently
$$K_2 = \frac{[V^{-r}]}{n^r[V^x]} = \frac{[V^{-r}]}{n^r[V^x]_{int}}. \tag{8.67}$$

Therefore
$$\frac{[V^{-r}]}{[V^{-r}]_{int}} = \left(\frac{n}{n_i}\right)^r. \tag{8.68}$$

Taking all possible charge states into account, the self-diffusion coefficient of silicon due to vacancies can be expressed as (Shaw 1975)

$$D_V^{SD}[V] = D(V^x)[V^x] + D(V^-)[V^-] + D(V^=)[V^=] + D(V^+)[V^+], \tag{8.69}$$

where
$$[V] = [V^x] + [V^-] + [V^=] + [V^+]. \tag{8.70}$$

For an intrinsic semiconductor, we can write the self-diffusion coefficient due to all the contributions from vacancy point defects as

$$D_V^{SD} = D_i^x + D_i^- + D_i^= + D_i^+, \tag{8.71}$$

where the symbol D_i^x denotes the contribution to diffusion coefficient from the vacancy of charge state x. However, if the semiconductor is extrinsic, the self-diffusion coefficient due to contributions from vacancies also depends on the electron and hole concentrations:

$$D_V^{SD} = D_i^x + D_i^- \left(\frac{n}{n_i}\right) + D_i^= \left(\frac{n}{n_i}\right)^2 + D_i^+ \left(\frac{p_i}{n_i}\right). \tag{8.72}$$

A similar relationship can be derived for interstitial defects. For example, in silicon the total self-diffusion coefficient is given by

$$D^{SD} = D_V^{SD}[V_{Si}] + D_I^{SD}[Si_I]. \tag{8.73}$$

The measurements of diffusion coefficients made by using tracer atoms require an additional factor called the correlation factor f that depends on the defect involved in diffusion and the crystal structure. Thus

$$D^{SD} = f_V\, D_V^{SD}[V_{Si}] + f_I\, D_I^{SD}[Si_I]. \tag{8.74}$$

The presence of two types of atoms and a variety of point defects having different charge states complicates the situation regarding self-diffusion in compound semiconductors. An additional complication is the high volatility of group V components. The loss of the volatile component alters the stoichiometry, which in turn affects the defect equilibria. As a result, the defect equilibria must be deduced from a series of equilibrium reactions, which must now include the vapor pressure of the volatile species.

According to Tan et al. (1991a), gallium self-diffusion in GaAs is dominated by triply negatively charged gallium vacancies, designated as V_{Ga}^{3-}, while in heavily p-doped materials it is controlled by the doubly positively charged gallium self-interstitials, designated as I_{Ga}^{2+}. Under thermal equilibrium and under intrinsic

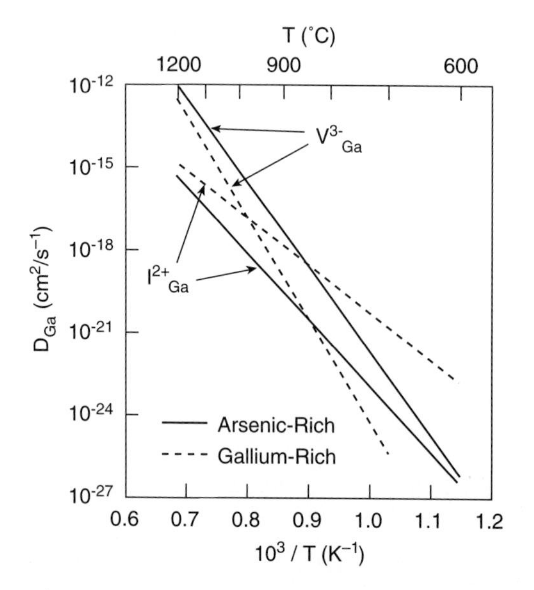

FIGURE 8.17
The contributions of V_{Ga}^{3-} and I_{Ga}^{2+} to self-diffusivity of Ga in As-rich and Ga-rich GaAs. (*After Tan et al. [1991a].*)

conditions, the best-estimated gallium self-diffusivity values contributed by V_{Ga}^{3-} and I_{Ga}^{2+}, labeled as D_{Ga}^{V} and D_{Ga}^{I}, are (Tan et al. 1991b)

$$\left.\begin{aligned}
D_{Ga}^{V}(n_i, 1\,\text{atm}) &= 2.9 \times 10^{8} \exp\left(-\frac{6\,\text{eV}}{k_B T}\right) \\
D_{Ga}^{V}(n_i, \text{Ga} - \text{rich}) &= 3.93 \times 10^{12} \exp\left(-\frac{734\,\text{eV}}{k_B T}\right) \\
D_{Ga}^{I}(n_i, 1\,\text{atm}) &= 6.05 \exp\left(-\frac{4.71\,\text{eV}}{k_B T}\right) \\
D_{Ga}^{I}(n_i \text{Ga} - \text{rich}) &= 4.46 \times 10^{-4} \exp\left(\frac{-3.37\,\text{eV}}{k_B T}\right)
\end{aligned}\right]. \tag{8.75}$$

We plot in Figure 8.17 the values of D_{Ga} as a function of the reciprocal temperature for the four cases included in Equation (8.75). For example, the self-diffusivity value due to V_{Ga}^{3-} in the arsenic-rich GaAs is higher than that in the gallium-rich material, an expected result because fewer gallium vacancies occur in the latter. This difference in the diffusivity values is particularly significant at lower temperatures. We can similarly rationalize the values of self-diffusivities due to the migration of I_{Ga}^{2+}.

8.7 DOPANT DIFFUSION IN SEMICONDUCTORS

We plot in Figure 8.18 the variations in the diffusion coefficients of several impurities in silicon as a function of reciprocal temperature (Gösele and Tan 1991). We can make several important observations. The activation energies associated with the

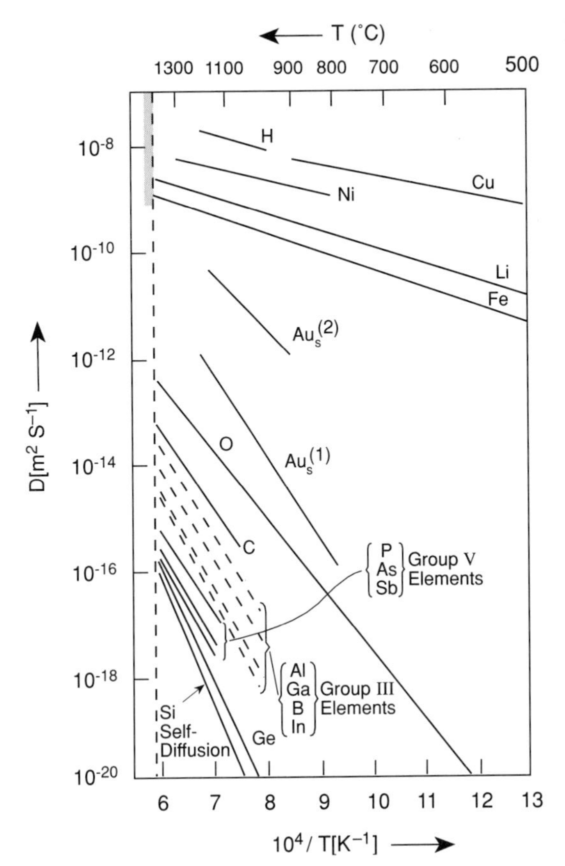

FIGURE 8.18
Diffusivities of various elements in silicon as a function of inverse absolute temperature.

diffusion of groups III and V dopants are lower than they are for self-diffusion. The activation energies for the diffusion of H, Cu, Ni, Li, and Fe are even lower, whereas the activation energies associated with the diffusion of Au fall between the two extremes. In the case of dopants, we attribute the observed effect to dopant–point defect interactions. We illustrate this effect using the diffusion of As, P, and B in silicon as examples.

At low concentrations of As ($\leq 10^{19}$ cm^{-3}), the diffusion of arsenic in silicon conforms to the simple theory. However, at higher concentration the concentration versus depth profile takes on a rectangular form, where a relatively large fraction of the arsenic atoms are electrically inactive due to their clustering. We assume that the formation of vacancy-arsenic complexes occurs according to the following reaction:

$$V_{Si} + 2As_{Si}^{+} = (V_{Si} - 2As_{Si}) \tag{8.76}$$

This reaction has been postulated to reduce diffusion. These complexes form only at high concentrations of arsenic. The diffusion of arsenic in silicon is assumed to occur predominantly by the vacancy mechanism, and the ($V_{Si} - 2As_{Si}$) complex appears to affect diffusion above 1050°C. Even in an oxidizing ambient where silicon interstitials are injected (see sections 7.4 and 7.6), the vacancies continue to control diffusion.

We can explain the reduction of diffusion coefficient of As(D_{As}) at high concentrations of As (C_T) as follows. Assuming that the total arsenic concentration consists of the following species, we can write C_T as

$$C_T = C(As_{Si}) + C(V_{Si} - As_{Si}) + C(V_{Si} - As_{Si}^{-}) + 2C(V_{Si} - 2As_{Si}), \quad (8.77)$$

where C denotes the concentration of the point defect complex indicated in the parenthesis. We can write the equilibrium concentration of each member of the defect pair in terms of the concentration of electrons n and the equilibrium constant for the appropriate point defect complex formation reaction as

$$C_T = n + K_a(T)n^2 + K_b(T)n^3 + K_c(T)n^4, \quad (8.78)$$

where n is equal to $C(As_{Si})$. The flux J is given by

$$J = -D_i \frac{C(As_{Si})}{n_i} \frac{\partial C(As_{Si})}{\partial x} = -D_i \frac{C(As_{Si})}{n_i} \frac{\partial C(As_{Si})}{\partial C_T} \frac{\partial C_T}{\partial x}. \quad (8.79)$$

The effective diffusion coefficient of As defined by $-J\left(\frac{\partial C_T}{\partial x}\right)^{-1}$ is given by

$$D_{As} = \frac{D_i(n/n_i)}{\dfrac{1 + 4K_c(T)n^3}{1 + 3K_b(T)n^2}} \quad (8.80)$$

where D_i denotes the diffusion coefficient in the intrinsic material. In Figure 8.19 we plot D_{As} as a function of C_T at a fixed temperature. D_{As} is clearly reduced at higher concentrations where clustering may occur.

The diffusion of P in silicon follows the expected complementary-error-function behavior when the surface concentration is less than the intrinsic carrier concentration at high temperature. We show a concentration versus depth profile, obtained at 1000°C, in Figure 8.20. There are three distinct regions in the observed profile. A plateau at the surface corresponds to a very high diffusion coefficient. Next is a high-gradient zone and a tail, which once again reflects high diffusivity. The first region is attributed to the difference between the total phosphorus concentration and the free carrier concentration at the surface of silicon shown in Figure 8.20. At high concentrations of phosphorus, the dominant diffusing species is thought to be ($P_{Si}^{+} - V_{Si}^{=}$) pairs, that is:

$$P_{Si}^{+} + V_{Si}^{=} = (P_{Si} - V_{Si}^{-}). \quad (8.81)$$

At lower levels of concentration of phosphorus, that is, deeper into Si, the ($P_{Si} - V_{Si}^{-}$) pairs start to dissociate according to the following reaction:

$$(P_{Si} - V_{Si}^{-}) = P_{Si}^{+} + V_{Si}^{-} + e^{-} \quad (8.82)$$

The resulting vacancies V_{Si}^{-} contribute to the diffusion. Fair and Tsai (1977) developed a theory based on these ideas that can effectively account for many of the features observed in Figure 8.20.

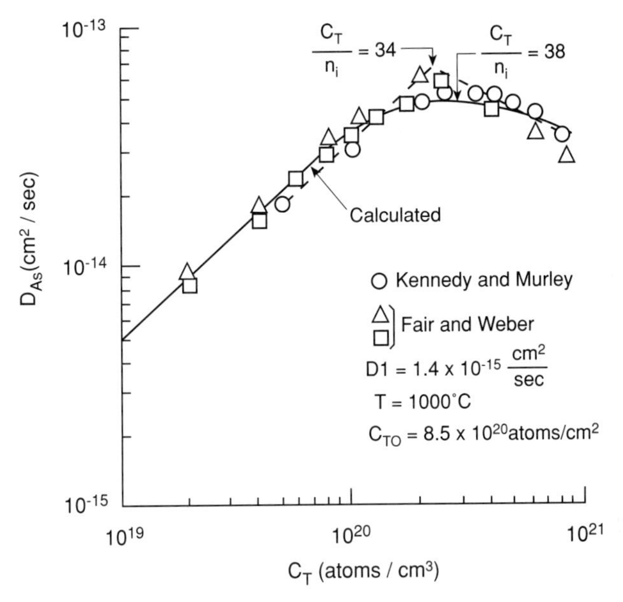

FIGURE 8.19
The effect of diffusivity of arsenic versus total concentration for diffusions into p-type silicon at 1000°C.

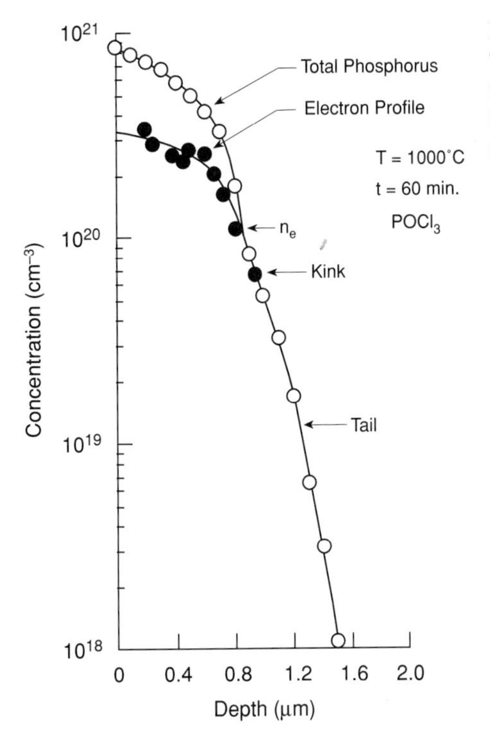

FIGURE 8.20
Total phosphorus and electron concentration profiles obtained by secondary ion-mass spectrometry and differential conductivity measurements.

Boron shows a rectangular-shape concentration profile after predeposition and is thought to diffuse by a vacancy mechanism under nonoxidizing conditions. Boron diffuses exclusively with V_{Si}^{+} vacancies. If $p > n_i$, the diffusivity of boron is enhanced, whereas it is reduced if $p < n_i$. The diffusivity of boron can be reduced by as much as a factor of 10 in highly doped n-silicon. We also know that boron diffusion can be enhanced in the presence of self-interstitial supersaturation, implying that interstitials play a role in B diffusion.

In III-V semiconductors the group II elements Be, Cd, Mg, Hg, and Zn act as p-type dopants. They substitute on the group III sublattice and show high diffusivities. Group VI elements such as Se, S, and Te are n-type dopants and reside on the group V sublattice. Group IV elements such as Si, C, Ge, and Sn can occupy either sublattice and are termed *amphoteric* behaving as either n- or p-type dopants. The choice of a particular element to be used as a dopant depends on several factors, such as level of diffusivity, atomic mass to ensure low implantation damage, the level of electrical activation, and solubility limit.

Si behaves as an amphoteric dopant in GaAs, so there is a limit to how many free carriers silicon can provide (2×10^{18} cm^{-3} electrons). Higher concentrations are not achievable by additional doping because of self-compensation. Silicon probably diffuses by a vacancy mechanism as determined from the dependence of diffusivity on arsenic vapor pressure. The concentration dependence of diffusion coefficient of silicon in GaAs appears to vary with the atomic mechanisms. Examples of dopant-diffusion species complexes that have been invoked to explain diffusion include $(Si_{Ga} - V_{Ga})$, $(Si_{As} - V_{As})$, $(V_{Ga}^{3-} - Si_{Ga})$, and $(Si_{Ga}^{+} - V_{Ga}^{3-})$. Be, which is a common p-type dopant used in GaAs, probably moves by the kick-out mechanism considered in section 8.1.1.

The diffusion of Zn in GaAs has been studied extensively because of its high solubility and fast diffusivity. Zn migrates via an interstitial-substitutional mechanism and behaves as a singly ionized acceptor in GaAs. Two mechanisms have been proposed to rationalize the diffusion of Zn in GaAs. Tuck and Kadhim (1972) postulate the following reaction:

$$Zn_i^{+} + V_{Ga} \longleftrightarrow Zn_s^{-} + 2h^{+}, \tag{8.83}$$

where an interstitial zinc ion (Zn_i^{+}) combines with a gallium vacancy (V_{Ga}) to form two holes and a substitutional zinc ion (Zn_s^{-}). An alternate suggestion is that zinc diffuses via a modified kick-out mechanism according to the following reaction:

$$Zn_i^{+} \longleftrightarrow I_{Ga} + Zn_s^{-} + 2h^{+}. \tag{8.84}$$

The diffusivity of zinc when the intrinsic point defects maintain thermal equilibrium is given by

$$D_{eff}^{(i)} = h\frac{D_i C_i^{eq}}{(C_s^{eq})^3} C_s^2, \tag{8.85}$$

where h is the electric-field enhancement factor. The diffusivity depends on arsenic vapor pressure according to

$$D_{eff}^{(i)} \propto P_{AS}^{1/4}, \tag{8.86}$$

as observed experimentally.

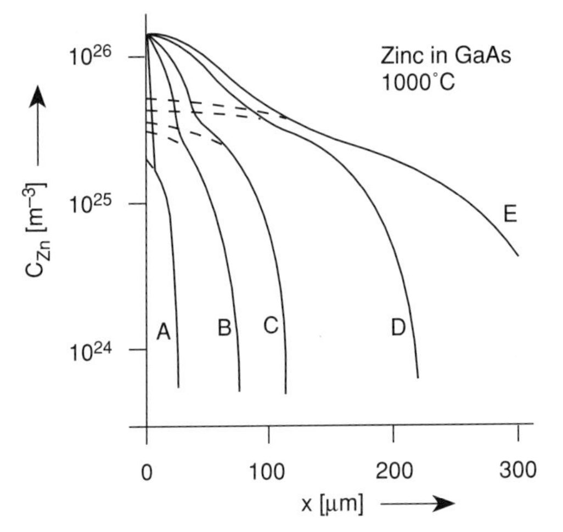

FIGURE 8.21
Concentration profiles of zinc diffusion in GaAs.

The concentration profiles of Zn in GaAs (see Figure 8.21) reveal a kink-and-tail structure. The enhanced tail is associated with an undersaturation of V_{Ga} or a supersaturation of I_{Ga}. This figure is consistent with the observation of interstitial-type dislocation loops in Zn-diffused GaAs at high concentrations.

8.8 DIFFUSION OF ELECTRICALLY ACTIVE CONTAMINANTS

Gold and platinum have a drastic influence on the minority-carrier lifetime in Si, since their energy levels are in the middle of the energy gap. These impurities are intentionally introduced into higher frequency Si devices to reduce the minority-carrier lifetime to enhance their frequency response. The impurities are considered contaminants in most of the devices.

Au and Pt dissolve substitutionally but diffuse as interstitials. We can accomplish this diffusion either by the kick-out mechanism or by the Frank-Turnbull mechanism discussed in section 8.1.1. It is instructive to compute the effective diffusion coefficient for each mechanism and then compare the answers to the experimental results.

Let us consider the case where atoms A diffuse by the Frank-Turnbull mechanism; the actual movement of A occurs by the interstitial A_I with a diffusivity D_{A_I}. The A atom is incorporated into the substitutional site A_S by consuming a vacancy V_S. It is convenient to denote the semiconductor atom site by the subscript S to retain generality of the analysis. Assuming local equilibrium ($A_I + V_S = A_S$), we can write

$$\frac{C_{A_I} C_v}{C_{A_S}} = \frac{C_{A_I}^{eq} C_V^{eq}}{C_{A_S}^{eq}}, \tag{8.87}$$

where C_{A_I}, C_v, and C_{A_S} are the actual concentrations of A_i, V_s, and A_s, respectively, and other terms can be interpreted in a similar manner. There are two processes of interest: one of supplying A_I and the other of establishing the equilibrium concentration of vacancies. The slower of the two processes determines the effective diffusivity. If C_v is close to C_v^{eq}, that is, the supply of A_I is the slower process, or $C_{A_I}^{eq} \ll C_{A_S}^{eq}$, the effective diffusivity of A_S is given by

$$D_{eff}^{A_S} = \frac{D_{A_S} C_{A_I}}{C_{A_I} + C_{A_S}} = \frac{D_{A_S} C_{A_I}^{eq}}{C_{A_S}^{eq}}, \tag{8.88}$$

where D_{A_S} is the diffusion coefficient of A_S. When the supply of vacancies limits the incorporation of A_S, that is, when the material is dislocation free, then the diffusion of A_S with an effective diffusion coefficient $D_{eff}^{V_S}$ must be governed by a flux balance that is given by

$$D_{eff}^{V_S} \frac{\partial C_{A_S}}{\partial x} = D_{V_S} \frac{\partial C_{V_S}}{\partial x}, \tag{8.89}$$

where D_{V_S} is the diffusivity of a vacancy. Hence

$$D_{eff}^{V_S} = \frac{D_{V_S} C_V^{eq}}{C_{A_S}^{eq}}, \tag{8.90}$$

where $C_{A_I} = C_{A_I}^{eq}$.

In order to determine which of the two diffusion coefficients is applicable, we introduce R_V where

$$R_V = \frac{D_{V_S} C_V^{eq}}{D_{A_S} C_{A_I}^{eq}}. \tag{8.91}$$

For $R_V \gg 1$, $D_{eff}^{A_S}$ is used, and for $R_V \ll 1$, $D_{eff}^{V_S}$ is used.

We can carry out a similar analysis for the kick-out mechanism. Here the incorporation of an A atom as A_S is associated with the generation of a self-interstitial. As before, we write the following expression that relates the concentrations of interstitials in various locations ($A_I = A_S + S_I$):

$$\frac{C_{A_I}}{C_{S_I} C_{A_S}} = \frac{C_{A_I}^{eq}}{C_{S_I}^{eq} C_{A_S}^{eq}}. \tag{8.92}$$

Introducing

$$R_I = \frac{D_{S_I} C_{S_I}^{eq}}{D_{A_S} C_{A_I}^{eq}}, \tag{8.93}$$

we observe for $R_I \gg 1$, the in-diffusion of A_I is much slower than the out-diffusion of the generated self-interstitial to the surface so that $D_{eff}^{A_S}$ holds. If $R_I \ll 1$, the incorporation of A_S is limited by the out-diffusion of the self-interstitial to the surface. We balance the flux of A_S with that of S_I, and because they are in opposite directions, we obtain

$$D_{eff}^{S_I} \frac{\partial C_{A_S}}{\partial x} = -D_{S_I} \frac{\partial C_{S_I}}{\partial x} \tag{8.94}$$

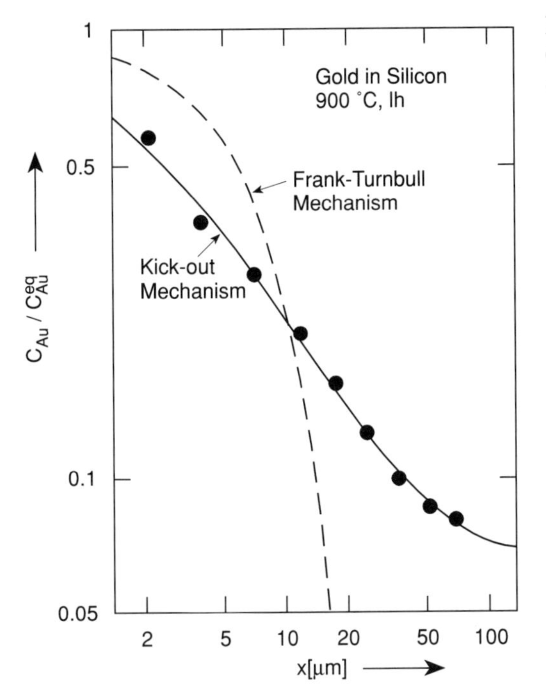

FIGURE 8.22
Gold concentration profile in dislocation-free silicon.

so that

$$D_{\text{eff}}^{S_I} = \frac{D_{S_I} C_{S_I}^{eq}}{C_{A_S}^{eq}} \left(\frac{C_{A_S}^{eq}}{C_{A_S}} \right)^2. \tag{8.95}$$

Note strong dependence of $D_{\text{eff}}^{S_I}$ on the local A_S concentration C_{A_S}. The shapes of the concentration versus distance profiles for both mechanisms are shown in Figure 8.22. The experimental results shown by black dots in Figure 8.22 for both indicate that the kick-out mechanism dominates the diffusion of Au in Si.

8.9 DIFFUSION OF CARBON, OXYGEN, AND HYDROGEN IN SILICON

We indicated in section 5.2.1 that carbon and oxygen are common impurities in as-grown Czochralski silicon crystals; their respective concentrations are 10^{16} and 10^{18} cm^{-3}. Hydrogen may be incorporated into silicon during the epitaxial growth (see section 6.1.2) and the etching of device structures (see Chapter 11).

Carbon dissolves substitutionally in the silicon lattice. Since the size of the carbon atom is very small (see Table 3.1), we expect it to diffuse interstitially. The supersaturation of self-interstitials is known to enhance carbon diffusion. The diffusion process involves a self-interstitial–carbon complex

$$C_{Si} + Si_I \longleftrightarrow (C_{Si} - Si_I), \tag{8.96}$$

where C_{Si} denotes the concentration of carbon on substitutional sites. The formation of the complex is favored by the strain-energy considerations. If we assume that the contribution of vacancies to the diffusivity of carbon is negligible, we can write

$$D_{eff} = \frac{D_{Cx} C_{Cx}}{C_S}, \tag{8.97}$$

where C_{Cx} is the solubility of the $(C_{Si} - Si_I)$ complexes and C_s is the solubility of substitutional carbon. We can see from Equation 8.97 that the higher the value of C_{Cx}, the higher the value of D_{eff}.

We discussed in detail the behavior of oxygen in silicon in section 5.6. We indicated that the oxygen atom occupies an interstitial site in the silicon lattice and diffuses by hopping between different interstitial sites. The formation of "thermal donors" in Czochralski silicon on annealing around 450°C that was covered in section 5.6.2.1 also requires oxygen atoms to have high diffusivity. Tan and Gösele (1982) attribute the high diffusivity to the formation of oxygen molecules according to

$$O_I + O_I \longleftrightarrow O_2, \tag{8.98}$$

where O_I is the interstitial oxygen. The O_2 molecules are assumed to diffuse extremely fast, since they are not bonded to silicon atoms. Therefore, we can write the effective diffusion coefficient of oxygen

$$D_{eff} = \frac{D_{OI} C_{OI} + 2D_2 C_2}{(C_{OI} + 2C_2)}, \tag{8.99}$$

where D_{OI} is the diffusivity of O_I and C_{OI} its concentration; D_2 and C_2 are the corresponding quantities for O_2 molecules.

Hydrogen, being the smallest element, is easily inserted in most materials. The number of structural and electronic defects introduced by hydrogen depends on the method used to incorporate hydrogen. It can appear as molecular hydrogen, as atomic hydrogen, or as a proton in semiconductors. H_2, H^0, and H^+ are present in p-type Si, whereas H_2 and H^0 are present in n-type Si. The relative amounts of H_2, H^0, and H^+ depend on temperature and doping levels.

8.10 SEQUENTIAL DIFFUSIONS

The fabrication of semiconductor devices is generally accomplished by sequential diffusions. For the sake of discussion, let us consider an n^+pn bipolar transistor. We can produce the collector, base, and emitter regions of the transistor by three sequential diffusions of P, B, and P. In addition to the difficulty in controlling the device dimensions due to the lateral diffusion discussed in section 8.2, we find that the diffusion of the base under the emitter is enhanced relative to the base diffusion outside the emitter. We refer to this phenomenon as the *emitter-push effect*, and we attribute it to the built-in electric field and the variations in point defect concentrations that occur during the sequential diffusions.

Let us first consider the built-in field effect. During the diffusion of the ionized donor impurities, electrons move along with the impurities. Since the electrons are much more mobile, they tend to move ahead of the donors during diffusion. This movement cannot happen indefinitely, because of the charge-neutrality requirements. In other words, a space-charge region forms that effectively retards the movement of the electrons and accelerates the migration of the donors. For the built-in field to have a significant effect on the kinetics of diffusion, the dopant concentration must exceed the intrinsic carrier concentration at the diffusion temperature.

The internal electric field ε_x can be expressed in terms of the electric potential $\phi(x)$ as

$$\varepsilon_x = \frac{\partial}{\partial x}\phi(x,t). \tag{8.100}$$

Considering a donor impurity, we can write

$$\phi(x,t) = \frac{E_{CB} - E_f}{q}, \tag{8.101}$$

where E_{CB} is the conduction-band edge, E_f is the Fermi level, and q is the charge of the electron. We note from Equation (1.87) that

$$np = n_i^2.$$

If all the donors are ionized, $N_D \cong n$. Hu and Schmidt (1957) have shown that

$$\varepsilon_x = \frac{k_B T}{q}\frac{\partial}{\partial x}\ln\left(\frac{N_D}{n_i}\right). \tag{8.102}$$

The diffusion flux is related to the two driving forces for diffusion

$$J_x = -D\frac{\partial N_D}{\partial x} - ZD\frac{q}{k_B T}N_D\varepsilon_x, \tag{8.103}$$

where Z is the charge state of the donor atoms. We can write Equation (8.103) as

$$J_x = -Dh\frac{\partial N_D}{\partial x} \tag{8.104}$$

where

$$h \equiv 1 + ZN_D\frac{\partial}{\partial N_D}\ln\left(\frac{N_D}{n_i}\right). \tag{8.105}$$

The quantity h is known as the *electric-field enhancement factor* and can be written as

$$h = 1 + \frac{N_D}{2n_i}\left[\left(\frac{N_D}{2n_i}\right)^2 + 1\right]^{-1/2}. \tag{8.106}$$

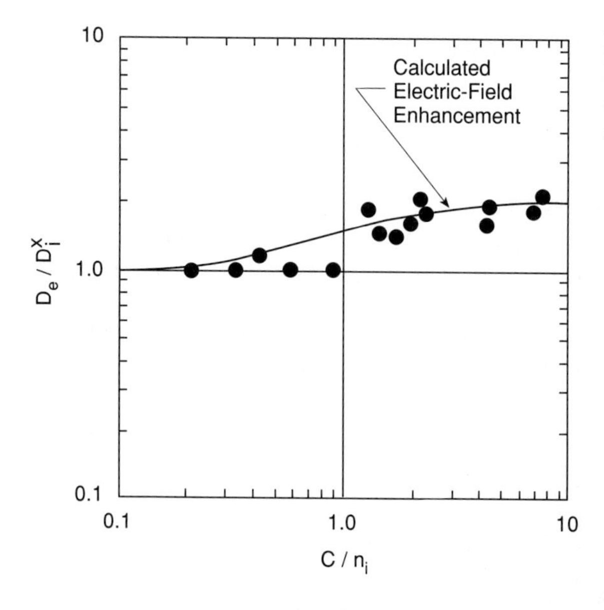

FIGURE 8.23
Electric-field enhanced diffusion of phosphorus in silicon at 900°C. (*After Fair [1979].*)

The maximum increase in diffusivity occurs when $N_D/2n_i \gg 1$, which gives $h = 2$. The electric-field enhancement effect observed during the diffusion of phosphorus in silicon appears in Figure 8.23. The effect becomes significant when the carrier concentration due to doping is considerably in excess of the intrinsic value at the diffusion temperature.

Fair (1979) has proposed that the enhanced diffusion of B in the base region during the fabrication of the emitter results from the dissociation of $P^+ - V^=$ complexes. This dissociation then increases the concentration of vacancies in the boron-doped base region, thereby increasing the diffusion rate in the region. Other experiments (Amigliato et al. 1977) have shown a net interstitial supersaturation below the phosphorous diffused layer, indicating that perhaps the interstitials play a role in the emitter-push effect.

8.11 OXIDATION-INDUCED DIFFUSION ENHANCEMENT OR RETARDATION IN SILICON

We discussed in sections 7.4 and 7.6 that silicon interstitials are generated during the oxidation of silicon. Since the covalent radii of the diffusant and the host lattice atoms differ, diffusion can introduce strain in the lattice (see section 3.3.1); we discuss this issue further in section 8.12. Depending on the nature of the point defect and the strain field, the generation of point defects may affect the diffusion behavior of impurities. We also expect doping to affect the oxidation behavior of silicon (see section 7.5).

The oxidation of silicon enhances the diffusion of B, In, Al, Ga, P, and As, whereas the diffusion of Sb is retarded. These effects are attributed to the generation of silicon self-interstitials during the oxidation process. As indicated in sections 7.4 and 7.6, the formation of self-interstitials results from the need to accommodate more than 100 percent volume expansion that occurs on the conversion of silicon to its oxide. The enhancement of diffusion of B occurs because it requires interstitials for its diffusion, whereas situations regarding P and As are more complex. On the other hand, since Sb diffuses by a vacancy mechanism, the formation of silicon interstitials during oxidation reduces the vacancy concentration, thus retarding its diffusion. We can study these phenomena using silicon nitride masks because they prevent the oxidation of silicon.

The diffusivity D^s_{per} of an impurity atom under conditions of nonequilibrium point defect concentration may be written as

$$D^s_{per} = D^s_I(n)\frac{C_I}{C_I^{eq}(n)} + D^s_v(n)\frac{C_v}{C_v^{eq}(n)}, \tag{8.107}$$

where C_I and C_v denote the perturbed point defect concentrations.

By defining a normalized diffusivity enhancement

$$\Delta^s_{per}(n) = \frac{D^s_{per}(n) - D^s(n)}{D^s(n)}, \tag{8.108}$$

where

$$D^s(n) = D^s_I(n) + D^s_v(n). \tag{8.109}$$

I and V emphasize all contributions associated with interstitial and vacancy defects, respectively. We can write a fractional interstitialcy component of diffusion as

$$\Phi_I(n) = \frac{D^s_I(n)}{D^s(n)}. \tag{8.110}$$

We can define the self-interstitial supersaturation ratio as

$$S_I(n) \equiv \frac{[C_I - C_I^{eq}(n)]}{C_I^{eq}(n)}. \tag{8.111}$$

Assuming the following mass action law is valid

$$C_I C_v = C_I^{eq} C_v^{eq}, \tag{8.112}$$

we can write $\Delta^s_{per}(n)$ in the form

$$\Delta^s_{per}(n) = [2\Phi_I(n) + \Phi_I(n)S_I - 1]\left[\frac{S_I}{1 + S_I}\right]. \tag{8.113}$$

The $\Delta^s_{per}(n)$ is given in Figure 8.24 for various values of S_I (Tan and Gösele 1985). $S_I < 0$, obtainable by thermal nitridation of silicon, corresponds to vacancy supersaturation. Oxidation at very high temperatures and in HCl atmospheres also generates a high degree of vacancy supersaturation. Under these conditions Sb diffusion is enhanced, whereas B and P diffusions are retarded. Oxidation of silicon produces conditions for which $S_I > 0$ so that P and B diffusions are enhanced and Sb diffusion is retarded. The value of Φ_I depends on the dopant and appears to be related to the ratio of the covalent radii of the diffusant and silicon atoms: The larger the ratio, the smaller the value of Φ_I.

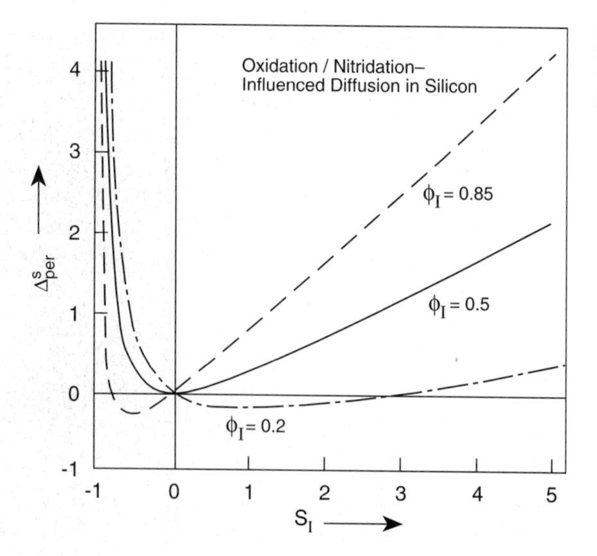

FIGURE 8.24
Diffusion enhancement versus self-interstitial supersaturation.

8.12 DIFFUSION IN POLYCRYSTALLINE SOLIDS

We discussed in section 3.3 that polycrystalline solids consist of grains that are separated from each other by grain boundaries. The *grain boundary,* a region where two crystals of differing orientation come together in some type of local equilibrium configuration, consists of intrinsic and extrinsic dislocations. As a result, the diffusion of atoms along the boundary is very rapid and is one of the dominant mechanisms in device failures involving polycrystalline materials.

When we diffuse atoms into a polycrystalline film, the atoms move simultaneously within the grains and along the grain boundaries. The interaction between the species migrating along the two diffusing paths determines the kinetics of atomic diffusion and hence the concentration versus depth profiles. It is customary to assign a width δ to the grain-boundary region and specify a diffusion coefficient D_b for diffusion along the grain boundary. The diffusion of atoms within the grains is given by the diffusion coefficient D_l. The schematic in Figure 8.25 shows the three kinetic regimes where grain boundaries are indicated by parallel vertical lines. In spite of the rapid diffusion of dopant atoms along the grain boundaries at high temperatures, the leakage of atoms into the grains, coupled with their bulk diffusion, produces an impurity profile as shown in Figure 8.25*a*. We refer to this diffusion kinetics as type A kinetics. At intermediate temperatures the atoms diffuse laterally into regions contiguous to the boundaries as depicted in Figure 8.25*b*, but the contribution of bulk diffusion to the dopant profile is relatively small. We refer to this situation as type B kinetics. In the low-temperature regime, the grain boundary diffusion dominates. The dopant atoms at the surface reach a distance of $(D_b t)^{1/2}$ along the grain boundary as shown in Figure 8.25*c*. This diffusion kinetics is called type C kinetics. We can obtain an estimate of the ratio R of the mass

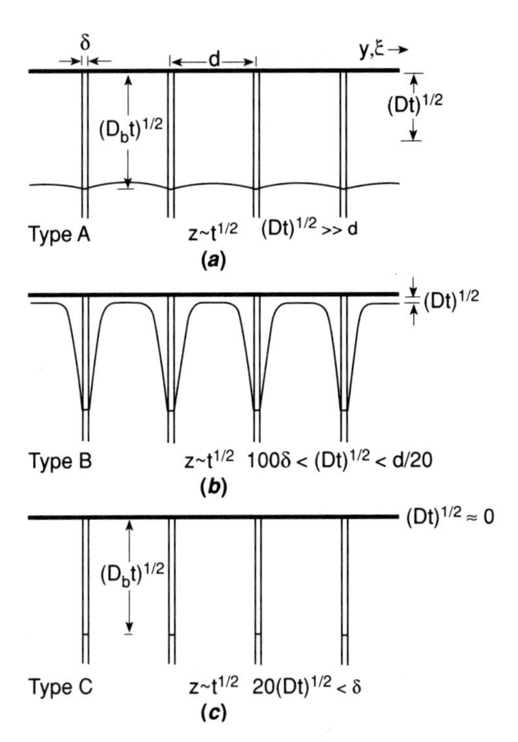

FIGURE 8.25
Three types of kinetics involving grain boundary diffusion at: (*a*) high (*b*) intermediate, and (*c*) low temperature regimes.

transport through the grain boundary and the grain from the relation

$$R = \frac{J_l A_l}{J_b A_b} = \frac{D_l \pi r^2}{D_b 2\pi r \delta}, \tag{8.114}$$

where r is the radius of the grain, δ is the width of the grain boundary, and A_l and A_b are the cross-sectional areas of the grain and the grain boundary. The thickness of the film usually indicates the grain size, and temperature influences the diffusion coefficient via an Arrhenius equation.

The determination of the kinetic regime requires the matching of concentration profiles to phenomenological solutions of the diffusion equations. Kaur and Gust (1988) provide extensive discussions of mathematical analysis and experimental determination of diffusion coefficients.

8.13 DIFFUSION-INDUCED DISLOCATION NETWORKS

Two different situations can develop when dopant atoms are diffused into substrates: (1) The concentration of the dopant is below its solubility limit in the substrate material, and (2) its concentration exceeds the solubility limit. Let us first consider the case where the concentration is below the solubility limit. As the dopant atoms are dissolved in the host lattice, the lattice parameter of the diffused region will either increase or decrease with respect to the substrate. The sign of the change depends on the difference between the tetrahedral radii of the host and the dopant atoms. This nature of the change is inferred because the replacement of host atoms by dopant atoms of a different size will change bond lengths in the tetrahedral arrangement. For example, consider diffusion doping of silicon by boron. Since the tetrahedral radius of B atoms is considerably smaller than the tetrahedral radius of silicon, the lattice parameter of the diffused region must be smaller than that of pure silicon. Prussin (1961) has given the following expression that relates the strain ε introduced in the material due to the incorporation of dopant to the concentration C:

$$\varepsilon = -\beta C, \tag{8.115}$$

where β is the solute lattice contraction coefficient. As a result of the strain, stresses are introduced in the material. The maximum stress that develops at the beginning of the diffusion process is given by the following expression:

$$\sigma_{max} = \beta C_S E/(1 - \upsilon), \tag{8.116}$$

where C_S is the surface concentration of the solute, E is Young's modulus, and υ is the Poisson's ratio of the substrate material.

A number of investigators (Queisser 1961; Schwuttke and Queisser 1962; Washburn et al. 1964; Levine et al. 1967) have shown that dislocations are introduced in the diffused region when the concentration of the dopant exceeds a certain value. X-ray topographs, reproduced from the work of Schwuttke and Queisser (1962) are shown as Figure 8.26. It is evident that the P-diffused (001) slice contains two sets of dislocations whose Burgers vectors lie in the wafer plane. Also the dislocations appear to lie in the same plane, implying that they are Lomer edge dislocations. Washburn et al. (1964) subsequently confirmed these results by using transmission electron microscopy to evaluate diffusion-induced dislocations. In addition, they observed helical dislocations, implying the occurrence of climb.

Levine et al. (1967) extended the above work of Washburn et al. (1964) and examined in detail dislocation structures in B- and P-diffused (110) and (111) wafers. Levine et al. did not observe crystallographically well-aligned dislocation structures. Figure 8.27 shows an example of the dislocation structure seen in B-diffused (111) silicon. Most of the dislocations have Burgers vectors that lie in the (111) plane and are indicated by short lines on the micrograph. In addition, dislocations whose Burgers vector is inclined to the substrate surface are visible, and they are delineated by arrows.

In the (110) B-diffused wafers, dislocation networks are quite complex. Figure 8.28 shows an example of the dislocation arrangement seen in such wafers (Levine et al. 1967). The network consists of long a/2 $[1\bar{1}0]$ edge dislocations aligned along the [001] direction. In addition, dislocations whose Burgers vectors are inclined to the (110) plane are visible, and they are delineated by short lines that represent the projections of their Burgers vectors onto the (110) plane.

Two mechanisms have been proposed to rationalize the observed dislocation structures. One model hypothesizes that diffusion-induced dislocations can be likened to misfit dislocations. These dislocations form to accommodate the lattice-parameters difference between the diffused and nondiffused regions. Mechanisti-

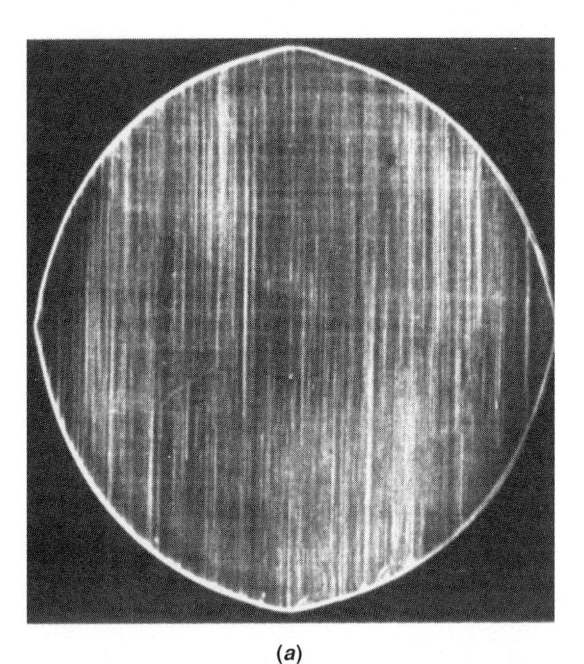

(*a*)

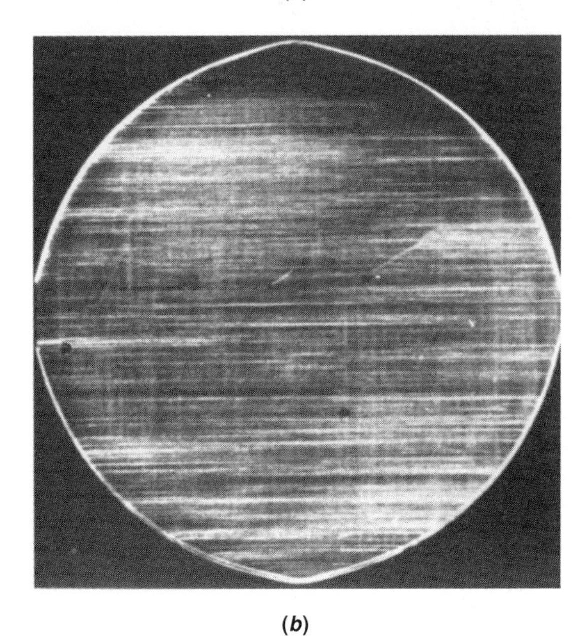

(*b*)

FIGURE 8.26
X-ray diffraction topographs obtained from a P-diffused Si sample where the concentration of P at the surface is $\sim 1 \times 10^{21}\text{cm}^3$. The operating reflections in (*a*) and (*b*) are $2\bar{2}0$ and 220, respectively. (*After Schwuttke and Queisser [1962].*)

cally, the lattice-parameter difference produces a misfit stress. Under the influence of this stress, dislocations undergo glide and interact with each other to form a network. Based on this model, the network should be located in the region where the diffusion-induced composition gradient is the largest. This assessment is not consistent with the experimental observations (Ravi 1981).

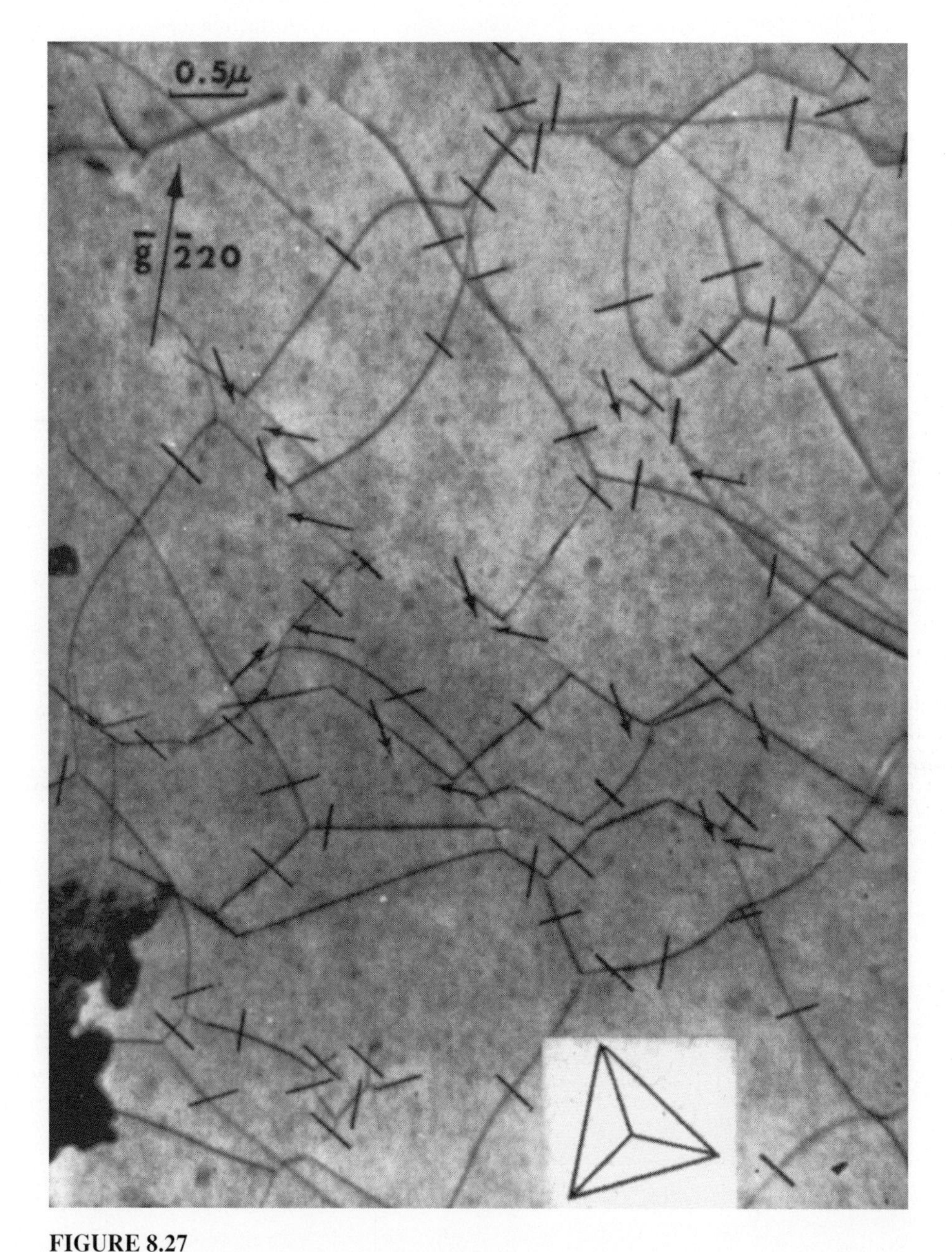

FIGURE 8.27
Dislocation structure observed in a B-diffused (111) Si sample; short lines delineate dislocations whose Burgers vectors lie in the (111) plane, whereas arrows mark dislocations with inclined Burgers vectors. (*After Levine et al. [1967].*)

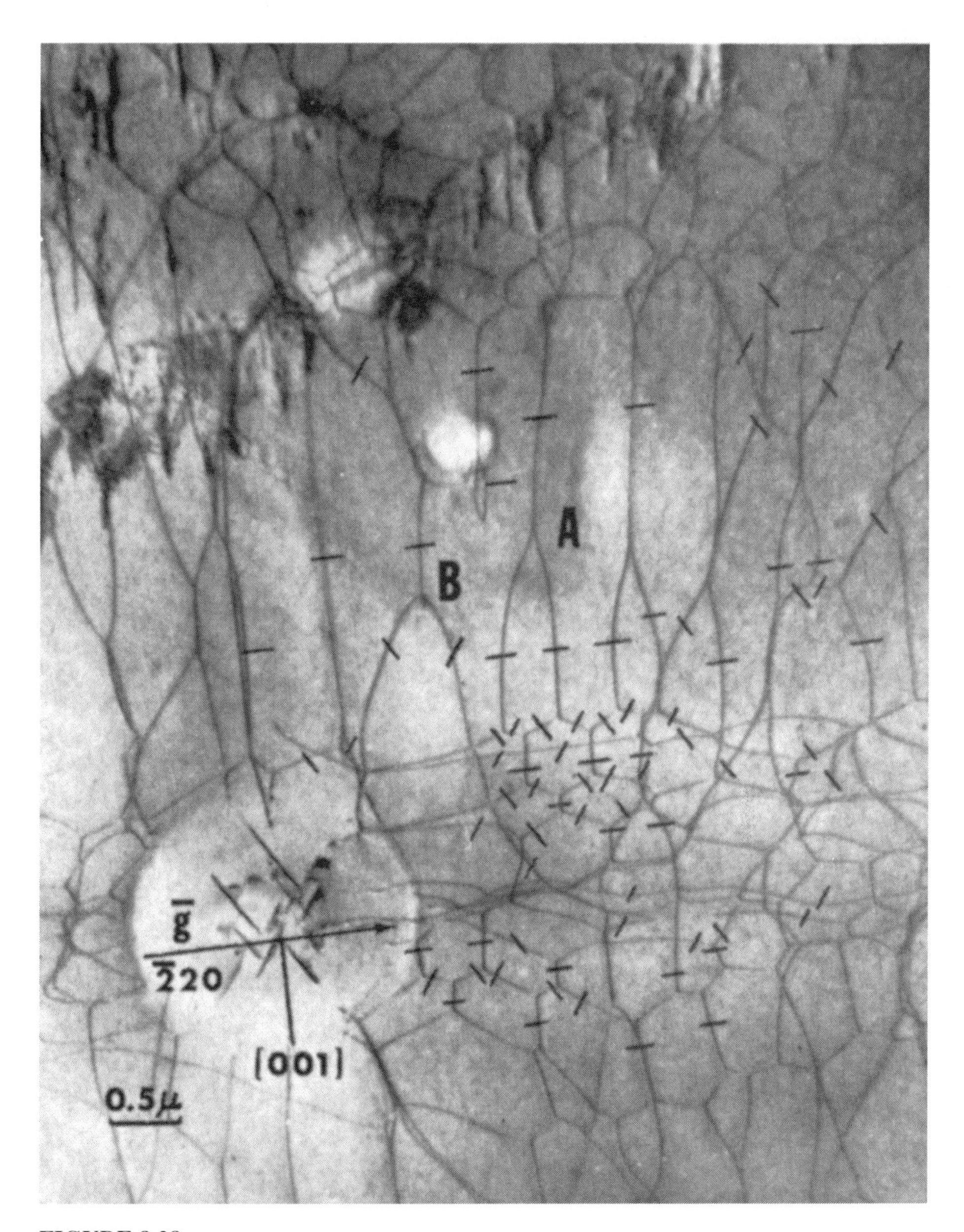

FIGURE 8.28
Dislocation networks observed in a B-diffused (110) Si wafer. Short lines delineate dislocations whose Burgers vectors are inclined to the (110) plane. (*After Levine et al. [1967].*)

Another suggestion is that under the influence of diffusion-induced stress dislocation loops nucleate from steps at the surface. This suggestion is reasonable because the stresses are the highest due to very high dopant concentration. Subsequently, these loops undergo glide on their respective slip planes. These loops will continue to glide until the diffusion-induced stresses fall below the critical resolved shear stress of the material at the diffusion temperature and interact with each other to form a network. In this case the network will be located somewhere in the diffused region and not at the interface between the diffused and nondiffused regions. This is indeed borne out by experimental results (Ravi 1981).

When the concentration of the diffusant in the solid exceeds its solid solubility limit at the diffusion temperature, the excess diffusant atoms could either cluster to form elemental precipitates or combine with atoms of the host lattice to form intermetallics. As the concentration of the diffusant is generally highest at the wafer surface, precipitation effects are confined to the surface and the near-surface regions. In the case of boron in silicon, precipitation requires a high degree of supersaturation. Most of the precipitates observed in boron-diffused silicon samples are in the form of platelets whose composition is SiB_x, where x varies from 2.89 to 4 (1977).

The precipitation of P in the form of silicon phosphide has been investigated extensively. Figure 8.29 shows a dark-field transmission electron micrograph of the silicon surface region obtained using a silicon phosphide spot. The needle-shape

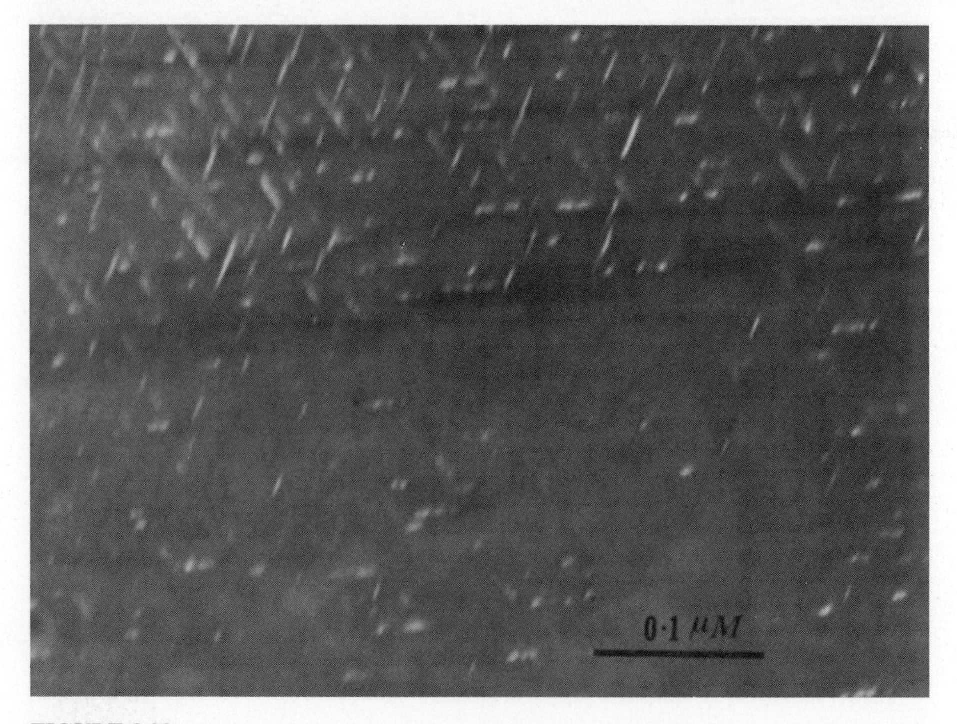

FIGURE 8.29
Dark-field transmission electron micrograph showing SiP precipitates in a P-diffused silicon wafer. (*After Armigliato et al. [1977].*)

precipitates align along the <110> directions. The structure of these precipitates is base-centered orthorhombic, and the precipitates compress the adjoining silicon matrix.

8.14 HIGHLIGHTS OF THE CHAPTER

The highlights of this chapter are as follows:

- The atomistic and phenomenological descriptions of diffusion are developed. Predepositions and drive-in diffusions involved in selective doping of semiconductors have been analyzed using the phenomenological approach.
- Self-diffusion in silicon may involve vacancies having many charged states and interstitials. The balance among the charged point defects may be shifted by doping, resulting in changes in diffusivity. This condition is referred to as the *Fermi-level effect.* The presence of charged vacancies also complicates the diffusion of dopants in silicon because many point defect–impurity interactions are possible. The diffusion of Ga in GaAs also involves charged point defects, that is, Ga-vacancies and interstitials.
- At high doping concentrations, diffusion-induced dislocation networks may form. The diffusion-induced stresses in the near-surface regions facilitate the formation of these networks.

REFERENCES

Armigliato, A.; D. Nobili; P. Ostoja; M. Servidori; and S. Solmi. In *Semiconductor Silicon,* ed. H. R. Huff and E. Sirtl. Pennington, NJ: Electrochem. Soc., 1977, p. 638.

Fair, R. B. *J. App. Phys,* 50, 860 (1979).

Fair, R. B., In *Impurity Doping Processes in Silicon,* ed. F. F. Y. Wang. New York: North-Holland, 1981.

Fair, R. B. and J. C. C. Tsai. *J. Electrochem. Soc.* 124, 1107 (1977).

Fick, A. *Ann. Phys.* 170, 59 (1855).

Ghandhi, S. K. *VLSI Fabrication Principles.* New York: John Wiley & Sons, 1983.

Gösele, U. *Ann. Rev. Mat. Sci,* 18, 257 (1988).

Gösele, U. and T. Y., Tan. In *Encyclopedia of Materials Science and Technology,* ed. W. Schröter. Weinheim: VCH, 1991, p. 197.

Grove, A. S.; O. Leistko; and C. T. Shah. *J. Appl. Phys.* 35, 2695 (1964).

Hu, S. M. and S. Schmidt. *Phys. Rev.* 107, 392 (1957).

Ikucher, T. *Soviet Phys. Solid State* 3, 401 (1961).

Kaur, I. and W. Gust, *Fundamentals of Gain and Interphase Boundary Diffusion.* Stuttgart: Ziegler Press, 1988.

Kennedy, D. P. and R. R. O'Brien. *IBM J. Res. & Dev.* 9, 179 (1965).

Kroger, F. A. *Chemistry of Imperfect Crystals.* New York: Interscience, 1964.

Levine, E.; J. Washburn; and G. Thomas. *J. Appl. Phys.* 38, 81 (1967).

Prussin, S. *J. Appl. Phys.* 32, 1876 (1961).

Queisser, H. J., *J. Appl. Phys.* 32, 1776 (1961).
Ravi, K. V. *Imperfections and Impurities in Semiconductor Silicon.* New York: John Wiley & Sons, 1981.
Schwuttke, G. H. and H. J. Queisser. *J. Appl. Phys.* 33, 1540 (1962).
Servidori, M. and A. Armigliato. *J. Mats. Sci.* 10, 306 (1975).
Shaw, D. *Atomic Diffusion in Semiconductors.* London: Plenum Press, 1973.
Shaw, D. *Phys. Stat. Sol.* 72, 11 (1975).
Shewmon, P. G. *Diffusion in Solids.* New York: McGraw-Hill, 1963.
Tan, T. Y. and U. Gösele. *Appl. Phys. Lett.* 40, 616 (1982).
Tan, T. Y. and U. Gösele. *Appl. Phys.* A37, 1 (1985).
Tan, T. Y.; U. Gösele; and S. Yu. *Critical Rev. Sol. State and Mater. Sci.* 17, 47 (1991a).
Tan, T. Y.; S. Yu; and U. Gösele. *J. Appl. Phys.* 70, 4283 (1991b).
Tuck, M. A. and M. A. H. Kadhim. *J. Mater. Sci.* 7, 585 (1972).
Van Vechten, J. A. In: *Lattice Defects in Semiconductors.* Bristol, England: Institute of Physics, 1975.
Washburn, J.; G. Thomas; and H. J. Queisser. *J. Appl. Phys.* 35, 1909 (1964).

PROBLEMS

8.1. Compute the distance traveled by oxygen in silicon at 1400 K after four hours. Compare this distance with that traveled by carbon in silicon under identical conditions.

8.2. Calculate the depth at which silicon changes from n-type to p-type if phosphorus is diffused into a p-type silicon containing 10^{17} acceptors per $(cm)^3$. Assume that the diffusion of phosphorus is carried out from a saturated surface at 1000°C for one hour.

8.3. Estimate the difference in time required to diffuse the dopant of your choice into silicon for a fixed distance of 10 μm below the surface if the diffusions are carried out at 1000° and 1400°C.

8.4. A silicon wafer has uniform p-type concentration of 10^{15} cm^{-3}. Phosphorus atoms are diffused into the wafer for two hours at 1100°C with a constant surface concentration. Calculate this concentration given that $\sqrt{D}$ for phosphorus is 0.27 μm 1 $hr^{1/2}$ at 1100°C and the junction depth is 2.1 μm.

8.5. A one-hour diffusion yields a junction depth of 1.5 μm. What would be the junction depth after a three-hour diffusion, all other conditions remaining the same?

8.6. Although diffusion coefficient should be independent of direction for silicon, during oxidation the diffusion depths of phosphorus can be different in the [111] and [100] directions. Explain why.

8.7. A p-n junction is to be fabricated in a semiconductor that has been doped n-type during growth. A p-type dopant is subsequently diffused in. Under what conditions can you reverse the process, that is, diffuse an n-type impurity in a p-type semiconductor? Explain your reasoning.

8.8. The diffusion experiment is carried out at a finite temperature T for a time t. In order to utilize the analytical expressions for solution of diffusion equation, the semiconductor wafer must conform to the semi-infinite approximation. This approximation is valid if the concentration falls to within 1 percent of the value at $x = 0$. At the normal temperatures used for diffusion in silicon, calculate the minimum thickness for the wafer.

8.9. Evaporation from the crystal can occur if diffusion is carried out at high temperature and from the vapor phase. The diffusion is one of solute from the vapor phase into the solid, when the boundary between the two is moving with a velocity v. Is there a maximum depth of penetration if the solute has a high evaporation rate and a low diffusion coefficient? Point out the implications for diffusing a solute into GaP, where the equilibrium vapor pressure of phosphorus close to the melting point is 50 atmospheres.

8.10. In epitaxial growth carried out at high temperatures, diffusivities of various impurities can be quite large. Impurities can diffuse out of the depositing layer into the substrate, and vice versa. What is the consequence of epitaxial growth on diffusion profiles?

CHAPTER 9

Ion Implantation

This chapter covers ion implantation, a technique used for selective doping of semiconductors. It emphasizes the underlying principles of ion implantation, ion channeling, ion implantation-induced damage, annealing behavior of the damage and process considerations.

9.1 ION RANGES AND IMPLANTATION PROFILES

Ion implantation refers to a process by which dopant ions having high kinetic energies are introduced into a semiconductor to change its carrier concentration and conductivity type. The kinetic energies range between 50 and 500 keV for most device applications. In the semiconductor technology ion implantation is primarily used to selectively dope surface regions of a wafer. Ion implantation is superior to chemical diffusion for this purpose because the lateral-diffusion effects discussed in section 8.3 are minimal. As a result, implantation has essentially replaced diffusion for selective doping in device processing.

When ions pass through a solid, they collide with nuclei and electrons, thereby losing energy. Eventually, they come to rest within the solid after some distance R, the range. The ion paths are not straight because of collisions, and thus R can have multiple values. The projection of R in the implantation direction is a more meaningful parameter because it determines the junction depth and is referred to as the *projected range* R_p. The relationship between R and R_p is shown in Figure 9.1*a*.

Some ions will collide less than the average, stopping beyond R_p; some will collide more and stop short of R_p. The statistical fluctuations in the ion concentration along the projected range are referred to as the *projected straggle* ΔR_p, as shown in Figure 9.1*b*. The ions are also scattered perpendicular to the incident direction. The resulting fluctuations in the ion concentrations in the transverse

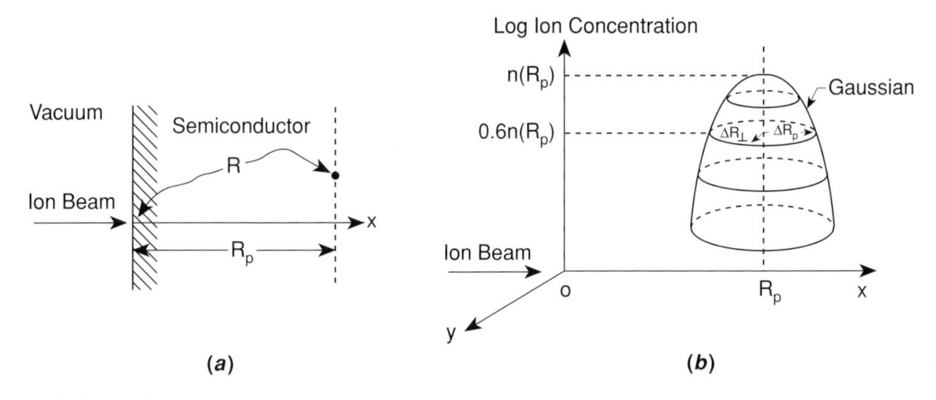

FIGURE 9.1
(*a*) Schematic of the ion range R and projected range R_p. (*b*) Two-dimensional distribution of the implanted atoms. (*From Sze [1985].*)

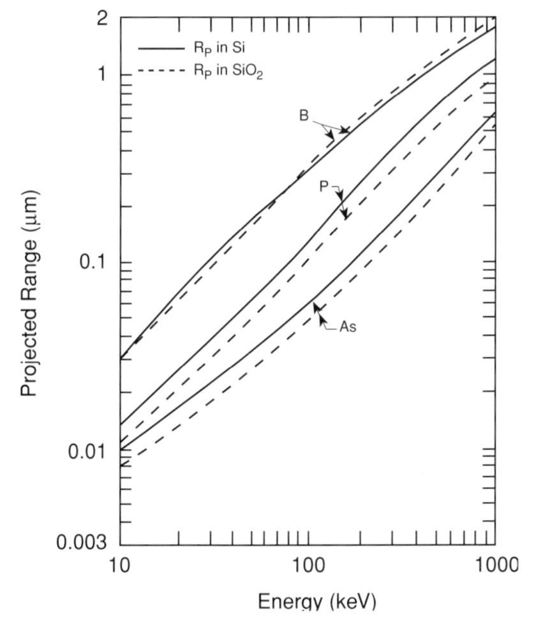

FIGURE 9.2
Projected range for B, P, and As in Si and SiO_2 at various energies. The results pertain to amorphous silicon targets and thermal SiO_2 (2.27 g/cm^3).

direction are called the *projected transverse* or *lateral straggle* $\Delta R_\perp$, as depicted in Figure 9.1*b*. The implications of the two straggles for device fabrication are that the vertical and lateral profiles of the implanted regions will not be sharp.

Figure 9.2 shows the projected ranges for B, P, and As in amorphous silicon and thermal SiO_2. As expected, for a given energy the lighter ion has a longer range than the heavier ion because the former have weak interactions. The calculated values of the ion projected straggle ΔR_p and the ion lateral straggle $\Delta R_\perp$ for B, P, and As in silicon, shown in Figure 9.3, follow the same dependence on the ion mass as depicted in Figure 9.2. For B ions $\Delta R_\perp$ exceeds ΔR_p at all energies.

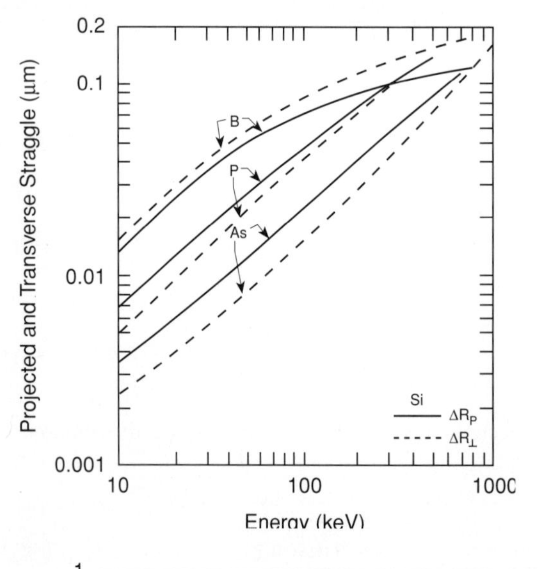

FIGURE 9.3
Calculated ion projected straggle ΔR_p and ion lateral straggle ΔR for As, P, and B ions in silicon.

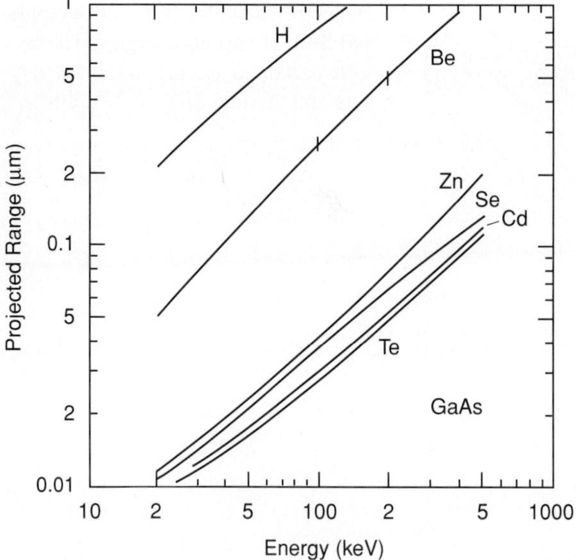

FIGURE 9.4
Projected range of H, Be, Zn, Se, Cd, and Te in GaAs.

The projected ranges of H, Be, Zn, Se, Cd, and Te ions into GaAs are shown in Figure 9.4. The lighter ions penetrate deeper than the heavy ions. As shown in Figure 9.5, H and Be ions also exhibit a much larger straggle than the Zn, Se, Cd, and Te ions.

9.1.1 Theory of Ion Stopping

Nuclear collisions and coulombic interactions with the electrons stop incident ions. By assuming that the energy losses due to the two mechanisms are mutually

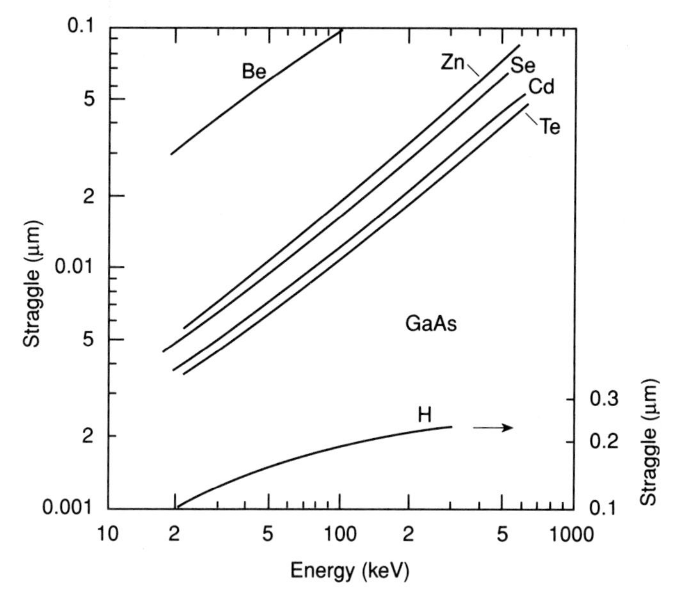

FIGURE 9.5
Straggle for H, Be, Zn, Se, Cd, and Te in GaAs.

independent and additive, Lindhard, Scharff, and Schiott (LSS) developed a theory for determining the range of an ion (Lindhard et al. 1963). The energy loss per unit distance due to the nuclear and electronic collisions is given by

$$\left(\frac{dE_{total}}{dx}\right) = \left(\frac{dE}{dx}\right)_{nuclear} + \left(\frac{dE}{dx}\right)_{electronic}, \tag{9.1}$$

where the nuclear and electronic losses depend on the ion energy. The range of the ions R is given by

$$R(E) = \int_E^0 \frac{dE}{\left(\frac{dE}{dx}\right)_{total}} = \frac{1}{N}\int_E^0 \frac{dE}{S(E)}, \tag{9.2}$$

where E is the energy of the incident ion, N is the number density of target atoms, and S(E) is the stopping power of the solid. If Sn(E) and Se(E) are, respectively, energy losses per unit distance due to nuclear collisions and coulombic interactions, then $S(E) = S_n(E) + S_e(E)$.

A description of the scattering process by a nucleus is provided in Figure 9.6. Assume that an incoming ion having energy E_1 and mass M_1 collides with a target atom whose mass is M_2. As a result of the collision, the incoming ion is scattered through an angle θ and the target atom is displaced from its stationary position as shown. The energy transferred T to the target atom is given by

$$T = \frac{4M_1M_2}{(M_1 + M_2)^2} E_1 \sin^2\left(\frac{\theta}{2}\right). \tag{9.3}$$

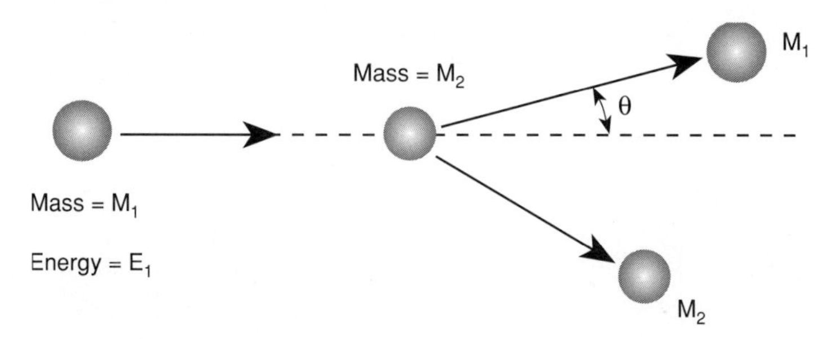

FIGURE 9.6
Schematic illustrating the collision between two hard spheres of masses M_1 and M_2. M_2 is at rest, and M_1 has energy E_1. θ is the scattering angle.

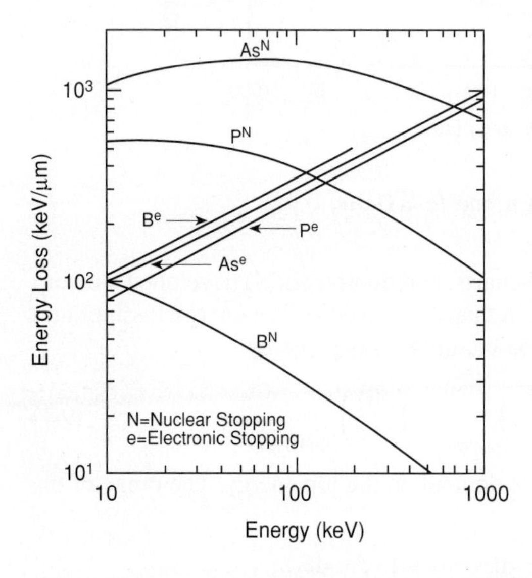

FIGURE 9.7
Calculated values of dE/dx for As, P, and B at various energies. The nuclear N and electronic e components are shown. Note the points at which nuclear and electronic stopping are equal.

The maximum energy is transferred in a head-on collision in which $\theta = 180°$. The scattering angle θ can be obtained by integrating the equation of motion for the scattering trajectory. The nuclear energy loss is given by

$$\left(\frac{dE}{dx}\right)_{\text{nuclear}} = N\int_{E}^{0} T d\sigma = NS_n(E), \tag{9.4}$$

where $d\sigma$ is the differential cross section for ion scattering. Lindhard et al. (1963) introduced a number of simplifications to make the above integration tractable. Using the LSS approach, Smith (1977) computed the values of $\left(\frac{dE}{dx}\right)_{\text{nuclear}} = S_n(E)$ for the implantation of As, P, and B into silicon, which are shown in Figure 9.7. In addition, the electronic losses are proportional to the velocity of the ion, that is, $\sqrt{E}$:

$$\left(\frac{dE}{dx}\right)_{\text{electronic}} = k_e\sqrt{E}. \tag{9.5}$$

The coefficient k_e is a relatively weak function of M_1 and M_2 and the atomic numbers of the incident ion and the stopping atom. Figure 9.7 also shows the computed values of $\left(\frac{dE}{dx}\right)_{electronic} = S_e(E)$ for implantation of As, P, and B into silicon. Note that in Figure 9.7 $S_n(E)$ increases with the mass of the implanted ion. Thus compared to light ions, heavy ions transfer much more of their energy through nuclear collisions. For B, $S_e(E)$ is the dominant loss mechanism over the whole energy range, while for P and As ions $S_n(E)$ dominates for energies up to 130 and 700 keV, respectively.

9.1.2 Implantation Profiles in Amorphous Solids

The concentration profile of the implanted ions in an amorphous solid is given by the following expression (Ghandhi 1983):

$$N(x) = \frac{Q_o}{\sqrt{2\pi}\,\Delta R_p} \exp\left[-\frac{1}{2}\left(\frac{x - R_p}{\Delta R_p}\right)^2\right], \tag{9.6}$$

where N(x) is the impurity concentration, Q_o is the dose (ions/cm^2), x is the distance from the surface in cm, R_p is the projected range in cm, and ΔR_p is the straggle in cm. Equation (9.6) ignores the effects of the transverse straggle $\Delta R_\perp$. This omission may introduce some error in determining the concentration of ions near the edges of a mask that is used for defining regions for selective doping.

The profile of the ion concentration given by Equation (9.6) is shown schematically in Figure 9.1*b*. The peak concentration occurs at R_p and falls off symmetrically on either side of R_p. Some dopants have shown considerable deviations from this Gaussian profile. These observations cannot be explained using Equation (9.6) because it is based on the simple range theory.

EXAMPLE 9.1. Calculate the ion concentration versus depth variations for 100 keV boron and phosphorous ions implanted into silicon crystals at ambient temperature. Assume that the dose in each case is 5×10^{16} ions/cm^2 and the effect of the transverse straggle on the ion concentration is negligible.

Solution. From Figure 9.2 we know that the projected ranges of the 100 keV B and P ions into a silicon crystal are 0.3 and 0.13 μm, respectively; the respective straggles are 0.07 and 0.045 μm. We can use Equation (9.6) to compute the values of N(x) for different values of x.

Boron

x in μm	N(x) (ions/cm^3)
0.20	1.06×10^{21}
0.22	1.48×10^{21}
0.24	1.97×10^{21}
0.26	2.44×10^{21}
0.28	2.74×10^{21}
0.30	2.85×10^{21}

Since the profile is Gaussian in character, the values of N(x) for x = 0.32, 0.34, 0.36, 0.38, and 0.40 mm would be the same as those for x = 0.28, 0.26, 0.24, 0.22, and 0.20 mm.

Phosphorus

x in μm	N(x) (ions/cm^3)
0.08	2.38×10^{21}
0.09	2.97×10^{21}
0.10	3.54×10^{21}
0.11	4.03×10^{21}
0.12	4.30×10^{21}
0.13	4.43×10^{21}

The values of N(x) will be identical to those listed above for $R_p + x$, where x = 0.14, 0.15, 0.16, 0.17, and 0.18. The preceding examples show that for the same ion energy, the boron implants are much deeper than those of phosphorus.

EXAMPLE 9.2. A GaAs wafer, 75 mm in diameter, is to be implanted with 100 keV Be ions to a dose of 5×10^{15} ions cm^{-2}. Determine the projected range, projected straggle, and peak concentration. Calculate also the required beam current if the implantation time is 90 sec.

Solution. We can obtain the projected range and projected straggle from Figures 9.4 and 9.5. The respective values are 0.25 and 0.1 μm. The peak concentration occurs at R_p and can be estimated from Equation (9.6).

$$N(x = R_p) = \frac{5 \times 10^{15}}{2.51 \times 0.1 \times 10^{4}}$$

$$= 1.99 \times 10^{20} \text{ ions/cm}^3.$$

$$\text{Total number of implanted ions} = \text{Area} \times \text{dose}$$

$$= \pi \times (7.5)^2 \times 5 \times 10^{15}$$

$$= 8.84 \times 10^{17} \text{ ions.}$$

$$\text{Beam current} = \frac{\text{total implanted charge}}{\text{time of implantation}}$$

$$= \frac{1.6 \times 10^{-19} \times 8.84 \times 10^{17}}{90}$$

$$= 0.157 \text{ mA.}$$

9.2 ION CHANNELING

Atoms in amorphous solids do not exhibit long-range positional correlations. However, some short-range correlations may exist. When the ions are incident on such a solid, the probability of the ions encountering atoms within the solid is extremely high. However, this situation does not generally occur with crystalline materials because the presence of three-dimensional atomic arrangements within the crystal creates open channels along certain crystallographic directions. This effect is illustrated in Figure 9.8*a*, which shows the perspectives of the ball models of the diamond-cubic lattice when viewed along the <100> and <110> directions. The "openness" that is observed along a specific direction is referred to as a *channel*. The channel width along a <110> direction is greater than the width along a <100> direction. If ions are implanted along a channeling direction, they will experience fewer nuclear collisions. They are primarily slowed down by the coulombic losses, as illustrated in Figure 9.8*b*. As a result, the ions can penetrate deeper into a crystalline solid than in an amorphous material. This effect is called *ion channeling*.

Implantations for the fabrication of devices are not carried out under channeling conditions, because controlling the concentration versus depth profiles of the implanted ions would be difficult. To make crystalline solids appear more like amorphous materials to the incident ions, the crystal is tilted away from a channeling direction. This approach ensures that the ions do not initially channel. However, they may subsequently align themselves along a channeling direction. As a result,

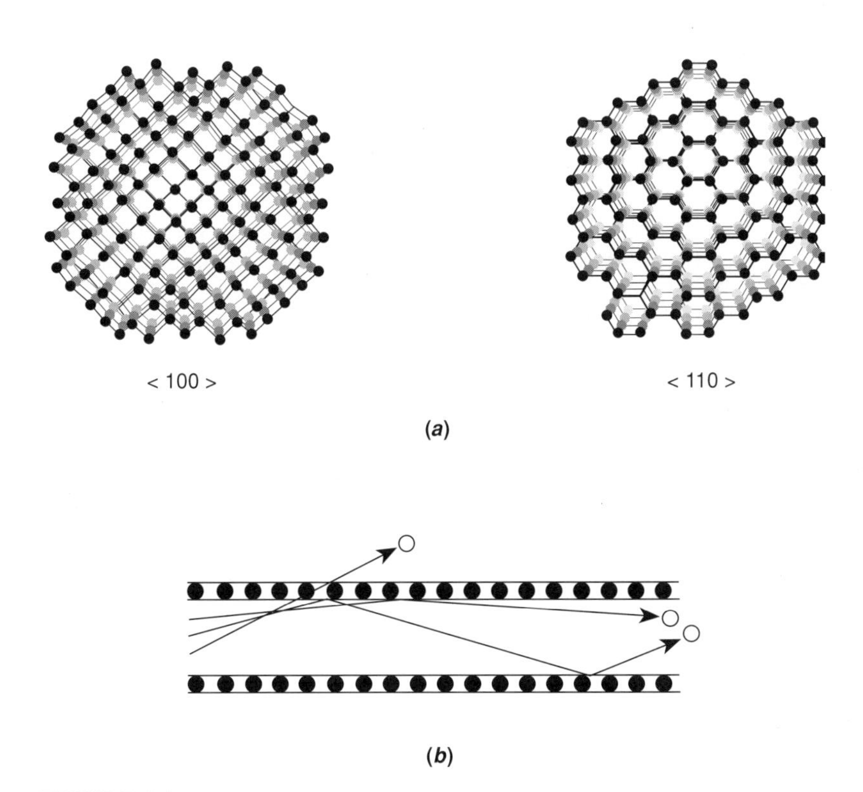

FIGURE 9.8
(*a*) Ball model showing relative degree of "openness" of the diamond-cubic(Si) lattice when traversing in <100> and <110> directions. (*b*) Schematic representation of ion trajectories in an axial channel for various entrance angles.

they may penetrate deeper than the range calculated from Equation (9.2). This effect may produce tails on the concentration versus depth profiles of the implanted ions. Experiments with phosphorus ions implanted into silicon indicate that the tail of the distribution is indeed due to channeling.

9.3 ION IMPLANTATION-INDUCED DAMAGE AND ITS ANNEALING BEHAVIOR

9.3.1 Damage

As discussed in the preceding section, ions undergo nuclear and electronic collisions when they enter a solid. If the energy transferred by the incoming ion to the host atom exceeds a limiting value, called the *displacement energy* E_d, the atom will be dislodged from its site. Figure 9.9 illustrates this situation. Since the displaced atom acquires energy as a result of nuclear collisions, it can in turn displace more atoms from their lattice sites. Likewise, the incident ions will continue to produce displacement damage until their energy falls to a level where the transferred energy during nuclear collisions is less than E_d. As a result of these multiple collisions, the displacement damage in ion-implanted solids can be quite extensive. The extent of the damage depends on incident-ion energy, ion dose, dose rate, mass of the ion, and temperature of implantation.

The distribution of the damage produced by an incident ion depends on whether the ion is lighter or heavier than the host atoms. Since the energy transferred in a collision is directly proportional to the mass of the ion (see Equation 9.3), a light ion will transfer a small amount of energy during each collision with the lattice atom. As a result, the incident ions will be scattered through large angles. The displaced lattice atoms will possess a small amount of energy and may not be able to produce additional atomic displacements. Furthermore, most of the energy of the incident ion is lost by electronic collisions so that relatively little crystal damage occurs. The ion range is comparatively large, and the damage will be spread out over a larger

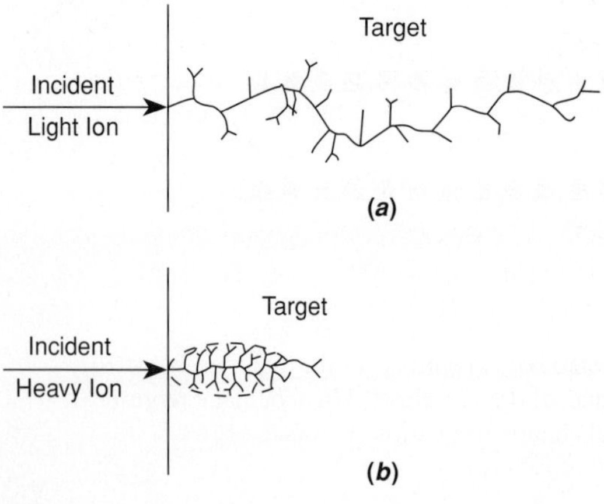

FIGURE 9.9
Damage due to (*a*) light ions and (*b*) heavy ions.

volume of the target. The damage produced by a single, light ion may thus take the form that is illustrated in Figure 9.9*a*.

The situation with heavy ions is quite different. In this case the energy transferred to the host atoms by nuclear collisions is quite substantial, implying that the displaced atoms in turn can produce displacement damage. The ions are also scattered through smaller angles, and the ion range is small. These factors localize the damage within a small volume as shown in Figure 9.9*b*.

In addition to the mass, the ion energy and dose affect the extent of the damage. In 1995, de la Rubia and Gilmer performed computer simulations of the damage produced in silicon by 5 keV silicon atoms. They observed that at the start of the collision process the energy is concentrated in relatively few atoms having large potential energies. The average potential energy of the atoms is greater than the latent heat of fusion of silicon, implying that collectively these atoms may have properties similar to that of liquid silicon. Subsequently the heat is extracted from the liquid-like cascade region by the surrounding silicon at an extremely rapid rate, resulting in amorphization of the region.

Figure 9.10 shows a disordered region observed in a (001) silicon crystal, implanted with 100 keV Si^+ ions to a dose of 1.0×10^{14} cm^{-2}, by high-resolution transmission electron microscopy (Narayan and Holland 1984). When the dose is increased, more disordered regions form and eventually begin to overlap, resulting in amorphous regions being surrounded by the crystalline material. One such region, observed in (001) silicon crystals implanted with silicon ions to 2×10^{14} cm^{-2}, appears in Figure 9.11 (Narayan and Holland 1984). With the added dose, the implanted volume may become amorphous as shown in Figure 9.12 for the case of self-ion damage in silicon (Narayan and Holland 1984). Figure 9.9 also shows that the crystalline-to-amorphous transformation, induced by ion implantation, would occur at lower doses for heavy ions. The amorphous regions have also been observed in (001) GaAs crystals implanted with 450 keV Se^+ ions to a dose of $10^{14}/cm^2$ (Sadana et al. 1985). The schematic in Figure 9.13 shows the resulting damage. The amorphous region (α-GaAs) is ~250 nm thick and extends all the way to the surface. In addition, a damaged volume containing channel cascades separates amorphous and crystalline portions of the crystal.

EXAMPLE 9.3. Estimate the number of silicon atoms displaced, both at the surface and for the peak value of the nuclear stopping, by the 300 keV, 10^{13} boron ions/cm^2.

Solution. The following simple expression relates the number of displaced atoms/cm^3 N with the displacement energy E_d, dose Q_O, and the nuclear stopping power $\left(\frac{dE}{dx}\right)_{nuclear}$:

$$N = \frac{Q_O}{E_d}\left(\frac{dE}{dx}\right)_{nuclear}.$$

The value of $\left(\frac{dE}{dx}\right)_{nuclear}$ can be obtained from Figure 9.7.

$$\text{N at the surface} = \frac{10^{13} \times 16 \times 10^7}{15} \quad \text{when } E_d = 15\,\text{eV}$$

$$\cong 10^{20}/\text{cm}^3.$$

$$\text{N at the peak disorder} = \frac{10^{13} \times 10^{2} \times 10^{7}}{15}$$
$$\cong 7 \times 10^{20}/\text{cm}^{3}.$$

9.3.2 Annealing Behavior

As indicated in section 9.3.1 ion implantation produces extensive damage. Since the point defects in semiconductors are electrically active (see section 3.4), as-implanted materials have poor electrical characteristics. Generally, the minority-

FIGURE 9.10
High-resolution transmission electron micrograph showing an amorphous cascade region in a (001) Si sample implanted with 100 keV Si^{+} ions to a dose of 1×10^{14} cm^{-2}. (*After Narayan and Holland [1984].*)

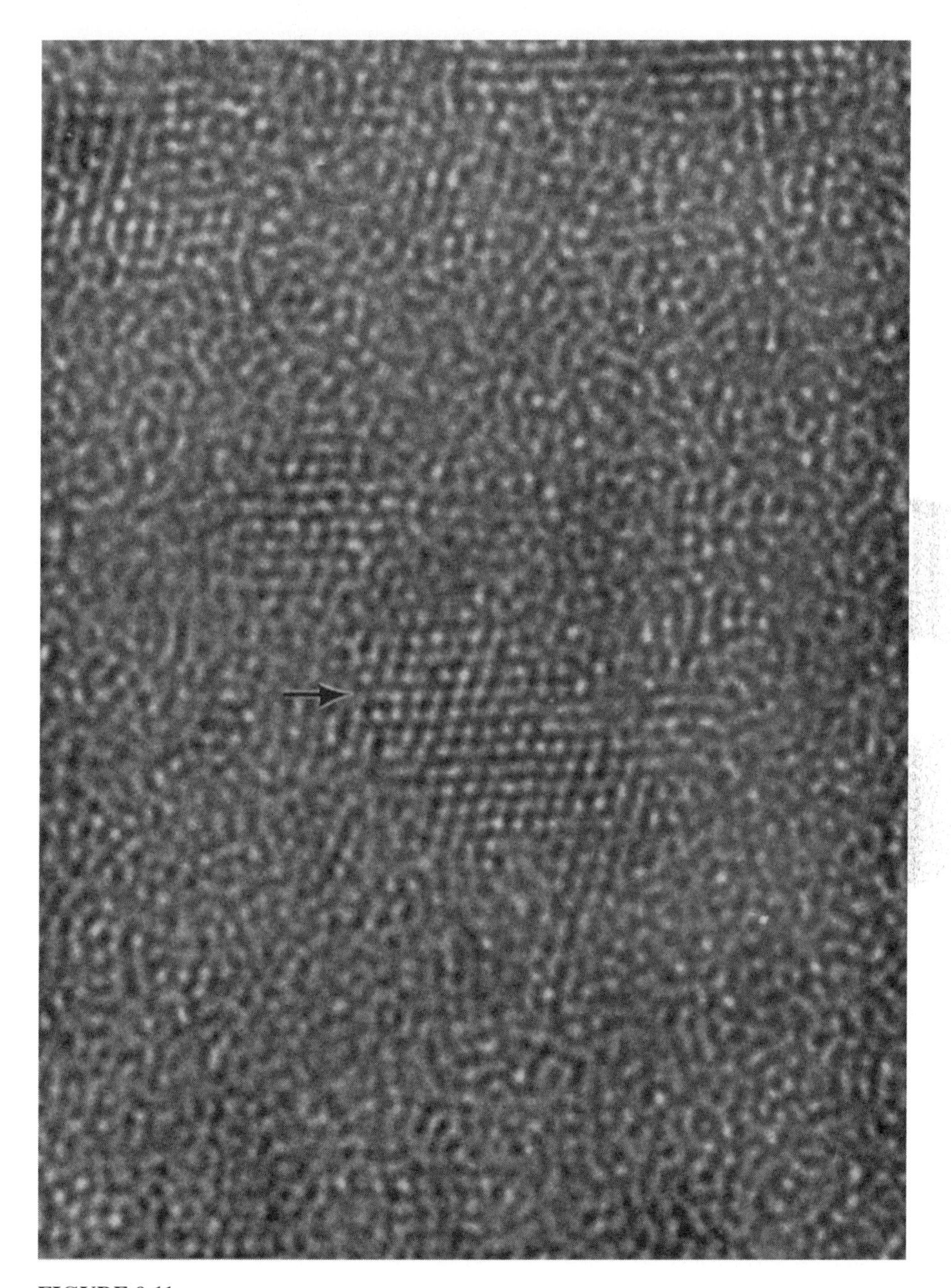

FIGURE 9.11
High-resolution transmission electron micrograph showing amorphous and crystalline regions (indicated by an arrow) in a (001) Si sample implanted with 100 keV Si^+ ions to a dose of 2×10^{14} cm^{-2}. (*After Narayan and Holland [1984].*)

FIGURE 9.12
High-resolution transmission electron micrograph showing the interface between the amorphous and crystalline Si. Note the presence of microcrystals in the amorphous regions that are indicated by an arrow. (*After Narayan and Holland [1984].*)

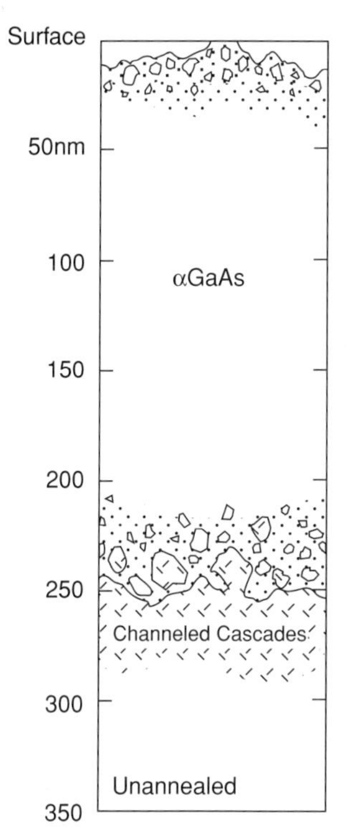

FIGURE 9.13
Schematics of damage distribution in (001) GaAs samples implanted with a dose of 10^{14} Se^{+}/cm^{2} at 450 keV. (*After Sadana et al. [1985].*)

carrier lifetime and mobility are severely degraded after ion implantation. Furthermore, only a fraction of the implanted ions are located on substitutional sites and contribute to the carrier concentration. To eliminate the detrimental effects of ion implantation, the material has to be annealed at elevated temperatures. This procedure serves two purposes. First, the point defect density is reduced because of the annihilation of some of the vacancies by the interstitials. Second, the implanted dopant atoms in interstitial sites could migrate to lattice sites and become electrically active.

We now discuss how damaged regions in ion-implanted materials respond to thermal annealing. In 1995, de la Rubia and Gilmer simulated the annealing behavior of the amorphous material in silicon produced by 5 keV silicon atoms. They found that on recrystallization only an extremely small fraction of the original volume is retained as damage in the form of self-interstitials, vacancies, and vacancy clusters.

We show in Figure 9.14 a micrograph obtained from silicon implanted with 5×10^{13} cm^{-2} 40 keV silicon ions and subsequently annealed at 740°C for 15 min (Eaglesham et al. 1995). Long defects, referred to as "rodlike" defects, consist of silicon interstitials. On additional annealing these defects either shrink by the emission of interstitials into the adjoining lattice or coalesce together to form larger ones, which may also dissolve on further annealing. The released interstitials may then participate in the transient enhanced diffusion of dopant atoms that is commonly observed in low-energy ion-implanted silicon (Stolk et al. 1995).

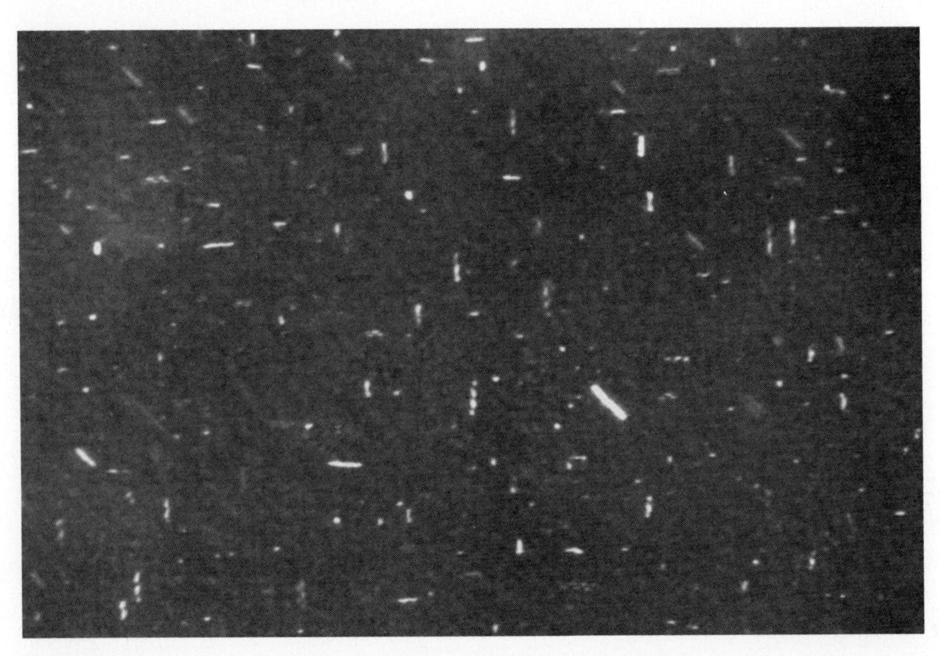

FIGURE 9.14
Micrograph obtained from a silicon sample implanted to 5×10^{13} cm^{-2} with 40 keV Si ions and subsequently annealed at 740°C for 15 min. (*After Eaglesham et al. [1995].*)

The rodlike defects in silicon implanted to 2×10^{14} cm^{-2} with 145 keV silicons ions and annealed at 950°C for 10 min tend to break up into dislocation loops. We illustrate this situation in Figure 9.15 (Eaglesham et al. 1995). The switch from the rod shrinkage and agglomeration to the loop formation at high energies could be caused by the presence of a high concentration of interstitials in the adjoining matrix. We envisage that the point defects constituting the rodlike defects may coalesce together to form extrinsic faulted dislocation loops that lie on {111} planes that are bounded by a/3 <111> Frank partials, as shown in Figure 9.16*a*. The driving force for the coalescence is the reduction in the surface energy of the defects; this issue is discussed in detail in section 5.5. To form such loops in III-V materials, groups III and V interstitials are required simultaneously (see section 3.3.3).

The faulted loops can grow on further annealing through the absorption of the interstitials at the cores of Frank partials. When the loops grow, their energies increase because the fault area and the length of the partial increase. At a certain size the energy of the faulted loop will become equal to that of the perfect loop bounded by ± a/2 <110> dislocations. The conversion of the faulted loops into the perfect loops is accomplished by the passage of Shockley partials across the fault planes (see section 5.5 for additional details). The resulting situation is shown in Figure 9.16*b*. If the implanted materials are still saturated with point defects, perfect loops can also expand by the absorption of point defects. During growth, loops may interact according to the following reaction to form a dislocation network

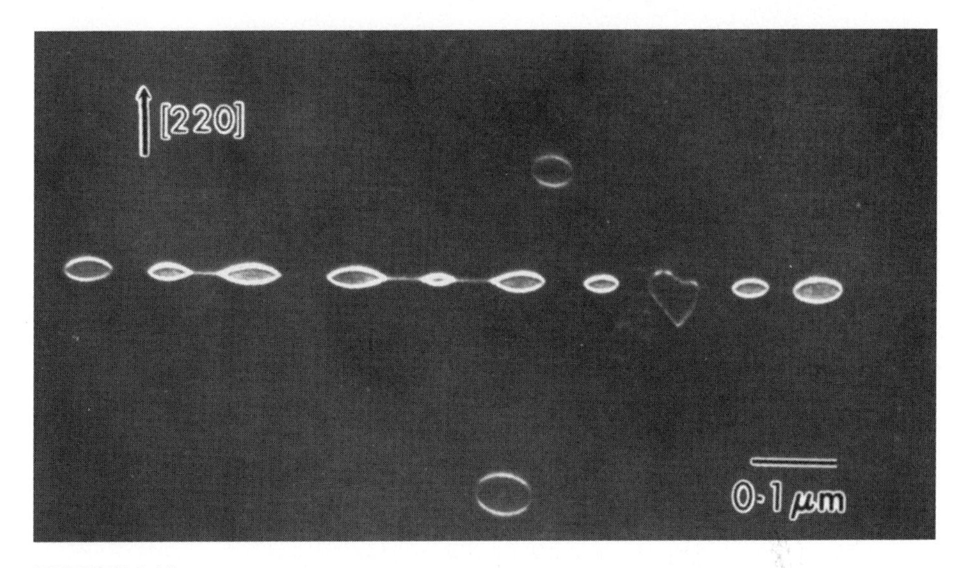

FIGURE 9.15
Dislocation loop formation from rodlike defects in silicon; a single defect gives rise to a row of Frank loops. 2×10^{14} cm^{-2} 145 keV Si ions plus annealing at 950°C for 10 min. (*After Eaglesham et al. [1995].*)

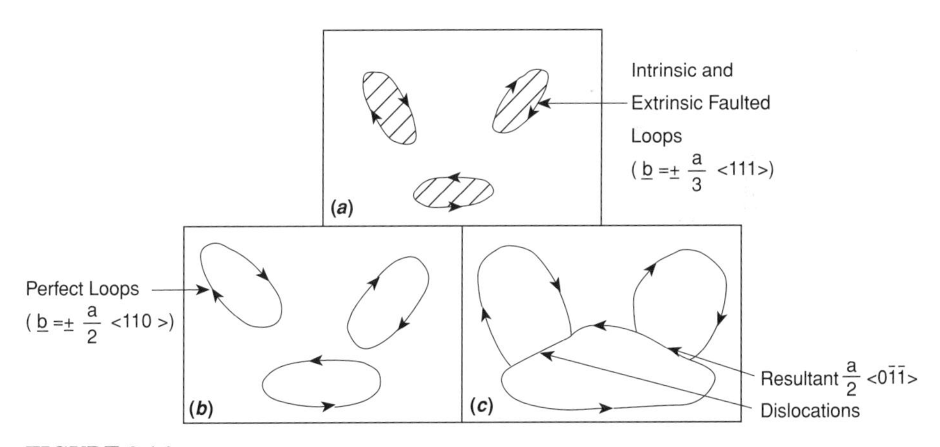

FIGURE 9.16
Schematics illustrating the annealing behavior of damage in implanted materials: (*a*) after short anneal showing faulted loops, (*b*) with additional anneal, faulted loops grow and transform into perfect loops, and (*c*) interaction between expanding perfect loops.

shown in Figure 9.16*c*:

$$\frac{a}{2} <1\bar{1}0> + \frac{a}{2} <\bar{1}0\bar{1}> \rightarrow \frac{a}{2} <0\bar{1}\bar{1}> \tag{9.7}$$

The resultant $a/2 < 0\bar{1}\bar{1} >$ dislocations are identified in Figure 9.16*c*.

Generally, the covalent tetrahedral radii of implanted ions and host atoms are different. Consequently, the occupation of substitutional sites by implants during annealing would lead to local strains. The overall strain energy of the system can be lowered by the migration of implanted atoms to dislocation loops and dislocations produced during the anneal because of favorable elastic interactions of the dislocation and the impurity strain fields.

The residual damage in materials like GaAs, InP, and so on after annealing may consist of dislocation loops, dislocations, faults, and twins. This situation is illustrated in Figure 9.17 for ~(001) GaAs implanted with 10^{14} Se^+ ions/cm^2 at 450 keV and subsequently annealed at 400°C (Sadana et al. 1985). Comparing Figures 9.13 and 9.17, it is clear that the extrinsic dislocation loops form in the channeled cascades, whereas isolated faults and dislocations are seen near the amorphous-crystalline interface. The high density of faults is observed in the amorphous region. In addition, the orientation of the crystalline region is replicated into the volume that evolves from the amorphous solid during annealing. This replication is possible only if the crystalline region in Figure 9.13 serves as the template during the amorphous-to-crystalline transition on annealing. This type of growth is referred to as *solid phase epitaxy*.

The fault bundles shown in Figure 9.17 are very likely growth faults. If amorphous-to-crystalline transformation occurs at a very rapid rate, growth mistakes in the {111} stacking sequence can occur, leading to stacking faults. The stresses that may develop in the lattice due to the incorporation of dopant atoms of different tetrahedral radii can further complicate this situation.

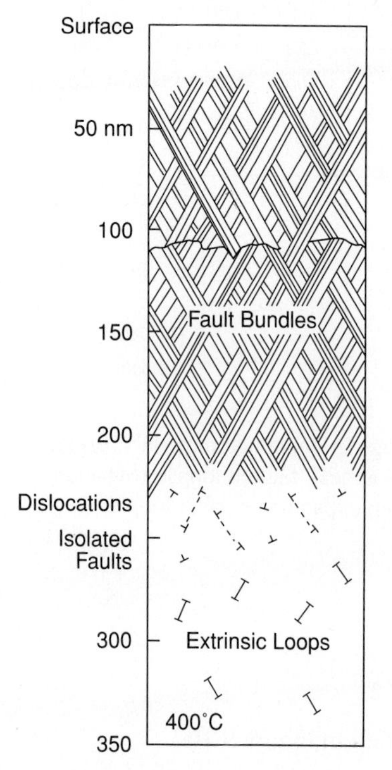

FIGURE 9.17
Schematic illustrating the damage distribution observed after annealing of GaAs samples implanted with Se^+ 10^{14}/cm^2 at 450 keV. (*After Sadana et al. [1985].*)

Let us now apply the above generic ideas on damage production and annealing to the following specific situations.

9.3.2.1 Isochronal annealing behavior of silicon implanted with boron ions

Figure 9.18 shows the isochronal annealing behavior of 150 keV boron ions implanted into silicon to three different doses (Seidel and MacRae 1971). The low-dose samples implanted to $8 \times 10^{12}/cm^2$ show a monotonic increase of free-carrier concentration, that is, P_{Hall} in Figure 9.18, with the annealing temperature, implying that the implanted ions are shifting to lattice sites on annealing. However, the annealing behavior of the samples implanted to higher doses is much more complex and can be divided into regions I, II, and III. In region I the free-carrier concentration increases with the annealing temperature. Extended defects, such as faulted dislocation loops, are not observed in this regime. The plausible explanation is that B atoms are migrating to the lattice sites during annealing. In region II the free-carrier concentration decreases with the increasing temperature. In addition, dislocation substructure is observed after annealing. It is likely that B atoms migrate to the dislocation cores because the B atoms are extremely small and will have strong elastic interactions with dislocation cores. Once the dopant atoms are removed from substitutional sites, they will not contribute to the free-carrier concentration. In region III the free-carrier concentration increases with the increasing temperature indicating that B atoms are returning to the lattice sites. This result suggests that the binding energy between B atoms and dislocations is overcome thermally. The released B atoms migrate to the lattice sites and increase the carrier concentration.

The preceding example shows that interactions between the implanted atoms and annealing-induced defect structure could complicate the activation of the dopants. To develop a full understanding of the behavior of the implanted dopants

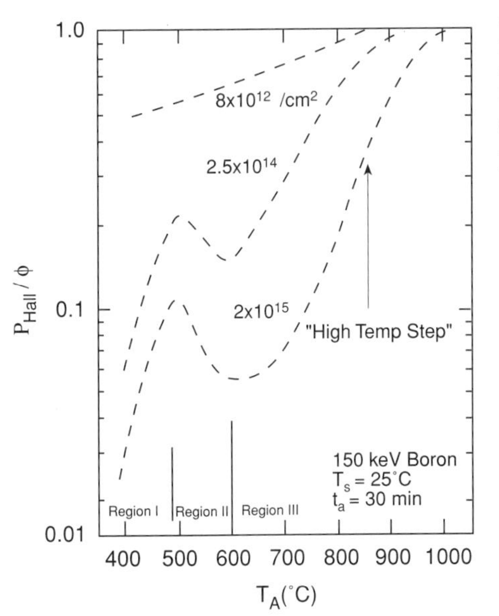

FIGURE 9.18
Isochronal annealing behavior of boron implanted into silicon. The ratio of free-carrier concentration P_{Hall} to dose ϕ is plotted against anneal temperature T_A for

during annealing, electrical measurements must be coupled with structural observations.

9.3.2.2 Isochronal annealing behavior of silicon implanted with phosphorous ions

Figure 9.19 shows the isochronal annealing behavior of phosphorus ions implanted into silicon at 250 keV and at six different doses (Crowder and Morehead Jr. 1969). Comparing Figures 9.18 and 9.19, we see the qualitative difference between the annealing behaviors of silicon implanted with B and P. When the dose is increased from 3×10^{12} to 3×10^{14} cm^{-2}, higher annealing temperatures are required to eliminate the progressively more complex damage. Amorphous layers are produced at doses of 1×10^{15} and 5×10^{15} cm^{-2}, and they extend to the surface. On annealing, the amorphous to crystalline transition occurs by solid phase epitaxy. During regrowth of the amorphous layers, the implanted atoms are incorporated into substitutional sites. Furthermore, after annealing, the carrier concentrations in the samples implanted to higher doses are lower than the concentrations in the low-dose specimens. It could be that after annealing the residual damage in the high-dose samples is much more extensive.

The annealing behavior of silicon implanted with the arsenic and antimony ions at room temperature is similar to the behavior of phosphorus ions except in which

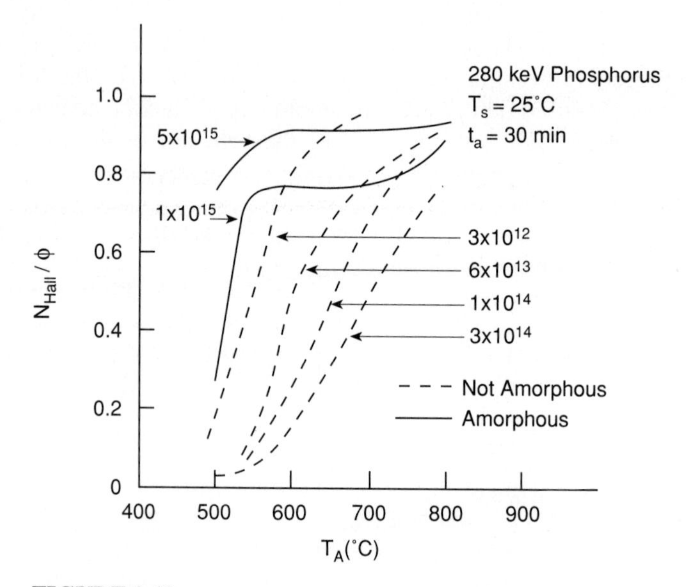

FIGURE 9.19
The ratio of free-carrier concentration to dose plotted against anneal temperature T_A for various phosphorus doses. The solid curves represent amorphous layers that anneal by solid phase epitaxy. The dashed curves represent implantation where the damage is not amorphous. (*After Crowder and Morehead Jr., [1960]*.)

lower doses are required for amorphization. This behavior occurs because As and Sb ions are considerably heavier than the P ions.

9.3.2.3 Isochronal annealing behavior of GaAs implanted with Be ions

Figure 9.20 shows the isochronal annealing behavior of GaAs implanted with the 250 keV Be ions to a dose of 1×10^{15} ions cm^{-2}, followed by a 30-minute anneal at a number of different temperatures, (Levige et al. 1977). The dopant profiles observed after the anneals at 600 and 700°C follow the Gaussian distribution computed using the LSS theory. On the other hand, the concentration profile seen after the 800°C anneal is fairly broad and is far from the Gaussian type because of the diffusion of Be atoms into the surrounding regions. The driving force for this diffusion is the difference in the activities of Be atoms in the implanted and unimplanted regions.

9.3.2.4 Annealing behavior of GaAs implanted with S and Se ions

Figure 9.21 shows the carrier-concentration profiles observed in GaAs implanted with 100 keV S^+ ions (dose-9×10^{12} cm^{-2}) and 400 keV Se^+ ions (dose-2×10^{12} cm^{-2}) and annealed at 850°C for 30 minutes (Eisen 1980). The doping profile for the S^+ ions deviates substantially from the Gaussian distribution, implying substantial out diffusion into the unimplanted regions. The situation regarding the Se^+ ions is totally different. The observed doping profile almost follows the profile calculated from the LSS theory, which suggests that the diffusion of Se in GaAs is considerably slower than that of S. The predictable characteristics of Se also make it a very useful implant for the fabrication of n-type regions in semi-insulating GaAs substrates required for field-effect transistors and integrated circuits.

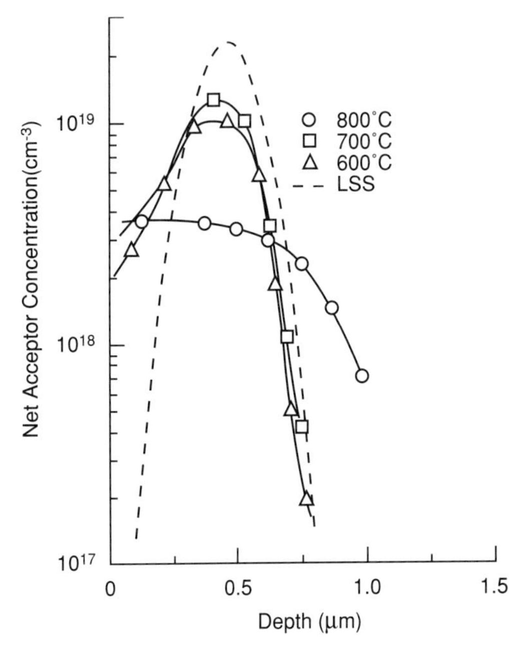

FIGURE 9.20
Annealing of beryllium implants in GaAs. (*After Levige et al. [1977].*)

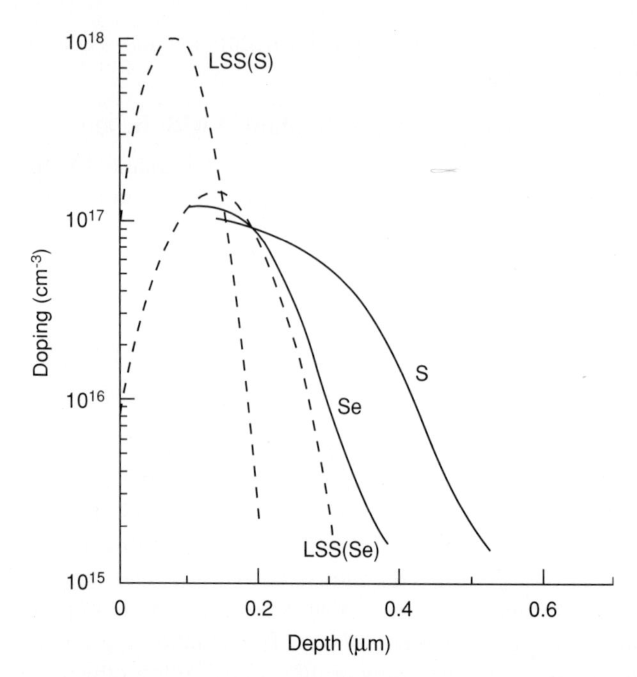

FIGURE 9.21
Annealing of selenium and sulfur implants in GaAs. (*After Eisen [1980].*)

9.3.2.5 Semi-insulating GaAs by the implantation of hydrogen ions

n-type and p-type GaAs behavior can be made semi-insulating by proton bombardment (Foyt et al. 1969). This behavior can be achieved with implants using 100 keV to 3 MeV H^+ ions to a dose of $10^{13}\ cm^{-2}$. The semi-insulating layers as thick as 20 μm can be fabricated in this manner. The carrier concentration in these layers is $10^{11}\ cm^{-3}$. The proton bombardment has also been used to isolate devices on a substrate and also to confine injected current in double-heterostructure GaAs/GaAlAs lasers. One drawback of the hydrogen ion-induced isolation is the lack of thermal stability; significant annealing of the damage can occur above 350°C.

EXAMPLE 9.4. The concentrations of the vacancies and interstitials in silicon crystal, grown under equilibrium conditions, have different values, whereas the concentrations of the two types of defects in implanted silicon are generally equal. Explain this statement.

Solution. It was shown in Chapter 3 that the concentration of vacancies n_v is given by the expression:

$$n_v = n \exp\left(-\frac{E_v}{k_B T}\right),$$

where n is the total number of atoms, E_v is the energy of formation of a vacancy, and k_B is the Boltzmann constant and T is the temperature. Likewise, the concentration of interstitials (n_i) is given by

$$n_i = n \exp\left(-\frac{E_i}{k_B T}\right).$$

Since E_v and E_i have different values, the concentrations of the two types of point defects will be different at a given temperature. In the case of ion implantation, the two types of defects are produced in pairs, even though later on they may separate spatially. Therefore, to a first approximation, the concentrations of the vacancies and interstitials will be equal.

EXAMPLE 9.5. *a*. Estimate the number of silicon atoms displaced for the peak value of the nuclear stopping by the 100 keV, 10^{13} phosphorous ions/cm^2.

b. Assuming that 50 percent of the point defects survive during the anneal at high temperature and that the edge of each loop is 10 nm, calculate how many extrinsic and intrinsic, hexagonal-shape Frank loops/cm^3 would form from the coalescence of residual point defects.

c. Calculate the separation between the loops.

Solutions. As indicated in Example 9.3, N can be related to the displacement energy E_d, dose Q_O, and $\left(\frac{dE}{dx}\right)_{nuclear}$ by the following expression:

$$N = \frac{Q_O}{E_d}\left(\frac{dE}{dx}\right)_{nuclear}.$$

The value of $\left(\frac{dE}{dx}\right)_{nuclear}$ for the peak value of the nuclear stopping is 525 keV/mm (see Figure 9.7).

$$\text{N at the peak disorder} = \frac{10^{13} \times 525 \times 10^7}{15}\ \text{cm}^{-3}$$

$$= 3.5 \times 10^{21}\ \text{cm}^{-3}.$$

Therefore, the number of interstitials produced is 3.5×10^{21} cm^{-3}. Since in the ion implantation process the number of vacancies produced is equal to the number of interstitials, the number of vacancies produced during the implantation is 3.5×10^{21} cm^{-3}.

b. Since 50 percent of the defects are annihilated during the anneal, the residual concentration of the vacancies and the interstitials is 1.75×10^{21} cm^{-3}. In calculating the number of point defects involved in the formation of faulted loops, we will follow the approach used in Chapter 3. It can be shown that point defects in one of the layers of the hexagonal loop are

$$n = \frac{3 \times L^2}{a^2},$$

where L is the length of the loop edge and a is the lattice parameter of silicon. Substituting the appropriate values of L and a, n is estimated to be 1017. Since the intrinsic and extrinsic loops contain two layers of point defects, the number of point defects consumed in forming a loop is 2034.

Therefore, the total number of the extrinsic loops

$$= \frac{1.75 \times 10^{21}}{2.03 \times 10^3}\ \text{cm}^{-3}$$

$$= 8.62 \times 10^{17}\ \text{cm}^{-3}.$$

The total number of the intrinsic loops

$$= 8.62 \times 10^{17}\ \text{cm}^{-3}.$$

c. The separation between the loops

$$\left(\frac{1}{8.62 \times 10^{17}}\right)^{1/3} \text{cm}$$
$$= 1.05 \times 10^{-6} \text{ cm}$$
$$= 10.5 \text{ nm}.$$

9.3.3 Diffusion of Implanted Impurities

As discussed in Chapter 8, diffusion of impurities in semiconductors is a complex phenomenon. This complexity is further compounded in the case of implanted materials because the point defect damage coexists with impurity concentration gradients. As an example, consider the annealing behavior of silicon implanted with 70 keV B ions to a dose of 10^{15} cm^{-2} that is shown in Figure 9.22 (Hofker and Philips 1975). The observed concentration profiles after the 1000 and 1100°C anneals are considerably different from the profile observed after implantation but can be rationalized in terms of the classical diffusion theory. At lower temperatures, however, the classical theory cannot predict the observed profiles. The boron profile after the 900°C anneal can be explained only if a diffusion constant is used that is about three times the value estimated from chemical diffusion at 800°C. As argued in section 9.3.2 point defects produced during implantation cause this transient enhanced diffusion.

This example illustrates that considerable care should be exercised in maintaining the as-implanted profiles during the activation of implanted species by thermal annealing. To produce electrical activity, implanted species must move from

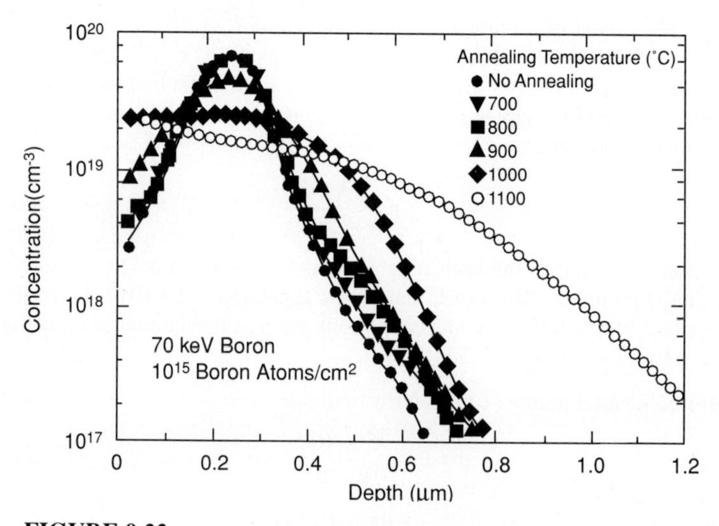

FIGURE 9.22
Boron atom concentrations as a function of annealing at various temperatures. The anneal time is 35 minutes. (*After Hofker [1975].*)

interstitial sites to contiguous, vacant substitutional sites, a fairly short distance in implanted materials. In principle, this process should be possible with a rapid thermal anneal at high temperatures. This approach is currently preferred for the activation of dopants in implanted semiconductors.

9.4 PROCESS CONSIDERATIONS

Production of device-worthy materials using ion implantation requires implantation systems, masks, and the facilities for thermal annealing. The discussion of the systems is beyond the scope of the book. The interested reader is referred to *VLSI Fabrication Principles* by Ghandhi (1983) and *Silicon Processing for the VLSI Era* by Wolf and Tauber (1986) for relevant information. Material issues pertinent to masks and thermal annealing are discussed below.

9.4.1 Materials for Masks

We can use ion implantation to dope regions of a wafer in a highly selective manner by using masks to localize implants to certain regions of a wafer. The desirable features of these mask materials are (1) their ion stopping power should be high so that thin layers can effectively block incoming ions, (2) the materials should be easily removable after implantation, and (3) the materials should be compatible with the photolithographic techniques. Materials such as SiO_2, Si_3N_4, polysilicon, metal films, photoresist, and polyimides are widely used for masking purposes.

The minimum thickness of the mask required to stop most of the implanted ions can be calculated from the range and the straggle of the ions in a mask. If the range parameters are R_p and ΔR_p, then a thickness of $R_p + 3\ \Delta R_p$ can provide a masking effectiveness of 99.99 percent (Wolf and Tauber 1986). The thicknesses of silicon dioxide and photoresist required for blocking this fraction of B, P, and As ions of different energies are shown in Figure 9.23. As expected, the lighter boron ions require thicker masks for effective stoppage.

Implantation of ions in a silicon dioxide mask produces displacement damage. To obviate this problem, the damaged mask can be annealed between 800 and 1000°C (Cass and Reddi 1973). If the SiO_2 mask is deposited on a photoresist, the adhesion between the silicon dioxide mask and the resist could be impaired. This impairment could lead to liftoff of the masking layer during subsequent processing. Another problem with the silicon dioxide mask is that at high implant energies the oxygen atoms could be dislodged from the mask and perhaps implanted into the underlying substrate. The metal masks also suffer from a metal ion implant problem that has severe deleterious effects because metals have deep energy levels in many kinds of semiconductors.

The polymers constituting photoresist masks could undergo cross-linking during ion implantation, thus making it difficult to remove the mask after ion implantation using standard procedure. A plausible solution could be to deposit a resist

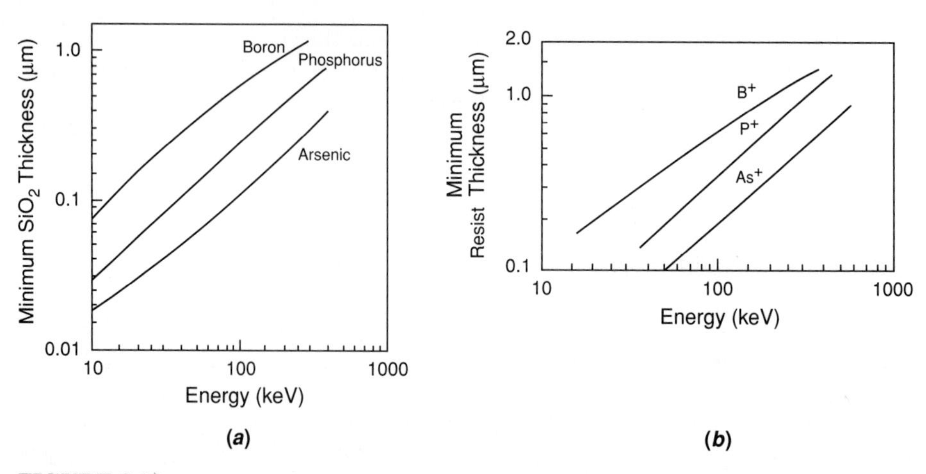

FIGURE 9.23
Minimum thickness to stop 99.99 percent of incident ions as a function of energy for (*a*) silicon dioxide and (*b*) photoresist. (*After Pickar [1975].*)

layer thicker than that required for the 99.99 percent masking effectiveness. After implantation, a thin, undamaged, non-cross-linked layer of the resist will be present adjacent to the substrate. This layer, together with the cross-linked layer, can then be removed in the usual manner.

9.4.2 Multiple Implants

Figure 9.24 shows that a fairly flat, deep profile of B in silicon can be obtained using multiple implants of different energies (Lee and Mayer 1974); such unique profiles cannot be obtained by chemical diffusion. The experimentally measured carrier concentrations and the profiles computed from the LSS theory are also shown. The principle of multiple implants is simple and involves shifting the ion distribution peak by changing the implant energy.

9.4.3 Annealing Setups

As indicated in section 9.3.2, ion-implanted semiconductors must be annealed for two reasons: (1) to provide thermal energy necessary to move implanted atoms from intersitital to substitutional sites and (2) to remove as far as possible ion-implantation-induced damage in the crystal.

Three types of annealing setups have been used to date: furnace annealing, laser annealing, and rapid thermal annealing (RTA). The furnace annealing setup is quite simple. However, this approach has a major drawback. As the annealing times are relatively very long, the profiles of the implanted impurities change considerably during the anneal. This difficulty can be obviated using laser annealing. Figure 9.25 compares the profiles of arsenic implanted into silicon and annealed with a continuous wave laser and in a furnace. The profile in the laser-annealed sample is

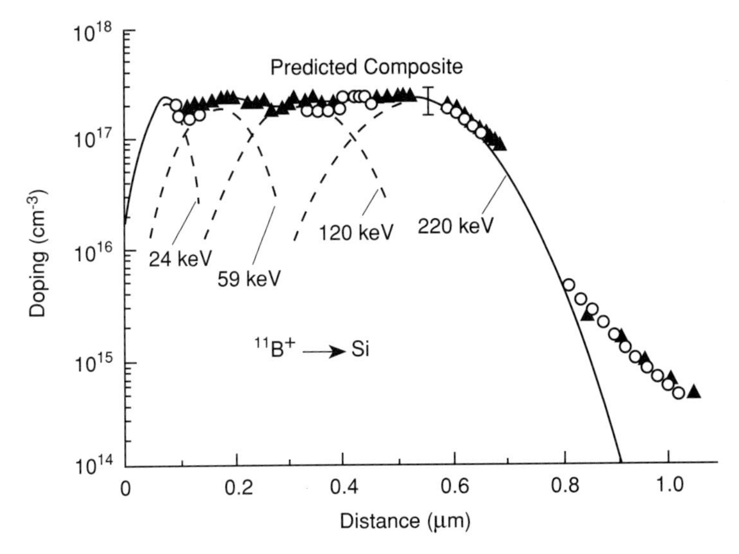

FIGURE 9.24
Composite doping profile using multiple implants. (*After Lee and Mayer [1974].*)

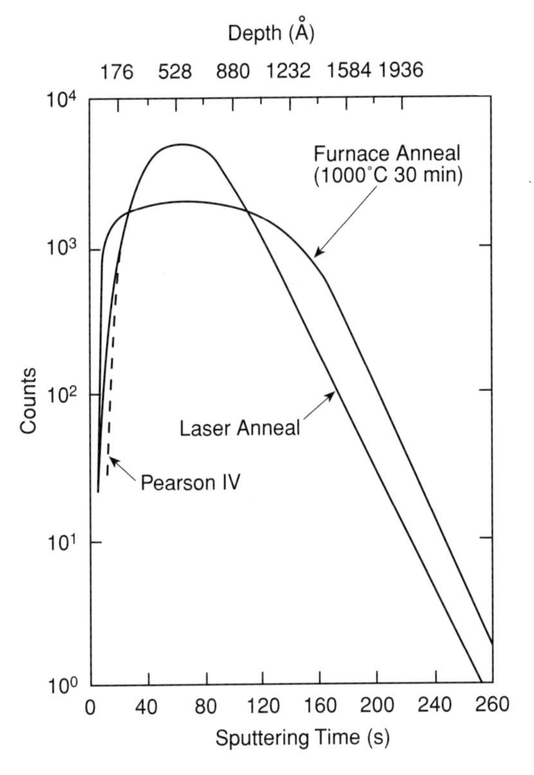

FIGURE 9.25
Profile of arsenic implanted into silicon and annealed both with a CW laser and with a standard thermal anneal. The as-implanted Pearson-IV distribution and the laser-annealed profiles are virtually identical.

close to the profile calculated from the theory. However, after the furnace annealing the profile is broadened due to bulk diffusion.

Depending on the energy density of the laser beams, the implanted material may either melt and then solidify or regrow by solid phase epitaxy. When melting occurs, the regrowth takes place very rapidly. Consequently, nonequilibrium concentrations of point defects are incorporated into the regrown layers. Since defects in semiconductors are electrically active, the electrical properties of the laser-annealed layers involving melting are poor.

The difficulties associated with furnace and laser annealings have been overcome in RTA setups. These setups consist of large-area incoherent energy sources that emit radiant light, which then heats the wafers and allows very rapid and uniform heating and cooling. During RTA, wafers are thermally isolated from each other so that radiative heating and cooling dominate. Temperature uniformity across the wafer is excellent, thus minimizing thermal gradients that can warp wafers. Various heat sources are used: arc lamps, tungsten-halogen lamps, and resistively heated, slotted graphite sheets. Most RTA is carried out in inert atmospheres (Ar and N_2) or in a vacuum.

The vapor pressures of silicon at temperatures (900 to 1100°C) where we anneal it to remove the ion-implantation-induced damage are very low. Consequently, the surfaces of implanted substrates do not deteriorate during annealing. On the other hand, the situation with III-V semiconductors is very different. The vapor pressures of group V species are very high at typical annealing temperatures (800 to 900°C for GaAs). If adequate precautions are not taken, surfaces of the implanted samples could undergo thermal decomposition. The application of RTA certainly reduces this problem. Dielectrics could also be deposited on implanted III-V surfaces prior to annealing. Alternatively, the implanted wafer could be covered with a dummy substrate during the anneal. This approach is called the *proximity anneal* and is quite effective in reducing the thermal decomposition-induced damage in III-V materials.

9.5 COMPARISON OF ION IMPLANTATION AND DIFFUSION FOR SELECTIVE DOPING

For selective doping, ion implantation offers the following advantages over diffusion:

1. Dopants can be introduced in a controlled manner at specific locations. This feature facilitates the fabrication of source and drain regions in MOS transistors.
2. The penetration depth of ions into a substrate increases with the increasing accelerating voltage. Therefore, by varying the voltage, the junction depth can be controlled. In addition, multiple implants at different accelerating voltages can produce through-thickness uniformity of the dopant concentration.
3. For silicon and III-V technologies, ion implantation is generally carried out below 673 K. Consequently, thermal-stability requirements on mask materials are less stringent than those for chemical diffusion; silica, silicon nitride, and different metallizations can be used as masks for selective doping.

TABLE 9.1
Some applications of ion implantation in semiconductor technology

Application	Device
Fabrication of junctions.	MOS and bipolar
Base formation.	Bipolar
Arsenic-implanted polysilicon emitter.	Bipolar
Emitter formation implantation.	Bipolar
Formation of source and drain regions.	CMOS
Silicon-on-insulator using oxygen and nitrogen implantations.	High-voltage devices

4. A wide range of ion doses, 10^{11} to $10^{17}/cm^2$, that is, from very low to very high, can be delivered to substrates, and the dose can be accurately controlled (± 1 percent). The lateral uniformity in ion concentration is also fairly good. In both these respects, ion implantation is superior to chemical diffusion.
5. Various ions can be mass separated to produce monoenergetic, highly pure, dopant ion beams.
6. Ion implantation is a nonequilibrium process. It is therefore possible to dope the substrate in excess of their solid solubility limits in the host lattice. However, for implanted species to remain in solution, the wafers should not be subsequently processed at high temperatures where excess dopant atoms become mobile and cluster together to form precipitates.

Ion implantation has some inherent disadvantages. The high-energy ions damage the lattice. Therefore, the implanted materials must be annealed at high temperatures to "heal" them. Since silicon is quite stable at high temperatures, the annealing step does not entail any problems. However, in implanted III-V materials, the group V atoms tend to boil off due to their high vapor pressures (see Chapters 5 and 6). Another complication is that the implanted profile may change during annealing because of bulk diffusion. Furthermore, the equipment for ion implantation is highly sophisticated, is very expensive, and has to be manned by experienced operators. However, a high degree of automation and fine control of the process are possible with ion implantation, and these features outweigh some of the disadvantages. As a result, ion implantation is used extensively in the semiconductor technology, and some of its applications are listed in Table 9.1.

9.6 HIGHLIGHTS OF THE CHAPTER

The highlights of this chapter are as follows:

- Implanted ions lose their energy by undergoing nuclear collisions and coulombic interactions with electrons of the host lattice atoms. Eventually ions come to rest within the solid after some distance R, the range. For a given energy the ranges of lighter ions in a solid are greater than those of heavier ions. Furthermore, along open channels in crystalline materials the ranges are higher because of fewer collisions.

- Ion implantation produces lattice damage. The computer simulation study indicates that at low ion energies, the damage in silicon is in the form of amorphous regions. Experimental studies suggest that with the increasing energy, these regions tend to overlap, leading to a continuous amorphous region.
- On annealing, the defect structure evolves in ion-implanted materials, whose complexity depends on the mass of the ion, dose, ion energy, and annealing temperature.
- During annealing, interactions between dopant atoms and defect structure can occur that could complicate the activation of the dopant.

REFERENCES

Cass, T. R. and V. G. K. Reddi. *Appl. Phys. Lett.* 23, (1973), 268.

Crowder, B. L. and F. F. Morehead Jr. *Appl. Phys. Lett.* 14, 313 (1969).

de la Rubia, T. D. and G. H. Gilmer. *Phys. Rev. Letts.* 74, 2507 (1995).

Eaglesham, D. J.; P. A. Stolk; H.-J. Gossman; T. E. Haynes; and J. M. Poate. *Nuclear Instruments and Methods in Physics Res.* B106, 191 (1995).

Eisen, F. H. *Rad. Eff.* 47, 99 (1980).

Foyt. A. G.; W. T. Lindley; C. M. Wolfe; and J. P. Donnelly. *Solid State Electron* 12, 209 (1969).

Gat, A.; J. F. Gibbons; T. J. Magee; J. Peng; V. R. Deline; P. Williams; and C. A. Evans Jr. *Appl. Phys. Lett.* 32, 276 (1978).

Ghandhi, S. K. *VLSI Fabrication Principles*. New York: John Wiley & Sons, 1983.

Hofker, W. K. and Philips, *Res. Repts.* Suppl. 8 (1975).

Lee, D. H. and J. W. Mayer. *Proc. IEEE* 623, 1241 (1974).

Levige, W. W.; M. J. Helix; K. V. Vaidyanathan; and B. G. Streetman. *J. Appl. Phys.* 48, 3342 (1977).

Lindhard, J.; M. Scharff; and H. Schiott. *Mat-Fys. Med. Dan Vid. Selsk,* 33, 1 (1963).

Narayan, J. and O. W. Holland. *J. Appl. Phys.* 56, 2913 (1984).

Pickar, K. *Ion Implantation in Silicon, Applied Solid State Science 5*. New York, Academic Press, 1975.

Sadana, D. K.; J. M. Zavada; H. A. Jenkinson; and T. Sands. *Appl. Phys. Lett.* 47, 691 (1985).

Seidel, T. E. and A. U. MacRae. *First Int. Conf. on Ion Implantation,* ed. by F. H. Eisen and L. J. Chadderton, Gordon, and Breach. New York: 1971.

Smith, B. "Ion Implantation Range Data for Silicon and Germanium Device Technologies," Research Studies, Forest Grove, OR: 1977.

Stolk, P. A.; H.-J. Gossmann; D. J. Eaglesham, and J. M. Poate; *Nuclear Instruments and Methods in Physics Research* B96, 187 (1995).

Sze, S. M. *Semiconductor Devices, Physics and Technology.* New York: John Wiley & Sons, 1985.

Wolf, S. and R. N. Tauber. *Silicon Processing for the VLSI Era.* Sunset Beach, CA: Lattice Press, 1986.

PROBLEMS

9.1. Project the diamond-cubic lattice, to scale, along [001], [110], and [111] directions. Calculate the radii of the circles that can be inscribed in the channels along the respective directions. Based on the value of the radii, comment on the channeling behaviors of ions along the three directions.

9.2. Compare the dopant profiles obtained by ion implantation and diffusion from a fixed source. Which approach would you prefer for fabricating a buried p-n junction and why?

9.3. A silicon slice is implanted with 200 keV P^+ ions. After implantation the wafer is annealed at a temperature and time where the value of $\sqrt{2Dt}$ is 0.05 μm. Assuming that there is no loss of dopant atoms during the anneal, calculate the final dopant profile.

9.4. A high-current ion implanter has a beam current of 25 mA and is used to implant oxygen ions into silicon substrates. If the wafer holder can accommodate 25 wafers, 100 mm in diameter, and the implant time is five minutes, calculate the dose received by each wafer.

9.5. A p-n junction is to be fabricated by implanting 150 keV As^+ ions into a p-type silicon (background concentration $\sim 5 \times 10^{14}$ cm^{-3}) through a window in a $Si0_2$ mask. (*a*) Estimate the thickness of the mask required for effective blockage of the ions. (*b*) Assuming a dose 5×10^{15} ions/cm^2, calculate the location of the resulting p-n junction.

9.6. (*a*) Estimate the number of gallium and arsenic atoms displaced, for the peak value of the nuclear stopping, by the 200 keV, 10^{13} tellurium ions/cm^2. Assume that E_d is 10 eV; $\left(\frac{dE}{dx}\right)_{\text{nuclear}} = 10^3$ keV/μm

(*b*) If the preceding displaced atoms coalesce together to form 5×10^{17} cm^{-3} circular, faulted loops lying on {111} planes that are bounded by a/3 <111> Frank partials, calculate the diameter of the loops.

9.7. A p-n junction is to be fabricated by ion implantation into p-type silicon doped to 5×10^{16} cm^{-3}. The implantation is carried out through a 25×25 μm window in a silicon dioxide mask. Assuming that 150 keV As^+ ions are used for implantation and that the maximum impurity concentration should not exceed 5×10^{19} cm^{-3}, calculate the total number of the implanted ions.

CHAPTER 10

Metallization

This chapter describes the techniques used for depositing thin conducting films that are used as contacts in the semiconductor technology, considers the evolution of microstructure in thin films, discusses some of the metallizations used for contacts, evaluates the ramifications of the contact-semiconductor interactions, and discusses the principles of diffusion barriers. It also covers interconnects and their electromigration behavior.

10.1 DEPOSITION OF THIN FILMS FOR CONTACTS AND INTERCONNECTS

Metallizations serve two functions in the semiconductor technology. Metallizations that allow an electrical signal to enter in and come out of a semiconductor are referred to as *contacts*. Metallizations that provide connections between different devices and components on a chip are termed *interconnects*.

We discussed in section 2.1.2 that metal-semiconductor junctions, that is, contacts, are of two types: Schottky and ohmic. The current-voltage characteristics of the Schottky contacts are very similar to those of p-n junctions. On the other hand, the current-voltage characteristics of ohmic contacts are linear over the entire range of voltages and the currents to which the contacts are subjected; that is, the ohmic contacts least interfere with the incoming and outgoing electrical signals.

Thin films of metals, alloys, and intermetallics are used for contacts and interconnects. The selective deposition of contacts and the desired pattern for interconnects are achieved by depositing thin films on appropriately masked regions. A variety of deposition techniques are currently available for producing such films: (1) physical vapor deposition (evaporation), (2) sputtering, and (3) chemical vapor deposition. Different principles govern the formation of a film in each technique, leading to differences in conformal or step coverage.

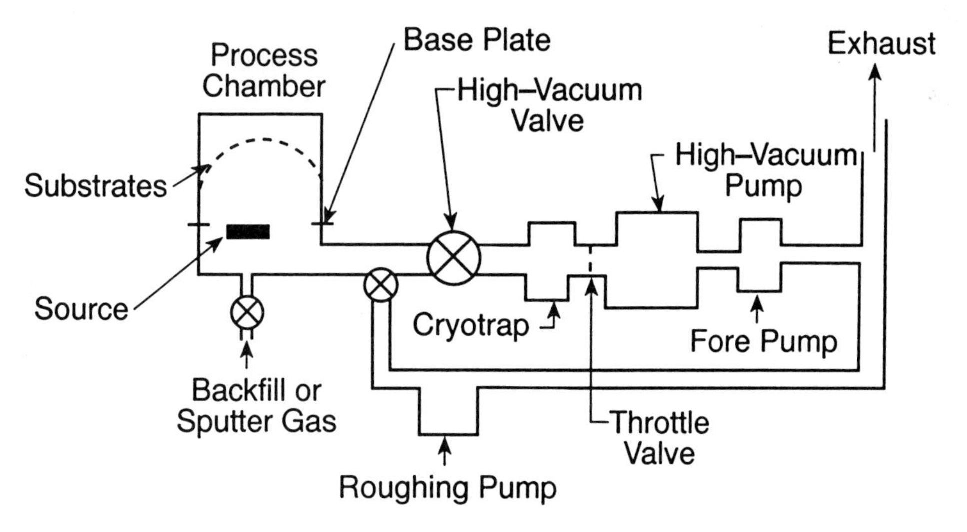

FIGURE 10.1
Schematic of a high-vacuum chamber with substrates mounted on a planetary substrate support above the source material. (*After Fraser [1983].*)

10.1.1 Physical Vapor Deposition (Evaporation)

The schematic in Figure 10.1 shows a typical system that is used for vacuum deposition of metallic thin films. The material to be evaporated is contained in an evacuated bell jar, which is identified as a process chamber in Figure 10.1. The source for the film can be melted using resistive or inductive or electron-beam heating. The substrate to be metallized is arranged on a planetary platten positioned directly above the source. Prior to the deposition of a film, the bell jar is evacuated to a base pressure of 5×10^{-7} torr using roughing and high-vacuum pumps. Subsequently the source is melted. In the high vacuum, atoms or atomic clusters leave the source and condense on the masked substrates, leading to the deposition of films. The temperatures of the source are chosen so as to give a vapor pressure of 10^{-2} torr, which results in reasonable growth rates.

For a given family of devices and chips, the thickness of the deposited films on all substrates undergoing processing simultaneously must be the same. The use of planetary plattens permits this uniformity. Figure 10.2 shows the location of a source and a receiver, that is, a substrate, on the surface of a spherical platten whose radius is r_o. The distance between the source and the receiver is r. The following expression gives the rate of mass loss per unit area R from the source (Glang 1970):

$$R = 4.43 \times 10^{-4} \left[\frac{M}{T}\right]^{1/2} P_e (\text{g/cm}^2\text{-sec}), \tag{10.1}$$

where M is the gram-molecular mass of the source material, T is the source temperature in K, and P_e is the equilibrium vapor pressure in Pa of the source at

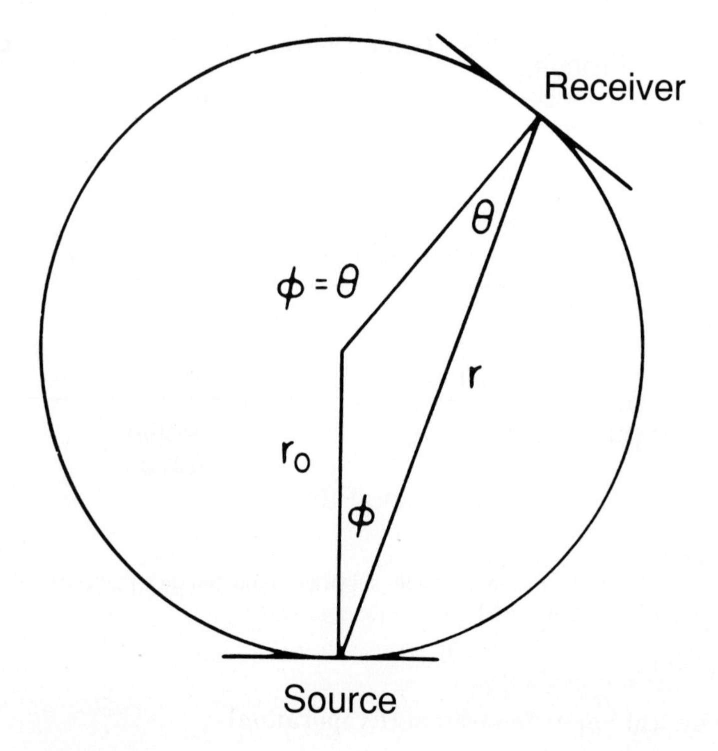

FIGURE 10.2
Schematic of vapor source and receiver mounted on a sphere of radius r_o. (*After Fraser [1983].*)

temperature T. The total loss of material per unit time R_T from the source may be found by integrating R over the source surface area:

$$R_T = \int R\, dA. \tag{10.2}$$

The flux of the source material at the receiver surface depends on the cosine of the angle between the normal to the source surface and the direction linking the source to the receiver ϕ. If θ is an angle between the center-receiver and source-receiver linkages as shown in Figure 10.2, then the deposition rate D is

$$D = \frac{R_T}{\pi r^2} \cos\theta \cos\phi. \tag{10.3}$$

Since the source and the receiver are located on the surface of a sphere, the following relations between ϕ and θ must hold:

$$\cos\phi = \cos\theta = \frac{r}{2r_0}. \tag{10.4}$$

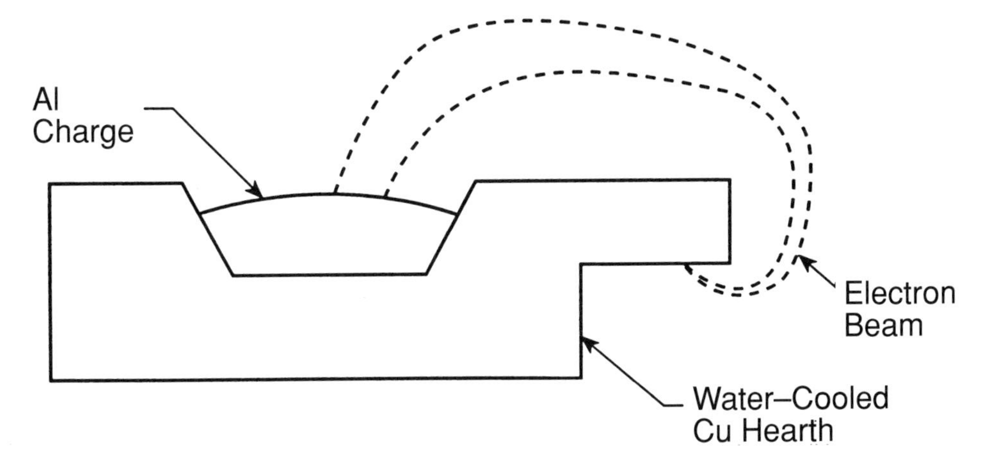

FIGURE 10.3
Electron-beam evaporation system. An imposed magnetic field deflects the electron beam so that the substrates are protected from the electrons. (*After Fraser [1983].*)

Substituting the values of cos ϕ and cos θ in Equation (10.3), we have

$$D = \frac{R_T}{4\pi r_0^2}. \tag{10.5}$$

Equation (10.5) implies that the deposition rate is the same for all points on a spherical surface when the source is also located on the same surface. However, steps on a substrate are not covered satisfactorily, and thinning and microcracks appear in the film because of shadowing effect of the step.

As indicated earlier the source material can be melted using resistive heating. This procedure is accomplished by either placing the material in a miniature basket fabricated from filaments of refractory metals or suspending the source material from coils of filaments. This approach is inexpensive. However, it has two main disadvantages. First, a possibility exists that the heater may contaminate the source material, resulting in a change in composition of the metallization. Second, when an alloy source is used, it is difficult to replicate its composition into the deposited film because of the differences in vapor pressures of atomic species constituting an alloy. The film will be richer in the constituent that has a higher vapor pressure.

High deposition rates can be achieved using induction heating and using boron nitride crucibles to contain the source materials. Dilute aluminum alloys as well as a variety of other source materials can be deposited by this approach.

Figure 10.3 shows an electron-beam evaporation source. Electrons are obtained from a hot filament and are then accelerated through a voltage of 10 keV before striking the source material. Appropriate provisions can be designed into the system so that impurities from the hot filament do not reach the melt. The electron beam can be scanned over the source surface to prevent nonuniform evaporation. Using a large source, thick films may be deposited without breaking the vacuum and recharging the source. By incorporating a number of sources in the deposition chamber, films of different compositions can be deposited sequentially. Depending on the source-substrate distance, deposition rates as high as 0.5 μm/min are feasible.

The major disadvantage of electron-beam heating is that for some metallizations, the 10 keV electrons generate the characteristic K-shell X-rays along with a continuum. These X-rays are able to penetrate into silicon substrates and cause radiation damage. This damage changes the metal-oxide-semiconductor capacitor characteristic of MOSFETs and must be annealed out by heating for high device reliability.

EXAMPLE 10.1. If the residual water vapor pressure is 2×10^{-5} Pa in an Al evaporator at 300 K, what is the oxygen content of a deposited Al film if it is deposited at a rate of 5 nm/sec? Assume that the Al-H_2O reaction results in the incorporation of Al_2O_3 in the film and that each water molecule has a reaction probability of 10^{-3}.

Solution. From the kinetic theory of gases, we can show that the bombardment rate of H_2O molecules per $(cm)^2$ per sec (N) is given by the following expression:

$$N = 6.4 \times 10^{19}\,(T)^{1/2}\,P,$$

where T is the absolute temperature and P is the pressure in Pa. Substituting the values of T and P in the preceding equation, we can calculate N:

$$N = 6.4 \times 10^{19} \times (300)^{-1/2} \times 2 \times 10^{-5}$$
$$= 7.4 \times 10^{13}\ \text{molecules/cm}^2\text{-sec.}$$

Assuming that the Al film is free from defects and voids, the number of atoms contained in $1(cm)^3 = 6.02 \times 10^{22}$.

Therefore, the number of atoms contained in a 5 nm thick Al film whose surface area is $1(cm)^2 = 5 \times 10^{-7} \times 6.02 \times 10^{22} = 3.01 \times 10^{16}$.

The Al atoms react with H_2O molecules to form Al_2O_3 according to the following reaction:

$$2Al + \frac{3}{2}H_2O \longrightarrow Al_2O_3 + \frac{3}{2}H_2.$$

Since the water molecules have a reaction probability of 10^{-3}, the total number of water molecules involved in the formation of $Al_2O_3 = 7.4 \times 10^{13} \times 10^{-3} = 7.4 \times 10^{10}$. Therefore, the oxygen content of 5 nm thick Al film $= 7.4 \times 10^{10}$ atoms/cm^2.

10.1.2 Sputtering

Sputtering refers to a process in which atoms are dislodged from the surface of a solid due to momentum exchange associated with collisions involving high-energy ions. This process is widely used for the deposition of a variety of thin films that find applications in the semiconductor technology. Films of refractory metal silicides and dielectrics can also be deposited using this technique.

The sputtering process can be explained qualitatively by referring to the schematic in Figure 10.4, which shows the sputtering of surface atoms produced by impinging inert gas atoms. As discussed in the case of ion implantation in sec-

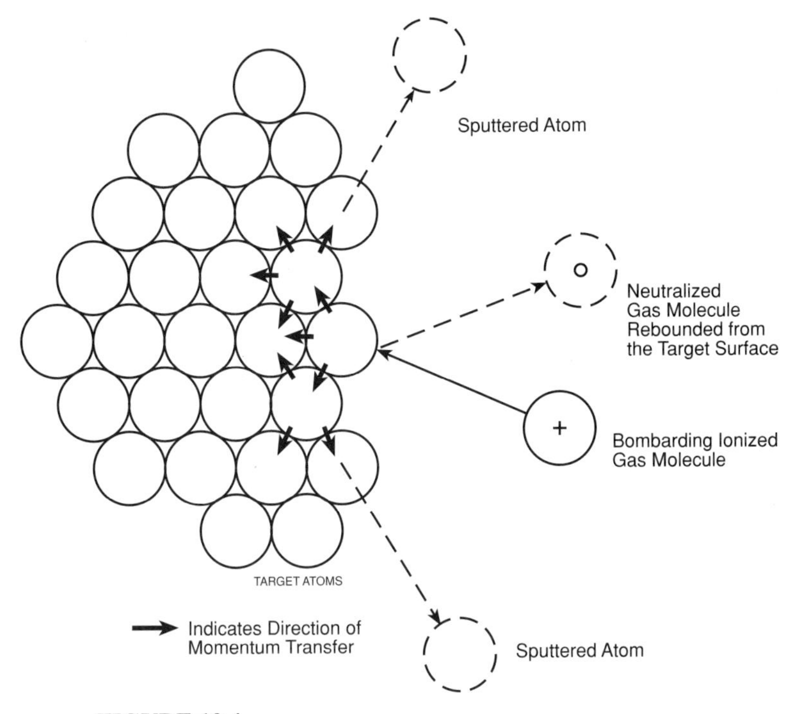

FIGURE 10.4
Collision process responsible for sputtering and generation of fast neutrals. (*After Wolf and Tauber [1986].*)

tion 9.1.1, the impinging ion imparts momentum to the surface atoms via elastic collisions. We can show from Equation (9.3) that for normal incidence the fraction energy (FE) transferred is given by

$$\mathrm{FE} = \left(\frac{\mathrm{T}}{\mathrm{E}_1}\right) = \frac{4\mathrm{M}_1\mathrm{M}_2}{(\mathrm{M}_1 + \mathrm{M}_2)^2}, \tag{10.6}$$

where T is the energy transferred to the target atom of mass M_2 and E_1 is the energy of the incident particle of mass M_1. The sputtering yield S can be expressed as

$$\mathrm{S(at\ constant\ FE)} = \frac{\alpha \mathrm{E}_1}{\mathrm{U}}\left(\frac{\mathrm{M}_2}{\mathrm{M}_1}\right), \tag{10.7}$$

where α is a function of $\left(\frac{M_2}{M_1}\right)$ and varies in a linear manner and U is the heat of sublimation of the target. Equation (10.7) is applicable to various materials.

We discussed in section 9.1.1 that the incident ion transfers its energy to the solid in two ways: electronic and nuclear collisions. The collisions with electrons result in excitation and ionization of atoms, whereas the collisions with nuclei lead to sputtering. The formation of collision cascades is envisioned in which the energy of the incident ion is transferred into atomic motion through a series of separate binary collision events. Each material has a threshold energy (about 5 to 40 eV) for sputtering that is approximately equal to the heat of sublimation of the solid. Sputtering yield increases with energy, reaches a maximum, and then decreases. The sputtering rate J is given by

$$J = 6.23j\ SM/\rho, \tag{10.8}$$

where j is the ion current density, M is the molecular weight, and ρ is the density of the target material. In general, S is a function of mass M_1 and atomic number Z_1 of the sputtering atoms and the kinetic energy E_1 of the incident ion, the angle of incidence θ, the mass M_2, and the atomic number Z_2 of the target atom, and the crystallinity and orientation of the target.

An interesting question is how to control the motion of the source atoms dislodged by sputtering so that they reach the masked substrates. This issue is addressed schematically in Figure 10.5. In Figure 10.5*a* a source and a substrate are contained in a partially evacuated chamber (pressure = a few to 100 m torr) containing argon atoms. If the source is at a negative potential with respect to the substrate, a plasma consisting of positively charged argon ions and electrons may be created between the source and the substrate, Figure 10.5*b*. The positively charged argon ions will move toward the source that is at a negative potential, and their impacts with the source surface will dislodge atoms of the source. The transferred mo-

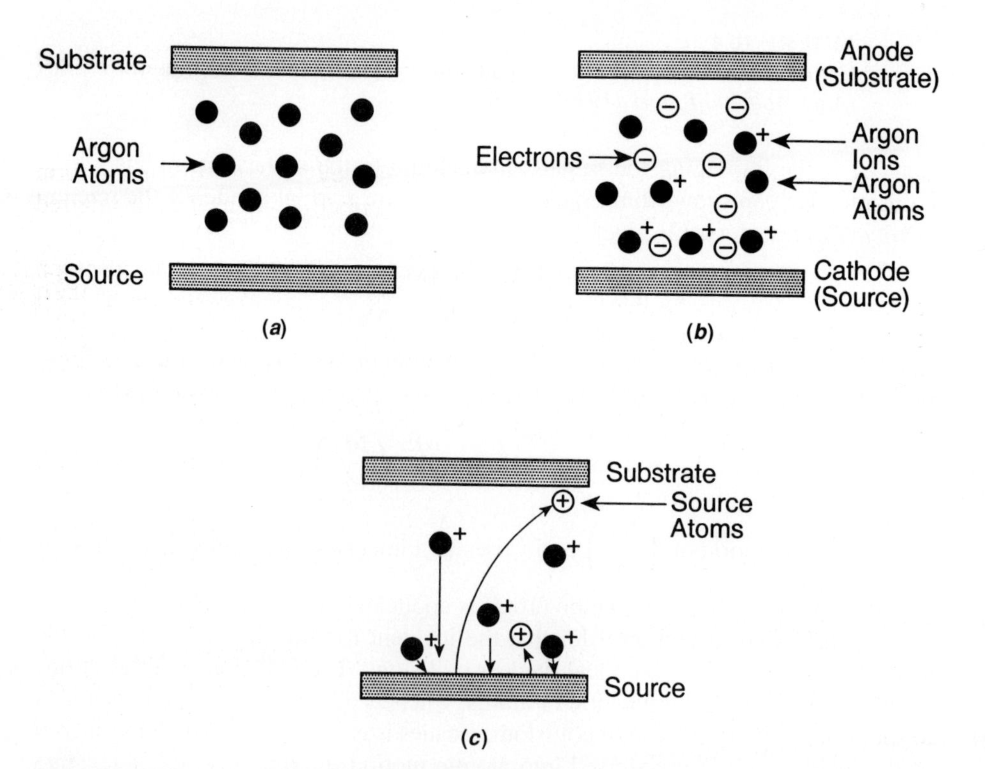

FIGURE 10.5
Schematics illustrating the principle of sputter deposition.

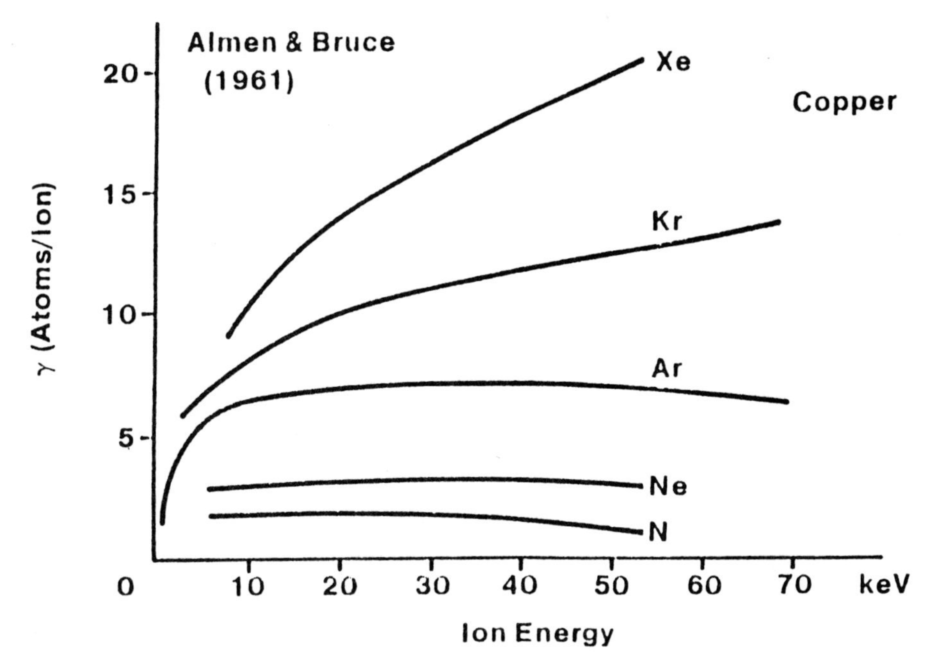

FIGURE 10.6
Sputtering yields of different noble gases from copper as a function of energy.
(*After Almen and Bruce [1961].*)

mentum will carry the atoms to the substrate where they coalesce together to form a thin film as shown Figure 10.5*c*. The process described in Figure 10.5 is referred to as *dc glow-discharge sputtering*.

The sputtering yield S defined as the number of atoms ejected per incident ion in this sputtering process depends on several factors: (1) direction of the incident ions, (2) target material, (3) mass of the bombarding ions, and (4) energy of the bombarding ions. In the energy range of 10 to 50 keV, the yield increases with ion energy and mass as indicated in Figure 10.6 for the sputtering of copper by various noble gas ions (Almen and Bruce 1961).

The dc sputtering cannot be used to sputter dielectric films because the voltage required between the target and the substrate would be extremely high. This problem can be obviated using RF sputtering; simplified dc and RF sputtering systems are compared in Figure 10.7. When an ac signal, below ~50 kHz, is applied to the electrodes, ions are mobile enough to establish a complete discharge at each electrode during each half-cycle; both electrodes alternatively behave as cathodes and anodes. Above 50 kHz two significant effects occur. First, electrons oscillating in the glow region acquire enough energy to cause ionizing collisions. Second, RF voltages can be coupled through some kind of impedance so that the electrodes need not be conductors.

RF sputtering is feasible because the target self-biases to a negative potential. Because electrons are considerably more mobile than ions, electrons have little dif-

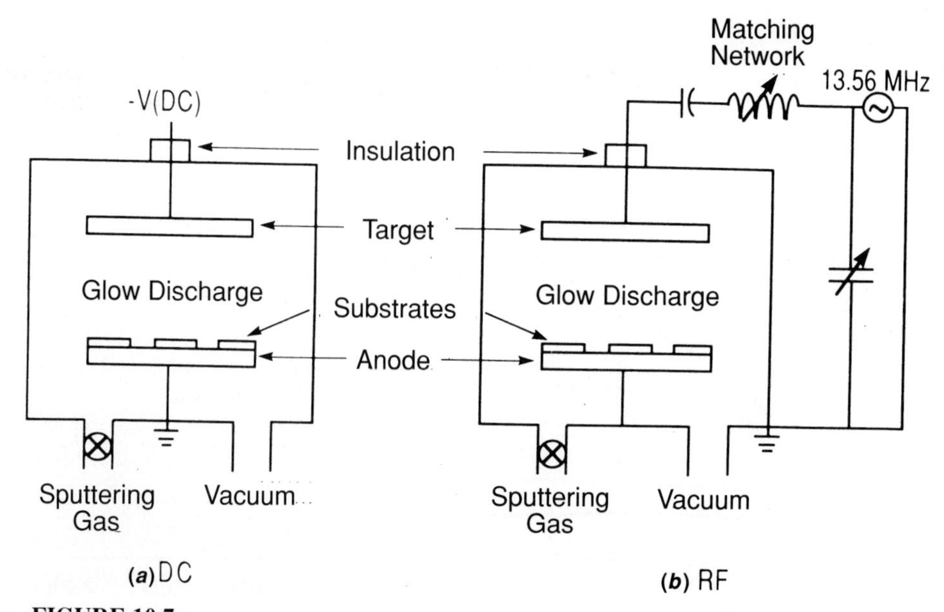

FIGURE 10.7
Schematics of simplified sputtering systems: (*a*) dc and (*b*) RF. (*After Ohring [1992].*)

ficulty in responding to the periodic change in the electric field. Once this periodic change is established, the target behaves like a dc target where the bombardment of positive ions sputter away target atoms. Furthermore, to avoid sputtering from a substrate, the sputter target must be an insulator and be capacitively coupled to an RF generator.

Table 10.1 shows dc sputtering yields for various metals by argon ions (Wolf and Tauber 1986). It is clear that the sputtering yields are fairly low, resulting in low

TABLE 10.1
Sputtering yields for metals in argon in atoms per ion

Target	Atomic Weight/ Density	100 eV	300 eV	600 eV	1000 eV	2000 eV
Al	10	0.11	0.65	1.2	1.9	2.0
Au	10.2	0.32	1.65	2.8	3.6	5.6
Cu	7.09	0.5	1.6	2.3	3.2	4.3
Ni	6.6	0.28	0.95	1.5	2.1	
Pt	9.12	0.2	0.75	1.6		
Si	12.05	0.07	0.31	0.5	0.6	0.9
Ta	10.9	0.1	0.4	0.6	0.9	
Ti	10.62	0.08	0.33	0.41	0.7	
W	14.06	0.12	0.41	0.75		

Source: Wolf and Tauber 1986.

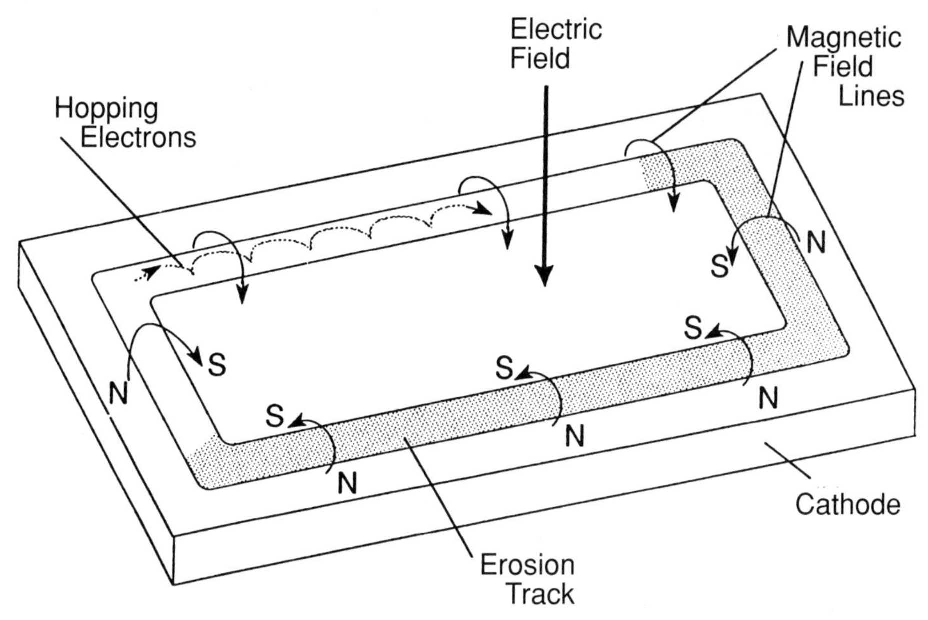

FIGURE 10.8
Applied fields and electron motion in a planar magnetron. (*After Ohring [1992].*)

film-growth rates. This situation can be substantially improved if during dc sputtering a magnetic field is applied parallel to the target and perpendicular to the electric field as shown in Figure 10.8. This process is referred to as *magnetron sputtering* and is achieved by placing bar or horseshoe magnets behind the target. The magnetic field lines first emanate normal to the surface, then are nearly aligned parallel to the surface, and finally return, completing the magnetic circuit. Electrons emitted from the cathode are initially accelerated toward the anode, but when they encounter a magnetic field they undergo a helical motion. In the region of the parallel magnetic field, they are bent in an orbit back to the target. As a result, the electron residence time in the plasma is prolonged and thus the probability of the creation of ions is enhanced. Appropriate orientation of target magnets helps define a track where the electrons hop around at high speed. Sputtering-induced erosion occurs within this track because ionization of the sputtering gas is most intense above it.

Magnetron sputtering is widely used commercially because high deposition rates are possible, for example, up to 1 μm/min for Al. They are an order of magnitude higher than deposition rates attained by conventional sputtering techniques. The electrons also do not bombard the substrate. As a result, substrate heating does not occur.

EXAMPLE 10.2. Electron-impact ionization and secondary emission of electrons by ions control J in a sputtering system; J is given by the following equation:

$$J = \frac{J_0 \exp \alpha d}{1 - \gamma[\exp(\alpha d) - 1]},$$

where J_0 = primary electron current density from an external source
α = number of ions per unit length produced by electrons
γ = number of secondary electrons emitted per incident ion
d = interelectrode spacing.

If the film-deposition rate during sputtering is proportional to the product of J and the sputtering yield S, calculate the proportionality constant for Cu in this system. Assume the deposition rate is 25 nm/min for 1 keV Ar ions, $\alpha = 0.1$ ion/cm, $\gamma = 0.08$ electrons/ion, d = 10 cm, and $J_0 = 100$ mA/cm^2.

Solution. Substituting appropriate values in the above equation, J is given by

$$= \frac{100 \times 10^{-3} \exp(0.1 \times 10)}{1 - 0.08[\exp(0.1 \times 10) - 1]}$$

$$= 0.316 \text{ A/cm}^2.$$

From Table 10.1 the sputtering rate S is 3.2 atoms per ion.

Since the deposition rate D = KJS, where K is the constant of proportionality, we can calculate K by substituting appropriate values for D, J, and S:

$$K = \frac{D}{JS}$$

$$= \frac{250 \times 10^{-8}}{0.316 \times 3.2}$$

$$= 2.5 \times 10^{-6} \left(\frac{\text{Ion cm}^3}{\text{min-Amp-atoms}} \right).$$

10.1.3 Chemical Vapor Deposition

Chemical vapor deposition (CVD) is a technique for depositing thin conducting films. It has several very attractive features: (1) the conformal nature of the deposit, that is, good step coverage, (2) the ability to deposit films simultaneously on a large number of substrates, and (3) the simplicity of the equipment required. Unlike physical vapor deposition, which suffers from shadowing effects and poor step coverage, low-pressure CVD can produce step coverage over a wide range of step profiles.

The principle of CVD is quite simple. As discussed in the case of the growth of epitaxial layers in Chapter 6, reactions among appropriate chemical species produce films of desired compositions. Figure 10.9 shows a simplified view of a low-pressure CVD reactor (Fraser 1983). In this case the reactor is surrounded by a furnace and is considered a “hot wall” system. However, if substrates are heated inductively, walls of the reactor remain cold, and the reactor is called a “cold wall” reactor. As shown in Figure 10.9, the reactants enter the reactor via the gas manifold. They react with each other in the vicinity of the hot substrates, resulting in thin films.

During the last decade the deposition of tungsten by CVD for IC applications has been pursued using tungsten hexafluoride as the source for W. Both thermal decomposition and reduction by hydrogen have been considered. For example, WF_6 can be pyrolyzed at high temperature to produce W films.

$$WF_6 + \text{heat} \longrightarrow W + 6HF \tag{10.9}$$

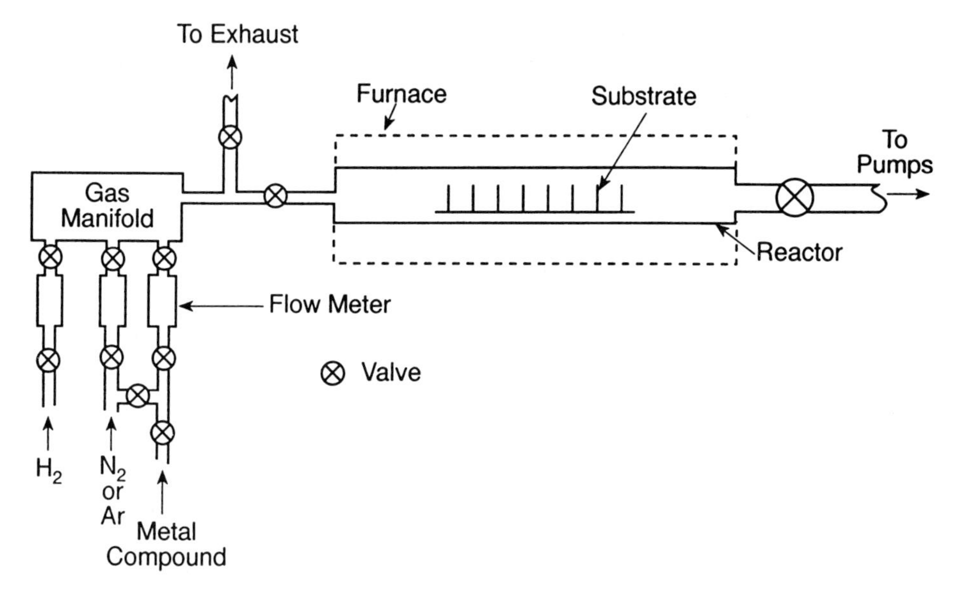

FIGURE 10.9
Schematic of a low-pressure CVD reactor system. To enhance reaction rates, the furnace could be augmented by a plasma source, intense light source, or other energy source. (*After Fraser [1983].*)

Alternatively, hydrogen gas may be used to reduce WF_6.

$$WF_6 + 3H_2 \longrightarrow W + 6HF \tag{10.10}$$

Deposition temperatures range from 60 to 800°C in various reactors. The reduction reaction in Equation (10.10) probably occurs at a much lower temperature than that for pyrolysis, given by Equation (10.9).

The deposition of Al films via the CVD route has been attempted using tri-isobutyl aluminum. These molecules appear to undergo two-stage decompositions (Fraser 1983).

$$\begin{aligned}[(CH_3)_2CH - CH_2]_3\,Al \xrightarrow{150^\circ C} & [(CH_3)_2CH - CH_2]_2\,AlH \\ & + (CH_3)_2\,C = CH_2\end{aligned} \tag{10.11}$$

$$[(CH_3)_2CH - CH]_2\,AlH \xrightarrow{250^\circ} Al + \frac{3}{2}H_2 + 2(CH_3)_2\,C = CH_2 \tag{10.12}$$

10.2 MICROSTRUCTURE OF THIN FILMS

To understand the evolution of microstructure in a thin film, the knowledge of various stages in its growth is essential. For the purpose of this discussion, we can divide growth into two stages: (1) nucleation of a film in the form of islands and (2) their subsequent growth leading to coalescence and a continuous film.

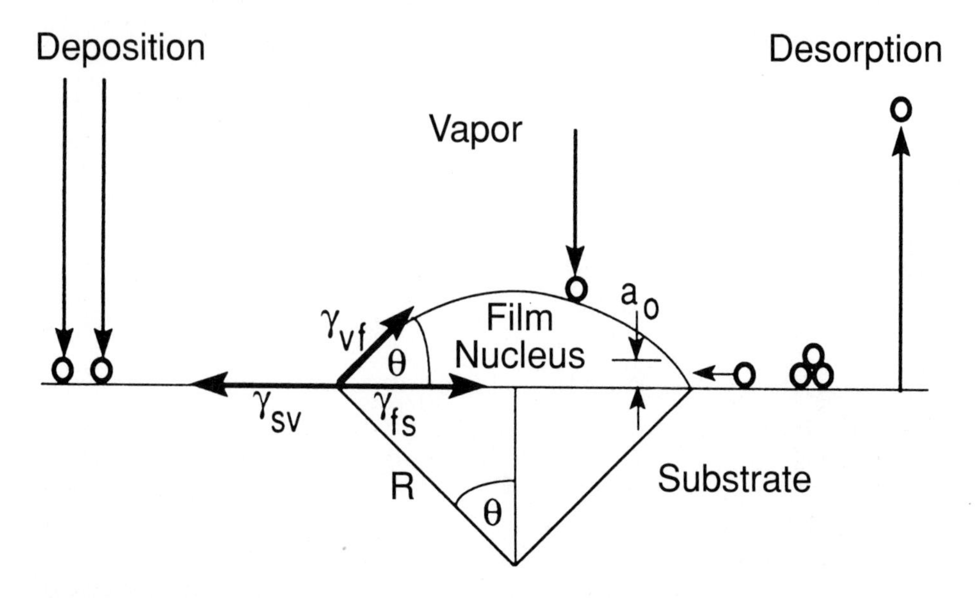

FIGURE 10.10
Schematic of atomic processes occurring on a substrate surface during vapor deposition. (*After Ohring [1992]*.)

The nucleation theory developed for the solid-solid transformations in materials science may be extended to understand the formation of nuclei during the source vapor-solid transformation. Figure 10.10 shows basic atomic processes occurring on the substrate surface during vapor deposition. Some of the incoming atomic species coalesce together at the substrate surface into a nucleus that is in the form of a hemisphere of radius R, and a fraction of the incoming atoms may undergo desorptions as shown in Figure 10.10. All the atoms required to form the nucleus do not coalesce together at once. The likely scenario is that the formation of the nucleus involves the migration of atoms on the surface.

The free-energy change accompanying the formation of the nucleus shown in Figure 10.10 is given by (Ohring 1992)

$$\Delta G_v = a_1 R^3 \Delta G_V + a_2 R^2 \gamma_{V,F} + a_3 R^2 (\gamma_{F,S} - \gamma_{S,V}), \qquad (10.13)$$

where ΔG_V is the volume free-energy change associated with the vapor-solid transition and $\gamma_{V,F}$, $\gamma_{F,S}$, and $\gamma_{S,V}$ are vapor-film, film-substrate, and substrate-vapor interfacial energies. For the nucleus shown in Figure 10.10, the volume ($a_1 R^3$), the curved surface area ($a_2 R^2$), and the projected circular area on the substrate ($a_3 R^2$) are involved, and the corresponding values of a_1, a_2, and a_3 are $\pi\ (2 - 3\ \cos\theta + \cos^3\theta)/3, 2\pi(1 - \cos\theta)$, and $\pi \sin^2\theta$, where θ is the contact angle.

The critical nucleus size R* can be determined by differentiating ΔG in Equation 10.13 with respect to R and then equating $\dfrac{d\Delta G_v}{dR} = 0$. Using this procedure, we can derive R* as

$$R^* = \frac{-2(a_2\,\gamma_{V,F} + a_3\,\gamma_{F,S} - a_3\,\gamma_{S,V})}{3a_1\,\Delta G_v} \qquad (10.14)$$

The nuclei whose sizes are smaller than R* will be unstable and will disintegrate. In addition, the nuclei will have a range of sizes exceeding R*.

The new nuclei will continue to form on the substrate surface as long as the internuclei spacings are larger than the surface migration distance of the atoms before they desorb back into the vapor. If the residence time of atoms on the surface is t_s and D_s is their surface diffusion coefficient, then the migration distance is $2\sqrt{D_s t_s}$. When the migration distance is larger than the internuclei separation, no new nuclei may form. Instead, the impinging atoms will migrate on the surface and will be adsorbed at ledges or steps present on the surfaces of the nuclei. Alternatively, some of the additional atoms, that is, the adatoms, may land directly on the existing nuclei, resulting in their growth.

The density of the nuclei depends on substrate temperature and deposition rate. When the substrate temperature is reduced, the driving force for the vapor-solid transformation is increased; that is, ΔG_V goes up. We see from Equation (10.14) that R* will be reduced, implying that the nuclei, which are unstable at higher substrate temperatures, can be stabilized by lowering the substrate temperature. Its ramification is that a larger number of smaller nuclei will form in the case of a low-temperature deposition. Likewise, the effect of deposition rate on R* can be analyzed using the following equation due to Ohring (1992):

$$\Delta G_V = \frac{k_B T}{\Omega}\left(\frac{\dot{x}}{\dot{x}_E}\right), \tag{10.15}$$

where k_B is the Boltzmann constant, T is the substrate temperature, Ω is the atomic volume, $\dot{x}$ is the deposition rate and $\dot{x}_E$ is the equilibrium evaporation rate from the nucleus at the substrate temperature. When $\dot{x}$ is increased, ΔG_V goes up. The effect is to stabilize smaller nuclei, an effect analogous to that of lowering the substrate temperature. In summary, a greater number of small nuclei will form when a thin film is deposited at either lower temperature or higher deposition rate, resulting in a finer grain size.

Once the film has been nucleated, several processes are involved in its evolution into a continuous film: (1) growth of the nuclei by the attachment of adatoms to steps or ledges, (2) Ostwald ripening, (3) sintering, and (4) cluster migration. These processes are illustrated in Figure 10.11. The incoming atomic species tend to attach to surface steps and kinks because of the higher density of bonding sites. As a result of these attachments, the islands grow laterally. The thermodynamic activities of adatoms at surfaces depend on their curvatures: the smaller the radius, the higher the activity. In Figure 10.11a, a_1 and a_2 refer to the activities of adatoms in the vicinity of nuclei of radii r_1 and r_2. Consequently, mass transport can occur from smaller to larger nuclei. Ultimately, some of the smaller nuclei may disappear, whereas the other nuclei may grow larger. The nuclei can also coalesce by sintering and lateral motion of smaller nuclei as illustrated in Figure 10.11. The driving force for both of these processes is the reduction in overall surface area of the nuclei. Figure 10.12, reproduced from the study of Pashley and Stowell (1966), shows successive electron micrographs of gold nuclei deposited on molybdenite at 400°C. Two nuclei in the center of each micrograph appear to coalesce by sintering.

Let us now imagine a situation in which thick films of a material are deposited at different temperatures. If the initial grain size is smaller than the film

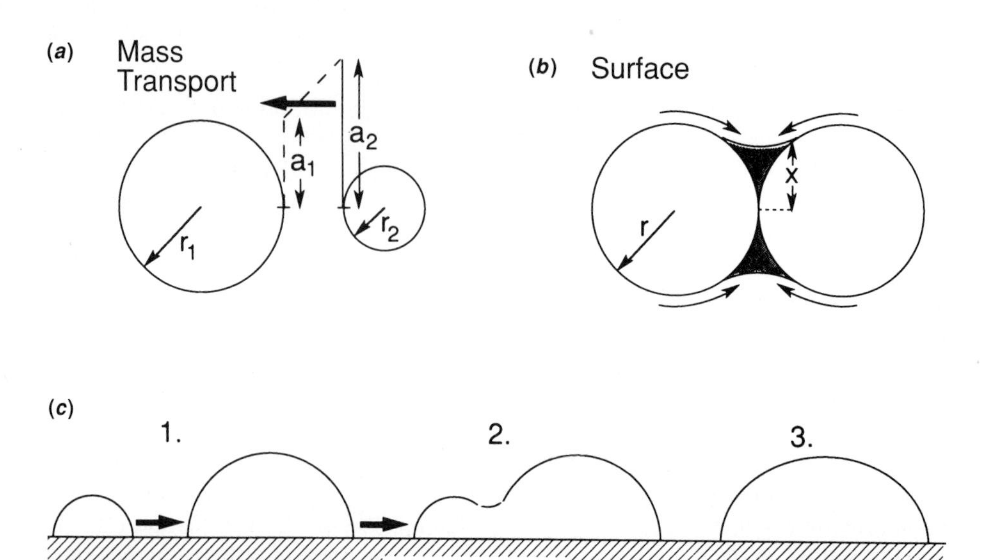

FIGURE 10.11
Coalescence of islands due to (*a*) Ostwald ripening, (*b*) sintering, and (*c*) cluster migration. a_1 and a_2 are adatom activities in the vicinity of islands of radii r_1 and r_2, respectively. (*After Ohring [1992].*)

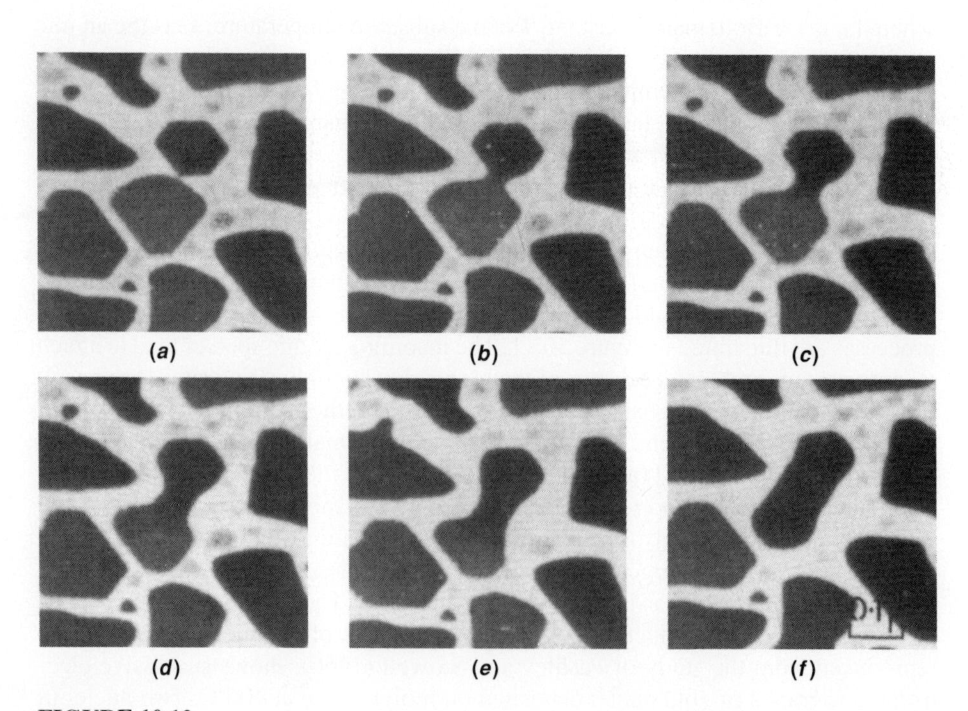

FIGURE 10.12
Successive electron micrographs of gold islands deposited on molybdenite at 400°C illustrating coalescence by sintering: (*a*) arbitrary zero time, (*b*) 0.06 sec, (*c*) 0.18 sec, (*d*) 0.50 sec, (*e*) 1.06 sec, and (*f*) 6.18 sec. (*After Pashley and Stowell [1966].*)

thickness, we can visualize the thick film to consist of a number of thin films, where each thin film may evolve by the formation of nuclei and their coalescence. Since the preceding layer can serve as a template for the deposition of the next layer, epitaxial effects may be observed in the microstructures of a thick film. In addition, at high deposition temperatures grain growth can occur following the formation of a film. Figure 10.13 illustrates the effect of deposition temperature T_S, normalized with respect to melting temperature of a film T_M, on resulting grain size. The resulting situation is referred to as a *zone model* for evaporated films. In zone I T_S/T_M is less than 0.2, and grains are equiaxed. This condition is likely to be due to the absence of boundary migration during the growth of the

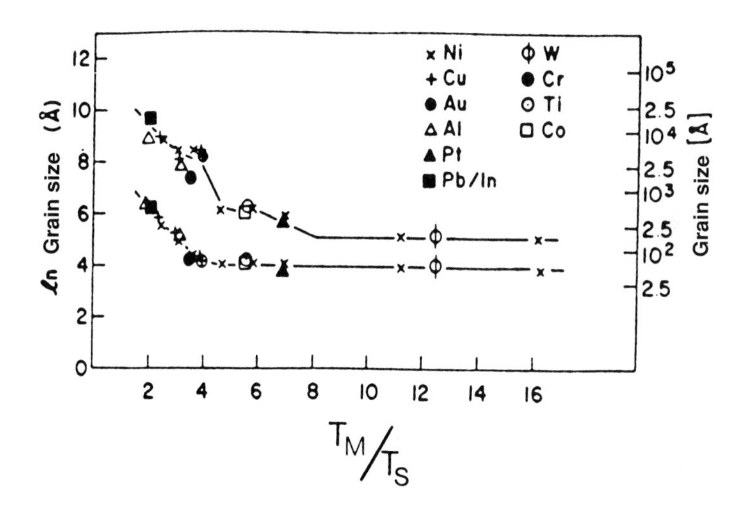

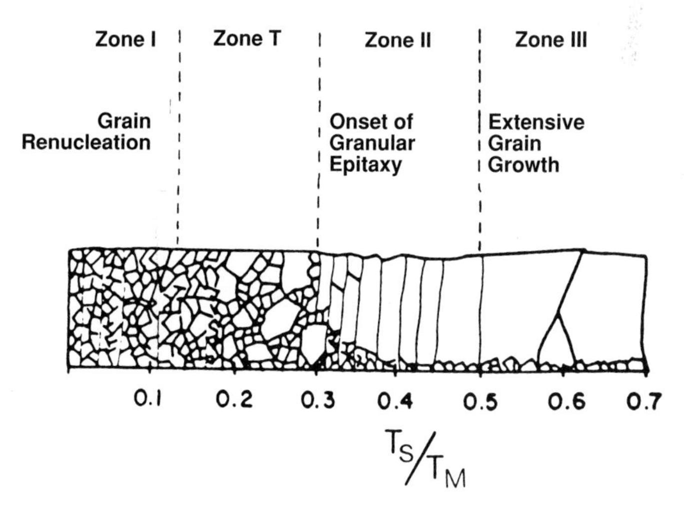

FIGURE 10.13
Zone model for evaporated metal films. (*After Hentzell et al. [1984].*)

film, a consequence of the low deposition temperature. In the range $0.2 < T_S/T_M < 0.3$, some grains are larger and are surrounded by small grains. This zone is called a *transition zone* T. The development of the bimodal grain-size distribution in this zone very likely occurs by the migration of highly mobile grain boundaries. In zones II and III, the effects of epitaxy are manifested. In addition, extensive grain growth occurs in zone III.

The preceding discussion assumes that the film is growing under normal incidence of source atoms. If this condition is not satisfied, the growing film columns may exert shadowing effects on their neighbors. In Figure 10.14 a computer-simulated microstructure of a Ni film during deposition shows this effect (Müller 1985). The principal effect of the oblique incidence is the incorporation of a large

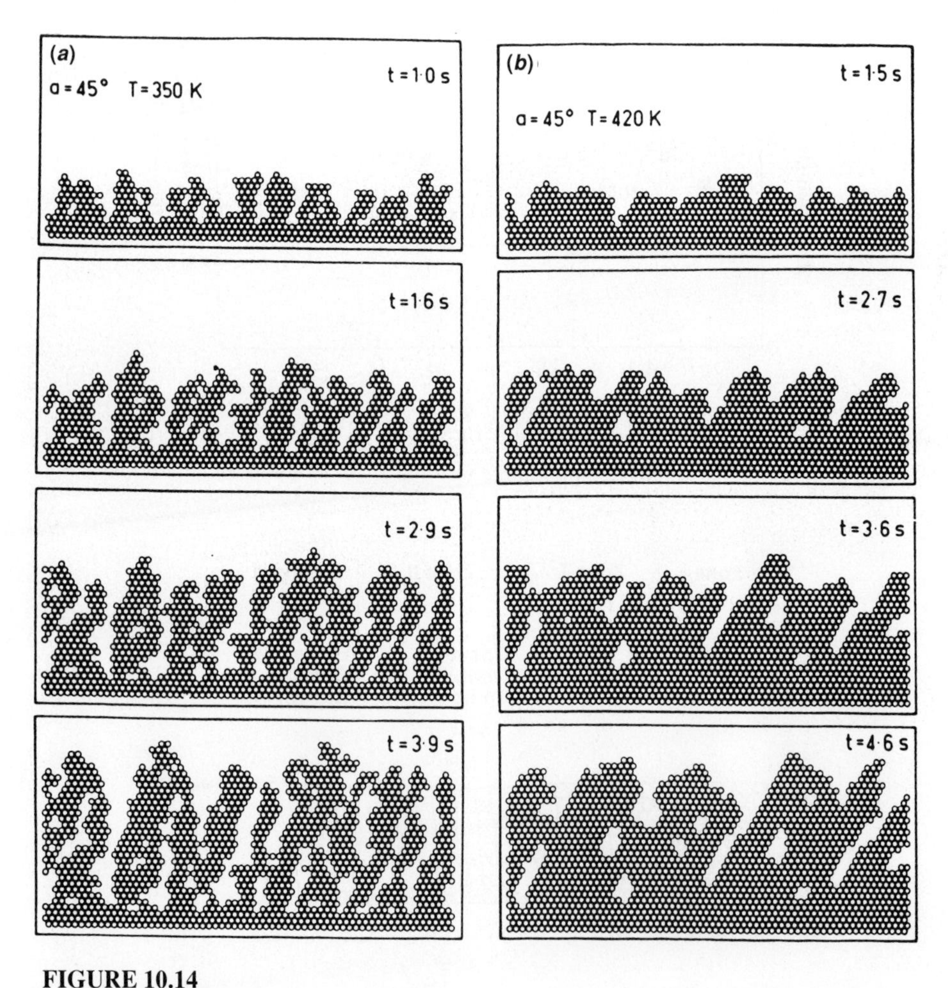

FIGURE 10.14
Computer-simulated microstructure of nickel film during deposition at different times for substrate temperatures of (*a*) 350 and (*b*) 420 K. The angle of deposition is 45°. (*After Müller [1985].*)

number of pores in the film. As shown in Figure 10.14, the volume of the pores is affected by the deposition temperature: the higher the temperature, the lower the pore volume.

Voids, whose sizes are considerably smaller than those of pores discussed above, are also observed in thin films deposited using a variety of techniques. Figure 10.15 shows an example of microvoids seen in an evaporated Au film (Nakahara 1979). Since as-deposited thin films can have excess vacancy concentrations of 10^{-2}, the vacancy supersaturation can be substantially reduced when the vacancies coalesce together to form voids. The driving force for this process is the reduction in energy associated with many isolated vacancies.

The microstructure of sputtered films has also been satisfactorily explained in terms of the zone model. In zone I, T/T_M is less than 0.3, and the grains are tapered with domed tops that are separated from each other by voids. There is a high density of dislocations, and the diameter of the crystals increases with T_s/T_M. In this zone the adatom diffusion is insufficient to overcome the shadowing effects. High points of the surface receive more flux than valleys, thereby inducing open

FIGURE 10.15
Transmission electron micrograph showing microvoids in an evaporated gold film. (*After Nakahara [1979].*)

boundaries. In zone II, $0.3 < T_s/T_M < 0.5$, columnar grains are separated by grain boundaries, and the surface is smoother. Grains grow with an activation energy about the same as for surface diffusion. In zone III, $0.5 < T_s/T_M < 1$, and we have equiaxed grains. The diameter of the grain grows at a rate governed by an activation energy that is close to the value for bulk diffusion.

Grain growth occurs both during deposition and postdeposition annealing of films. Grain growth results in an increase in the average size of the grain and also brings about changes in the orientation of the grains. In columnar grain structure the average grain size in the plane of the film is considerably smaller than the average grain size perpendicular to the film. Equiaxed grains follow a normal grain growth behavior:

$$r^m - r_0^m = \alpha t, \tag{10.16}$$

where r is the average size of the grain at time t, r_0 is the initial size at time $t = 0$, and α is a constant. The exponent m is generally around two and follows an Arrhenius-type dependence on temperature. Deviations from Equation (10.16) are expected for columnar grains. Normal grain growth terminates once the average grain size is comparable to the film thickness. Hence when columnar grains intersect both the top and bottom of the deposit, normal grain growth stops. Grain growth in thin films is a function of multiple variables: temperature, substrate and its surface features, stress, ion bombardment, and composition of films.

EXAMPLE 10.3. Two spherical nuclei of radii R_1 and R_2 coalesce in the gas phase to form a spherical nucleus of radius R_3. Given that the mass is conserved in the coalescence process and the surface energy is γ, calculate the energy reduction.

Solution. Total volume of the two nuclei $= \frac{4}{3}\pi R_1^3 + \frac{4}{3}\pi R_2^3$.

Mass of the two nuclei $= \frac{4}{3}\pi\rho(R_1^3 + R_2^3)$, where ρ is the density of the material.

Mass of the resulting nucleus $= \frac{4}{3}\pi\rho R_3^3$.

Since the mass is conserved during coalescence, the two masses must be equal.

Therefore
$$\frac{4}{3}\pi\rho R_3^3 = \frac{4}{3}\pi\rho(R_1^3 + R_2^3)$$

that is
$$R_3 = (R_1^3 + R_2^3)^{1/3}.$$

The total surface energy of nuclei prior to coalescence $= 4\pi\gamma(R_1^2 + R_2^2)$.

The total surface energy after coalescence $= 4\pi\gamma(R_1^3 + R_2^3)^{2/3}$.

Reduction in surface energy $= 4\pi\gamma\left[(R_1^2 + R_2^2) - (R_1^3 + R_2^3)^{2/3}\right]$.

EXAMPLE 10.4. If the density of 10 nm diameter voids in an aluminum film deposited at 50°C is $5 \times 10^{14}/cm^3$, calculate the number of vacancies required for their formation. Compare the required number to the thermal equlibrium concentration of vacancies in 1 gm-mol aluminum at 50°C.

From Table 3.1 the tetrahedral radius of an Al atom = 0.126 nm.

The number of vacancies required to form a 10 nm diameter

$$\text{void} = \frac{(50)^3}{(1.26)^3} = 6.25 \times 10^4.$$

The total number of vacancies required to form

$$\text{the voids} = 5 \times 10^{14} \times 6.25 \times 10^4$$
$$= 3.1 \times 10^{19}/\text{cm}^3.$$

The equilibrium concentration of vacancies in 1 gm-mol of Al (N), that is, the Avagadro number, can be computed using Equation (3.11)

$$n_D/N = \exp\left(\frac{-E_D}{k_B T}\right).$$

$E_D = 0.5$ eV, $k_B = 8.6 \times 10^{-5}$ eV/K, and T = 323 K.

$$n_D/N = \exp\left(\frac{-0.5}{8.6 \times 10^{-5} \times 323}\right)$$

$$n_D/N = 1.5 \times 10^{-8}$$

Substituting $N = 6.022 \times 10^{23}$ in the preceding expression, we have

$$n_D = 1.5 \times 10^{-8} \times 6.022 \times 10^{23}$$
$$= 9.03 \times 10^{15}.$$

This concentration is much smaller than that required to form the voids.

10.3 CONTACT METALLIZATIONS

We indicated earlier that there are two types of contacts: Schottky barrier and ohmic. The underlying physics of the two types of metal-semiconductor junctions and their behaviors under equilibrium and forward and reverse biases are discussed in section 2.1.2. The role of Schottky diodes is firmly established in the semiconductor technology because they perform functions that no other junction devices can accomplish. Ohmic contacts are used in a very large number of devices, such as field-effect transistors, bipolar junction transistors, photodetectors, light-emitting diodes, double-heterostructure lasers, and solar cells. Therefore, this type of contact is an integral part of all sophisticated semiconducting devices.

The contacts should have good mechanical properties, demonstrate superior corrosion resistance, and must adhere well to the semiconductor during device processing and operation. Stresses at the contact-semiconductor interface must be low so that metallizations do not lift off. High interfacial stresses may also affect the properties of the underlying semiconductor.

10.3.1 Films for Schottky Contacts

The following are desirable characteristics of Schottky contacts: (1) the film must adhere to the semiconductor after deposition, during subsequent device processing, and

TABLE 10.2
Barrier heights of some metals and silicides on n-type silicon

Metal	Barrier height (eV)
Al	0.72
Cr	0.61
Mo	0.68
Ni	0.61
Pt	0.90
CoSi	0.68
$CoSi_2$	0.64
NiSi	0.66
$NiSi_2$	0.70
PtSi	0.84
Pd_2Si	0.72–0.75
$TaSi_2$	0.59
$TiSi_2$	0.60
WSi_2	0.65

also in service; (2) the film should not introduce excessive stress in the semiconductor because this stress may alter the electronic characteristics of the contacts; and (3) the Schottky barrier height should be maintained during the device operation.

Table 10.2 lists the barrier heights of some of the metals and silicides on n-type silicon. This table shows that Schottky barrier heights, ranging from 0.60 to 0.90 eV, can be easily accessed. However, the listed values can be achieved only if the metal-semiconductor interface is clean. Therefore, the need for surface cleanliness during the fabrication of a Schottky contact cannot be overemphasized.

Single metals, such as Al, Pt, Ni, and so on, are thermodynamically unstable on silicon; that is, they tend to react with each other. For example, Al and its alloys react with silicon during high-temperature processing, transforming an Al-Si Schottky diode into a p^+-n junction. Likewise, Pt reacts with silicon at high temperature to form PtSi. This reaction lowers the barrier height by $\sim$0.06 eV. On the other hand, various silicides are thermodynamically stable on silicon; that is, the activities of silicon in the silicides and in the silicon are essentially the same at the silicide-silicon interface. We can thus fabricate Schottky diodes whose current-voltage characteristics do not change during high-temperature processing. Consequently, silicides such as $TaSi_2$ and $TiSi_2$ have emerged as preferred metals for Schottky diodes in the silicon technology. An additional advantage is that producing silicide in situ by diffusing the metal into silicon ensures the cleanliness of the silicide-silicon interface.

Basically, three approaches are used to form silicides: (1) deposition of the pure metal on single crystal or polycrystalline silicon, (2) coevaporation of the silicon and the refractory metal from two separate sources, and (3) sputter deposition of the

TABLE 10.3
Methods of silicide formation

Method of formation	Advantages	Disadvantages
Direct metallurgical reaction. M + xSi ⇒ M Si_x Metal deposited by evaporation, sputter, or CVD.	Selective etch possible.	[M]/[Si] depends on phase formed; sensitive to sintering environment; rough surface.
Coevaporation from an independent Si and M source.	Smooth surface; sintering environment not as critical.	[M]/[Si] control difficult but possible; no selective etch possible; poor step coverage.
Cosputtering from independent Si and M targets.	Good control of [M]/[Si]; smooth films; sintering environment not as critical; deposition of sandwich possible.	Difficult calibration to achieve [M]/[Si] control.
Sputtering from a composite MSi_x target.	Excellent [M]/[Si] control if correct target chosen; good step coverage.	Contamination from target.
Chemical vapor deposition: atmospheric, low pressure, or plasma enhanced.	High throughput; excellent step coverage.	Rough surface; [M]/[Si] control difficult but possible; possible poor adhesion.

Source: Wolf and Tauber 1986.

silicide either from a composite target or by cosputtering or layering. In most cases the deposition is followed by an anneal to form the silicide. Table 10.3 lists the advantages and disadvantages of the respective approaches.

Table 10.4 lists the barrier heights of some metals on n-type GaAs. Au and Ag produce excellent Schottky diodes, but they form various intermetallics with GaAs during high-temperature processing. As discussed in the case of silicon, no single metal can yield thermally stable contacts. To obviate this problem, Sands, Palstrøm, and co-workers (1990) devised stable, epitaxial Schottky contacts consisting of intermetallics of transition-metal group III metal and rare earth arsenides. Table 10.5 lists the values of barrier heights for some intermetallics and arsenides on n-type III-V semiconductors. The Sands-Palmstrøm approach for producing stable contacts for III-Vs is conceptually identical to the silicide approach for silicon.

TABLE 10.4
Barrier heights of some metals on n-type GaAs

Metal	Barrier height (eV)
Ag	0.88–0.93
Al	0.80
Au	0.9–0.95
Pt	0.86–0.94

TABLE 10.5
Schottky barrier heights of some intermetallics and arsenides on n-type GaAs

Metal	Barrier height (eV)
NiAl	0.73
CoAl	0.76
CoGa	0.76
NiAs	0.73
ScAs	0.81
ErAs	0.84

Source: Sands et al. 1990.

EXAMPLE 10.5. Calculate the thickness of the metal film required to produce a 0.1 μm thick $TiSi_2$ layer on the surface of silicon by reacting a Ti layer with the underlying silicon. Assume that the surface area of Si is $1(\text{cm})^2$.

Solution. Volume of $TiSi_2 = 10^{-5} \times 1 = 10^{-5}\ \text{cm}^3$.

Mass of the film $= \rho_{TiSi_2} \times 10^{-5}$ gms.

Molar mass of $TiSi_2 = 104$.

Total number of Ti and Si atoms in the film

$$= \frac{6.02 \times 10^{23} \times \rho_{TiSi_2} \times 10^{-5}}{104}$$

$$= 5.79 \times 10^{16} \rho_{TiSi_2}.$$

Number of Ti atoms in the film $= 1.93 \times 10^{16} \rho_{TiSi_2}$.

Mass of Ti consumed in the formation of $TiSi_2$.

$$= \frac{48 \times 1.93 \times 10^{16} \rho_{TiSi_2}}{6.02 \times 10^{23}}$$

$$= 15.39 \times 10^{-7} \rho_{TiSi_2}.$$

$$\text{Volume of the Ti film} = \frac{1.54 \times 10^{-6} \rho_{TiSi_2}}{\rho_{Ti}}.$$

Since the area of the film is $1(\text{cm})^2$, thickness of the Ti film consumed =

$$t_{Ti} = \frac{1.54 \times 10^{-6} \rho_{TiSi_2}}{\rho_{Ti}}.$$

10.3.2 Films for Ohmic Contacts

The desirable characteristics of ohmic contacts are identical to those of Schottky contacts. In addition, the contact resistance should be very low and I-V characteristics should be linear.

The deposition of a metal film on a clean semiconductor surface generally results in a Schottky barrier. Therefore, to produce ohmic contacts, approaches must be developed to lower the barrier height substantially so that the current-voltage characteristics become linear. The diagrams in Figure 10.16 show some of the approaches that have been developed to lower the barrier height on n-GaAs (Murakami et al. 1989, and Sands et al. 1990). If a thin layer of the semiconductor surface is heavily doped, the metal-semiconductor barrier can be overcome by tunneling. The resulting contact is referred to as a *tunneling contact*. The barrier can

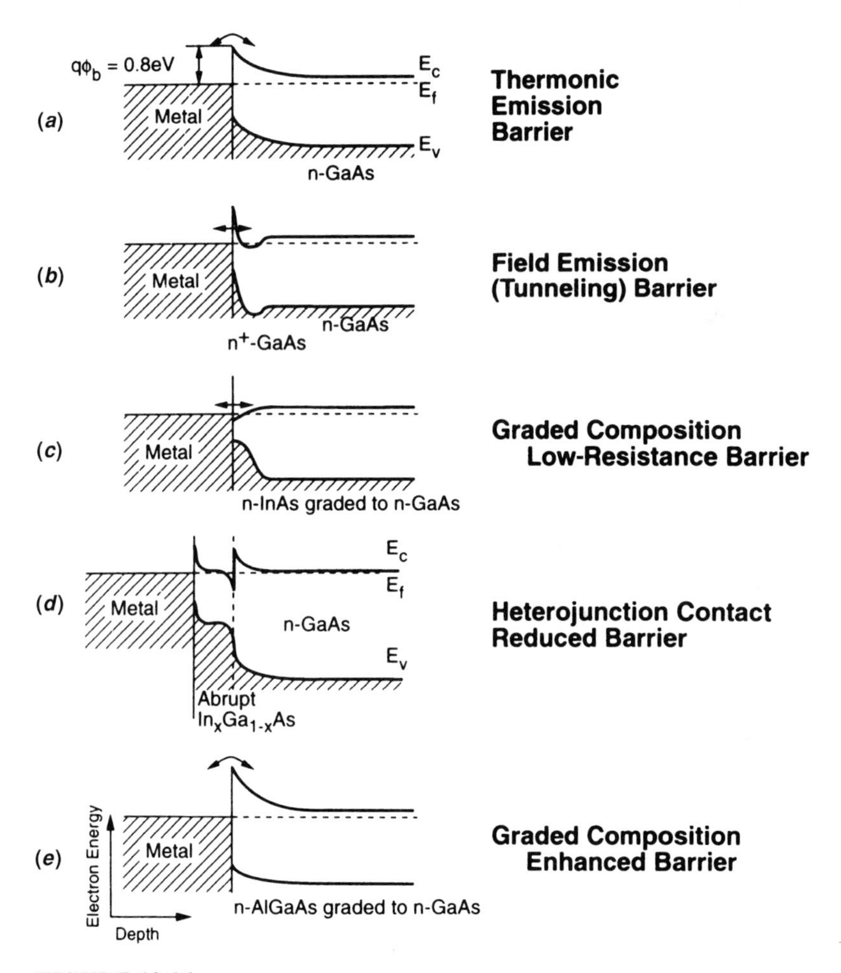

FIGURE 10.16
Schematic band diagram of metal-semiconductor contacts illustrating the metallurgical approaches to barrier control. The top panel shows the 0.8 eV thermionic emission barrier conventionally observed for metals on GaAs. The accompanying panels illustrate decreased barrier heights via tunneling, graded composition, and heterojunction in compound interfaces as well as increased barriers (bottom panel) resulting from larger band-gap interface compounds. (*After Murakami et al. [1989] and Sands et al. [1990].*)

also be reduced by composition grading so that the metal is in contact with a material having a lower band gap. Even though Figure 10.16 illustrates the situation for GaAs, the preceding approaches are generic in nature.

In the silicon technology aluminum is used for the formation of ohmic contacts. These contacts are formed by annealing an evaporated aluminum film on a silicon substrate. The composite is subsequently annealed between 450 and 550°C. During the anneal aluminum diffuses into silicon producing a p^+ surface layer, whereas silicon may out-diffuse into aluminum to form an aluminum-silicon alloy layer. The contact is formed between the alloyed layer and the p^+ region and is ohmic in nature due to tunneling. We can also form tunneling-type ohmic contacts to highly doped source and drains of MOSFETs using silicides.

Tunneling-type contacts have also been developed for III-V materials. An important example is Au-Ge-Ni contacts to n-GaAs. When multilayer structures consisting of GaAs/Au/Ge/Ni are annealed around 500°C, the contact becomes ohmic and has low resistance. We envisage that during annealing, Ge in-diffuses into GaAs. As a result, the surface region is heavily doped, and tunneling occurs between this region and a NiAs film that is observed at the metal-semiconductor interface (Murakami et al. 1989). Using a similar approach, Pd-based ohmic contacts have also been developed for n-GaAs (Wang et al. 1990, 1991, and 1992).

10.4 CONSEQUENCES OF METAL-SEMICONDUCTOR INTERACTIONS

We infer from section 10.3.2 that to form an ohmic contact, metal-semiconductor composites must be annealed at relatively high temperatures. Since chemical activity differences between the atomic species constituting the metal-semiconductor contact exist, species must interdiffuse until their respective activities are equal. Some of the consequences of interdiffusion are discussed below.

Junction spiking. As indicated in section 10.3.2, the processing of deposited Al films for contacts in the silicon technology typically involves a 400°C anneal. This anneal enables the aluminum to reduce the thin native SiO_2 film and adhere to silicon, resulting in lower contact resistance. Referring to the Al-Si phase diagram shown in Figure 10.17, we can see that at this temperature Si dissolves into Al up to 0.3 wt%. During annealing, Si from the substrate diffuses into Al via grain boundary paths to satisfy the solubility requirement. Concomitantly, Al migrates into Si in an attempt to balance the activity differences. If we assume that the native oxide has pin holes, the penetration of Al into Si will be deeper in regions that are contiguous to the pin holes. When enough Al penetrates at a particular location, a conducting metal filament shorts the underlying p-n junction, and junction "spiking" is said to occur. Figure 10.18 illustrates various events leading to spiking.

A plausible solution for the preceding problem is to use Si-Al alloy films containing about 1 wt % Si. The presence of Si in the film eliminates the need for out-diffusion of silicon from the substrate. However, another complication arises during processing of these films. During the heating and cooling cycles, Si is first held in solution in Al but then precipitates on grain boundaries as the Si solubility is reduced

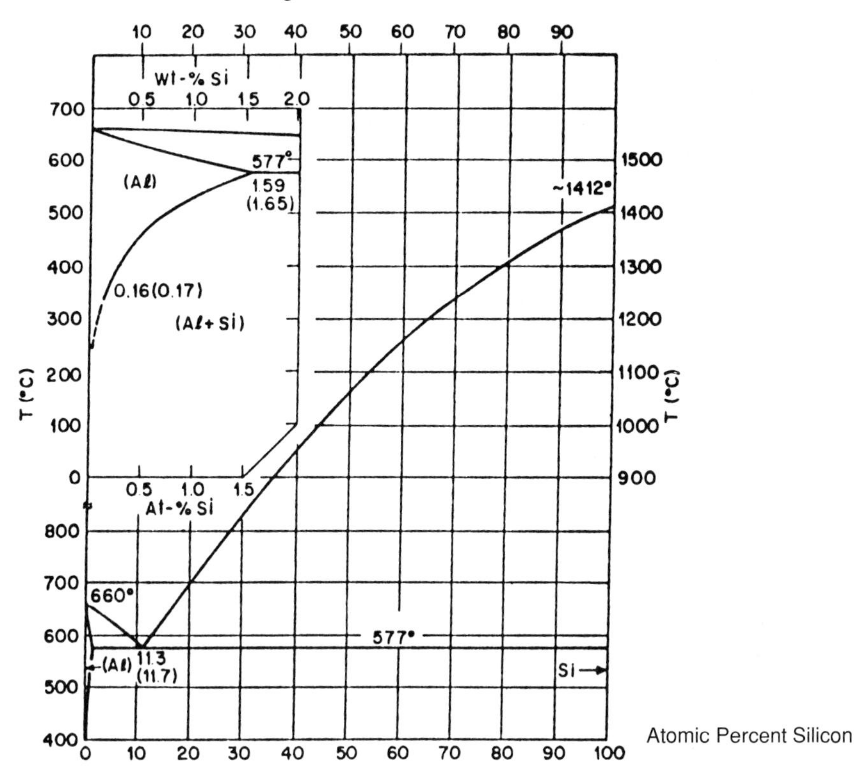

FIGURE 10.17
Aluminum-silicon phase diagram.

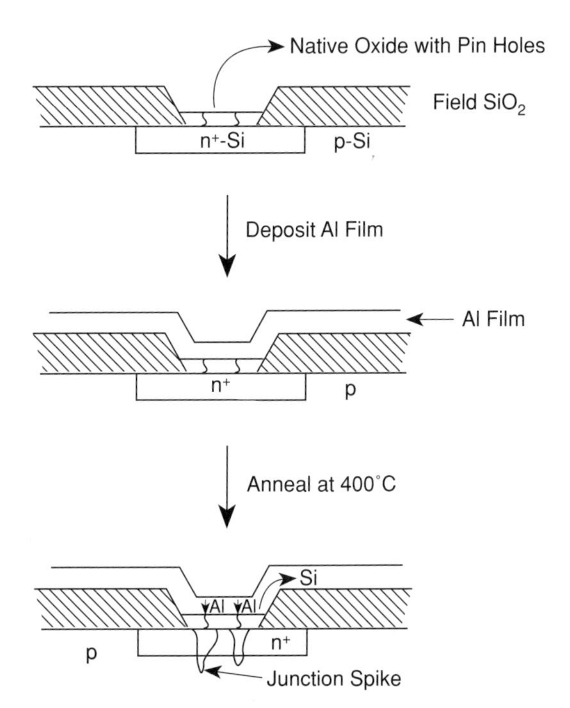

FIGURE 10.18
Schematic illustrating the formation of junction spikes due to Al-Si interdiffusion reactions.

during cooldown. These precipitates grow, may reach the contact, and affect its electrical properties.

Kirkendall porosity. The preceding discussion of junction spiking assumed that a differential does not exist between the diffusivities of Al into Si and Si into Al. Generally, there is a differential in the diffusivities. Consider a diffusion couple consisting of A and B species and assume that the diffusivity of B into A is larger than that of A into B. When this couple is annealed at high temperatures, more B atoms would diffuse into A than A atoms would diffuse into B. As a result, vacancies are created in the B component. These vacancies could cluster together to form pores that are referred to as *Kirkendall porosity.* If the porosity is contiguous to the metal-semiconductor junction, adhesion of the contact as well as its electrical characteristics may be affected. The excess B atoms in A may also come together to form precipitates.

Formation of intermetallics during device aging. To assess the long-term reliability of light-emitting devices (LED) and double-heterostructure lasers, devices are aged around 200°C under current injection. A typical InP/InGaAsP LED is shown in Figure 10.19; note that metallizations are gold-based alloys. When this device is aged for about 2000 hours, dark spot defects (DSD) develop in the active layer; examples of DSDs observed in aged LEDs are shown in Figure 10.20. A number of studies have shown that DSDs are gold-based intermetallics and are associated with dislocations and stacking faults (Mahajan et al. 1984).

We visualize that DSDs result from in-diffusion of Au into the active layer where it combines with the host lattice atoms to form intermetallics. This suggestion

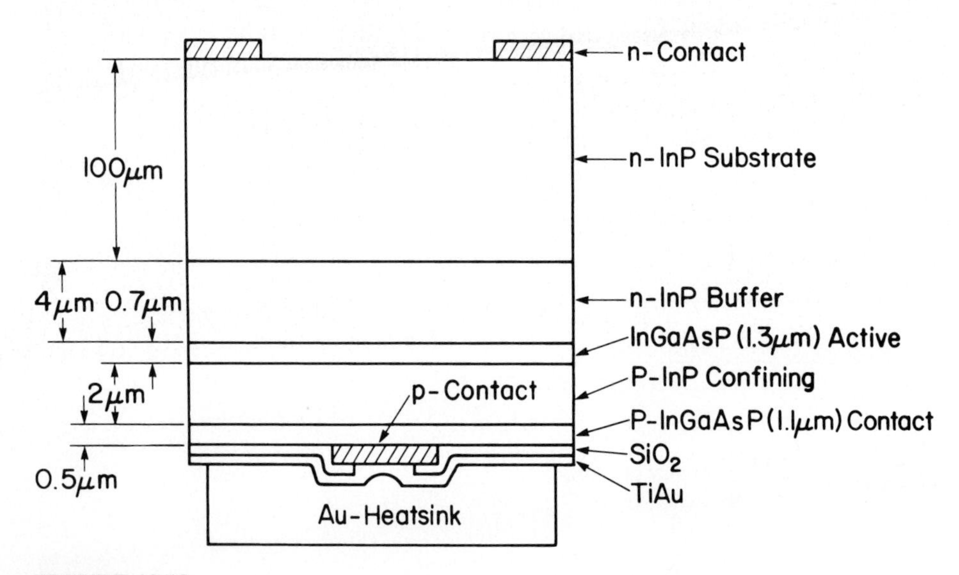

FIGURE 10.19
Schematic of a dielectrically isolated, front-emitting InP/InGaAsP LED. (*After Mahajan [1989].*)

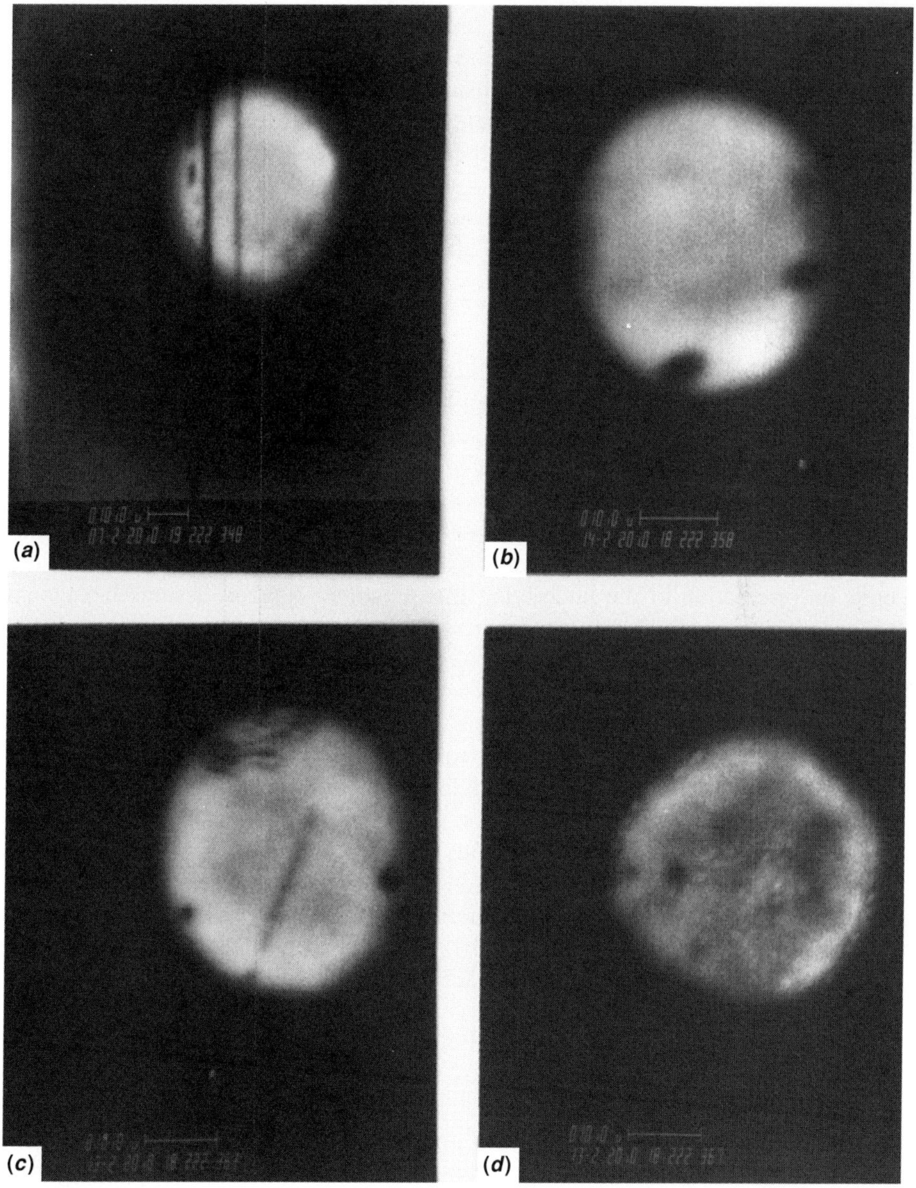

(a) (b) (c) (d)

FIGURE 10.20
(a, b, c and d) Spatially resolved cathodoluminescence images of the light-emitting region observed through the p-InP confining layers of various LEDs. (*After Mahajan et al. [1984].*)

is reasonable in view of the fact that the diffusivity of Au into InP at 400°C is $\sim 10^{-8}$ cm^{-2}/sec. Since the volumes of the intermetallics formed are likely to be different from those of the host lattice consumed in their formation, dislocations and stacking faults may form to accommodate these volume differences. DSDs also affect current injection during the device operation (see section 2.2.2) because they are electrically conducting, leading to local heating. This effect, together with the possibility of nonradiative recombinations at dislocations and stacking faults, may lead to the nonluminescent regions in Figure 10.20. An obvious solution to this problem is to eliminate Au from the device metallization. Titanium-based metallizations have been developed for this purpose.

Contact spreading. In LEDs current confinement is effected using extremely small circular p-contacts ($\sim$25 μm in dia.), which spread during the contact anneal and are no longer circular. The morphology observed after annealing of a circular gold contact on a (001) InP surface is shown in Figure 10.21 (Elias et al. 1987). The assigned crystallography is consistent with the assumption that the origins of In- and P-sublattices are at 0, 0, 0 and 1/4, 1/4, 1/4. After annealing, the contact has assumed a rectangular shape. Elias et al. (1987) have shown that the final shape depends on orientation of the underlying InP substrate: contacts assume hexagonal shapes on $(111)_{In}$ and $(\bar{1}\bar{1}\bar{1})_P$ planes, whereas a distorted hexagon is seen on a (110) surface.

Elias et al. (1987) have attempted to rationalize the preceding observations as follows. When an Au-InP couple is annealed, Au diffuses into the semiconductor and forms reaction pits that are bounded by $(111)_{In}$, $(11\bar{1})_{In}$ and $(\bar{1}11)_P$, $(1\bar{1}1)_P$ facets. Since Au has much stronger driving force to react with the $\{111\}_{In}$ facets than with the $\{\bar{1}\bar{1}\bar{1}\}_P$ facets, the recession rate of the In-facets is much greater than that of the P-facets. This anisotropy leads to wider spreading of the contact along the [110] direction.

The above observations may have ramifications in metal semiconductor field-effect transistors (MESFETs) based on III-V materials. On a (001) surface there are two possible orientations for a gate contact, that is, [110] and $[1\bar{1}0]$ directions. If the longer dimension of the metal contact is along the [110] direction, its spreading

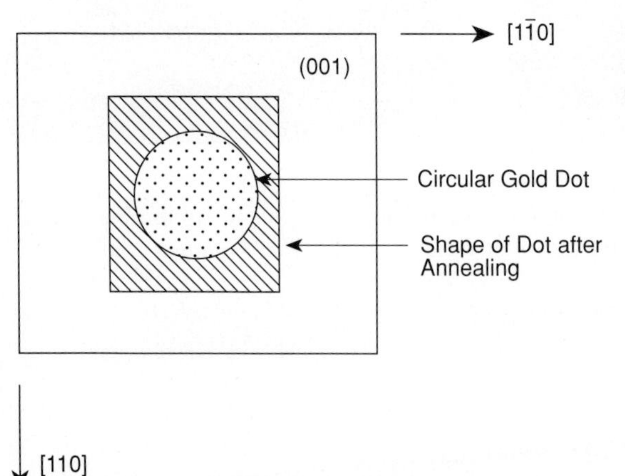

FIGURE 10.21
Schematic showing morphology of a circular gold contact on a (001) InP surface after annealing. (*After Elias et al. [1987].*)

during the anneal would be very small. As a result, the designed separations between the source-gate and gate-drain contacts will be maintained. If on the other hand, the longer dimension is along the $[1\bar{1}0]$ direction, control of the designed separations will not be possible because of extensive spreading during the contact anneal.

EXAMPLE 10.6. Suppose gold-based metallizations are used as contacts in the LED structure shown in Figure 10.19. What is the maximum time the structure can be aged at 200°C without the migration of gold into the active layer?

Solution. Assume that the active layer is separated from the top contact by 3 μm and the diffusivity of gold D into InP is 10^{-8} cm^2/sec.

The distance x that gold migrates during annealing $= 2\sqrt{Dt}$, where t is the annealing time.

$x = 3\ \mu m$ and $D = 10^{-8}\ cm^2/sec.$

$$\text{Therefore } t = \frac{9 \times 10^{-8}}{4 \times 10^{-8}}$$

$$= 2.25 \text{ sec}$$

10.5 DIFFUSION BARRIERS

We can conclude from the examples discussed in section 10.4 that inserting thin film diffusion barriers between metals and semiconductors could either reduce or eliminate the effects associated with interdiffusion. Ideally, a barrier layer B sandwiched between two materials A and C should have the following attributes (Nicolet 1978):

- B should constitute a kinetic barrier to the movement of A and C atoms; that is, the diffusivity of A and C in B should be small.
- The solubility of B in A and C should be small, and B should be thermodynamically stable with respect to A and C at the highest processing temperature.
- B should adhere well to both A and C, form low-resistance interfaces, and have high electrical and thermal conductivities.

Researchers have explored the use of many materials as barriers between silicon and Al contacts. These materials include refractory metals, transition-metal compounds, silicides, and dual-layer barriers, such as refractory metal-silicide and transition-metal–compound-silicide combinations. Table 10.6 lists some of these materials and the reaction products.

To select a barrier layer for a particular diffusion couple, three approaches have been proposed: (1) barriers having blocked grain boundaries, (2) passive compound barriers, and (3) sacrificial barriers. Barriers having blocked grain boundaries suppress rapid diffusion along grain boundaries. The marked improvement in Mo and Ti-W films containing small amounts of N or O as diffusion barriers is attributed to this mechanism.

Transition-metal nitrides, carbides, and borides exhibit nearly ideal barrier characteristics because they are chemically inert and have negligible mutual solubility and diffusivity. TiN films have been extensively explored for device applications.

TABLE 10.6
Aluminum-diffusion barrier—Silicon contact reactions

Diffusion barrier	Reaction temperature (°C)	Reaction products	Failure mechanisms
Cr	300	Al_7Cr	C (EC = 1.9 eV)
V	450	Al_3V, Al-V-Si	C (EC = 1.7 eV)
Ti	400	Al_3Ti	C (EC = 1.8 eV)
Ti-W	500		D
ZrN	550	Al-Zr-Si	C
PtSi	350	Al_2Pt, Si	C
Pd_2Si	400	Al_3Pd, Si	C
NiSi	400	Al_3Ni, Si	C
$CoSi_2$	400	Al_9Co_2, Si	C
$TiSi_2$	550	Al-Ti-Si	D
$MoSi_2$	535	$Al_{12}Mo$, Si	D
Ti-Pd_2Si	435	Al_3Ti	C
W-$CoSi_2$	500	$Al_{12}W$	C
TiN-PtSi	600	AlN, Al_3Ti	C
TiC-PtSi	600	Al_4C_3, Al_3Ti	C
TaN-NiSi	600	AlN, Al_3Ta	C

Note: C = compound formation; D = diffusion.
Source: after Wittmer 1984.

In the sacrificial barrier approach, B reacts with both A and C and is totally consumed. The resulting products are thermodynamically more stable with respect to each other. A Ti layer inserted between Al and Si films is an example of this type of barrier. On annealing, Ti reacts with Al and Si forming $TiAl_3$ and $TiSi_2$.

In summary, the inherent instability of a metal-semiconductor contact can be alleviated by the incorporation of a diffusion barrier layer. This approach certainly ensures reliability but entails additional processing steps.

10.6 FILMS FOR INTERCONNECTS

In integrated circuits various components on a chip, that is, transistors, diodes, resistors, capacitors, and inductors, need to be connected to each other. The components are linked with each other using thin conductor lines, that is, interconnects, that are separated from each other and the underlying semiconductor by an insulating layer.

The interconnects must also have good mechanical properties, high conductivity, and excellent corrosion resistance. They must adhere well to dielectrics separating them from underlying semiconductors. This adhesion should be maintained during the lifetime of a chip. The metallizations used for interconnects should also be metallurgically compatible with the Schottky and ohmic contacts; that is, interactions between the contacts and the interconnect should be minimal. Otherwise, the occurrence of the reactions would lead to the deterioration of the electrical properties of the contacts.

Interconnects should be deposited readily, should be patternable easily with high resolution, should be relatively soft and ductile so that they can withstand temperature cycling during service, must have high electrical conductivity, and should be capable of carrying high current densities. Aluminum films satisfy most of these requirements. Therefore, they are extensively used as interconnects in the silicon technology.

Aluminum films bond well to the SiO_2 and Si_3N_4 insulating layers after a relatively short anneal because Al reacts with the insulating layers to form thin layers of Al_2O_3 and AlN. The bonding occurs between Al_2O_3 and SiO_2 or AlN and Si_3N_4. In addition, Al films can be deposited readily using physical vapor deposition and magnetron sputtering. They are readily wet-etched into fine-line patterns without attack on the underlying insulating layers. Finally, these interconnects are compatible with the Al contacts discussed in section 10.3.2.

The Al interconnects have some disadvantages. First, since Al is electronegative in nature, it is prone to corrosion in acidic and basic environments (Learn 1976). It is also soft and can be easily scratched during wafer and chip handling. Both of these problems can be circumvented by coating the processed wafer with a protective film of either phosphosilicate glass or Si_3N_4. In addition, these interconnects are susceptible to a failure mode known as electromigration at high current densities. We discuss this topic in section 10.7.

In MOSFETs interconnection delays depend on the so-called RC time constant, where R and C are, respectively, the effective total resistance and capacitance of the device at the gate and interconnection level. The higher the value of RC, the slower

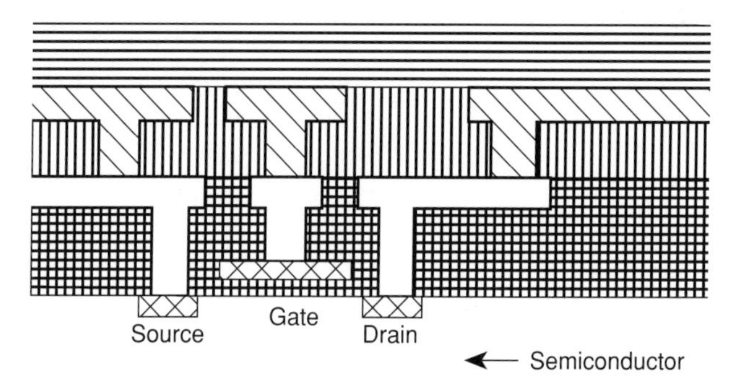

FIGURE 10.22
Schematic showing multilevel interconnects in a MOSFET. The cross-hatched regions represent the source, drain, and gate contacts. Heavy-bordered regions represent different levels of metal interconnects isolated by the insulator films. (*After Murarka [1995].*)

the speed of the device. We can write RC as (Murarka 1995)

$$RC = \frac{\rho}{t}\frac{L^2\varepsilon_{ox}}{t_{ox}}, \tag{10.17}$$

where ρ, L, and t are, respectively, the resistivity, length, and thickness of the interconnect and ε_{ox} and t_{ox} are the permittivity and thickness of the oxide that separates the interconnect from the semiconductor. With decreasing device dimensions and increasing circuit complexities, the length L of the interconnects is becoming larger. To keep RC low, we would like to reduce L. We can effectively reduce L if interconnections are carried out in more than one plane, separated from each other by insulating layers. We refer to this approach as a *multilevel interconnection scheme* (Murarka 1993). The schematic in Figure 10.22 illustrates this concept.

10.7 ELECTROMIGRATION IN INTERCONNECTS

Electromigration refers to mass transport in thin film conductors when they are subjected to high, direct current densities. In silicon technology involving Al interconnects, electromigration is manifested as hillocks on film surfaces as shown in Figure 10.23*a*, as bridges between two conductor lines as depicted in Figure 10.23*b*, and as a discontinuity in a conductor line as shown in Figure 10.23*c*. As the tech-

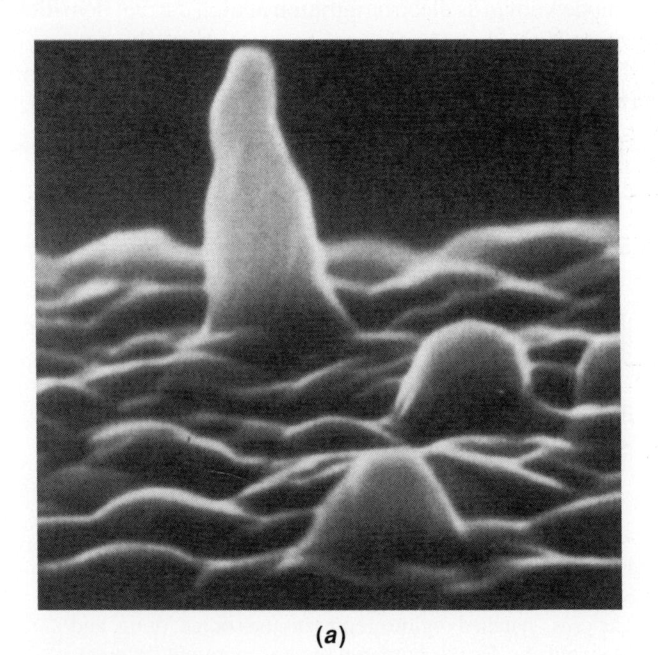

(*a*)

FIGURE 10.23*a*
Manifestations of electromigration damage in aluminum films: hillock formation. (*After Vaidya et al. [1980].*)

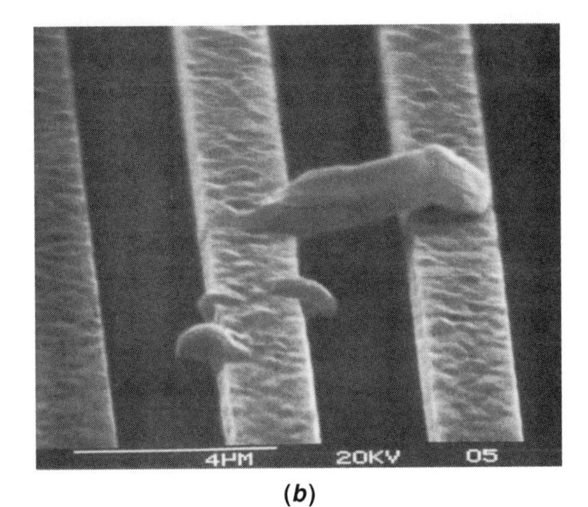

(*b*)

FIGURE 10.23*b*
Manifestations of electromigration damage in aluminum films: whisker bridging between two conductor lines. (*After Vaidya et al. [1980].*)

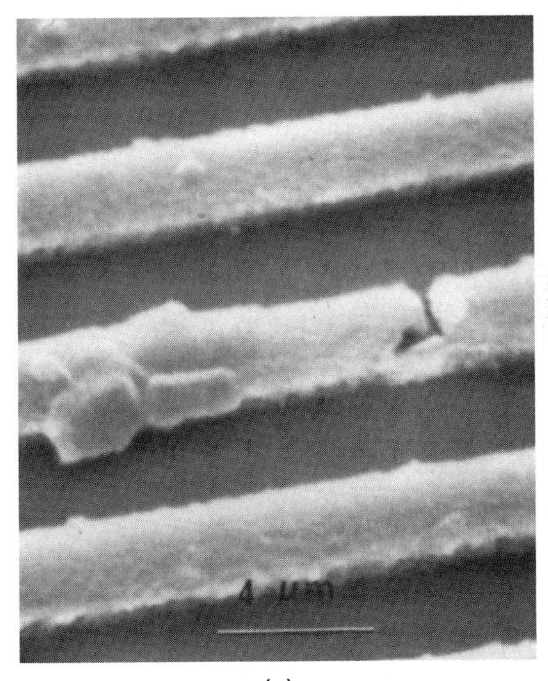

(*c*)

FIGURE 10.23*c*
Manifestations of electromigration damage in aluminum films: mass accumulation and depletion. (*After Vaidya et al. [1980].*)

nology progresses from VLSI to ULSI, the need to enhance electromigration resistance is becoming a serious technological challenge.

The mechanism governing electromigration is not fairly well understood. At high current densities, electrons can impart sufficient momentum to metal ions to propel them toward the anode. This electron "wind" force acts in a direction opposite to that due to an applied electric field ε. Therefore, a net force F acting on the ions is given by

$$F = Z^{*}q\varepsilon = Z^{*}q\rho J \tag{10.18}$$

where Z^* is an "effective" ion valence, q is the electronic charge, ρ is the electrical resistivity of the metal, and J is the current density. This force results in a net flux of metal ions toward the anode. In polycrystalline interconnects, flux divergences occur at grain-boundary triple points, leading to either local accumulation or local depletion of mass. These local mass changes lead to the formation of hillocks and voids. If hillocks grow sufficiently, they can flip onto their side and short the contiguous conductor lines. On the other hand, the continued growth of voids could lead to discontinuities in a conductor. The activation energies associated with electromigration have also been measured in a number of interconnects. These measurements indicate that the electrotransport in films occurs by the movement of atoms along grain boundaries, that is, by grain-boundary diffusion.

During cooldown after deposition, tensile stresses are generated in aluminum interconnects. These stresses can lead to the formation of voids and delamination of dielectric layers. The voids are generally under a passivation layer and appear to occur in lines with widths less than 2 μm when temperatures are in the range of 150° to 180°C during processing. The confinement of the metal by the substrate and the dielectric layer appears to play a crucial role in stress-induced voiding. However, its mechanism has not been clearly established, and the climb of dislocations might be involved in void formation.

The prevention of electromigration in the semiconductor technology is a formidable challenge to materials engineers. To enhance the electromigration resistance, researchers have investigated the influence of various materials parameters on electromigration characteristics of Al interconnects. Since the occurrence of flux divergences at grain-boundary triple points produces electromigration-induced damage, reducing the number of triple points per area, that is, increasing the grain size, should enhance the electromigration characteristics. Of course, single-crystal films would make ideal interconnects because the damage sites are eliminated, but it is not feasible to deposit them on amorphous insulators.

The introduction of <111> texture in Al films increases their electromigration resistance (Vaidya et al. 1980). Even though this observation is not well understood, we can develop a plausible explanation. The textured films probably contain triple points consisting of special grain boundaries where the propensity for flux divergence is small.

Now imagine a situation where grain boundaries run normal to the sides of an interconnect. This so-called bamboo structure can be achieved when the film width is of the order of the grain size. This structure should have a high electromigration resistance. This explanation is indeed borne out by the study of Vaidya (1981) in which she examined the influence of line width on mean time to failure (MTF) of Al-0.5 percent Cu interconnects; the results are reproduced as Figure 10.24.

The electromigration resistance can also be improved by alloying. In the middle 1970s IBM scientists significantly reduced the electromigration problem in Al interconnects by adding small amounts of copper. The controlling mechanism is not clear, but it is likely that impurities segregate to grain boundaries and affect the grain-boundary diffusion behavior. The tendency for grain-boundary segregation will be

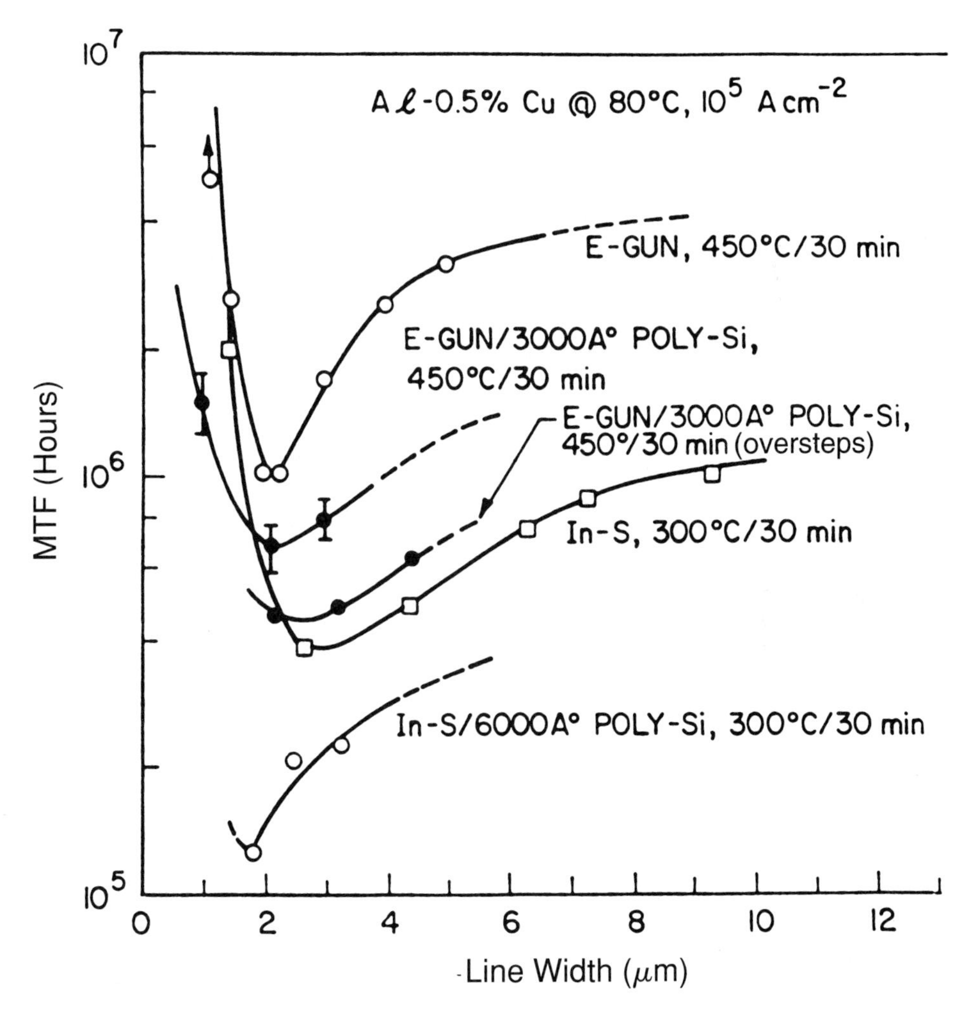

FIGURE 10.24
Mean time to failure as a function of stripe line width for evaporated (E-gun) and sputtered (S-gun and In-S) Al films. E-gun, S-gun, and In-S refer to electron-beam evaporation, sputter deposition, and induction-source evaporation, respectively. (*After Vaidya [1981].*)

higher for impurities whose atomic sizes are larger than the host lattice atoms because of the strain-energy considerations. This hypothesis is indeed consistent with the observed effect of copper on electromigration resistance of Al interconnects.

A serious effort is underway to replace the Al interconnects with copper in the ULSI technology (Murarka and Hymes 1995). To promote adhesion between Cu and SiO_2, substitutional impurities that segregate at the surface are added to the films. These impurities reduce SiO_2, and the resulting oxides adhere well to SiO_2. Furthermore, the scratch resistance of Cu is higher than that of Al, and the electromigration characteristics of Cu interconnects are superior to the electromigration characteristics of Al. These interconnects are also compatible with the Al contact technology. Furthermore, the activation energies for grain-boundary diffusion scale well with the melting point of a metal; that is, metals having higher melting points have higher activation energies and therefore are less prone to electromigration.

10.8 HIGHLIGHTS OF THE CHAPTER

The highlights of this chapter are as follows:

- We discuss the principles of various techniques that are available for depositing thin conducting films for Schottky and ohmic contacts. Compared to physical vapor deposition and sputtering, chemical vapor deposition provides better step coverage.
- The characteristic microstructural features of thin films are grain boundaries and voids. The densities of these features depend on the deposition temperature and rate.
- The Schottky and ohmic contacts based on pure metals are thermodynamically unstable on the semiconductor surface. As a result, they undergo reactions with the semiconductor. To avoid interdiffusion, diffusion barriers have been developed.
- The phenomenon of electromigration observed in aluminum interconnects is considered. Material parameters that can enhance electromigration resistance are discussed.

REFERENCES

Almen, O. and G. Bruce. *Nucleation Instrumentation Methods* 11, (1961), pp. 257–.

Brillson, L. J. In *Handbook on Semiconductors,* ed. P. T. Lansberg. Amsterdam: Elsevier Science Publishers, 1992, p. 281.

Elias, K. R.; S. Mahajan; C. L. Bauer; and A. G. Milnes. *J. Appl. Phys.* 62, 1245 (1987).

Fraser, D. B. In *VLSI Technology,* ed. S. M. Sze. New York: McGraw-Hill, 1983, p. 347.

Glang, R. *Handbook of Thin Film Technology,* ed. L. I. Maissel and R. Glang. New York: McGraw-Hill, 1970, p. 1.

Hentzell, H. T. G.; C. R. M. Grovenor; and D. A. Smith. *J. Vac. Sci. Tech.* A2, 218 (1984).

Learn, A. J. *J. Electrochem. Soc.* 123, 894 (1976).
Mahajan, S. *Progress in Mats. Sci.* 33, 1 (1989).
Mahajan, S.; A. K. Chin; C. L. Zippel; D. Brasen; B. H. Chin; R. T. Tung; and S. Nakahara. *Mater. Lett.* 2, 184 (1984).
Müller, K. H. *J. Appl. Phys.* 58, 2573 (1985).
Murakami, M.; H. J. Kim; Y.-C. Shih; W. H. Price; and C. C. Parks. *Appl. Surf Sci.* 41/42, 195 (1989).
Murarka, S. P. *Metallization Theory and Practice for VLSI and ULSI.* Boston: Butterworth, 1993.
Murarka, S. P. *Intermetallics* 3, 173 (1995).
Murarka, S. P. and S. W. Hymes. *Critical Reviews in Solid State and Materials Science,* 20, 87 (1995).
Nakahara, S. *Thin Solid Films* 64, 149 (1979).
Nicolet, M.-A. *Thin Solid Films* 52, 415 (1978).
Ohring, M. *The Materials Science of Thin Films.* New York: Academic Press, 1992.
Pashley, D. W. and M. J. Stowell. *J. Vac. Sci. Tech.* 3, 156 (1966).
Sands, T.; C. J. Palmstrøm; J. P. Harbison; V. G. Keramidas; N. Tabatabaie; T. L. Cheeks; R. Ramesh; and Y. Silberberg. *Mats. Sci. Eng.*(R) 5, 99 (1990).
Vaidya, S. *Appl. Phys. Lett.* 39, 960 (1981).
Vaidya, S.; T. T. Sheng; and A. K. Sinha. *Appl. Phys. Lett.* 36, 464 (1980).
Wang, L. C.; Y. Z. Li; M. Kappes; S. S. Lau; D. M. Hwang; S. A. Schwarz; and T. Sands. *Appl. Phys. Lett.* 60, 3016 (1992).
Wang, L. C.; X. Z. Wang; S. N. Hsu; S. S. Lau; P. S. D. Lin; T. Sands; S. A. Schwarz; D. L. Plumton; and T. F. Kuech; *J. Appl. Phys.* 69, 4364 (1991).
Wang, L. C.; X. Z. Wang; S. S. Lau; T. Sand; W. K. Chan; and T. F. Kuech. *Appl. Phys. Lett.* 56, 2129 (1990).
Wittmer, M. *J. Vac. Sci. Tech.* A2, 273 (1984).
Wolf, S. and R. N. Tauber. *Silicon Processing for the VLSI Era.* Sunset Beach, CA: Lattice Press, 1986.

PROBLEMS

10.1. If a 1 μm thick layer of $TaSi_2$ is to be produced by the sequential deposition of Ta and Si layers and followed by annealing, calculate the required thicknesses of the respective layers. Assume that the area of the film is 1 cm^2.

10.2. *a.* What is the maximum energy that can be transferred to the target atoms when an aluminum target is bombarded by 2 Ar ions? [Hint: Use Equation (9.3).]
b. What is the mean energy of the struck target atom if the threshold sputtering energy for the Al target is 13 eV?

10.3. Giving mechanistic details, compare and contrast the step-coverage characteristics of physical vapor deposition, sputtering, and chemical vapor deposition.

10.4. *a.* A 1 cm^2 clean surface is covered with 6×10^{10} hemispherical titanium clusters divided equally between 20 and 40 nm clusters. If during the ripening process the small clusters are consumed by the large ones, calculate the reduction of the surface energy of the clusters, given that the surface energy of titanium is 1650 ergs/cm^2.

b. The pressure P above a cluster of radius r is given by the Gibbs-Thompson equation:

$$\frac{P}{P_o} = \exp\left(\frac{2\gamma\Omega}{rkT}\right),$$

where P_o is the equilibrium pressure above the planar surface in the absence of a cluster, γ is the surface energy, Ω is the volume per atom, and T is the temperature. The transport of material from the clusters with smaller radii to larger-size clusters can be rationalized from this equation. If the clusters in (*a*) are transformed into parellopipeds, will they coalesce into each other to reduce the total surface energy of the clusters? Explain the mechanism of coalescence.

10.5. *a.* Explain the formation of voids in as-deposited films. Can these voids be eliminated by a high-temperature anneal? What are the likely sinks for the resulting vacancies in polycrystalline thin films?

b. A polycrystalline film of silicon contains a void density of 5×10^{14} cm^{-3}, and the void diameter is 50 nm. Calculate the total number of vacancies involved in the formation of these voids.

10.6. *a.* Assuming that silicon is slightly soluble in aluminum and the surface energy of {111} planes of silicon is the lowest, show schematically the shapes of molten aluminum-induced dissolution pits on the (001) and (111) surfaces of silicon; label all directions and planes.

b. Suppose the droplet on the (001) surface is subjected to a temperature gradient along the [001] direction. Assuming temperature at the top surface is lower than that at the bottom, discuss the movement of the droplet. This process is referred to as *thermomigration*. Can this approach be used to make p-n junctions in silicon?

10.7. A diffusion couple is formed between two FCC metals A and B and their respective lattice parameters are 0.35 and 0.4 nm. If B atoms diffuse faster into A, vacancies are created in B, whereas interstitials are produced in A. If the concentration of vacancies in B is 5×10^{12}/cm^3, calculate the void density if the average diameter of the voids is 50 nm. In addition, discuss plausible scenarios regarding the fate of the B interstitials in A.

10.8. Suppose a circular gold contact is produced on the (112) surface of GaAs. During subsequent annealing, the contact spreads. Assuming that the interdiffusion-induced reaction pits are bounded by $\{111\}_{Ga}$ and $\{\bar{1}\bar{1}\bar{1}\}_{As}$ facets, sketch the shape of the contact after annealing. Label all directions.

10.9. Using the criteria outlined for diffusion barriers in section 10.6, suggest three barriers for the following situations: (1) between InP and Au, (2) between Si and Ti, (3) between GaAs and Ti, (4) between GaN and Au, and (5) between SiC and Ti. Give reasons for your choices.

10.10. Assuming that the grain-boundary diffusion controls the electromigration-induced damage in thin film interconnects, discuss the following situations:

a. Which three alloying additions would you use to enhance the electromigration resistance of Al thin films and why?

b. Explain the role of precipitates in reducing electromigration

c. Explain the role of bamboo structure on the electromigration behavior.

CHAPTER 11

Lithography and Etching

This chapter covers lithography and etching, the two complementary processing steps used in semiconductor technology. The chapter considers the three main components of lithography, that is, resist materials, masks, and radiation sources used for the transfer of a pattern. It also emphasizes the mechanisms of wet and dry etching.

11.1 LITHOGRAPHY

We discussed in Chapters 6 through 10 the processes that are used in semiconducting technology. An interesting question is, How are these processes integrated for the fabrication of circuit elements consisting of bulk resistors, diffused resistors, capacitors, transistors, diodes, and so forth? The designs for electronic circuits contain geometrical patterns of all the circuit elements, their connections, and their electrical characteristics. The main challenge is to reproduce these patterns on the surface of a semiconductor. The current technology accomplishes this task with lithography and etching.

Lithography is the key to the transfer of the circuit pattern onto the surface of a semiconductor wafer. Its implementation entails masking of a radiation-sensitive film deposited on the semiconductor surface and exposing the film to a suitable radiation, followed by chemical/physical removal of the exposed or unexposed regions of the film. The radiation sensitive film is referred to as a *resist*. The three main components of lithography are resists, masks, and radiation sources.

11.1.1 Resists

Resists are photosensitive polymeric materials and are of two types: positive and negative. In a positive resist the exposed areas of a resist are removed during development, leaving a positive image in the resist. On the other hand, in a negative resist the unexposed areas are removed during development, leaving a negative image in the resist. Positive resists are made up of polymeric materials, which on exposure to radiation undergo breakup of molecular chains, resulting in a local reduction of molecular weight. As a result, the exposed regions become chemically more active and can thus be selectively dissolved in a developer. Negative resists are also polymeric materials, but in this case the molecular chains cross-link on exposure to radiation. In a suitably chosen developer, the cross-linked regions are insoluble. The schematics in Figure 11.1 illustrate the transfer of patterns using positive and negative resists.

The resists used in optical lithography have a number of components. The basic matrix material, called a *resin,* has the requisite mechanical properties and also serves as a binder. It can withstand both chemical and plasma etchings. Since the resin may not be sensitive to optical radiation, a sensitizer is added. In a positive resist, the sensitizer is photosensitive, absorbs optical radiation, and helps in the breakup of the polymeric chains. In a negative resist, the sensitizer facilitates the transfer of optical energy to the resin, resulting in cross-linking of the polymeric chains. A solvent is incorporated into the resist to make it fluid so that it is easier to deposit the resist by spin coating. Other components of the resist are adhesion promoter and thinner. As the name implies, the adhesion promoter (hexamethyl disilazene) facilitates adhesion to the surface being coated, whereas the thinner keeps the resin in the liquid state until it is applied to the substrate.

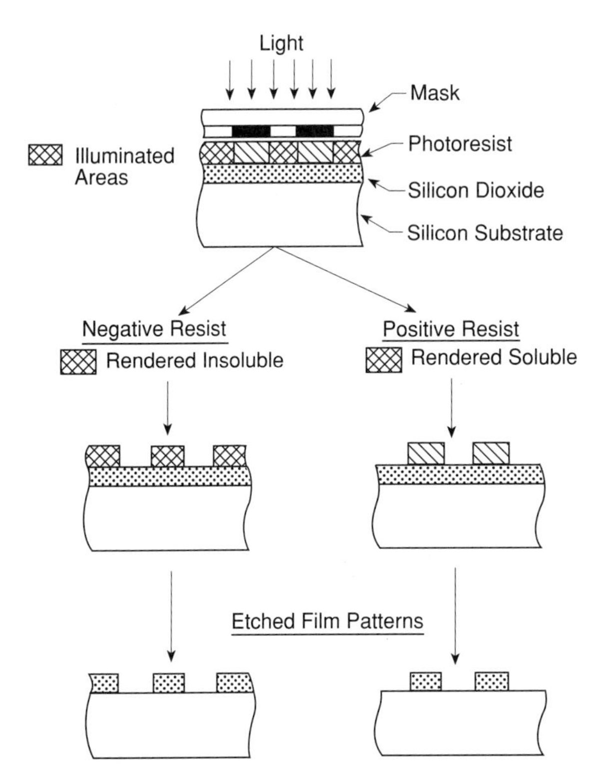

FIGURE 11.1
Schematics illustrating pattern transfer using positive and negative resists.

A number of properties of the resist are common to all lithographic technologies (Thompson et al. 1983): sensitivity, contrast, resolution, optical density, etching resistance, purity, solubility, adhesion, spectral response, ease of processing, and toxicity. These properties can be achieved by the manipulation of polymer structure, molecular properties, and methods of synthesis. The specific chain groups in the polymer resin or the sensitizer, molecular weight, molecular weight distributions, and glass transition temperature are important characteristics. During lithography we control the uniformity and thickness of the resist, prebake and postbake conditions, conditions of development, exposure dose, and substrate.

The sensitivity of a resist is a measure of the energy required for it to develop after an exposure to an optical radiation and is expressed in coulombs/m^2. To ascertain the sensitivity, we expose the resist of known thicknesses to varying doses of radiation. The exposed layers are subsequently developed, and the remaining resist thicknesses are measured. We normalize the remaining thicknesses with respect to the original thicknesses and label the ratio as p. We plot p as a function of log D, where D is the dose. The resulting curve is called the *response* or *sensitivity curve,* and we show this curve for a positive resist in Figure 11.2*a*. Initially p is equal to one until a dose D_o is reached where the breakup of the polymeric chains in the resin begins to occur. The sensitivity of a positive resist is defined as the dose D_c required

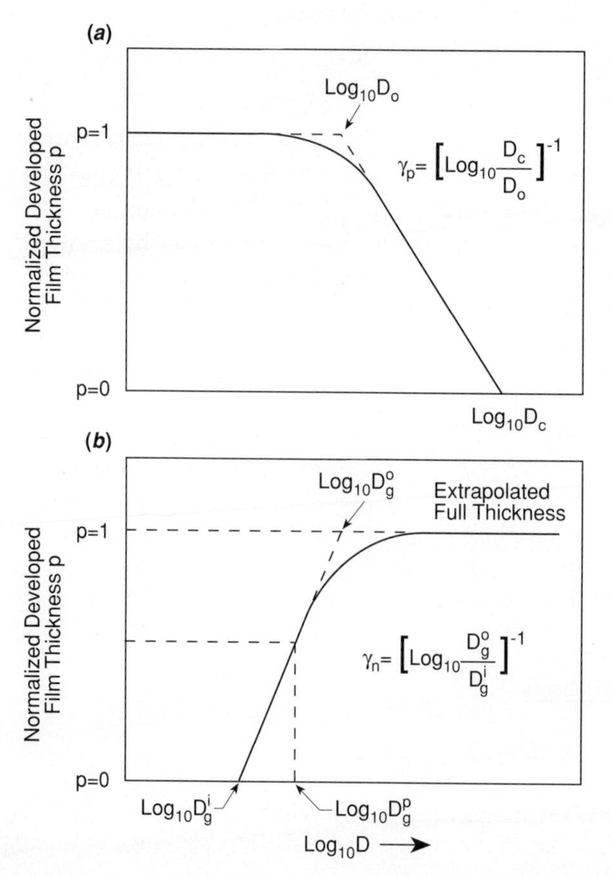

FIGURE 11.2
Developed thickness normalized to initial thickness p for (*a*) positive and (*b*) negative resists.

to effect complete solubility of the exposed regions, given that under the same conditions the unexposed region is completely insoluble. For a negative resist, sensitivity is defined as the dose D_g^1 at which a lithographically useful image is formed and is equal to $[\alpha h\nu I(z)]$, where α is the linear absorption coefficient of the resist, $h\nu$ is the energy of the radiation, and $I(z)$ is the intensity. We show the sensitivity curve for a negative resist in Figure 11.2*b*. Sensitivity thus measures the energy required for the resist to develop properly and influences the speed with which the exposure has to be carried out. There are practical limits to the sensitivity: The low values of the sensitivity may not be acceptable from throughput considerations. On the other hand, the high values of the sensitivity may imply that polymeric chains in the resist undergo chemical reactions at room temperature, and the resist does not have an acceptable shelf life.

In Figure 11.2*a*, p is equal to one until $\log D_o$ is reached and then it attains a value of zero at $\log D_c$. The sharpness of the transition from D_o to D_c is a measure of the resist contrast. The contrast (γ_p) of a positive resist is given by

$$\gamma_p = \frac{1}{\log D_c - \log D_o} = \log\left[\frac{D_c}{D_o}\right]^{-1}, \tag{11.1}$$

where D_o is the dose at which developer begins dissolving the irradiated film. Likewise, the contrast (γ_n) of a negative resist is given by

$$\gamma_n = \frac{1}{\log D_g^o - \log D_g^1} = \log\left[\frac{D_g^o}{D_g^1}\right]^{-1}, \tag{11.2}$$

where D_g^1 is the dose below which no image is formed and D_g^o is the dose at which the thickness of the imaging film equals that of the resist prior to exposure. The higher the resist contrast, the more vertical the resist profile after dissolution.

The *resolution* of a resist refers to the smallest feature that can be resolved using the particular resist. The resolution therefore represents the minimum feature size that can be replicated into a resist. The line width after etch is the basic dimensional parameter of interest and requires a careful control. In particular, the feature that is most difficult to resolve should govern the practical limit for resolution. Ideally, if the resist profile were vertical, the line width is a good measure of the dimension in the mask. This correlation is only possible if there is infinite contrast. Since the energy delivered to the resist is dissipated in a more diffuse fashion due to diffraction and scattering, the profile of the resist after development will exhibit some slope. The angle of the resist profile at the point where the resist thickness falls to zero is directly proportional to γ and the resist thickness. For an exposure system to satisfactorily print a given feature, the modulation transfer function (MTF) for the feature must be greater than a critical MTF (CMTF) value $CMTF_{resist}$ for the resist used, where $CMTF_{resist}$ is given by

$$CMTF_{resist} = \frac{10^{1/\gamma} - 1}{10^{1/\gamma} + 1}. \tag{11.3}$$

The thicker layers of the positive resists with higher values of γ have resolution comparable to that of the thin layers of the negative resists. The advantage of using thicker layers is that they provide better step coverage, more defect protection, and greater dry-etch resistance.

A resist will be effective if, after absorbing the initial energy, its top portions become transparent to further irradiation. In this way the upper regions of the resist, which are exposed first, do not prevent the radiation from going through the resist and reaching its unexposed portions. The difference between the absorbance of radiation of the unexposed and exposed resists is called *actinic absorbance*.

During the pattern transfer the resist has to survive contact with an etchant or a plasma etching process. The resistance to etching of a resist is merely reported in relative terms by comparing the etch rate of a given resist with that of a conventional photoresist. The etching resistance of resists to wet etching is generally much better than their resistance to dry etching. If the etch resistance is such as to cause less than a 10 percent change in line width after etching, the resist is acceptable.

The resists can be deposited on a variety of materials, such as silicon, chromium, gold, silicon dioxide, silicon nitride, phosphorus doped silica, and polysilicon. In each case the resist must have adequate adhesion to the underlying substrate. Extensive undercutting, loss of resolution, and destruction of film can result from poor adhesion. On occasions, adhesion promoters have been utilized. Elevated-temperature postbake cycles also promote adhesion. Dehydration bakes prior to coating improve adhesion.

A typical positive photoresist used in optical lithography consists of a matrix component. The current compositions consist of a low molecular weight novalic resin. The sensitizer is diazonaphthaquinones, which makes the resist insoluble in aqueous developing solutions. The solvents that facilitate spinning of resists into thin films are typically mixtures of n-butyl acetate, xylene, and cellulose acetate. On the absorption of light, the quinonediazide undergoes a photochemical rearrangement known as the *Wolff rearrangement* shown in Figure 11.3. This procedure is

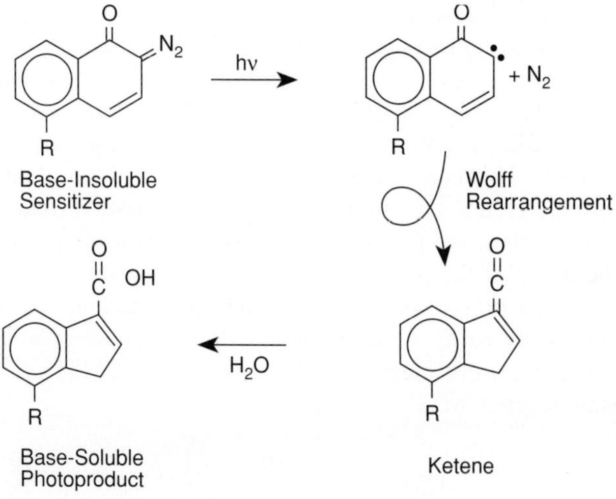

FIGURE 11.3
Photochemical transformations of the quinonediazide sensitizer.

followed by hydrolysis to yield indenecarboxylic acid. This material is readily soluble in the basic solutions used as a developer. Since the developer does not permeate the unexposed film regions, the positive resists retain their size after exposure and even after immersion in the developer, as we show in Figure 11.4.

The common negative resists utilize cyclized synthetic rubber resin as the matrix material. They are extremely soluble in toluene and xylene. Upon exposure to light (400 nm wavelength), the bis-azide is excited and forms a highly reactive species called nitrene :N-Ar-N:, where Ar is an aromatic group. This reactive species acts as a cross-linking agent to connect polyisoprene molecules in the synthetic rubber. The absorption of light eliminates nitrogen, and an extremely reactive nitrene intermediate, shown in Figure 11.5, forms. Subsequent reactions produce a variety of products including the formation of amines. Polymer dissolution occurs first by swelling of the matrix, followed by chain disentanglement. Light that is scattered off the projection optics can also cross-link a thin layer of negative resist.

Now let us examine how the topology of the surface upon which the image must be printed affects the simultaneous achievement of good line-width control, high resolution, and good step coverage. Even though resolution considerations are specified in only two dimensions, the IC fabrication requires faithful reproduction

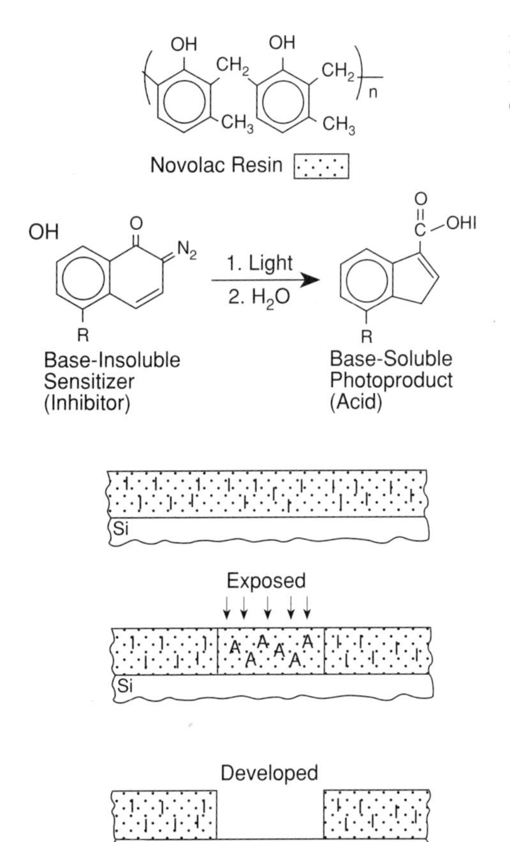

FIGURE 11.4
Mechanism of positive resist action in quinomediazide-novolac resists.

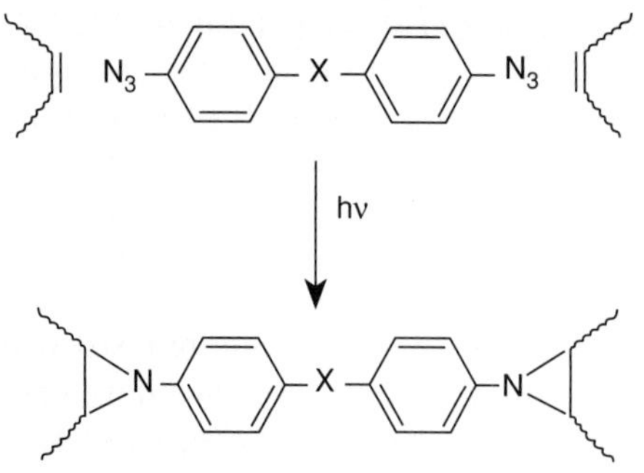

FIGURE 11.5
Photochemical transformations of a bisazide sensitizer in negative resists based on cyclized polyisoprene.

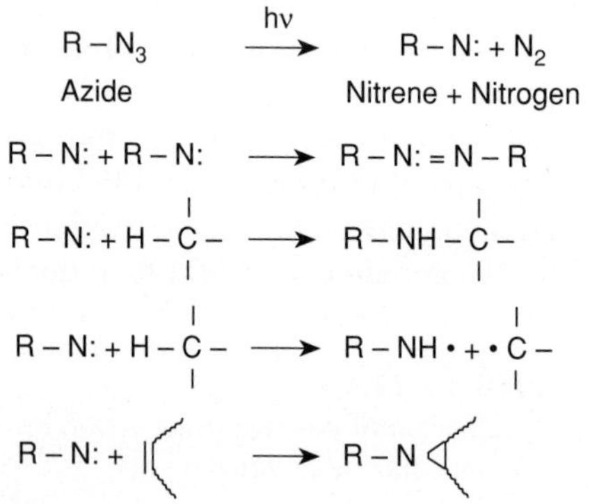

$$R-N_3 \xrightarrow{h\nu} R-N: + N_2$$

Azide — Nitrene + Nitrogen

$$R-N: + R-N: \longrightarrow R-N:=N-R$$

$$R-N: + H-\overset{|}{\underset{|}{C}}- \longrightarrow R-NH-\overset{|}{\underset{|}{C}}-$$

$$R-N: + H-\overset{|}{\underset{|}{C}}- \longrightarrow R-NH\bullet + \bullet\overset{|}{\underset{|}{C}}-$$

of three-dimensional structures. Ideally, a flat surface and a thin resist are necessary to obtain excellent resolution. However, good step coverage requires thick films. Only when development of exposed resist can be controlled to produce images with absolutely vertical side walls can the images in both thick and thin areas of a resist be developed in proper dimensions.

During IC fabrication a number of nonplanar topographical features are present. As a result, the resist layers have varying thicknesses, which limits the achieveable resolution. Even though the spin-coated resist films tend to planarize or smooth out the bumps, the extent of planarization is not ideal (LaVergne et al. 1985). The lack of planarization contributes to the loss of resolution. If the development does not produce vertical walls, the time required to develop an image at the proper dimensions at the substrate in a thin area is insufficient to develop an image at the substrate in the thick area. If overdevelopment occurs, the thin areas will give line widths that are too wide.

The topographical feature has a serious effect on the modulation of light. Light transmitted through the resist is reflected from the substrate. The two light waves interfere with each other, producing standing waves. In addition, the material composition and the scattering of light from topographical features produce linewidth variations that are particularly severe when highly reflective metals and silicides are patterned. The minimization of these problems is possible using several approaches: image-reversal processing, contrast-enhancement lithography, and multilevel-processing schemes.

The advantages of positive and negative resists are used effectively in the image-reversal processing. In this technique a positive resist is exposed to radiation through a patterned mask. The latent image of the mask is produced in those areas where the resist has undergone photoinduced polarity changes. The film is then treated chemically and thermally to transform the latent-image region to an insoluble but photoinsensitive compound. The next step is to flood the entire resist with radiation so that the originally unexposed regions can be dissolved in a developer. The final developing process yields the image of a negative photoresist. This procedure has been carried out successfully using diazoquinone resists. One of the additional advantages of this process is in controlling the wall profile. For example, the latent image produced in the resist has a positive slope, whereas the developed image following image reversal can have positive, negative, or vertical slope depending on the conditions of processing and developing. The positive wall slope is due to an insoluble area at the top, which determines the line width in the negative mode, whereas the foot of the resist structure determines the width of the image in the positive mode.

To improve the achievable resolution limit in a given optical lithographic system, it is useful to practice contrast-enhancement lithography. Organic dyes (nitrones, diazonium salts, polysilanes) are spin coated onto the resist prior to exposure. The exposure to light bleaches the dye layer, and the central areas are bleached more than the edges because they receive less light in a given lithographic system. The transmitted intensity that exposes the resist is modified so as to improve the contrast. The resolution limit can thus be optimized, but the throughput is affected.

The simplest multilevel-processing scheme involves depositing an antireflection coating on the substrate prior to coating the resist. The coating absorbs the light before it reaches the substrate and thus reduces the distortions due to the standing wave patterns discussed earlier in section 11.1.1. Multilevel processes that consist of at least two coating steps and an image-transfer step provide a generic solution to the topography problem. A multilevel scheme is illustrated in Figure 11.6. The wafer is coated with a thick layer of opaque material whose principal purpose is to attain the maximum possible planarization. A thin layer of the resist is then placed on the planarized layer. High-resolution images are formed by exposing the resist layer to light. The resulting image is transferred in some anisotropic etching process through the planarizing layer to yield high aspect ratios with vertical walls. Since the walls of the images in the planarizing layer can be vertical, the process is made insensitive to variations in the coating thickness of the planarizing layer. A number of modifications to the multilevel process have been investigated to ensure vertical etching and to simplify processing steps (Wilson et al. 1988).

The limitations associated with the positive and negative photoresists become severe as the resolution requirements shift toward submicron levels. The physical limits associated with resolution in optical lithography have prompted the drive toward lower wavelength radiation in lithography. Significant changes in lithographic equipment and development of novel resists must occur before using lower wavelength radiation for device fabrication.

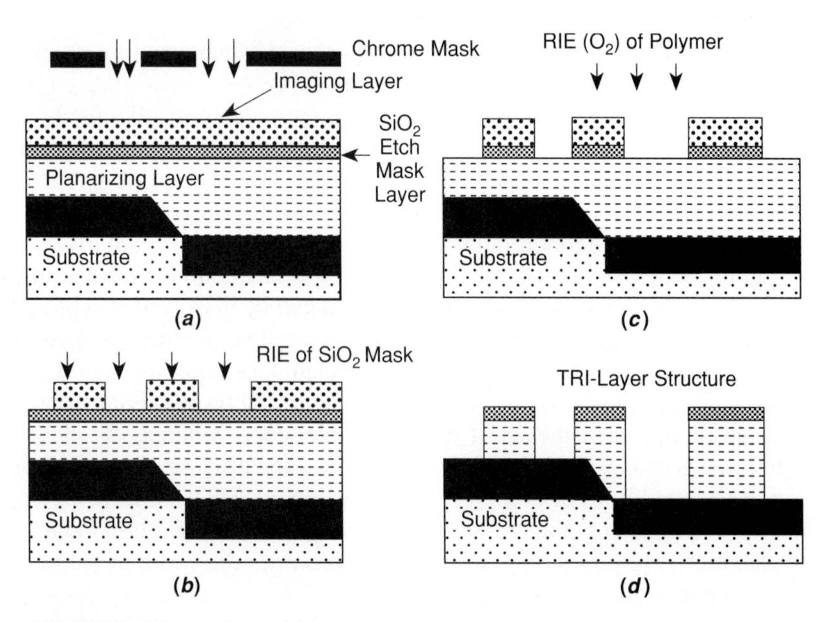

FIGURE 11.6
Multilayer process proposed for line-width control and good step coverage. RIE refers to reactive ion etching.

Conventional positive photoresists do not function adequately in deep ultraviolet lithography, but some improvement can be achieved by redesigning the dissolution/inhibitor/matrix system. Polymethylmethacrylate (PMMA) is sensitive to deep ultraviolet radiation and can be used as a resist (Lin 1975). It has high resolution, excellent film-forming characteristics, and wide processing latitude and is readily available. However, it is an insensitive resist and requires 0.5 to 1.0 J/cm^2 dose for workable processing. The sensitivity can be improved using several approaches: copolymers used as resists, incorporation of aromatic groups, and substituted systems. Negative resists have low resolution but offer high sensitivity, ease of handling, and wide process latitude, but they suffer from swelling phenomenon. Swelling in negative resists during developing can be minimized to some extent by structure manipulation, but cannot be completely eliminated as long as the development of the resist depends on solubility differences. Attempts are thus being made to develop dry resists that do not depend on the differential solubility due to cross-linking. The dry developable resists have two components: a monomer and a low plasma-resistant base polymer. The monomers are grafted to the base polymer and/or polymerized to a homopolymer on exposure to radiation. The graft polymer has a lower plasma etch rate than the base polymer. The monomer can be sublimed or vaporized from the unexposed area by baking prior to plasma development. Polysilanes have been extensively investigated because of their high stability to O_2-reactive ion etching. An additional benefit of this process is the potential to conduct etching in a clean vacuum environment.

The polymer resists used for electron, x-ray, and ion-beam lithographies are also organic polymers, but do not contain additives. The energies involved are such that the polarity change in the molecules is not responsible for the solubility differences. The polymers chosen for such resists either cross-link with one another like negative resists or undergo breakup of chains when exposed to radiation like positive resists.

Electron resists utilize the dependence of solubility and dissolution rate on molecular weight and its distribution, since the resists do not have additives. Three types of positive electron-resists have been explored: PMMA, polyolefin sulfones, and cross-linked PMMA. Negative electron resists that have been explored are epoxide containing polymers, polystyrene-based polymers, and copolymers of PMMA with side chains containing unsaturated bonds. Electrons on entering the resist suffer both elastic and inelastic collisions. The inelastic collisions involve the transfer of the kinetic energy of electrons to the energies of the electrons in the molecule. As discussed in section 9.1.1, the angle of scattering is very small so that the trajectory of the incoming electron is relatively unaffected. On the other hand, the elastic collisions occur when the electrons collide with nuclei of atoms. As a result, the paths of the electrons deviate considerably from their initial directions. These deviations broaden the area over which the resist is exposed as compared to the cross section of the incident beam. We refer to the phenomenon as *proximity effect.* The transfer of high-resolution patterns requires the elimination of the proximity effect. One way to achieve this goal is to find the relationship between the line width and spot diameter by lengthy computation and then account for it in the design. We can also use higher-energy electrons because they are scattered very little in a thin resist layer and there is a sharper concentration of energy dissipation around the point of impact. Low-energy electrons can also be used for lithography if we can deposit ultrathin layers of resist. However, thin layers have poor etching resistance, high pinhole density, and unreliable planarization. We can also use PMMA films produced by the Langmuir-Blodgett technique for lithography.

X-rays penetrate deep into the resist without undergoing scattering. Therefore, high-resolution patterns are possible in thick resists that facilitate pattern transfer. When a resist molecule encounters an X-ray photon, photoelectrons are generated by inelastic collisions. The photoelectrons in turn modify the properties of the resist by either cross-linking molecules or breaking up polymeric chains. The penetration range of photoelectrons is quite small, enabling high-resolution patterns to be obtained. Unlike photoresists, the sensitivity of X-ray resist is significantly dependent on the polymer structure, which requires the determination of the number of chain-breaking or cross-linking events per unit energy of absorbed radiation. Increased sensitivity is achieved in positive resist polymethyl methacrylate by having a narrow molecular weight distribution (Lai et al. 1976). Polymers with a low glass-transition temperature are unstable and cause a noticeable change in resist profiles. Developments to synthesize high-temperature resists ($\sim$200°C) (Toukhy et al. 1984) or processing by ultraviolet (Hiraoka et al. 1981) or plasma (Ma 1982) exposure to enhance thermal stability are reported.

Resists for ion-beam lithography suffer chemical changes as the ions travel through the resist. Since the mass of the ions is large, the amount of energy deposited per unit penetration distance of an ion beam is also large. The sensitivity of the resists is therefore expected to be high. Lateral spread of ion beam is minimal. PMMA is being explored as a possible ion resist.

EXAMPLE 11.1. Obtain an expression for the energy absorbed by a resist at a distance z when photons of energy $h\nu$ penetrate the resist. Assume the dose of photons is D_0 and the linear absorption coefficient of the resist is α.

Solution. The depth-dose function of a resist is related to the rate of energy dissipation during penetration of photons into the resist film. The dose at a depth z in the resist is D(z) and is given by

$$D(z) = I(z)\,\tau,$$

where I(z) is the intensity if the radiation at depth Z and τ is the exposure time. The intensity I(z) is given by

$$I(z) = I_0F(z),$$

where I_0 is the incident intensity of the photons and F(z) is the depth-dose function. The energy absorbed per unit volume of the resist polymerizes or breaks the bonds in the resist. The rate of energy adsorption at a depth z is

$$\frac{dE}{dt} = h\nu\frac{dI}{dz} = \alpha\, h\nu\, I(z) = \alpha\, h\nu\, I_0F(z),$$

where $h\nu$ is the photon energy and α is the absorption coefficient. At a distance z the energy absorbed per unit volume is obtained by integration and is given by

$$E(z) = \alpha\, h\nu\, D_0F(z),$$

where D_0 is the incident dose and is equal to $I_0\tau$, I_0 being the incident intensity.

11.1.2 Masks

Masks serve two functions in the fabrication of semiconducting devices and circuits: (1) preparation of the master pattern and (2) permanent transfer of the pattern onto a semiconductor surface. A mask can be fabricated by depositing a thin chromium film on a glass plate, followed by the layer of a photosensitive resist material. The pattern is transferred onto the film by exposing the resist to light of a suitable wavelength. The masks for high-density circuits are prepared using electron-beam lithography. The exposed composite is developed, followed by etching to remove both the chromium and the photoresist from those regions where light has to go through the mask. However, the regions where chromium is required are protected during etching by the resist. The remaining resist is then removed, resulting in a patterned glass plate called a *reticle*.

Soda lime glass plates were originally used for the masks. Since their thermal expansion coefficient is large and is different from that of silicon, they have been replaced by borosilicate glass plates. Based on thermal expansion considerations alone, fused silica appears to be an ideal choice for the mask material. Chromium is

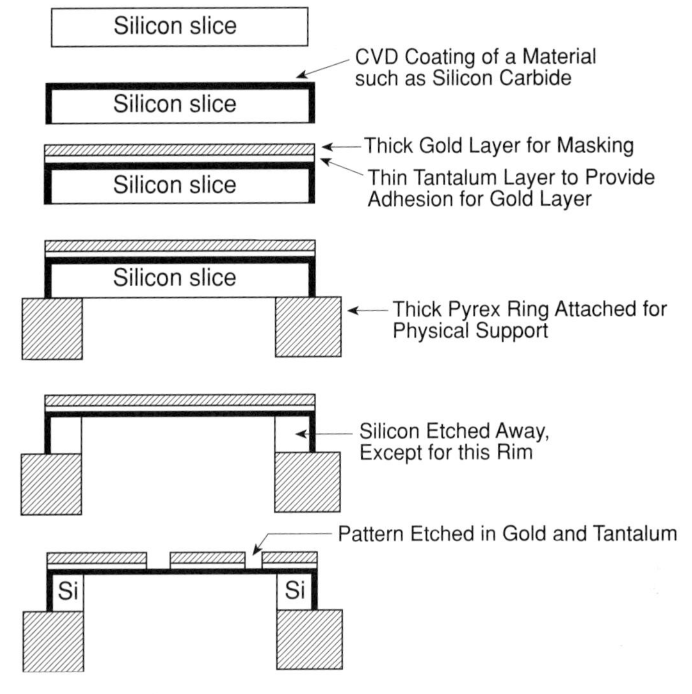

FIGURE 11.7
A method of fabricating a mask for X-ray lithography.

the most commonly used coating for pinhole-free masks. Antireflection coatings are sometimes employed to reduce reflection from the chromium surface.

The masks for X-ray lithography are thin due to reduced penetration of X-rays. In addition, mask durability, dimensional stability, and ruggedness are important concerns. The materials currently being used as membranes for masks are polyimide, parylene, SiC, Si_3N_4, BN, and so forth, and their thicknesses are determined by their mass absorption coefficients. Figure 11.7 shows the typical steps involved in the fabrication of a mask for X-ray lithography. First, a film of silicon carbide is deposited on a silicon substrate. Layers of Ta and Au are then sequentially deposited. The function of the Ta layer is to facilitate the adhesion of the Au film to the SiC layer. After attaching a pyrex ring to the composite for mechanical support, the silicon substrate is removed by etching from the back side. The Au and Ta layers are subsequently patterned to produce a mask. Surface roughness and stresses in the film must be kept at low levels.

The processing-induced defects in masks can arise for various reasons, and their origins are illustrated in Figure 11.8 (Runyan and Bean 1990). The existence of pinholes in a mask is generally attributed to pinholes in the metal films. Whenever a printing or developing problem occurs, one sees bridges in a mask. The missing element is due to either poor adhesion or the poor data fed into the pattern generator. Errors in resist exposure, difficulties in etching the metal films, and errors

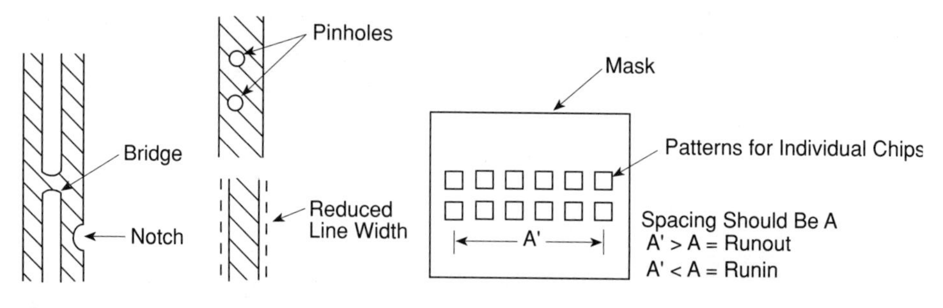

FIGURE 11.8
Typical defects in masks. (*After Runyan and Bean [1990].*)

in describing the size may contribute to the wrong geometries. Poor handling can result in notches and scratches. Particulates should be avoided during mask making. Runout/runin are defects attributed to the actual spacing A′ between patterns for individual chips when compared to the spacing A set by design. If $A' > A$, the defect is said to be a *runout,* and if $A' < A$, the defect is said to be a *runin*. Temperature differences between the center and the periphery of a wafer and the wafer bowing can contribute to runout errors.

Generally many masks are needed to complete the fabrication of devices and integrated circuits. Therefore, various masks must be aligned with respect to each other so as to produce the circuit elements at the desired locations. However, errors do occur. The errors in positioning of one pattern with respect to another are called *overlay errors*. The mismatch can arise because of the difference between the coefficients of thermal expansion of the mask and the wafer. High-intensity light used during exposure can cause a difference in temperature between the center and the periphery of the wafer. Reflectivity differences in the metal films can cause temperature differences in a wafer during exposure. If air is used to cool the wafer, its uneven flow can lead to temperature variations. Overlay problems can also arise from the random placement of a feature within a die or from a random misplacement of a die within the whole mask array. Elaborate patterns have been designed to obtain statistical data for determining level-to-level pattern-printing misregistration. Device designs should be tolerant of alignment errors. Figure 11.9 illustrates some of these errors (Runyan and Bean 1990). In electron-beam lithography the mask-placement errors are one of the main factors that determine the limitation of lithographic equipment. The factors that contribute to this error are beam-position drift and detector sensitivity, drift of the exposure system itself, loading errors, substrate height variations, mechanical distortion of the mask by the cassette, and static and dynamic temperature variations.

11.1.3 Radiation Sources and Lithographies

Lithography can be carried out using three types of radiation sources: (1) optical, (2) X-rays, and (3) electrons and ion beams. Currently, the optical sources are widely used in the semiconductor technology and cover a wavelength range span-

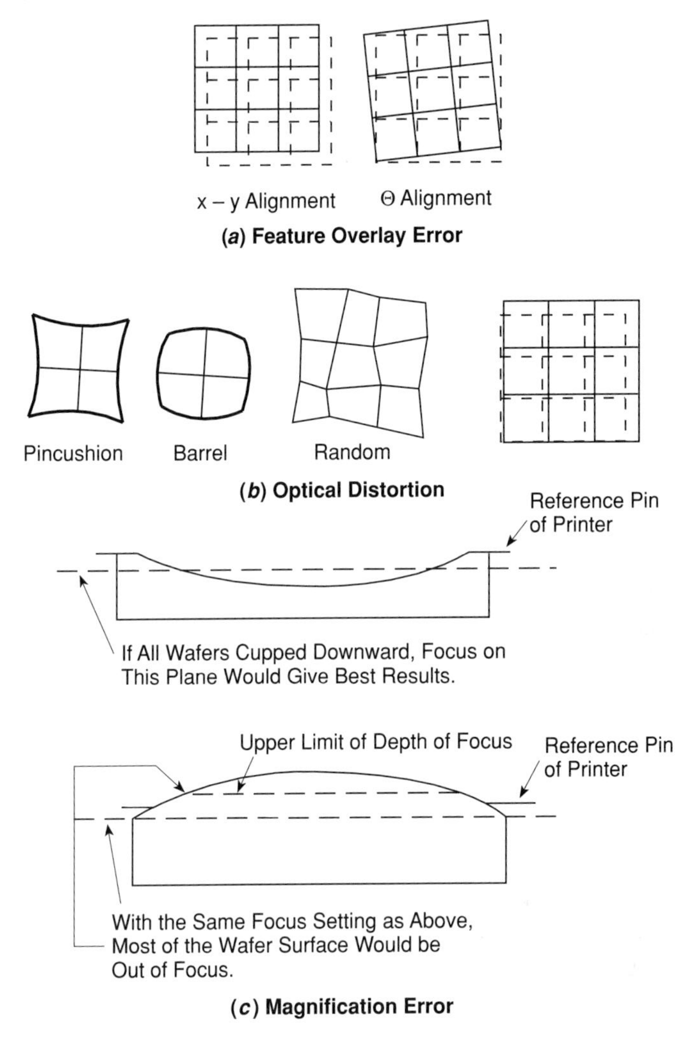

FIGURE 11.9
Alignment errors. (*Adapted from Runyan and Bean [1990].*)

ning deep and near ultraviolet. As the feature sizes are further reduced to below ~0.1 μm, alternative radiation sources such as X-rays and electrons and ion beams are being developed.

11.1.3.1 Optical and X-ray lithographies

The simplest method for transferring a desired pattern onto a resist is to hold the mask in contact with the resist and expose the resist to photons. As a result, we obtain an image in the resist that is defined by the shadow of the desired set of shapes that are present in the mask. This technique is called *contact printing*. The presence of particles on the surface, spikes in epitaxial layers, and defects caused by the

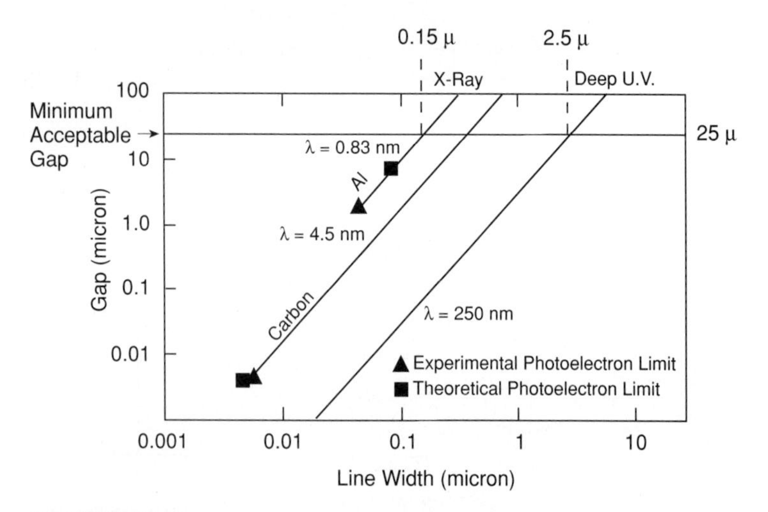

FIGURE 11.10
Line widths achievable in proximity printing for different photon wavelengths. (*After Broers in* Microelectronic Materials and Processes *[1989]*.)

adhesion of mask and resist due to physical contact limits the utility of contact printing. The reproducibility of this method is poor because the mask degrades with each use.

To avoid damage to the mask, a wide separation between the mask and the resist is desirable. In proximity printing the mask and the resist are separated from each other and are parallel. Figure 11.10 shows the smallest width of a line, that is, line width, that can be achieved by proximity printing. Since the desired level of accuracy is $<2\ \mu m$, this technique is inadequate for the current generation of devices.

If photons are produced from a point source and if they behave as particles, the shadow produced by a straight edge or a slit would indeed be sharp, as shown in Figure 11.11*a*. However, the wave nature of photons causes spreading of radiation into regions that are not directly in the line of sight of photons. We show this effect in Figure 11.11*b*. The intensity distribution depends on the distance between the mask and the resist. When this distance is small in comparison to the wavelength of photons, diffraction occurs that is known as *near-field,* or *Fresnel diffraction.* The *far-field,* or *Fraunhofer diffraction,* takes place when the distance is large in comparison to the wavelength of the incident photons. The minimum line width W_m that can be satisfactorily printed is given by

$$W_m = (d\lambda)^{1/2}, \tag{11.4}$$

where λ is the wavelength of photons and d is the mask-wafer separation.

The source of photons in the optical lithography systems is high-pressure mercury lights operating around 35 to 40 atm pressure. Three wavelengths emitted by the source at 436 nm (G-line), 405 nm (H-line), and 365 nm (I-line) are of particular interest. Optical sources operating at shorter and shorter wavelengths are desirable. Deep ultraviolet excimer laser sources, emitting in the 230 to 260 nm

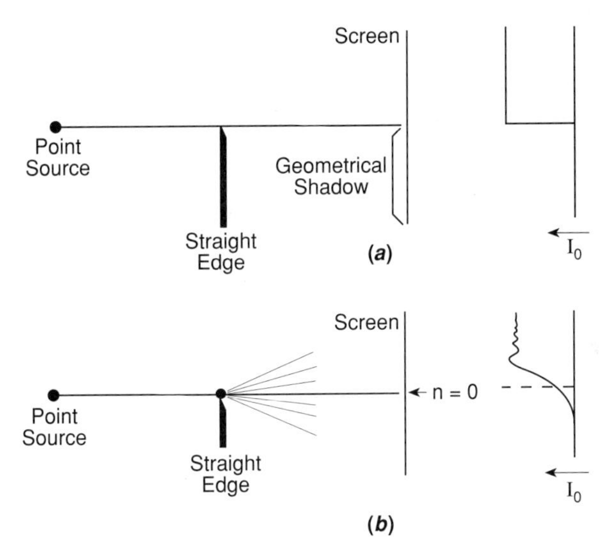

FIGURE 11.11
The distribution of intensity of a spatially coherent light source as it passes by a straight edge: (*a*) according to geometrical optics and (*b*) according to wave optics.

range, have been developed and are expected to play an important role in the fabrication of line widths below 0.25 μm.

A synchrotron provides well-collimated, high-intensity X-rays. The need to keep the source far from the wafer surface affects resolution in X-ray lithography using synchrotons. The decrease in resolution results from the leakage of X-rays through the edges of a mask, shadows due to the finite size of an X-ray source and the finite size of mask. These effects are illustrated in Figure 11.12. Line widths of about 0.2 μm can be printed using X-rays (4 to 50Å range) by contact or proximity

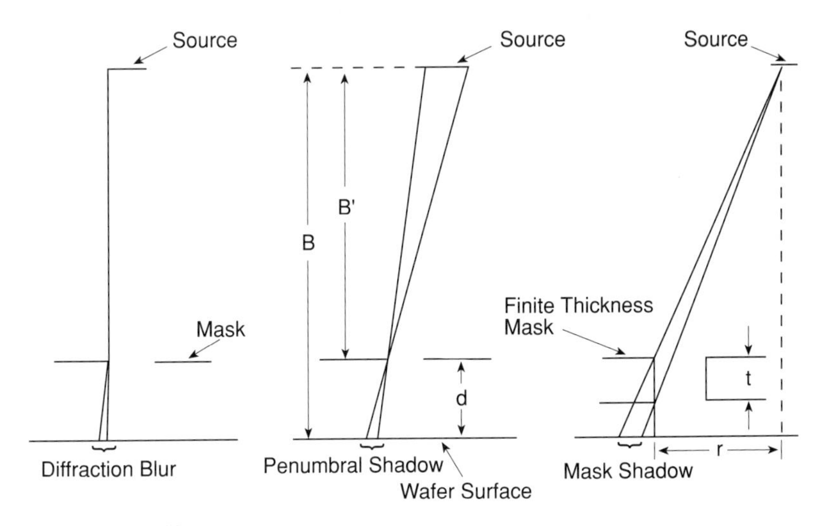

FIGURE 11.12
Factors that affect the resolution in X-ray lithography. Various parameters B, B′, d, t, and r are defined in the figure.

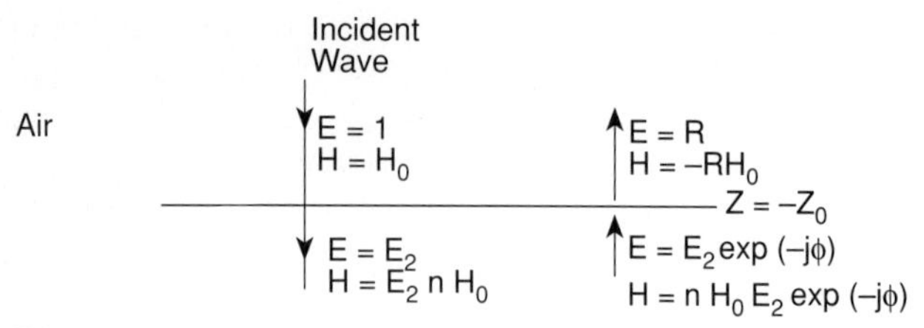

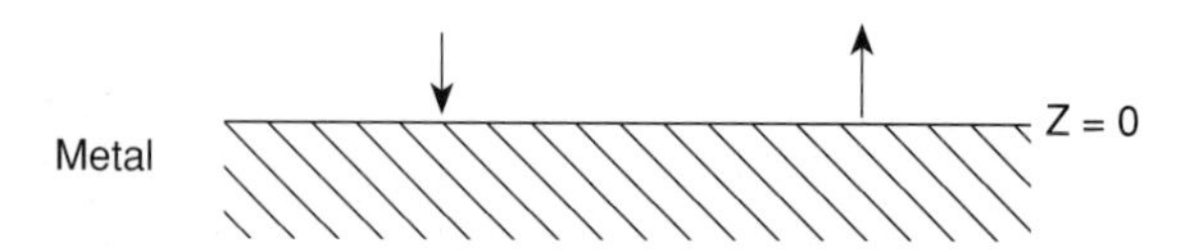

FIGURE 11.13
Schematic showing interference between incident wave and reflected wave.

printing. These types of printing are necessary because of the absence of suitable mirrors or lenses to project or form images.

As mentioned earlier, during exposure the standing wave patterns due to interference between the incident and reflected waves are created in the resist. The incident light undergoes reflection from the medium surface whose index of refraction is different from that of the resist. We show this situation in Figure 11.13. The incident and the reflected beams interact to create a standing wave. As a result, the light intensity will vary in a periodic fashion normal to the wafer surface, and has an undulated intensity pattern in the resist on development. To avoid the periodic variations in intensity, the use of highly monochromatic sources and reflecting surfaces such as silicon and metal is discouraged.

Proximity printing involving optical sources has poor reproducibility, but can be carried out effectively using X-rays because their short wavelength would permit much greater separation between the mask and the substrate. The diffraction effects can also be ignored in X-ray lithography for separation on the order of a few micrometers.

To avoid contact printing–induced damage to masks and to improve the resolution of optical-proximity printing, projection printing has been developed. In projection lithography, a mask containing the desired pattern is demagnified by an optical system and is projected onto a resist. The demagnification allows the mask to have dimensions larger than those required on the wafer, which makes it easier to fabricate, inspect, and repair the mask. In projection lithography the image is built using reflective optics. As a result, it is subject to a series of aberration errors that can blur and distort the image. However, the excellent design of the available optical systems has reduced many of these errors, resulting in an extensive use of the projection lithography.

In an optical projection system, the transmitted and diffracted light from the mask in Figure 11.13 has to be collected and projected onto the wafer as shown in

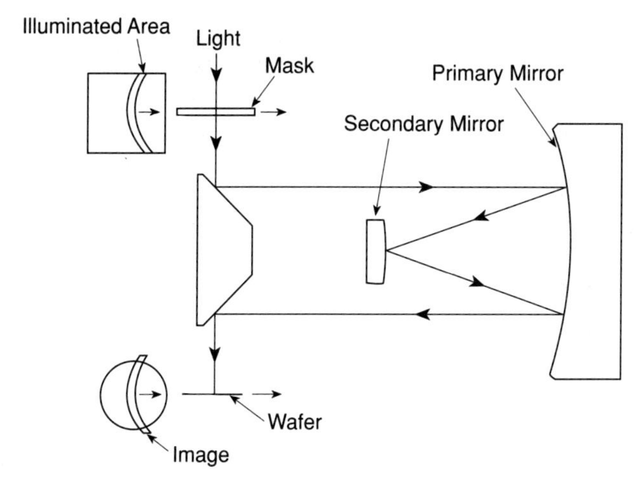

FIGURE 11.14
Sketch of a scanning projection printer.

Figure 11.14. The numerical aperture (NA) of the objective lens determines the ability to collect this light and to project it as an image. NA is in turn related to the refractive index n of the medium around the lens and the half angle α of the two most divergent rays of light that can pass through the lens by

$$\mathrm{NA} = \mathrm{n} \sin \alpha. \tag{11.5}$$

NA is typically in the range 0.16 to 0.40 in projection aligners.

The ability of a system to form separate images of closely spaced objects is known as *resolution* S_r. A typical estimate of resolution may be obtained from

$$S_r = \frac{0.6\lambda}{\mathrm{NA}}. \tag{11.6}$$

The higher resolution possible with a larger NA is only obtained at the expense of the depth of focus. The optical image degrades when the system is defocused, and the amount of defocusing that can be tolerated is called the *depth of focus* δ. An estimate of δ is given by

$$\delta = \frac{\lambda}{2(\mathrm{NA})^2}. \tag{11.7}$$

The depth of focus must be sufficient to cover the flatness errors (typically $\pm$ 10 μm) common in wafers after high-temperature processing.

A practical way to improve the resolution while retaining the depth of focus is to employ shorter wavelength photons. The shortest wavelength used in conventional optics is about 193 nm obtained from an excimer laser source. The current optical sources use emissions from mercury vapor at 436 nm (G-line) and 405 nm (H-line). The use of wavelengths shorter than 193 nm is limited by absorption in optical materials. In any case optical lithography may not be useful for line widths smaller than 100 nm. Projection X-ray systems are currently not available because of the difficulties in fabricating mirrors for X-rays.

The increased diameter of a silicon wafer makes it impractical to design a projection system that can simultaneously expose the whole wafer. The systems employing refraction of light are designed to project a portion of the mask onto a corresponding part of the wafer. The field is then stepped and repeated across the wafer so that the wafer size is not a limitation in the lithographic process. In addition to higher resolution, refraction corrections for wafer distortion, and overlay accuracy lead to improved resolution, but these benefits occur at the expense of low throughput.

The size of the image field over which a high-resolution exposure can be made in projection printing is determined by the focal number F (≠) and the diameter D of the entrance pupil. The focal number is given by

$$F(\neq) = \frac{f}{D}. \tag{11.8}$$

At a given resolution the contrast is typically measured by modulation transfer function (MTF). For a given incident light energy on a plane surface, the contrast M can be represented by

$$M = \frac{I_{max} - I_{min}}{I_{max} + I_{min}}, \tag{11.9}$$

where I_{max} is the image intensity at the center of the bright area and I_{min} is the image intensity at the center of the dark pattern. The quantity M can be evaluated at the plane of the mask M_{mask} and at the plane of the image M_{im}. Then

$$MTF = \frac{M_{im}}{M_{mask}}. \tag{11.10}$$

If $M_{mask} = 1$, then MTF is given by M_{im}. MTF varies with the mask feature size and the spatial coherence of the source, as shown in Figure 11.15.

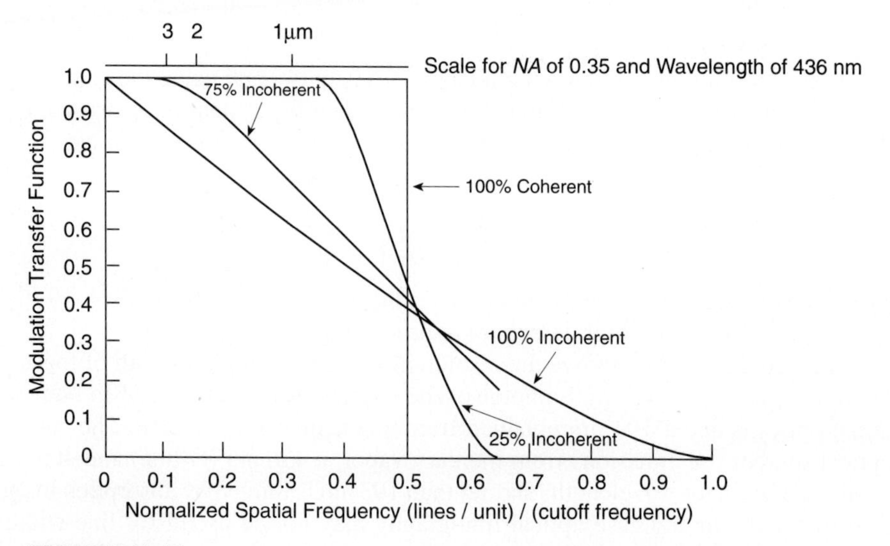

FIGURE 11.15
Modulation transfer function versus normalized frequency for different values of source coherence.

The coherence of a light source is a measure of the degree to which the light from a given source is emitted in phase. Phases of the wave can differ from one another because of differences in the time of emission of the photon or in the path traversed due to separate origination points for the emission of light. Any light source that is not a point source loses spatial coherence. The degree of coherence depends on the size of the source, angular range of light waves from the source allowed to pass through an aperture, and the distance of the image plane from the source. Since an ideal point source has zero brightness, most light sources will have partial coherence. To obtain high contrast, reasonable image fields, and depths of focus, partial coherence (60 percent) is desired. Furthermore, if MTF is known as a function of the frequency, the intensity distribution of the pattern can be obtained from the intensity distribution of the mask. Multiplying by the exposure time gives the dose of photons across the surface of the resist film.

EXAMPLE 11.2. How is it possible to avoid the effect of standing waves in contact printing?

Solution. For thin resist films on silicon dioxide over silicon, standing wave effects may be produced by monochromatic contact printing. The standing waves are produced by the interference of incoming light with the light reflected from the silicon surface. Some of the process issues are edge roundings, exposure time, and scalloping of resist lines. To avoid these problems, it should be possible to maintain the optical thickness of oxide plus resist as an odd multiple of the quarter wavelength of the exposure light.

The number of quarter wavelengths in the optical thickness at 3650Å is given by

$$\frac{(\text{oxide thickness} \times \text{index of refraction}) + (\text{resist thickness} \times \text{index of refraction})}{(3650/4)}.$$

Other means used to counteract standing wave effects are postexposure bake, use of polychromatic light source, and control of developing rate of photoresist.

11.1.3.2 Electron- and ion-beam lithographies

Electron and ion beams offer several advantages over photon lithography. Besides giving a better resolution because of small wavelength, they can be focused directly to produce the pattern on the resist without the need for a mask. However, the engineering problems involved in designing equipment, its cost, and very low throughput will restrict their use to situations in which they are the only option for lithography.

To expose the substrate coated with resist, it is necessary to subdivide the resist into a grid of addressable locations. The smallest elementary area to be exposed in the resist is called a *pixel*. Each pixel must receive a certain number of electrons or ions so that the exposed resist is significantly different chemically from the unexposed pixel. Thus the exposure dose must exceed a minimum value

$$\frac{Q}{q} \geq \frac{N_m}{\rho_p^2}, \tag{11.11}$$

where Q is the dose delivered to the resist (Coulombs/cm^2), q is the charge of the electron or the ion, N_m is the minimum number of electrons or ions to strike and lose their energy in each pixel, and ρ_p is the linear dimension of the pixel. The length of time it takes to write a full pattern is an important parameter in the design of electron- and ion-beam systems. To decrease the exposure time without compromising resolution, the ready availability of electron sources and the high quality of electron optics have led to the development of alternative beam shapes, such as Gaussian round beams, fixed-shape beam configurations and variable-shape aperture beams.

The electrons emitted from a source have an energy distribution that depends on the temperature of the source. The distribution of velocities and directions of emission of electrons imply that they cannot be brought to a point focus when they pass through a magnetic lens. A maximum current J_m can be brought to focus through a convergence angle α, and J_m can be approximately given by

$$J_m = J_0\left(\frac{eV}{k_B T}\right)\alpha^2, \tag{11.12}$$

where J_0 is the current emitted from the gun cathode, V is the beam-accelerating voltage, T is the temperature of the electron source, and k_B is the Boltzmann constant.

The maximum current that can be drawn from an electron gun is limited by the negative space charge surrounding the filament. Since all the emitted electrons do not pass through the center of the lens, some of them are subject to spherical aberration. The spread in energy of the emitted electrons contribute to chromatic aberration. These factors limit the smallest diameter that we can obtain in an electron-beam system (Pease 1981).

Electron-beam lithography is mask free. Registration between the substrate and the desired pattern can be obtained by analyzing the back-scattered electrons from the surface topography. This analysis permits control and correction for level-to-level misalignments produced by wafer distortion and temperature differences.

Focused ion beams of 0.1 μm diameter can be produced and can be scanned to fabricate patterns of any desired shape. The size of the ion source and quality of optics available limit the use of ion-beam lithography. Bright ion sources based on liquid-metal field emitters have been used to repair optical and X-ray masks.

11.2 ETCHING

Conceptually, the fabrication of ICs involves the addition of materials at some locations and the removal of materials from other regions. As discussed in section 11.1.1, selective addition is carried out using lithography. The removal of unwanted materials is accomplished by etching. If etching involves aqueous solutions, it is termed *wet etching*. On the other hand, *dry etching* entails the use of gaseous species. Figure 11.16 shows the two basic methods of pattern transfer. In the subtractive method the planar film is removed from the areas not protected by the mask. In the additive method the film is deposited over the mask pattern, and the mask with the undesired film deposit is removed later.

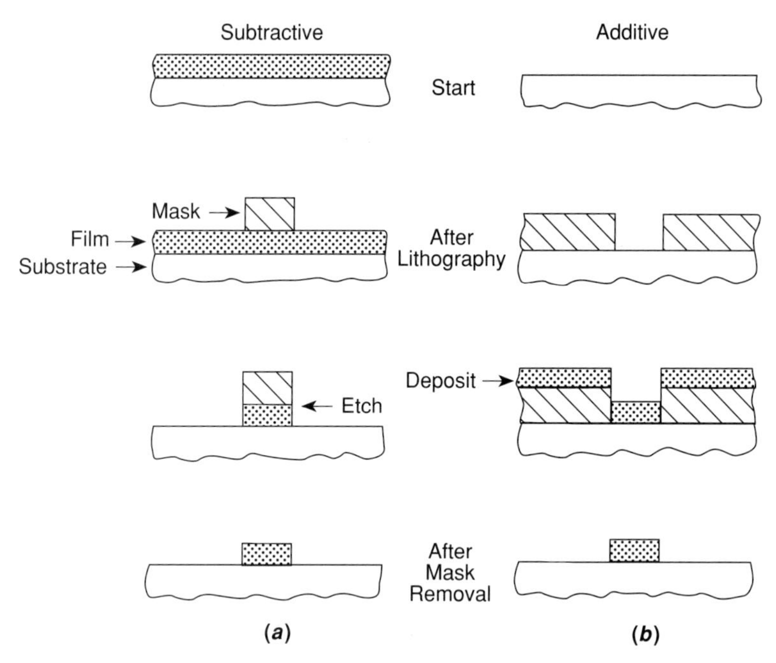

FIGURE 11.16
Process steps in (*a*) subtractive and (*b*) additive pattern transfer.

11.2.1 Wet Etching

Films are removed using chemical methods. Resists are stripped by a solution consisting of H_2SO_4 and H_2O_2. Organic and some metallic films can be dissolved in H_2O: NH_4OH: H_2O_2 : : 5: 1: 1 solution. Any remaining oxide film can be removed by HF.

A variety of etchants are used in IC fabrication. A solution consisting of HF: HNO_3: CH_3COOH : : 8: 75: 17 by volume serves as a nonpreferential etch for silicon. The etch rate is 5 μm/min at 25°C. Preferential etchants have also been developed. For example, in a solution containing 50 gms of KOH and 150 gms of water, maintained at 80°C, the etch rate of {110} silicon is 700 times faster than that for {111} silicon.

During wet etching an etchant is simultaneously in contact with a variety of materials, and only one of them is to be removed. To maintain the dimensions of the original pattern, the etchant must be highly selective, particularly with respect to the mask material. Wet etching has a number of problems. The etched material reveals ragged edge forms. The adherence of the film to the substrate is extremely important; without adherence the dimensional control is poor. Bubbles can grow during etching and act as localized masks preventing the proper removal of film. The wet etching is isotropic so that the line width is difficult to control, as shown in Figure 11.17.

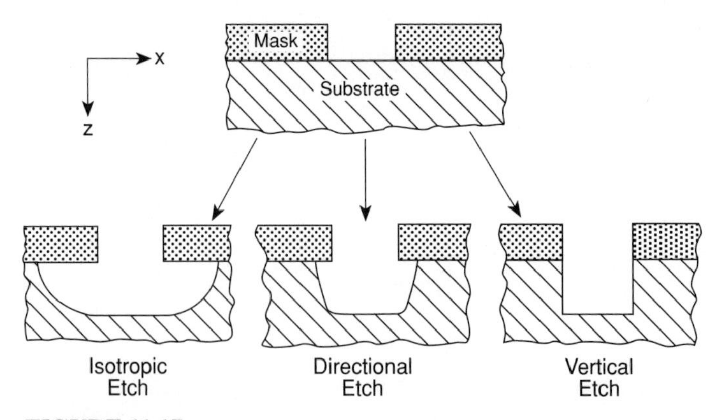

FIGURE 11.17
Profiles of an etched pattern.

11.2.2 Dry Etching

Wet etching exhibits high selectivity and is easy to carry out. However, when feature dimensions are in the 1 μm range, the reproducible transfer of patterns becomes difficult. Dry etching obviates this problem. One of the earliest applications of dry etching was the removal of hardened photoresist using plasma ashing. This process was accomplished by a glow discharge consisting of oxygen radicals.

The extension of plasma techniques to etch materials other than hardened resists has been prompted by a number of limitations of the wet-etching technique. As indicated above wet etching is not suitable for small feature sizes, 0.5 to 1.0 μm, because surface tension of etch solutions produces bridges between stripes and precludes etching of the film.

The fabrication of high-resolution patterns requires a method of material removal that provides high anisotropy in etching, is very selective in etching, and provides reasonable rates for material removal. At the present time dry-etching techniques are the only methods available that satisfy these requirements. A number of different techniques for dry etching have been developed. They have been classified in a convenient manner and are shown in Figure 11.18. The simplest technique utilizes a reaction between a gas and the exposed areas of resist to form a volatile species, as in reactive gas etching. Alternatively, light from a laser source can be focused either to provide heat in a localized region or to promote selective reactions so that vapor species are generated and removed from the material to be etched. All other methods of etching involve the use of ions and/or plasma.

The plasma is considered as a partially ionized gas consisting of an equal number of positive and negative charges and a large number of neutral molecules, excited molecules, and radicals. The plasma is typically generated by applying an electric field to a gas maintained at low pressures in a chamber. The plasmas that are of interest in microfabrication are called *cold plasma* and are characterized by

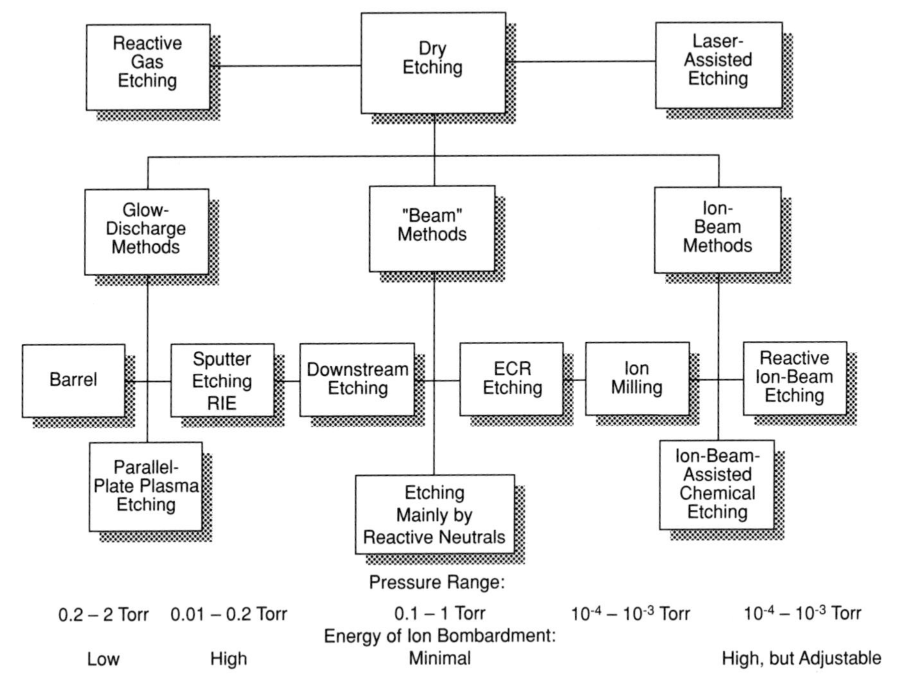

FIGURE 11.18
Dry-etching methods and typical energy and pressure ranges.

high values of E/p or E/n, where E is the strength of the electric field, p is the pressure in the plasma chamber, and n is the particle number density. Under these conditions electrons can achieve high energies compared to the ions, which can be at ambient temperatures. Due to their low mass electrons can transfer very little, if any, of their kinetic energy to the kinetic energy of atoms and molecules with which they collide. The collision of electrons with atoms and molecules is essentially elastic in nature. On the other hand, electrons acquire sufficient kinetic energy from the applied electric field and can transfer it to the atoms and molecules as potential energy during inelastic collisions. The ions, however, collide with the other species of comparable mass and can transfer kinetic energy efficiently from one to another. Whatever energy they gain from the electric field is shared quickly with all other particles, causing them to come to equilibrium quickly. The net result is that atoms and molecules are at equilibrium with the temperature of the container of the plasma, whereas electrons start expending their energies in creating a variety of reactions with the atoms and molecules, thereby changing their nature and reactivity.

A stable plasma is possible if the generation and neutralization rates of ions are balanced. Table 11.1 lists the types of reactions that can occur when an electron collides with an atom or molecule. The reactions that occur between electrons and ions

TABLE 11.1
Electron impact reactions

Excitation (rotational, vibrational, electronic)

$$e + 2x \rightarrow x_2^* + e$$

Dissociative attachment

$$e + x_2 \rightarrow x^- + x$$

Dissociation

$$e + x_2 \rightarrow 2x + e$$

Ionization

$$e + x_2 \rightarrow x_2^+ + 2e$$

Dissociative ionization

$$e + x_2 \rightarrow x^+ + x + 2e$$

*Indicates an excited state.

TABLE 11.2
Reactions at surfaces

Ion-surface interactions

- Neutralization and secondary electron emission
- Sputtering
- Ion-induced chemistry

Electron-surface interactions

- Secondary electron emission
- Electron-induced chemistry

Radical or atom-surface interactions

- Surface etching
- Film deposition

with atoms on the surface of the plasma chamber are also very important, and a number of these are listed in Table 11.2. The interactions of ions with other molecules are listed in Table 11.3. The type of reactions that occur between ions and other ions, molecules, and so forth resulting in a loss of charge are given in Table 11.4. Both positive and negative ions are produced in a plasma, and a number of reactions that produce negative ions are listed in Table 11.5. The plasma thus consists of a variety of atoms, molecules, and radicals, and these have been investigated extensively (Massey et al. 1969). We see from these tables that the plasma is much more reactive than a gas from which it is derived. The specific types of particles in a plasma, their densities, and the factors on which they depend are a complex

TABLE 11.3
Interactions between atoms

Metastable-neutral collisions (penning ionization)

$$A^* + G \rightarrow G^+ + A + e$$

Metastable-metastable ionization

$$Ar^* + Ar^* = Ar + Ar^+ + e$$

Charge transfer

$$A + A^+ \rightarrow A^+ + A$$

Asymmetric charge transfer

$$B^+ + C \rightarrow B + C^+$$

*Indicates an excited state.

TABLE 11.4
Deionization reactions

Radiative recombination	Ion-ion recombination
$A^+ + e \rightarrow A^* + h\nu$	$A^+ + B \rightarrow AB + h\nu$
Dielectronic recombination	Mutual neutralization
$A^+ + e \rightarrow A^{**}$	$A^+ + B^- \rightarrow A^* + B^*$
Dissociative recombination	Three-body recombination
$AB^+ + e \rightarrow A^* + B^*$	$A^+ + B^- + M \rightarrow AB + M$
Three-body recombination	
$A^+ + e + e \rightarrow A^* + e$	

*Singly excited atom or molecule.
**Doubly excited atom or molecule.

function of the cross sections for the reactions mentioned in the preceding tables and are not listed in the tables.

The gaseous plasma is conducting. Therefore, the voltage drop across the plasma is very small and is called the *plasma potential* V_P. The applied field, however, shows the most voltage drop in regions that are adjacent to the electrodes. The

TABLE 11.5
Negative ion formation

1. Radiative attachment

   $e + A \Leftrightarrow A^- + h\nu$

2. Three-body collision

   $e + A + B \rightarrow A^- + (B + W_k)$

3. Associative detachment and dissociative attachment

   $e + XY \Leftrightarrow X^- + Y$

4. $e + (XY) \Leftrightarrow (XY)^{*-}$

   $(XY)^{*-} + A \Leftrightarrow (XY)^- + A + W_k + W_p$

5. Ion-pair production

   $e + XY \Leftrightarrow X^+ + Y^- + e$

6. Charge transfer in heavy-particle collision

   $A + B \rightarrow A^+ + B^-$

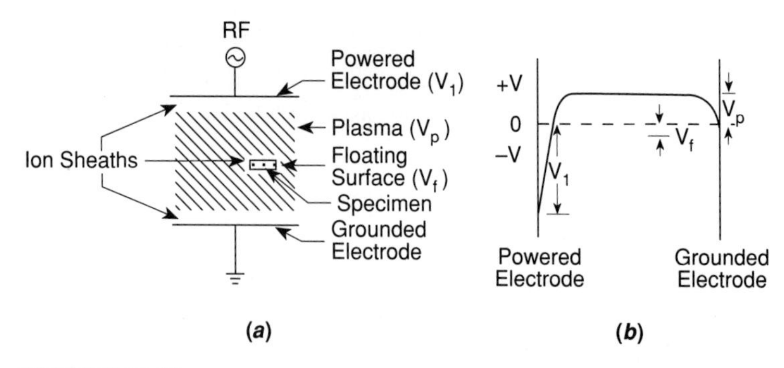

FIGURE 11.19
Schematic of (*a*) an RF glow discharge plasma and (*b*) voltage distribution.

regions are termed *plasma sheaths* and are shown in Figure 11.19. These regions are also deficient in electrons because of the encounter of plasmas with electrode walls, and the sheaths form due to the difference in the mobility of electrons and ions. The number density of charged particles in the plasma body is typically 10^4 to 10^6 and is smaller than the neutral species in the plasma. These neutral species are of primary interest in a plasma.

The plasma etching may be envisaged to consist of seven primary processes, as shown in Figure 11.20. The reactive species must form first in the plasma. They then diffuse to the surface to be etched. Once they reach the surface, the species should be chemisorbed on the surface, diffuse on the surface, and react with the surface atoms to form a reaction product. The product must then be desorbed from the surface. Finally, the desorbed product must be transported away from the surface.

Most of the gases commonly used in dry etching do not react with the material to be etched. The generation of reactive species by electron/molecule bombardment is a vital step for plasma etching to be initiated. For example, CF_4 does not etch

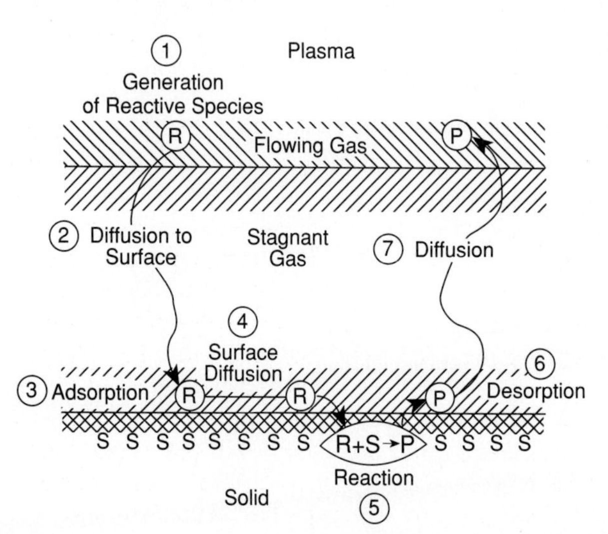

FIGURE 11.20
Schematic representation of the seven basic steps in plasma etching.

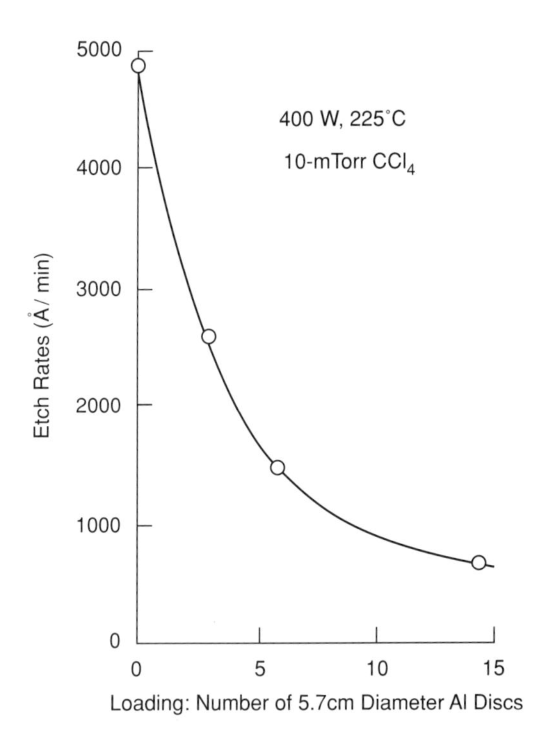

FIGURE 11.21
Loading effect observed in plasma etching of Al discs using CCl_4.

silicon until it is dissociated by electron collision to form fluorine atoms. The flow rate of the etch gas is extremely important in delivering the active species to the region to be etched, especially since the gas is consumed in the process of etching. One of the consequences of the inadequate flow is the so-called loading effect in which the etch rate decreases with the increase in area of the etchable material exposed to the plasma. The loading effect is clearly illustrated in Figure 11.21 for the etching of Al in CCl_4 plasma. One way to obviate this problem is to fully load the system by providing a large area of the material to be etched. The gas must be chemisorbed on the surface to be etched, forming a chemical bond with the surface atoms in order for etching to occur. For example, it has been determined experimentally that CF_4 does not chemisorb on silicon, whereas the CF_3 radical does. Once chemisorbed on the substrate, the radical has to move on the surface until the conditions are favorable for the formation of a reaction product, which generally depends on temperature. Chlorine, for example, chemisorbs on silicon, dissociates, and moves around on the surface, but chlorine is unable to form the reaction product $SiCl_4$; therefore, no etching can take place. Even if a reaction product were to form, it should have sufficient volatility to desorb from the surface. For example, aluminum surface exposed to fluorine atoms readily forms AlF_3, which is essentially nonvolatile at ambient temperatures, so that etching comes to a halt. The desorbed reaction product must go far enough to the effluent gas stream to be disposed of. If the reaction product is not stable, it might decompose on entering the plasma region and deposit all over the reaction chamber. A consequence of this behavior is the so-called memory effect; that is, the reactor retains the memory of the previous etching step, which makes it difficult to control the process in manufacturing.

TABLE 11.6
Etch gases used to dry etch films

Films	Gases
Si	CF_4, CF_4/O_2, CF_3Cl, CCl_4, Cl_2, SiF_4/O_2, NF_3, ClF_3
Si_3N_4	CF_4/O_2, C_2F_6, C_3F_8, CF_4/H_2
Organic materials	O_2, O_2/SF_6, O_2/CF_4
Al	CCl_4, CCl_4/Cl_2, $CHCl_3/Cl_2$, BCl_3, BCl_3/Cl_2, $SiCl_4$
W, WSi_2, Mo,	CF_4, C_2F_6, SF_6, CCl_2F_2,
$MoSi_2$, $TiSi_2$, $TaSi_2$	NF_3, CCl_4 (O often with O_2)
Ti	CF_4, $CClF_3$, $CBrF_3$
Au	$C_2Cl_2F_4$, Cl_2
Cr	Cl_2/O_2, CCl_4/O_2

One of the major considerations in plasma etching is the selection of gas that can be dissociated in a discharge to form the reactive gas species. For example, a CF_4 plasma discharge has abundant CF_3^+ ions, CF_3 radicals, and F atoms. Etching of Si in this discharge is due to F atoms. CF_3 radicals will chemisorb on Si but will not etch because of the presence of carbon on the surface from the breakup of CF_3 radicals. The selection of gases to etch materials is therefore a difficult task. Table 11.6 lists several of the gases used and the materials that they etch. We can use analytical techniques, such as emission spectroscopy, laser-induced fluorescence spectroscopy, mass spectrometry, and Fourier transform-infrared spectroscopy, to assess potential gases for dry-etching purposes.

Plasma etching for a given combination of gas and materials to be etched must provide high etch rates, accomplish uniformity in etching, be very selective in the materials that will be etched when exposed simultaneously to different materials, show strong anisotropy in etching, and provide some means by which the endpoint of etching can be recognized so that the etching process can be terminated at the appropriate time. In addition, one should avoid the radiation damage to the devices, remove sources of contamination, and safely dispose of all toxic gases in the effluent stream. These needs are not simultaneously achievable in all cases so that considerable experimentation precedes successful implementation of dry etching.

Plasma etching is highly directional. The presence of sheaths in a plasma propels positive ions to bombard the surface at normal incidence, resulting in an anisotropic flux at the plasma-solid boundaries. The ions may stimulate the action of the neutral gas-solid reactions. A classic illustration of this phenomenon is provided by the etching of silicon by XeF_2, which we show in Figure 11.22. The etch rates of silicon exposed to XeF_2 molecules alone and of Ar^+ ion beam alone are quite low. However, XeF_2 in the presence of Ar^+ ions shows a 12-fold increase in the etch rate of silicon. The effects of neutrals and ions in etching are clearly synergistic with the resulting material removal rate exceeding the sum of the substrate chemical etching

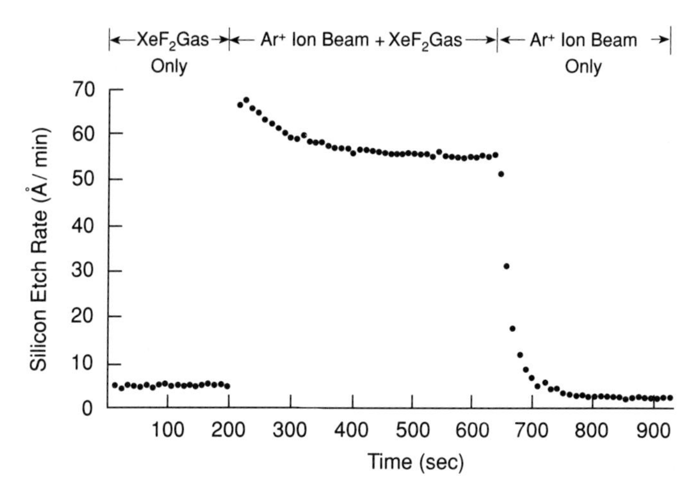

FIGURE 11.22
Synergistic role of Ar^+ ion beam and XeF_2 gas in plasma etching of silicon.

or ion-beam etching acting alone. Researchers assume that the ions supply kinetic energy and disturb the surface being etched, and the sheath potential provides the directionality to the path of the ions before striking the surface. The interrelationship between the anisotropy obtainable in etching and its implications in lithography has been extensively discussed by Mogab (1981).

Selectivity in plasma etching is defined as the ratio of etch rates between different materials when exposed to the same plasma environment. The importance of selectivity in feature size control, performance, and yield of devices is self-evident. Some selectivity is possible by maintaining the bombardment energies of ions above the threshold energy for one material and below that of the other. The predominance of exploiting only this feature can result in angled features in etching, stemming from the angular dependence of sputtering rates. A much more common approach to attain selectivity in etching is the manipulation of the plasma chemistry by the introduction of gases. In addition to maximizing or minimizing a particular reactive radical in a plasma, this procedure introduces another mechanism for selective etching. This mechanism entails interaction of ions with an etching product to form nonvolatile films that coat the substrate and prevent etching reactions from taking place. However, the flux of the ions keeps the horizontal areas clear from this undesirable film, but the vertical features are protected and untouched by the ions. For example, the density of fluorine atoms increases as oxygen is added to fluorocarbon discharges, and gases that are introduced to affect plasma chemistry, such as hydrogen, tend to consume fluorine atoms. The etch rates of silicon, silicon dioxide, and resist can be manipulated by these additions, as shown in Figure 11.23. The selectivity is improved if fluorocarbons decompose to form carbon fibers. The accumulation of carbon is less on SiO_2 than it is on Si because of the direct reaction between carbon and oxygen resulting from SiO_2 producing CO, CO_2, and so forth. In extreme cases the carbon could polymerize and basically stop the etching process.

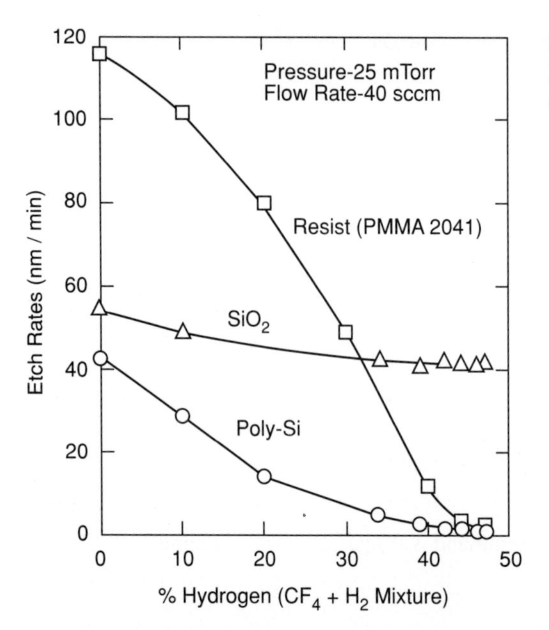

FIGURE 11.23
Effect of hydrogen addition on CF_4 etching of SiO_2, poly-si, and resist.

Uniformity of etching, surface quality, and reproducibility in etching are highly desirable features in fabrication processes. Process parameters, such as gas-flow rate, distribution of the etchant gas, and temperature variations in the reactor are important. Many of these factors are a function of the reactor design. To overcome and optimize a number of requirements necessary in electronic microfabrication, specially designed reactors have been constructed. Figure 11.24 shows some of these reactors.

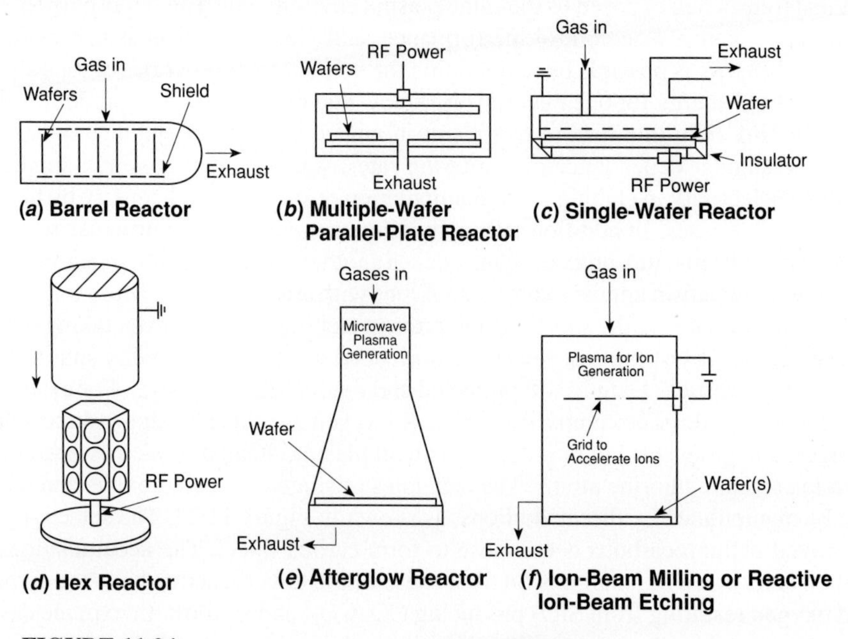

FIGURE 11.24
Various reactor configurations used for plasma etching.

The detection of the etching endpoint is important in the plasma-etching process, since on many occasions overetching must be allowed to accommodate the variations in the topography of the film. Monitoring pressure changes, mass spectrometry, laser interferometry, and reflectance measurement are some of the tools used to detect the endpoint of etching and control the time of etching.

Currently, macroscopic models are not available to simulate etching profiles for various combinations of gas phase and reactor variables; therefore, painstaking statistical design of experiments is necessary to optimize etching conditions. Reactor parameters such as frequency, excitation power, discharge gas, pumping speed, geometrical factors (area), and gas-flow rates, cannot be varied sufficiently to meet all fabrication requirements. Therefore, many different techniques of dry-etching reactions have been specially designed. Detailed considerations of these are constantly emerging and can be found in the literature.

EXAMPLE 11.3. Given that in a plasma chamber electrodes are 0.02 m apart and the frequency used to generate the plasma is 13.56 MHz, calculate the voltage at which efficient transfer of energy occurs.

Solution. An electron subjected to an electric field ε gains an energy E where

$$E = q\varepsilon\lambda,$$

where λ is the average distance traversed by the electron before collision and q is the charge on an electron. An electron, by virtue of its small mass, can lose only a small fraction of the energy that it acquires by the electric field. The electron thus has a small drift velocity due to the electric field that is superimposed on its random thermal-velocity distribution.

In an alternating field the electron oscillates according to the equation of motion:

$$m\,\frac{d^2x}{dt^2} = q\xi \sin \omega t,$$

where w is the angular frequency and $\frac{d^2x}{dt^2}$ = acceleration. We have

$$\text{velocity} = \frac{dx}{dt} = \frac{q\xi}{m\omega}\cos(\omega t).$$

Therefore, the maximum energy gained from the field is

$$\frac{1}{2}\,m\left(\frac{dx}{dt}\right)^2 = \frac{1}{2m}\left(\frac{q\xi}{m\omega}\right).$$

If a voltage V_o is applied between the two electrodes separated by a distance d, then $\varepsilon = \frac{V_o}{d}$. An electron can be trapped in the interelectrode region if its mean free path λ is greater than x_{max}. We obtain

$$x = \frac{q\varepsilon}{\omega^2 m}\sin \omega t$$

So that

$$x_{max} = \frac{q\varepsilon}{\omega^2 m} = \frac{qV_o}{d\omega^2 m}.$$

Hence the electron will be trapped if

$$\omega^2 > \frac{qV_0}{d^2m}.$$

When the electron is trapped, eventually an ionizing collision occurs so that efficient transfer of energy occurs from the RF field. Since the typical frequency used for generating an RF plasma is 13.56 MHz, the voltage that one has to apply between electrodes that are 0.02m apart is

$$V_0 = \frac{d^2m\omega^2}{q} = \frac{4 \times 10^{-2} \times 9.11 \times 10^{-31} \times (2\pi \times 13.56 \times 10^6)^2}{1.6 \times 10^{-19}}$$

$$= 1663 \text{ Volts.}$$

EXAMPLE 11.4. Calculate the relative etch rate of SiO_2, Si_3N_4 and polycrystalline Si in a plasma etcher using CF_4 with a small amount of oxygen.

Solution. The equations that represent etching are similar regardless of the compound etched. For example, we can write the etching of SiO_2 as $SiO_2 + [O/F] = [SiF_4, SiOF_2, Si_2OF_6] + O_2 + F_2 + CO_2 + H_2O$ and so on. As a requirement in plasma etching, all the etch products are gaseous. Therefore, the etch rates must vary as a function of the atomic percent of silicon, even though absolute etch rates depend upon many variables. Therefore, we have the following:

Compound	Atomic percent Si	Relative etch rate
SiO_2	33.3%	1.0
Si_3N_4	42.8	4.0
Si	100	–

Given the relative etch rates of SiO_2 and Si_3N_4 to be 1.0 and 4.0, respectively, the relative etch rate of silicon is

$$= \frac{(4.0 - 1.0)}{(42.8 - 33.3)} \times (100 - 33.3) = 21.$$

11.3 HIGHLIGHTS OF THE CHAPTER

The highlights of this chapter are as follows:

- Using lithography, we can transfer any geometrical pattern onto the surface of a substrate. The three main components of lithography are resists, masks, and radiation sources.
- Resists are either positive or negative and are polymeric materials. The pattern is transferred by exposing the suitably masked areas of the resists to one of the following radiation sources: optical, X-rays, electron beams, or ion beams.
- Unwanted materials can be removed from the wafer surface using wet and dry etchings. When high selectivity is desired during etching, plasma and reactive-ion etchings are better than wet etching.

REFERENCES

Electron Beam Technology in Microelectronic Fabrication, ed. G. R. Brewer. New York: Academic Press, 1980.

Hiraoka, H. and J. Pacansky. *J. Vac. Sci. Technol.* 19, no. 4, 132 (1981).

Introduction to Microlithography, ed. L. F. Thompson; C. G. Wilson; and M. J. Bowden. ACS Symposium # 19. Washington, DC: American Chemical Society, 1983.

Lai, J. H. and L. T. Shepherd. *J. Apply. Polym. Sci.* 20, 2367 (1976).

LaVergne, D. and D. C. Hofer. *Proc. SPIE,* 2, 539 (1985).

Lin, B. J. *J. Vac. Sci. Technol.* 12, 1317 (1975).

Lin, B. J. In *Fine Line Lithography,* ed. R. Newman. New York: North-Holland, 1980.

Ma, W. H. *SPIE Submicron Lithogr.* 333, 19 (1982).

Massey, H. S. W.; E. H. S. Burhop; and H. B. Gilbody. *Electronic and Ionic Impact Phenomens.* 2nd ed. London: Oxford University Press, 1969.

Materials for Lithography, ed. L. F. Thompson; C. G. Wilson; and J. M. J. Frécht, ACS Symposium #266. Washington, DC: American Chemical Society, 1984.

Microelectronic Materials and Processes, ed. R. A. Levy. Dordecht: Klüwer Academic Publishers, 1989.

Moreau, W. M. *Semiconductor Lithography: Principles, Practice and Materials.* New York: Plenum Press, 1988.

Murarka, S. P. and M. C. Peckerar. *Electronic Materials Science and Technology.* Boston: Academic Press, 1989.

Particle Control for Semiconductor Manufacturing, ed. R. P. Donovan. New York: Marcel Dekker, 1990.

Pease, R. F. W. *Contemporary Physics,* 22, 265 (1981).

Runyan, W. R. and K. E. Bean. *Semiconductor Integrated Circuit Process Technology.* Reading, MA: Addison-Wesley, 1990.

Semiconductor Materials and Process Handbook, ed. G. E. McQuire. Noyes Publications, 1988.

Toukhy, M. A. and J. A. Jech. *SPIE Adv. Resis. Technol.* 159, 469 (1984).

Wilson, C. G. and M. J. Bowden. In *Electronic and Photonic Applications of Polymers, Advances in Chemistry Series 218,* Washington DC: American Chemical Society, 1988.

Wolf, S. and R. N. Tauber. *Silicon Processing.* vol. 1, Sunset Beach, CA: Lattice Press, 1986.

PROBLEMS

11.1. A mask is in physical contact with a resist. If W is the opening in the mask, what is the corresponding opening in the photoresist if the resolution is diffraction limited? The wavelength is λ.

11.2. How does the answer to problem 11.1 change if there is a gap s between the mask and the photoresist?

11.3. The refractive index of a photoresist is given by 1.68–0.02 i. What is the amount by which the incoming light of wavelength 404.7 nm is attenuated?

11.4. What is the transmission factor of X-rays when they go through a mask of thickness z? What is the amount of unattenuated X-rays when they go through a support material for the mask of thickness z. What is the condition to obtain maximum contrast?

11.5. A photoresist is deposited on a wafer by spin coating. Develop an expression for the thickness of the resist as a function of the spinning rate.

11.6. A positive resist is formulated such that during development in a solvent the dissolution kinetics are (*a*) linear and (*b*) nonlinear. Use a schematic sketch to explain how the kinetics can affect the contrast.

11.7. What defect density is permissible to obtain 70 percent yield for a silicon chip of area 500 μm on a side? By how much should the density of defects decrease if the chip area is 1.5 cm^2? If there are 12 process levels, what is the maximum number of fatal defects per cm^2 per step that can be tolerated?

11.8. The etch rates of silicon and silicon dioxide for fluorine atoms is given by the expression:

$$\text{Etch rate} = A n_F T^{1/2} \exp(-E_A/k_B T) \text{A}/\text{min},$$

where the constants A and E_A for Si and SiO_2 are, respectively, as follows:

Film	A	E_A (kcal/mol)
Si	2.86×10^{-12}	2.48
SiO_2	0.614×10^{-12}	3.76.

For $n_F = 3 \times 10^{15} cm^{-3}$, plot the ratio of etch rates between the temperatures −50 to 2000°C. Determine the temperature at which a selectivity ratio of 95:1 is reached.

CHAPTER 12

Challenges in Growth and Processing of Semiconductors

The semiconductor industry is under continued, relentless pressure to develop products that are better and cheaper than their predecessors. Currently, there are three main consumer-driven electronics markets: delivery of entertainment, computing services, and communications. These services have many common characteristics. In the future the electronic equipment required for these services will be essentially identical, leading to the convergence of the three technologies.

The silicon-based microelectronics play a major role in all three markets. As this technology advances, the feature size in integrated circuits (ICs) will become smaller. We highlight some of these developments in Table 12.1. We also show in Figure 12.1 that by the year 2000 the gate length of MOSFETs may be reduced to 0.2 μm, gate-oxide thickness to 4 nm, and junction depth to 0.04 μm (Chang and Sze 1996). These reductions would lead to faster ICs. We expect similar changes in bipolar junction transistors.

Figure 12.2 shows the evolution of circuit complexity versus year (Chang and Sze 1996). It is evident that ICs incorporating MOSFETs have the highest

TABLE 12.1
Performance projection

	Year		
	1960	**1991**	**2000**
Minimum feature length (μm)	25	0.7	0.2
Component density (devices/cm^2)	1	8×10^6	0.03
Gate delay (ns)	500	0.1	0.03
Wafer size (mm)	25	200	250

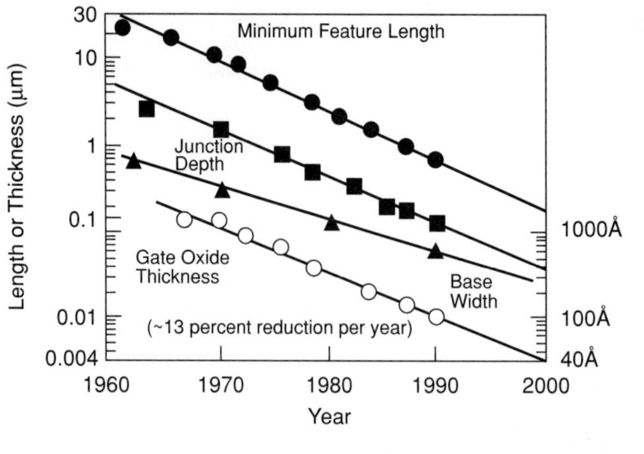

FIGURE 12.1
Dimensional scaling of MOSFETs and bipolar junction transistors versus time. (*After Chang and Sze [1996].*)

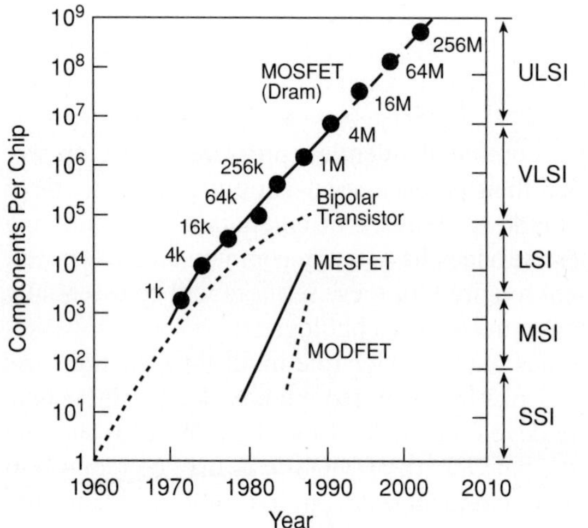

FIGURE 12.2
Evolution of VLSI circuit complexity. (*After Chang and Sze [1996].*)

complexity. The momentum for bipolar transistors is leveling off. MESFETs and modulation-doped field-effect transistors (MODFETs) are still in the development stage, but their potential is very high. In addition, the integration of highly complex silicon-based devices with the high-speed capability of III-V devices will create novel applications for the semiconductor technology in the three markets.

To achieve extremely fast and sophisticated devices, we must further refine on several fronts the science and technology of growth and processing of semiconductors. We highlight some of these challenges in this chapter.

12.1 GROWTH OF BULK CRYSTALS OF III-N MATERIALS

AlN, GaN, and InN are scientifically interesting and technologically relevant materials. Their respective band gaps are 6.2, 3.39, and 1.89 eV. Devices emitting in the blue-green region of the spectrum can be fabricated using these materials. They also have applications in ultraviolet (UV) detectors. Furthermore, we can use AlN and GaN to fabricate devices operating at high temperatures because the concentration of thermally generated carriers is very low due to their wide band gaps. Currently, we grow III-N epitaxial layers on sapphire, spinel, SiC, and Si substrates. Since the mismatch between the overgrowth and the substrate is significant, high densities of misfit and threading dislocations are observed in the layer (Mahajan 1996). To achieve long-term device reliability, we must improve the perfection of the layer. We can achieve this objective via either homoepitaxy or lattice-matched heteroepitaxy. For homoepitaxy we require bulk crystals of III-N materials.

We can use two approaches to grow the above crystals: (1) vapor transport and (2) liquid encapsulated Czochralski (LEC) growth under extremely high pressures. The vapor of species from the source to the growth region is achieved by imposing a temperature gradient in an evacuated vessel, a technique that is being currently used to grow SiC crystals. The growth of bulk AlN crystals by this method has been reported (Balkas et al. 1996).

The pressures of nitrogen over GaN and InN melts are extremely high. Therefore, we need to impose very high pressures during the LEC growth. Baranowski (1996) has attempted to grow bulk GaN crystals using LEC. The as-grown crystals are relatively small, but their perfection is adequate (Hu et al. 1996). We anticipate that with an additional effort, we may be successful in growing modest size crystals.

12.2 GROWTH OF QUANTUM WELLS, WIRES, AND DOTS

We showed in section 1.2.1 that when electrons are confined in an infinite-potential well, they acquire discrete energy levels that are given by Equation (1.21). For example, the lowest level E_1 corresponds to $n = 1$ and is given by

$$E_1 = \frac{\pi^2 \hbar^2}{2ma^2}, \tag{12.1}$$

where $\hbar = \dfrac{h}{2\pi}$, h being Planck's constant, m is the mass of the electron, and a is the width of the well. The advent of the epitaxial growth techniques discussed in Chapter 6 allows us to grow structures in which carriers in an active layer can be confined by barrier layers. Depending on the dimensionality of the resulting confinement, we can obtain quantum wells, wires, and dots.

Using growth on patterned substrates (see section 6.2), we can fabricate quantum wells and wires (Kapon 1994). Let us imagine a situation in which we sequentially deposit two materials, differing in their band gaps, on a nonplanar substrate. The resulting quantum wells are shown in Figure 12.3. The lateral patterning of the layers occurs because the crystal facets exposed on a patterned substrate have different growth rates. The difference in the incorporation rate of adatoms and lateral

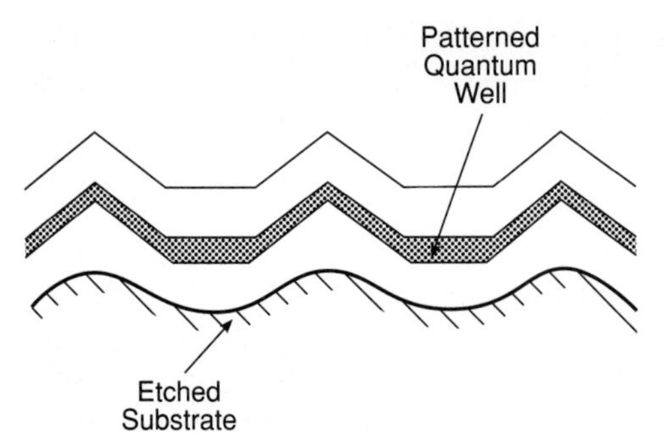

FIGURE 12.3
Schematic cross-sectional view of a patterned quantum-well heterostructure formed after growth on a nonplanar substrate. (*After Kapon [1994].*)

variation in flux of the growth species due to geometrical and/or surface diffusion effects cause the variation in growth rate.

Quantum wires having lateral dimensions in the 10 nm range can also be fabricated by epitaxial growth on nonplanar substrates. The schematic in Figure 12.4 shows the GaAs-GaAlAs quantum-wire heterostructures grown by organometallic vapor phase epitaxy (OMVPE) in a $[01\bar{1}]$-oriented channel etched in a (100) GaAs substrate (Kapon 1994). The growth of the GaAlAs layers sharpens the corner between the exposed $\{111\}_A$ planes of the grooved substrate. Subsequent growth of a thin GaAs quantum-well layer increases the corner radius slightly because the migration distance of gallium is greater than that of aluminum, resulting in a crescent-shape GaAs quantum wire. We can regain the sharpness of the corner by growing a thick GaAlAs barrier layer as shown in Figure 12.4. Using sequential growths of GaAs and GaAlAs, we can obtain multiple quantum wires and quantum wells within a single groove. Figure 12.5 shows an example of such an array, taken from the study of Kapon (1994). This figure depicts a cross-sectional electron micrograph in which the GaAs quantum wires and quantum wells, embedded in the high band-gap material GaAlAs, are clearly visible.

That different regions of the quantum wires shown in Figure 12.5 have different effective band gaps can be discerned from Figure 12.6 that shows a cathodoluminescence (CL) spectrum of the array (Christen et al. 1992). The CL

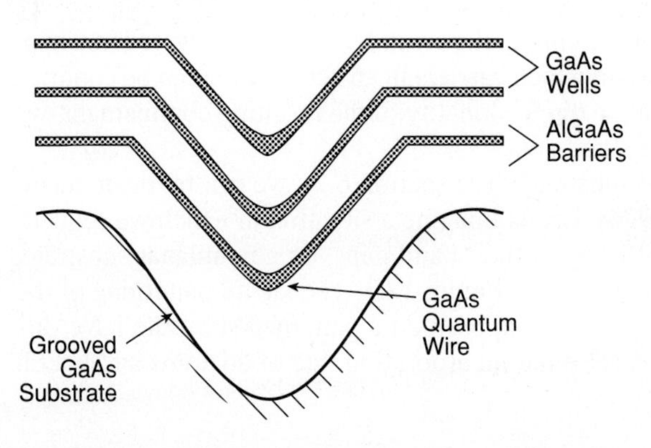

FIGURE 12.4
Schematic cross section of a GaAs-GaAlAs heterostructure, grown by OMVPE in a $[01\bar{1}]$ groove in a (100) GaAs substrate, showing the formation of a crescent-shape quantum wire. (*After Kapon [1994].*)

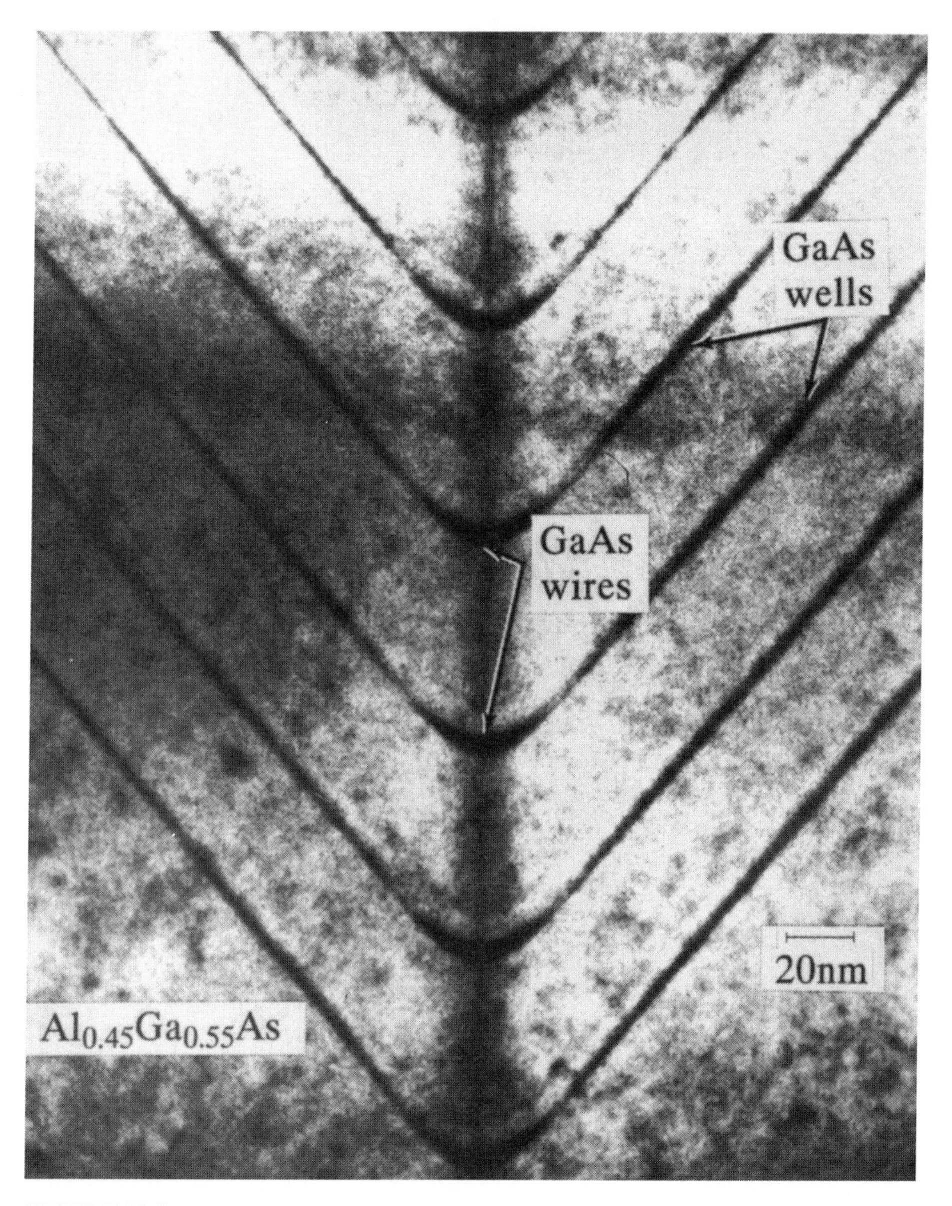

FIGURE 12.5
Cross-sectional electron micrograph showing a vertical array of GaAs-GaAlAs quantum wires formed in a groove. (*After Kapon [1994].*)

emissions from the quantum wires (QWRS) and quantum wells (QWLS) occur at 770 and 660 nm, respectively, whereas the underlying GaAs substrate emits at 820 nm.

Let us now consider a situation in which nanoclusters of a semiconductor are buried in an insulating matrix. In this case the carriers within the semiconductor are confined in all three directions. We refer to these nanocrystals as *quantum dots*

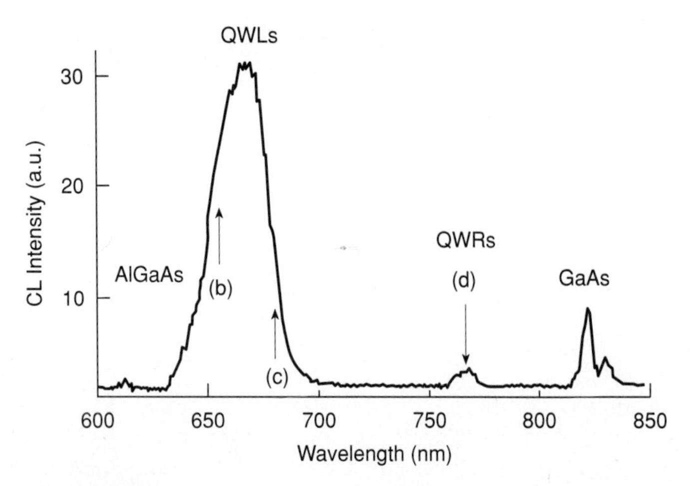

FIGURE 12.6
Cathodoluminescence spectrum obtained from Figure 12.5. (*After Christen et al. [1992].*)

(Steigerwald and Brus 1989). The size of the dots may range from 1 to 20 nm. Figure 12.7 shows an electron micrograph of CdS quantum dots that are confined in a glass matrix (Liu and Risbud 1992). Like the quantum wires and the quantum wells, the dots have unusual optical properties.

Currently, there is a considerable interest in light emitters operating in the blue-green region of the spectrum. Two types of wide band-gap materials, that is, ZnSe and

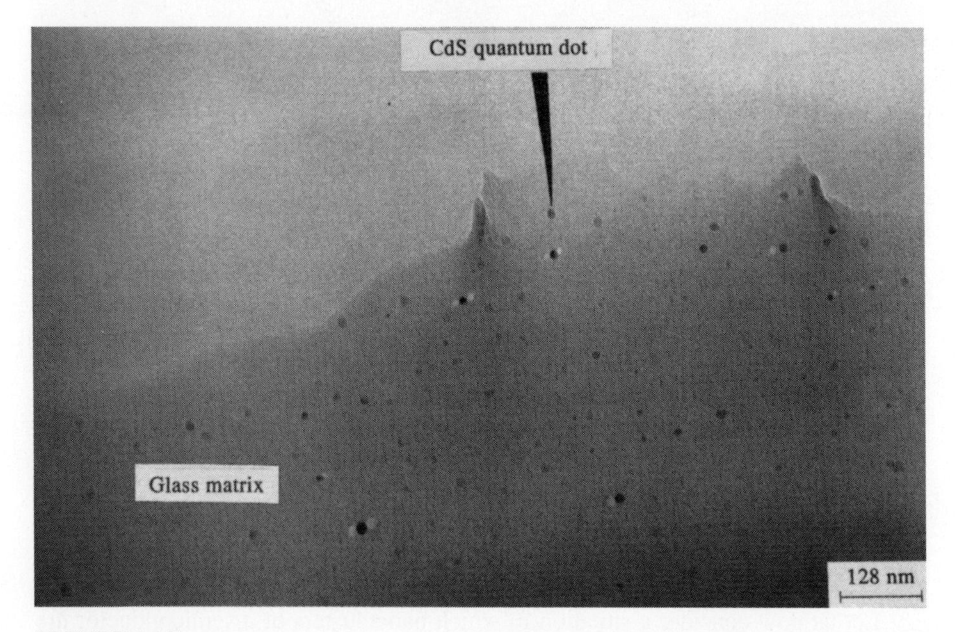

FIGURE 12.7
Electron micrograph showing CdS quantum dots dispersed in the glass matrix. (*After Liu and Risbud [1992].*)

GaN, have been used to fabricate such devices (Haase et al. 1991; Nakamura 1996). These developments represent significant achievements in the field of material engineering because the fabrication of these emitters required controlled n- and p-type doping. With the availability of the advanced epitaxial growth techniques, such as OMVPE and MBE, it has been possible to achieve these devices in ZnSe, GaN, and related materials.

12.3 THIN DIELECTRICS

As dimensions of the feature sizes shrink in the silicon technology, a concomitant reduction in the gate oxide thickness is required to prevent short-channel effects. We illustrate this trend for complementary MOS (CMOS) technology in Figure 12.8 (Kwong 1996). For example, an excessive reduction in channel length of a MOSFET without an adequate thickness scaling can result in threshold voltage instabilities. Therefore, to minimize the undesirable short-channel effects, we need to reduce the oxide thickness, resulting in a device that behaves in a more long-channel fashion.

Thin gate oxides (<10 nm) of MOS structures in ultralarge scale integration (ULSI) should have the following characteristics: (1) low defect density, (2) good barrier properties against impurity diffusion, (3) high-quality Si/SiO_2 interface with low interface state density and fixed charge, and (4) stability under irradiation. To meet these stringent requirements, we pay considerable attention to preoxidation cleaning procedures (Kwong 1996). Tay et al. (1987) used high-pressure oxidation of silicon to obtain 12 nm thick oxide at 700°C. The properties of the resulting oxide are reasonable. The high-pressure approach is attractive in relationship to the

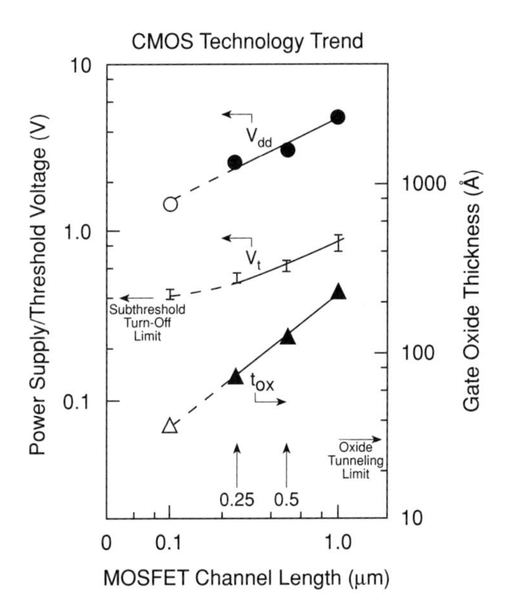

FIGURE 12.8
Illustrating the variation in the thickness of the oxide with the channel length. (*After Kwong [1996].*)

thermal budget, but its applicability to ULSI MOS processing needs to be explored further.

Researchers have recently grown thin oxides by rapid thermal oxidation of silicon in oxygen and N_2O ambients (Ting et al. 1991; Green et al. 1994). We compare the growth rates in the two ambients at 1000°C in Figure 12.9 (Ting et al. 1991). It is clear that the oxide grows at a faster rate in the O_2 environment. The lower growth rate in the N_2O atmosphere has been attributed to the presence of a SiO_xN_y layer in contact with the silicon surface (Ting et al. 1991; Green et al. 1994). We show the resulting diffusion-couple situation in Figure 12.10. According to the schematic, for oxidation to occur even at small thicknesses, the silicon atoms must diffuse through the oxy-nitride layer. This slow process reduces the growth rate of the oxide (Ting et al. 1991; Green et al. 1994).

The oxy-nitride layer is also an excellent barrier to the diffusion of boron atoms (Ting et al. 1991). Du et al. (1990) attribute this effect to its denser atomic structure

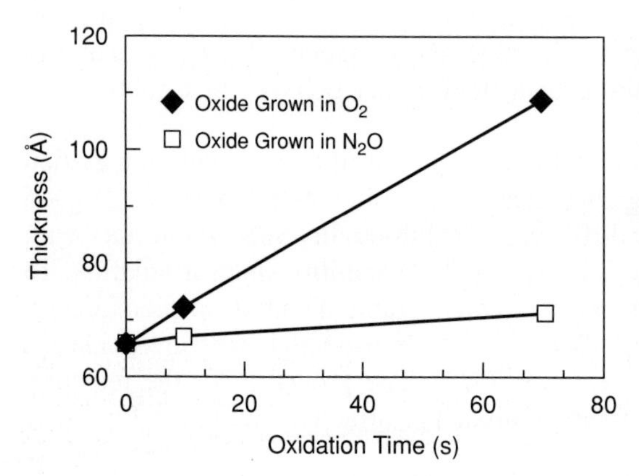

FIGURE 12.9
Illustrating the variation in thicknesses of two oxide samples subjected to rapid thermal oxidation in O_2 at 1000°C. The two samples were originally oxidized at 1050°C, one in N_2O and the other in O_2, both to a thickness of 6.5 nm before the 1000°C oxidation. (*After Ting et al. [1991].*)

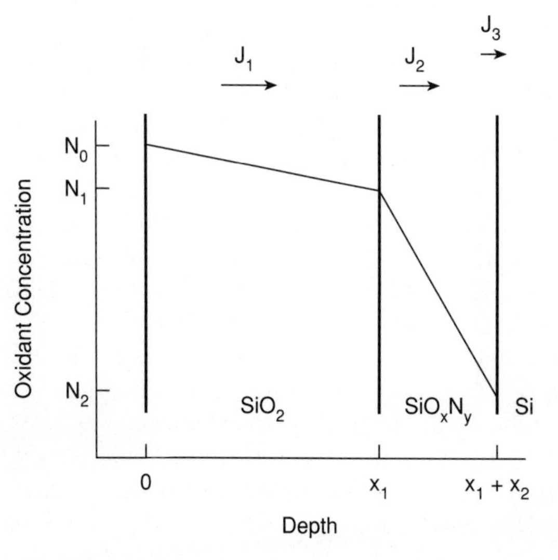

FIGURE 12.10
Schematic showing the layer structure that develops on the silicon surface during oxidation in N_2O. (*After Ting et al. [1991].*)

in comparison to that of the oxide. Alternatively, Fair (1996) argues that this effect is due to the reduced concentration of the peroxy linkage defects (plds), that is, $\equiv$ Si-O-O-Si $\equiv$, in the oxide in the presence of nitrogen. Fair suggests that plds react with nitrogen according to the following reaction:

$$N^+ + \equiv \text{Si-O-O-Si} \equiv \longleftrightarrow \equiv \text{Si-O-N-O-Si} \equiv \tag{12.2}$$

He labels the resulting species as a nitrogen-related defect (nrd). The plds appear to participate in the diffusion of As, B, P, and Ga in the oxide. We can understand the retardation of B diffusion because the O-O bond (bond energy = 0.44 eV) is weaker than the Si-O bond (bond energy = 3.82 eV). The formation of nrds according to reaction (12.2) reduces the concentration of plds in the oxide. As a result, the diffusion of boron in the oxide is slowed.

III-V and II-VI materials do not have suitable native oxides. Therefore, developing dielectrics for devices based on compound semiconductors is a challenging task. For most applications the dielectrics SiO_xN_y, Si_xN_y, and SiO_x are used. The composition of the dielectric is optimized to enhance the device properties. For high-speed devices, it is desirable to have a low-dielectric constant so that the RC time constant will have a reduced value; see Equation (10.17).

12.4 FORMATION OF SHALLOW JUNCTIONS

For high-performance silicon technology, we require shallower source/drain junctions. Junctions with relatively high surface dopant concentrations, ultrashallow depths, low contact and sheet resistances, and low-junction leakage currents are critical for the advanced CMOS technologies. We project that junctions with depth less than 50 nm will be required, and the surface dopant concentration is expected to decrease from 10^{18} cm^{-3} to 5×10^{17} cm^{-3}.

Several approaches have been developed for the formation of shallow junctions: (1) low-energy ion implantation, (2) diffusion from doped deposited layers, (3) gas-immersion laser doping, and (4) plasma-immersion ion implantation (Kwong 1996). This chapter covers only the first two approaches, which are in widespread use.

We indicated in Chapter 9 that we can control the junction depth by varying the energy of the implanted ions. This approach may work for heavier ions, such as As and Sb. However, in the case of B it is difficult to control junction depth using this approach because of the channeling tail (Kwong 1996; Hong et al. 1988). Three processes have been developed to overcome the channeling problem. Instead of the B ions, we can implant ionized BF_2 molecules. Since these molecules are much heavier than the B ions, the extent of the channeling tail is reduced. However, there is a drawback because the projected ranges at the lowest available energies ($\sim$10 keV) are still too large (Kwong 1996). As a result, it is difficult to achieve junction depths below 60 nm by the implantation of BF_2 molecules. We can also reduce

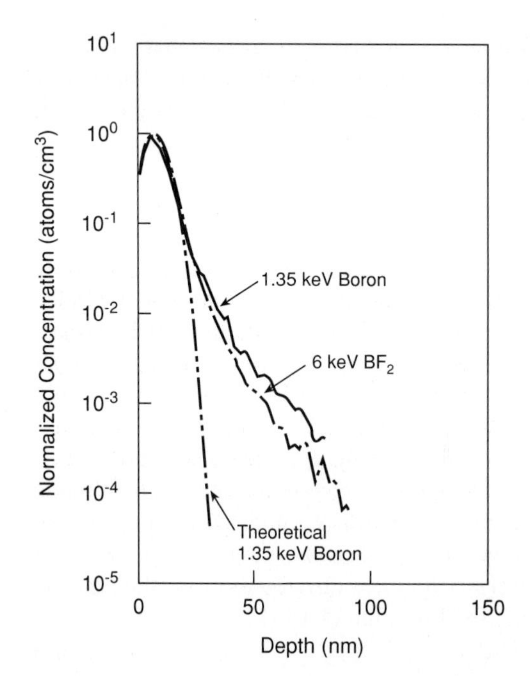

FIGURE 12.11
Normalized as-implanted boron profiles in a crystalline silicon for a 1.3 keV B implant with the modified system and a 6 keV BF_2 implant with a conventional implanter. (*After Hong et al. [1988].*)

the channeling tail by modifying an implanter (Hong et al. 1988). Figure 12.11 compares the normalized as-implanted boron profiles in crystalline silicon for a 1.35 keV (modified implanter) and a 6 keV BF_2 implant (Hong et al. 1988). It is clear that the profiles in the two cases are very similar.

The channeling tail can also be reduced by implanting dopants into an amorphous Si surface layer, which is in turn produced by implanting isoelectronic species, such as Si, Ge, and Sn (Ruggles et al. 1989; Bousetta et al. 1991). The tail is eliminated because the crystalline region is absent. The leakage current of the resulting p-n junction is very sensitive to the postannealing residual damage and its location relative to the junction. Furthermore, the position of the original amorphous-crystalline interface with respect to the junction determines the amount of leakage current (Sedgwick et al. 1988). We attribute leakage current to the electronic effects associated with the defects discussed in section 3.5.

We argued in Chapter 9 that a high-temperature anneal following implantation serves two purposes: (1) to activate the dopant atoms and (2) to reduce the implantation-induced structural damage. We can obviate these difficulties in fabricating shallow junctions by using diffusion from doped deposited layers (Kwong 1996). The advantage of this approach is that the dopant in the substrate has a high surface concentration and has a sharp diffusion profile.

Various materials have been used as diffusion sources for shallow-junction formation: polycrystalline or epitaxial Si, polycrystalline $Si_{1-x}Ge_x$, silicides, and spin-on oxides. The presence of high-diffusivity paths along the grain boundaries in polysilicon keeps the dopant distribution essentially uniform within the polysilicon.

When we anneal the silicon–doped silicon composite, the dopant atoms migrate into the substrate because of the differences in the activity of the dopant atoms over the source and the substrate. We can control the depth of penetration of the dopant atoms by varying the time and temperature of the diffusion anneal. (See Chapter 8 for relevant details.) In this process the cleanliness of the silicon–diffusion source interface is very important. As expected the presence of an interfacial oxide hinders the diffusion of dopant atoms into the substrate (Raicu et al. 1990).

Hsieh et al. (1990) and Georgiou et al. (1990) successfully fabricated shallow n^+-p and p^+-n junctions using the polysilicon diffusion source. Grider et al. (1991) used $Si_{1-x}Ge_x$, an alternative diffusion source, to produce shallow p^+-n junctions. The advantage of the $Si_{1-x}Ge_x$ source over polysilicon is that $Si_{1-x}Ge_x$ shows more selectivity of deposition between Si and SiO_2.

An alternative approach entails the use of silicide contacts as diffusion sources for dopants. In this process the silicide is formed by reacting metals, such as titanium and cobalt, with the underlying silicon. The formation of the silicide can be facilitated by ion-beam-induced interface mixing that helps to break up the native oxide (Ku et al. 1990). Dopants are then implanted into the silicide, followed by rapid thermal-annealing drive-in to diffuse the dopant atoms into the substrate.

We show in Figure 12.12 SIMS profiles for diffusion of boron and arsenic into Si obtained using various implanted silicides (Kwong 1996); the depth scale of the profiles originates at the silicide-Si interface. It is clear that out-diffusion of B and

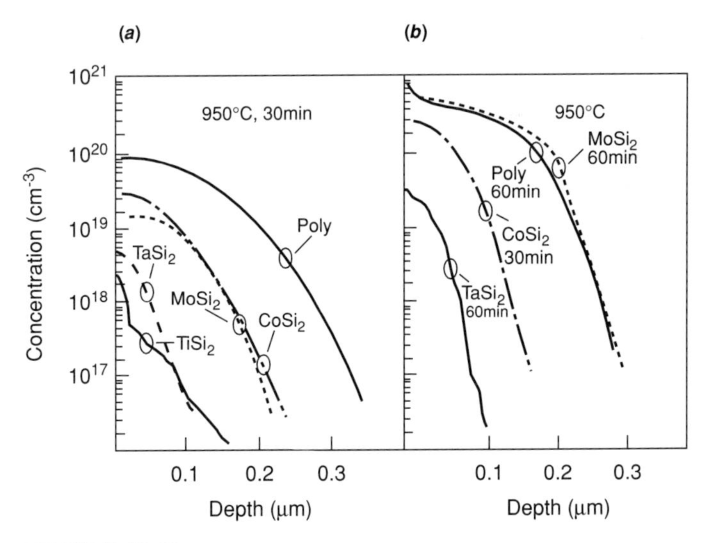

FIGURE 12.12
SIMS profiles in Si after diffusion of (*a*) B and (*b*) As from various diffusion sources upon furnace annealing. The origin of the depth axis is at the silicide-silicon interface. (*After Kwong [1996].*)

As from $CoSi_2$ and $MoSi_2$ occurs much more readily than from $TiSi_2$ and $TaSi_2$. In the case of $TiSi_2$, the formation of TiAs and TiB_2 precipitates in the silicide could be responsible for the observed behavior (Probst et al. 1988).

A major advantage of spin-on films as dopant sources is that these films can be deposited at ambient temperature, thereby eliminating one thermal cycle involving deposition. Usami et al. (1992) used spin-on polymeric-boron doped films and phosphorus doped oxides to fabricate very shallow junctions. By controlling various process parameters, such as the amount of dopant in the film, annealing time, and temperature, shallow junctions 50 nm deep and less were produced.

12.5 MULTILEVEL INTERCONNECTIONS FOR THE ULSI AND GIGABIT SCALE INTEGRATION (GSI) ERAS

We indicated in section 10.6 that the value of the RC time constant increases with decreasing device dimensions and increasing circuit complexities. However, RC remains low if we spread out interconnections in more than one level. In addition, the reduced cross section of the interconnects increases the current density. As discussed in section 10.7, this increases the probability of electromigration-induced damage. Therefore, our push toward ULSI and GSI poses several challenging issues regarding interconnects: (1) the choice of metals for interconnects, (2) the choice of metals for connecting interconnects at different levels, and (3) the choice of dielectrics separating interconnects at various levels.

Murarka (1997) compares the properties of various potential metals for multilevel-interconnect applications that are listed in Table 12.2. Even though the aluminum interconnects and their derivatives have performed well over the years, we may need to replace them with metals that are more electromigration resistant. Murarka (1997) argues that copper offers the most viable alternative, based on several factors: (1) it is a near-noble metal, (2) it has good electrical conductivity, and (3) its activation energy for self-diffusion is higher than those for Ag, Au, and Al. The one drawback of the copper interconnects is their poor adhesion to silicon dioxide that separates metallizations located at different levels. As argued in section 10.7, adding small amounts of aluminum to the interconnects can improve the adhesion (Murarka and Hymes 1995).

In Figure 10.22 we show that the interconnects at different levels are connected to each other by metallizations that go through holes in interlevel dielectrics. These extremely small holes are referred to as *vias*. W is the metal of choice for filling the vias, that is, the vertical interconnect, and is deposited by chemical vapor deposition; see Equations (10.9) and (10.10).

Silicon dioxide has served well as an interlayer dielectric for silicon ICs. However, as the feature size is being reduced, there is a need for interlayers with lower dielectric constants to reduce the RC time constant. Fluorine-doped SiO_2 films, containing 2 to 10 at percent F, have a dielectric constant in the range of 3.0 to 3.7, a

TABLE 12.2
Comparison of properties of possible interlayer metals

	Metal				
Property	**Cu**	**Ag**	**Au**	**Al**	**W**
Resistivity ($\mu\Omega$-cm)	1.67	1.59	2.35	2.66	5.65
Young's modulus ($\times 10^{-11}$) dyn/cm^2	12.98	8.27	7.85	7.06	41.1
Thermal conductivity (Wcm^{-1})	3.98	4.25	3.15	2.38	1.74
Coefficient of thermal expansion $\times 10^6$ (°C)$^{-1}$	17	19.1	14.2	23.5	4.5
Melting point (°C)	1085	962	1064	660	3387
Specific heat capacity (J Kg^{-1}K^{-1})	386	234	132	917	138
Corrosion in air	Poor	Poor	Excellent	Good	Good
Adhesion to SiO_2	Poor	Poor	Poor	Good	Poor
Deposition					
Sputtering	Yes	Yes	Yes	Yes	Yes
Evaporation	Yes	Yes	Yes	Yes	Yes
Chemical vapor deposition	Yes	?	?	Yes (?)	Yes
Etching					
Dry	?	?	?	Yes	Yes
Wet	Yes	Yes	Yes	Yes	Yes
Delay (ps/mm)	2.3	2.2	3.2	3.7	7.8
Thermal stress per degree for films on Si (10^7 dyn/cm^2 °C)	2.5	1.9	1.2	2.1	0.8
Self-diffusion					
Activation energy (eV)	2.19	1.97	1.81	1.48	5.47
Prefactor (Do) (cm^2/s)	0.78	0.67	0.091	1.71	0.04
Mean free path of electrons (Å)	390	520	380	150	50 (+6)* 81 (+3)*

*For two different valences of tungsten.

Source: After Murarka 1997.

value significantly lower than the value for SiO_2 films. Only polymeric materials have dielectric constants close to 2. Aerogels, containing air or gas trapped in a porous structure, could be organic or inorganic and have dielectric constants in the range of 1 to 2 (Labadie et al. 1992; Hrubesh 1995).

One of the potential applications of wide-gap semiconductors, such as SiC, GaN, and AlN, is in high-temperature electronics. The contact properties not only have to be attractive but also should be retained over a long period during the high-temperature operation—a very challenging task indeed!

12.6 DIRECT WRITING

Currently, electron-beam lithography is widely used for making masks. It has a strong position for patterning prototype devices and advanced application-specific ICs, which are produced in small quantities (Newman et al. 1992).

Direct write electron-beam lithography is highly flexible because it does not require a mask. However, it has several drawbacks: low throughput due to serial writing and relatively high cost. We need to address the serial-writing issue because it is a bottleneck in the current technology.

The following examples illustrate the potential of this technique. We show in Figure 12.13 a scanning electron micrograph of a bipolar custom LSI in which the interconnects were directly written using an electron-beam lithography system (Nakamura et al. 1985). Two levels of aluminum metallizations and contact holes show excellent patterning and excellent accuracy. The carriers and the edges of the wiring patterns are smooth.

FIGURE 12.13
A scanning electron micrograph showing two levels of metallizations that have been directly written by electron-beam lithography. (*After Nakamura et al. [1985].*)

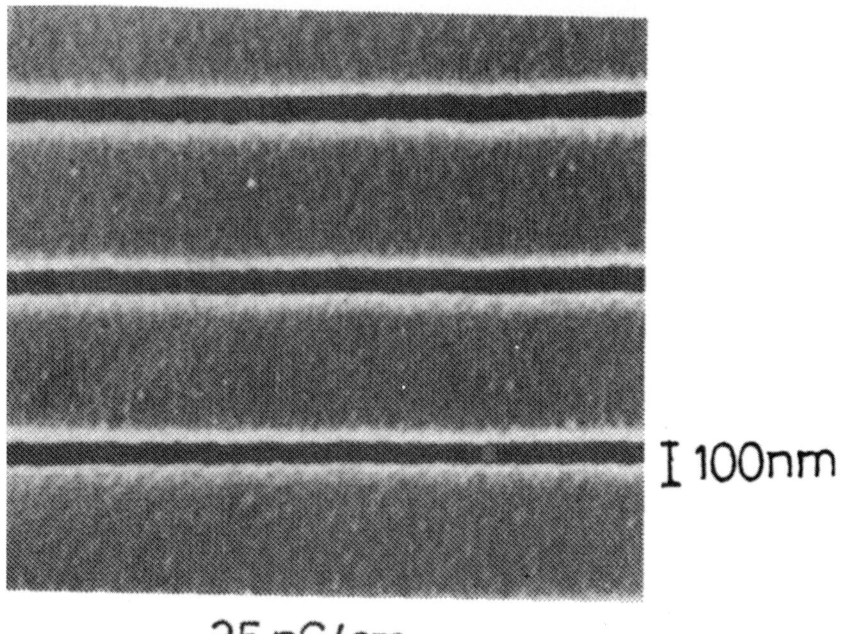

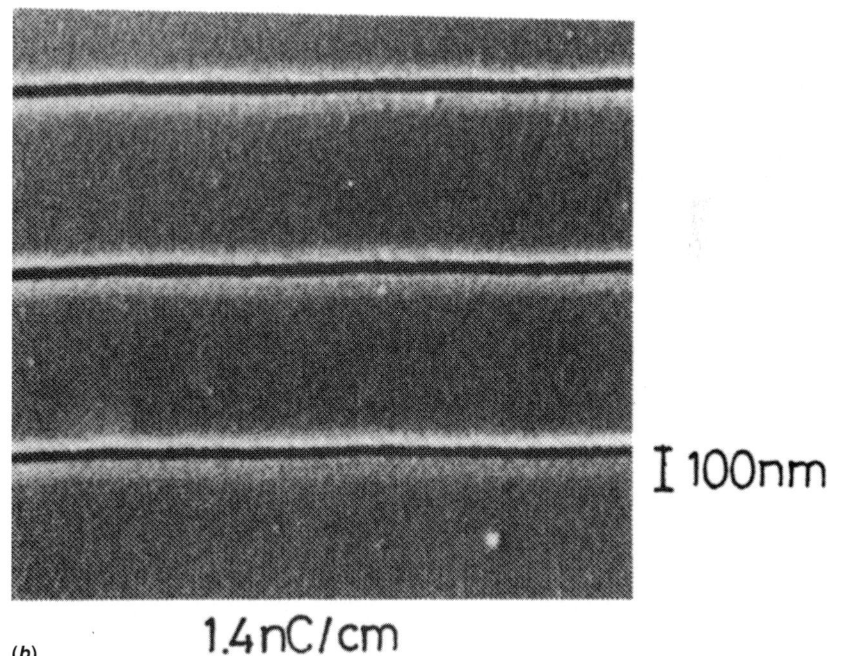

FIGURE 12.14
200 nm thick resist pattern delineated on GaAs substrates using 40 keV electrons and two different doses: (*a*) 3.5 nC/cm and (*b*) 1.4 nC/cm. (*After Gamo et al. [1995].*)

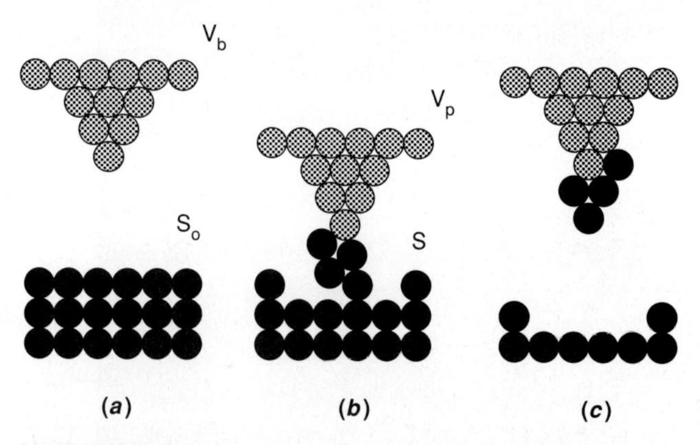

FIGURE 12.15
Direct patterning by atomic manipulation with an STM. (*a*) Situation before removing atoms from sample surface. The tip voltage is V_b, and an automatic control maintains an average tip-sample separation at S_o. (*b*) Situation during atom removal. The automatic control is disengaged, the tip is moved toward the sample, and the tip voltage is pulsed to value V_p. (*c*) Situation after atoms have been removed from the sample to create a nanoscale pit. (*After Calling et al. [1995].*)

We show in Figure 12.14 line patterns written into a 200 nm thick resist spun on GaAs; the beam energy was 40 keV, and the respective doses in Figure 12.14*a* and 12.14*b* were 1.4 and 3.5 nC/cm (Gamo et al. 1995). The line widths in the two cases are 20 and 40 nm, implying a strong dependence of the width on the dose.

In the last five years, another direct-writing technique has emerged. It is based on scanning tunneling microscopy (STM) and can be used to remove and deposit materials at specified locations. We can illustrate its operation using a schematic shown in Figure 12.15 (Calling et al. 1995). Figure 12.15*a* shows an extremely sharp metal tip at voltage V_b that is separated from the sample to be patterned by a distance S_o. In Figure 12.15*b* we move the tip toward the specimen, and the tip voltage is pulsed to a value V_p. This movement removes atoms from the specimen surface, and these atoms accumulate on the tip as shown in Figure 12.15*b*. The tip is subsequently retracted from the sample, and the sequence is repeated. Calling et al. (1995) used this technique to produce trenches in silicon. Figure 12.16 shows a 10 nm wide trench in a germanium island on (001) Si that has a uniform depth of 0.42 nm. This result is remarkable!

In summary, over the years we have made great strides in the growth and processing of semiconductors. We are on the threshold of moving from the ULSI era to the GSI era. This development entails many challenges, and we have elaborated on some of them in this chapter.

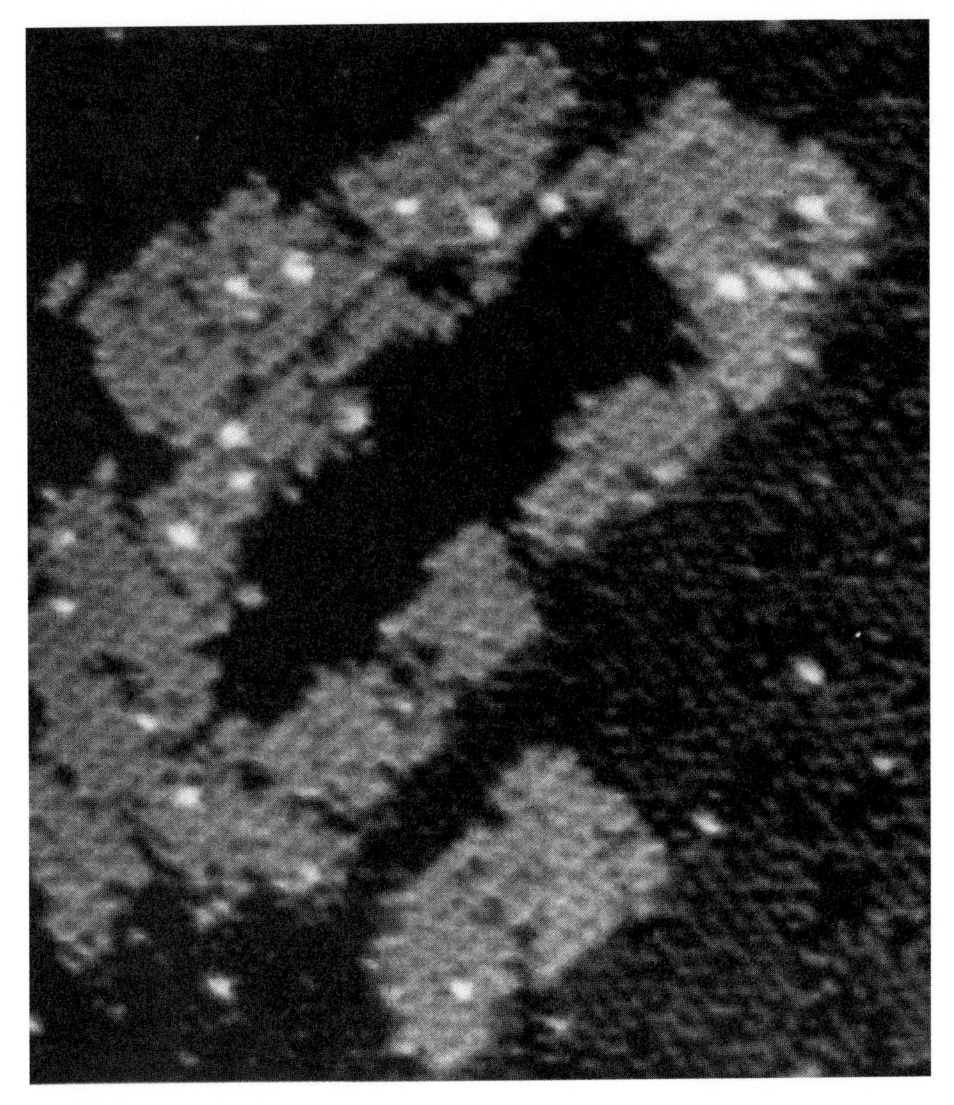

FIGURE 12.16
An STM image of a 10 nm wide trench in a germanium island on (001) silicon; the 0.42 nm deep trench was fabricated using the pulsing technique depicted in Figure 12.15. (*After Calling et al. [1995].*)

REFERENCES

Balkas, C. M. presented at the Fall MRS Meeting, Boston (1996).
Baranowski, J. M. presented at the Fall MRS Meeting, Boston (1996).
Bousetta, A.; J. A. van den Berg; and D. G. Armour. *Appl. Phys. Lett.* 58, 1626 (1991).

Calling, C. T.; I. I. Kravchenko; and M. G. Lagally. *J. Vac. Sci. Technol.* B13, 2828 (1995).
Chang, C. Y. and S. M. Sze. In *Materials Science and Technology:* Silicon Device Structures; ed K. A. Jackson. Weinheim: VCH, 1996, vol. 16, p. 327.
Christen, J.; E. Kapon; E. Colas; D. M. Hwang; L. M. Schiavone; M. Grindman; and D. Brimberg. *Surface Science* 267, 257 (1992).
Du, H.; R. E. Tressler; K. E. Spear; and C. G. Pantano. *J. Electrochem. Soc.* 136, 1527 (1990).
Fair, R. B. *The Physics and Chemistry of SiO_2 and Si-SiO_2 Interface-3,* ed H. Z. Massovd, E. H. Poindexter, and C. R. Helms. Pennington, NJ: Electrochemical Society, 1996, vol. 96-1, p. 200.
Gamo, K.; K. Yamashita; F. Emoto; S. Namba; N. Samoto; and R. Shimizu. *J. Vac Sci. Technol.* B3, 117 (1995).
Georgiou, G. E.; T. T. Sheng; J. Kovalchick; W. T. Lynch; and D. Malm. *J. Appl. Phys.* 68, 3707 (1990).
Green, M. L.; D. Brasen; K. W. Evans-Lutterodt; L. C. Feldman; K. Krisch; W. Lennard; H.-T. Tang; L. Manchanda; and M.-T. Tang. *Appl. Phys. Lett.* 65, 848 (1994).
Grider, D. T.; M. C. Ozturk; and J. J. Wortman. *Proc. of 3rd Int. Symp. ULSI Science and Technology.* Pennington, N.J: Electrochem. Society, 1991, p. 259.
Haase, M. A.; J. Qiu; J. M. DePuydt; and H. Cheng. *Appl. Phys. Lett.* 59, 1272 (1991).
Hong, S. N.; G. A. Ruggles; J. J. Paulos; J. J. Wortman; and M. C. Ozturk. *Appl. Phys. Lett.* 53, 1741 (1988).
Hrubesh, L. W. *Mats. Res. Soc. Symp. Proc.* 381, 267 (1995).
Hsieh, T. Y.; H. G. Chun; D. L. Kwong; and D. B. Sprate. *Appl. Phys. Lett.* 18, 1779 (1990).
Hu, C. unpublished research, Carnegie Mellon University (1996).
Kapon, E. *The Encyclopedia of Advanced Materials,* ed D. Bloor, R. J. Brook, M. C. Flemings, and S. Mahajan. Oxford: Pergamon Press, 1994, p. 2121.
Ku, Y.; S. K. Lee; and D. L. Kwong. *J. Electrochem. Soc.* 137, 728 (1990).
Kwong, D. L. *Materials Science and Engineering,* Silicon Device Processing; ed K. A. Jackson. Weinheim: VCH, 1996, vol. 16, p. 395.
Labadie, J. W.; J. L. Hedrick; V. Wakharkar; D. S. Hofer; and T. P. Russell. *IEEE Compons. Hybrids Manuf. Technol.* 15, 925 (1992).
Liu, L. C., and S. H. Risbud "Nanosize Silicon and GaAs Particles Left After Stopping Dissolution in Molten Glass," Mat. Res. Soc. Vol. 272, 1992, p. 35.
Maex, K., et al. *ULSI Science and Technology,* (1991), p. 254.
Mahajan, S. *Mats. Res. Soc. Symp.* 410, 3 (1996).
Murarka, S. P. accepted for publication in *Mats. Sci. Eng.* (R) (1997).
Murarka, S. P. and S. W. Hymes. *Critical Reviews in Solid State and Materials Science,* 20, 87 (1995).
Nakamura, S. presented at the Fall MRS Meeting, Boston (1996).
Nakamura, K.; Y. Sakitani; T. Konishi; T. Komoda; N. Saitou; and K. Sugawara. *J. Vac. Sci. Technol.* B3, 94 (1985).
Newman, T. H.; P. J. Coane; M. G. R. Thomson; and F. J. Hohn. *Microlith. World* 1, 16 (1992).
Probst, V.; H. Schaber; P. Lippens; L. Van den Hove; and R. De Keersmaecker. *Appl. Phys. Lett.* 52, 1803 (1988).
Raicu, B.; W. A. Keenan; M. Current; D. Mordo; and R. Brennan. *Proc. SPIE Symp. on Rapid Thermal and Related Processing Techniques,* 1393, 161 (1990).
Ruggles, G. A.; S. N. Hong; J. J. Wortman; M. Ozturk; E. R. Myers; J. J. Hren; and R. B. Fair. *MRS Symp. Proc.* 128, 611 (1989).

Sedgwick, T. O.; A. E. Michel; V. R. Deline; S. A. Cohen; and J. B. Lasky. *J. Appl. Phys.* 63, 1452 (1988).
Steigerwald, M. L. and L. E. Brus. *Annu. Rev. Mats. Sci.* 19, 471 (1989).
Taur, Y., et al. *IEDM Tech. Digest,* New York: IEEE, 1993, p. 127.
Tay, S. P.; J. P. Ellul; J. J. White; and M. I. H. King. *J. Electrochem. Soc.* 134, 1484 (1987).
Ting, W.; H. Hwang; J. Lee; and D. L. Kwong. *J. Appl. Phys.* 70, 1072 (1991).
Usami, A.; M. Ando; M. Tsunekane; and T. Wada. *IEEE Trans. Electron Devices* 39, 105 (1992).

APPENDIX A

Physical constants

Quantity	Symbol/Unit	Value
Angstrom unit	Å	1Å = 10^{-1} nm 10^{-4} μm = 10^{-8} cm = 10^{-10}m
Avogadro constant	N_{AVO}	6.02204×10^{23} mole^{-1}
Bohr radius	a_B	0.5917 Å
Boltzmann constant	k_B	1.38066×10^{-23} J/K(R/N_{AVO})
Elementary charge	q	1.60218×10^{-19} C
Electron rest mass	m_0	0.91095×10^{-30} kg
Electron volt	eV	1 eV = 1.60218×10^{-19} J = 23.053 kcal/mole
Gas constant	R	1.98719 cal/mole − K
Permeability in vacuum	μ_0	1.25663×10^{-8} H/cm ($4\pi \times 10^{-9}$)
Permittivity in vacuum	ϵ_0	8.85418×10^{-14} F/cm ($4\mu_0 c^2$)
Planck constant	h	6.62617×10^{-34} J − s
Reduced Planck constant	$\hbar$	1.05458×10^{-34} J − s ($h/2\pi$)
Proton rest mass	M_p	1.67264×10^{-27} kg
Speed of light in vacuum	c	2.99792×10^{10} cm/s
Standard atmosphere		1.01325×10^{5} Pa
Thermal voltage at 300 K	k_BT/q	0.0259 V
Wavelength of 1 − eV quantum	λ	1.23977 μm

Conversion of energy units

	ERG	JOULE	eV	CAL_{TH}*
1 erg	1	10^{-7}	6.2418×10^{11}	2.3901×10^{-8}
1 joule	10^{7}	1	6.2412×10^{18}	0.23901
1eV	1.6021×10^{-12}	1.621×10^{-19}	1	3.8291×10^{-20}
1cal_{th}	4.1840×10^{7}	4.1840	2.6116×10^{19}	1

*cal_{th} = thermochemical calory

APPENDIX B

Properties of important semiconductors at 300K

Semiconductor		Lattice constant Å	Bandgap (eV)	Band[a]	Mobility[b] ($cm^2/V - s$) μ_e	μ_p	Dielectric constant
Element	Ge	5.64	0.66	I	3900	1900	16.0
	Si	5.43	1.12	I	1450	450	11.9
IV–IV	SiC	3.08[c]	2.99	I	400	50	10.0
III–V	AlSb	6.13	1.58	I	200	420	14.4
	GaAs	5.63	1.42	D	8500	400	13.1
	GaP	5.45	2.26	I	110	75	11.1
	GaSb	6.09	0.72	D	5000	850	15.7
	InAs	6.05	0.36	D	33000	460	14.6
	InP	5.86	1.35	D	4600	150	12.4
	InSb	6.47	0.17	D	80000	1250	17.7
II–VI	CdS	5.83	2.42	D	340	50	5.4
	CdTe	6.48	1.56	D	1050	100	10.2
	ZnO	4.58	3.35	D	200	180	9.0
	ZnS	5.42	3.68	D	165	5	5.2
IV–VI	PbS	5.93	0.41	I	600	700	17.0
	PbTe	6.46	0.31	I	6000	4000	30.0

[a]I = Indirect, D = Direct.

[b]The values are for drift mobilities obtained in the purest and most perfect materials available to date.

[c]Silicon carbide crystallizes in the wurtzite structure.

INDEX